ADVANCES IN CHEMICAL PHYSICS

VOLUME 110

ADVANCES IN
CHEMICAL PHYSICS

Edited by

I. PRIGOGINE

Center for Studies in
Statistical Mechanics and
Complex Systems
The University of Texas
Austin, Texas
and
International Solvay
Institutes, Université
Libre de Bruxelles
Brussels, Belgium

AND

STUART A. RICE

Department of Chemistry
and
The James Franck Institute
The University of Chicago
Chicago, Illinois

VOLUME 110

AN INTERSCIENCE® PUBLICATION
JOHN WILEY & SONS, INC.
NEW YORK • CHICHESTER • WEINHEIM • BRISBANE • SINGAPORE • TORONTO

For ordering and customer service, call 1-800-CALL WILEY

Library of Congress Catalog Number: 58-9935

ISBN 0-471-33180-5

Printed in the United States of America.

10 9 8 7 6 5 4 3 2 1

CONTRIBUTORS TO VOLUME 110

Koji Ando, University of Tsukuba, Institute of Materials Science, Tsukuba, Ibaraki 305, Japan

A. D. O. Bawagan, Ottawa-Carleton Institute, Carleton University, Ottawa, Ontario K1S 5B6, Canada

Ernest R. Davidson, Indiana University, Department of Chemistry, Bloomington, IN

G. D. Fletcher, Iowa State University, Ames, IA

Kiyokazu Fuke, Kobe University, Department of Chemistry, Kobe, 657-8501 Japan

M. S. Gordon, Iowa State University, Ames, IA

Kenro Hashimoto, Tokyo Metropolitan University, Computer Center, Minami-Ohsawa, Hachioji, 192-0397 Japan

James T. Hynes, University of Colorado, Department of Chemistry and Biochemistry, Boulder, CO

Suehiro Iwata, Computer Center, Institute of Molecular Science, Myodaiji, Okazaki, 444-8585 Japan

T. Kurz, Universität Göttingen, Drittes Physikalisches Institut, D-Göttingen, Germany

W. Lauterborn, Universität Göttingen, Drittes Physikalisches Institut, D-Göttingen, Germany

Xiangzhu Li, University of Waterloo, Department of Applied Mathematics, Ontario, Canada

Ruth McDiarmid, BCST, National Research Council, Washington, DC

R. Mettin, Universität Göttingen, Drittes Physikalisches Institut, D-Göttingen, Germany

C. D. Ohl, Universität Göttingen, Drittes Physikalisches Institut, D-Göttingen, Germany

Joe Paldus, University of Waterloo, Department of Applied Mathematics, Ontario, Canada

M. W. Schmidt, Iowa State University, Ames, IA

INTRODUCTION

Few of us can keep up any longer with the flood of scientific literature, even in specialized subfields. Any attempt to do more and be broadly educated with respect to a large domain of science has the appearance of tilting at windmills. Yet the synthesis of ideas drawn from different subjects into new, powerful, general concepts is as valuable as ever, and the desire to remain educated persists in all scientists. This series, *Advances in Chemical Physics*, is devoted to helping the reader obtain general information about a wide variety of topics in chemical physics, a field that we interpret very broadly. Our intent is to have experts present comprehensive analyses of subjects of interest and to encourage the expression of individual points of view. We hope that this approach to the presentation of an overview of a subject will both stimulate new research and serve as a personalized learning text for beginners in a field.

I. Prigogine
Stuart A. Rice

CONTENTS

CHAPTER 1

A CRITICAL ASSESSMENT OF COUPLED CLUSTER METHOD IN QUANTUM CHEMISTRY

JOSEF PALDUS

*Max-Planck-Institute for Astrophysics, Karl-Schwarzschild Str. 1,
Postfach 1523, 85740 Garching, Germany,
and
Department of Applied Mathematics,* Department of Chemistry,
and Guelph-Waterloo Center for Graduate Work in Chemistry,
University of Waterloo, Waterloo, Ontario N2L 3G1, Canada*

XIANGZHU LI

*Department of Applied Mathematics, University of Waterloo, Waterloo,
Ontario N2L 3G1, Canada*

CONTENTS

---

*Permanent address.

Advances in Chemical Physics, Volume 110, Edited by I. Prigogine and Stuart A. Rice.
ISBN 0-471-33180-5. © 1999 John Wiley & Sons, Inc.

Abstract

Methods based on the coupled-cluster (CC) *Ansatz* and most widely applied for the computation of molecular properties and electronic structure are reviewed. The applications of each method are presented, and its performance is evaluated. Following introductory remarks and a brief historical overview of CC methodology and its applications, we first outline the scope of our review, establish the required notation, and recall the foundations and origins of the CC *Ansatz*. We begin with single-reference CC approaches, which have greatly matured since the first *ab initio* study in early seventies, and are currently used to solve many diverse problems. The handling of nondegenerate closed-shell states being nowadays routine, we focus on open-shell systems and the important role of spin adaptation, as well as on methods extending or improving the ubiquitous CC method with singles and doubles (CCSD). Multireference CC approaches still defy routine usage and are therefore given only a cursory treatment. Their main purpose is to provide a basis for the so-called state-selective, or state-specific, CC approaches that employ a single, yet possibly multideterminantal, reference. Perturbative and the so-called quadratic configuration interaction methods are briefly addressed from the viewpoint of CC theory. We also make an attempt to assess and interrelate numerous CC-based approaches to molecular properties, ranging from those related to the shape of the potential-energy surfaces (geometry, harmonic force fields, etc.) to properties characterizing the interaction with electromagnetic fields (static moments, polarizabilities, etc.). We then conclude the methodo-logical part of our review with a few comments concerning the computational aspects. The subsequent part, devoted to applications, presents various examples that illustrate the scope, efficiency, and reliability of various CC methods, particularly those of recent provenance. We focus on the general account of correlation effects, potential-energy surfaces and related properties, ionization potentials and electron affinities, vertical electronic excitation energies, and various static and dynamical properties. The final section of this review attempts to summarize the status quo of the CC methodology and its applications, as well as to ponder its future prospects.

It is the glory of science that no establishment could withstand the overwhelming impact of Heisenberg's discovery of quantum mechanics in 1925. Once precise numbers can be calculated, the dam wall breaks. But it is rare for the younger generation to make such an immense stride all in one move. Usually, many smaller steps are needed before the decisive step becomes possible, and it is the smaller steps, none of them decisive in itself, that the refereeing system is ideally designed to prevent. $\cdots$ The worst mistakes of the past had been in observation, not in theory.

Fred Hoyle, *Home Is Where the Wind Blows*
University Science Books, Mill Valley, CA, 1994

I. INTRODUCTION

Ever since the first explicit formulation of the basic single-reference (SR) coupled-cluster (CC) equations by Čížek [1, 2] more than three decades ago, we have witnessed a steady development of the theory and computer implementation of this formalism. An increased need for accurate and reliable quantum mechanical descriptions of molecular electronic structure that account for many electron correlation effects has resulted in more and more frequent exploitation of this formalism. Although the importance of the exponential CC *Ansatz* for a proper formulation of a size-extensive theory was first realized by nuclear physicists [3, 4], it is in the modeling of many-electron systems where this formalism found its most comprehensive and widespread use. The first *ab initio* pilot study [5] clearly indicated the remarkable ability of the approach to accurately describe correlation effects, at least in the nondegenerate closed-shell (CS) ground states of simple molecular systems. The development of generic codes [6, 7] for the CC method truncated at the pair cluster level, facilitated by a rapidly growing computer technology, then permitted extensive testing and subsequent exploitation of the basic SR CC method (cf. [8–14]).

Nowadays, a plethora of various CC methods is routinely used in practical applications, and several standard approaches are available in most commercially distributed quantum chemistry software packages, such as ACES II [15], CADPAC [16], COMENIUS [17], DIRCCR12-95 [18], GAUSSIAN [19], MOLCAS [20], MOLPRO [21], OPENMOL [22], PSI [23], TITAN [24], and others [25, 26]. Applications range from small molecular systems involving 4–20 electrons [9–14, 27–32], where large basis sets can be employed and yield highly accurate results, to relatively large systems, such as benzene dimers [33], nucleic acid bases [34], or DNA base pairs [35]. Even these latter systems have been modeled using valence-double-zeta (VDZ) and triple-zeta (VTZ) correlation-consistent (cc) basis sets [36]. The range of systems and properties handled via CC approaches is thus steadily expanding [30–32]. Nevertheless, while the CC theory is nowadays well understood and effectively implemented for the nondegenerate ground states of CS systems, the study of open-shells—with the exception of the simplest ones—is still in its infancy. Despite the vast literature that exists on the subject (for reviews see, e.g., [14, 27–32, 37]) and both past [38–43] and very recent [32, 44–53] methodological advances, our grasp of the problem remains rather limited. While a number of very useful applications in an increasing variety of systems (particularly those of high spin variety) exists today, no standard software is available that can ubiquitously handle any open-shell (OS) system and its properties.

Our present objective is a critical assessment of CC methodology and its applications. We will therefore focus on OS or quasi-degenerate systems. In view of the existence of numerous reviews on various aspects of CC theory and its applications, and in view of the complexity of some theoretical developments that still await practical implementation, we shall concentrate on SR-like state-selective, or state-specific (SS), approaches and pay only cursory attention to genuine multireference (MR) approaches, however great their theoretical appeal and desirability, because they are still too demanding for a generic implementation and widespread exploitation.

Following a brief historical overview of theoretical and practical developments in the next section, we will outline the scope of this review in Section I.B. The methodologically oriented aspects will then be addressed in Section II, while actual computation of various molecular properties will be described in Section III. Finally, in Section IV we will present some thoughts on the status quo and future developments.

A. Historical Outline

The origins of the CC approach to many-fermion systems (for personal accounts, see [54–56]) can be traced back to the early fifties, when the first attempts were made to understand the correlation effects in an electron gas [57, 58] and nuclear matter (cf. [59]). For both of these extended systems, it was absolutely essential that the method employed scale linearly with the particle number, or, as we say today, be *size extensive.** The first methodological achievement in this direction was made by Brueckner [60], who observed that up to the fourth order of Rayleigh–Schrödinger perturbation theory, the size-nonextensive unlinked terms cancel out. His conjecture that this result holds in all orders of Rayleigh–Schrödinger perturbation theory—today often referred to as the many-body perturbation theory (MBPT), or the Møller–Plesset (MP) PT when the Hartree–Fock (HF) operator is employed as the unperturbed Hamiltonian—was proved by Goldstone [61], who, almost simultaneously with Hubbard [62] and Hugenholtz [63], introduced mathematical techniques of quantum field theory, particularly those based on Feynman-like diagrams, into the MBPT. It was especially the work of Hubbard [62] that clearly showed the exponential structure of the wave operator W in terms of connected (in contrast to linked!) diagrams (see Section 3.5 of [14]), which directly implied the exponential coupled-cluster *Ansatz* $W = \exp(T)$. (See, e.g., [55].) Stimulated by this result, Coester and Kümmel [3, 4] endeavoured to exploit the exponential form of the wave operator and to calculate the connected-cluster amplitudes directly rather than via MBPT. They formulated general equations for this purpose, but did not work out their explicit form until much later [64].

In the meantime, on the atomic and molecular electronic structure front, the need for a proper account of many electron correlation effects was realized and the correlation problem clearly defined, primarily by Löwdin [65]. Goldstone's MBPT was first exploited in atomic systems by Kelly [66], while Sinanoğlu's [67] approach, directed toward molecular problems, was already based on the cluster-type expansion with an approximate treatment of pair clusters. The desirability of the exponential *Ansatz*—an idea that

---

*The term was coined much later [7, 8], in order to make a clear distinction from other approximate methods, such as limited-configuration interaction, which do not have this property. Because methods that are not size extensive are useless for handling extended systems, this property was automatically required of any many-body theory without being mentioned explicity.

originated a long time ago in statistical mechanics (cf. [68, 69])—for the study of many-electron correlation problems, was also pointed out at that time by Primas [70].

Motivated by these earlier developments, as well as by the work of Tolmachev [71], Čížek [1, 2] succeeded in deriving the *explicit* form of coupled-cluster equations for the cluster amplitudes that were written down earlier in their *implicit* form by Coester and Kümmel [3, 4]. We must emphasize here that although these equations were actually written down only for the most important case of pair (i.e., doubly excited) cluster components (i.e., for the approximation $T \approx T_2$), referred to originally as coupled-pair many-electron theory (CPMET [72], or CCD in today's terminology), the diagrammatic formulation of Čížek [1, 2] is completely general, making it straightforward to obtain coupled-cluster equations at any level of approximation. Clearly, the required labor increases with the excitation level at which one truncates the cluster expansion (cf., e.g., [73]), but, should one require a high-order formalism, the procedure can be easily automated (cf., e.g., [74]). However, while it is easy to generate the required equations, it is not at all easy to exploit them computationally in practical applications in view of the rapidly increasing dimensionality of higher order cluster amplitudes.

From a theoretical viewpoint, the importance of this development lies in the fact that the CC approach bridges the Gell-Mann and Brueckner electron gas theory for high densities [58], characterized by ring diagram contributions, and Brueckner's theory of nuclear matter [60], relying on ladder diagrams. Hence, the CC theory incorporates not only both ladders and rings, but also their possible combinations, thus properly accounting for intercluster interactions. Moreover, it does so in a size-extensive fashion, whatever the selected truncation scheme. In fact, the CC equations can be regarded as a kind of recursive relations for the generation of MBPT diagrams [1, 2, 75], as will be pointed out in greater detail in Section II.C.1.

Initially, little attention was paid to these developments by the majority of quantum chemists. This was partially due to very limited computing facilities—certainly by today's standards—that were available at the time, but also due to the general preference for conceptually much simpler variational approaches in the spirit of Hartree–Fock (HF) or self-consistent-field (SCF) and configuration interaction (CI) methods. The fact that the initial formulation of CC theory relied heavily on field-theoretical math-ematical techniques, which at the time were not appreciated by most quantum chemists, contributed to its undeserved initial neglect. To bridge this latter obstacle and to provide another viewpoint on the essence of the CC *Ansatz*, the basic CCD equations were rederived using the standard wave function (i.e., the first quantization) formalism [76]. Soon afterwards, pilot calculations using simple semiempirical as well as *ab initio* models were carried out [5, 75], which, together with the cluster analysis of some model wave functions [77], provided convincing evidence of the potential effectiveness and usefulness of CC approaches. Likewise, a pedagogical exposition of the time-independent version of the second quantization-based diagrammatic techniques [78] significantly contributed to the general appreciation and dissemination of MBPT-based methodologies. Yet another viewpoint was provided by an algebraic derivation of CC equations based on the well-known (see, e.g., [79]) expansion of $e^{-A}Be^{A}$ in terms of commutators [80, 81].

Although CC theory had existed for over a decade, its widespread exploitation awaited the emergence of first minicomputers in the late seventies, when the first general-purpose codes implementing the CC method, truncated at the pair cluster level (CCSD method), were developed by Pople's and Bartlett's groups [6, 7]. The CCSD methodology is nowadays often used when an accurate and reliable account of many-electron correlation effects is called for and the standard spin-orbital-based, and even spin-adapted, CCSD codes form a part of many software packages, as was already mentioned.

Unfortunately, the situation is much less satisfactory for general OS systems, particularly those requiring a genuine MR description. Undoubtedly, impressive progress has been achieved in the development of a general theory, and various controversies have been resolved. Yet, our knowledge of the cluster structure of general OS wave functions can hardly be regarded as being definitely established: While we can successfully handle many important OS situations, no general-purpose code enabling us to model multiple OS states with arbitrary spin multiplicity exists.

Part of the difficulty stems from a certain arbitrariness in generalizing the CS cluster *Ansatz*. While this *Ansatz* is unique when one relies on a nondegenerate spin-orbital single-determinantal reference or when one considers CS systems, that is no longer the case for MR generalizations. This nonuniqueness opens up several possibilities not only for various SR-based extensions, particularly in their spin-adapted or approximately spin-adapted forms, but even for genuine MR generalizations.

In order to classify the existing approaches to OS systems that are either fully or partially based on the CC *Ansatz*, we can distinguish two basic categories of methods, namely, those which formulate or postulate a new kind of exponential *Ansatz* that is suitable for handling OS situations (possibly of a restricted class) and those which bypass this problem by relying on the standard CS or spin-orbital SR cluster *Ansatz*, while describing the related OS situation (mostly for states resulting by an excitation of closed-shell or simple OS ground states) via the additional CI-type linear expansion. We can thus say that the former approaches rely on a *full* (or *complete*, or *integral*) cluster *Ansatz*, while the latter ones use a *partial* (or *incomplete*) *Ansatz*.

To the first category belong the two genuine types of MR approaches: the so-called *Fock space*, or *valence universal*, methods [37–39, 41–43, 82–84] (see also [14, 29, 37]) and the *Hilbert space*, or *state universal*, methods [40, 85–94] (see also [29, 37]). These methods have been tested in various open-shell or quasidegenerate situations, but one cannot speak of a systematic exploitation (see Section II.B), the main drawback stemming from the intruder state problems and multiplicity of solutions (cf., e.g., [90–92, 95]), but also from the complexity of the formalism, restricting its applicability.

Thus, it is the third class of approaches relying on the full CC *Ansatz*, referred to as *one-state*, *state-selective*, or *state-specific* (SS) methods, that have been most vigorously pursued during the past decade. This rather heterogeneous group of methods relies on different types of cluster *Ansätze*, with either full, partial, or no spin adaptation, and involves different truncation schemes for higher excited cluster amplitudes. In all cases, however, it uses effectively an SR formalism, even though the reference employed may consist of several Slater determinants (with a priori fixed weights). The general characteristic of these methods, which we could also refer to as *quasi-MR*, or *pseudo-MR*, approaches, is the fact that, in contrast to genuine SR formalism, where we partition the spin orbitals into the two distinct subsets of occupied and unoccupied (or virtual) ones, in these MR SS approaches, just as in genuine MR CC approaches, we always rely on a spin-orbital partitioning into three subsets: occupied, active, and virtual. Based on this partitioning, a suitable single-, but possibly multideterminantal (or effectively multideterminantal) reference is constructed and used.

The important SS approaches that are currently intensely pursued were already alluded to. They either rely on the standard SR CC formalism and account for nondynamical correlation effects by considering suitable subsets of higher-than-pair cluster amplitudes [96–99], thus providing a viable alternative to theoretically desirable, but practically unaffordable, high-order SR CC approaches considering triples and quadruples (CCSDT [100–102], CCSDTQ [103]), or approximate the higher-than-pair clusters by exploiting various external sources, as in *externally corrected* CCSD procedures [47–50].

We shall also pay attention to the second category of methods employing partial or incomplete cluster *Ansätze*. Their exploitation and performance were reviewed rather recently [28, 30, 31], so that we shall primarily address their relationship with other methods. A prototype of these approaches that employed a simple linear *Ansatz* acting on a CC wave function was introduced a long time ago [104, 105]. The imposed requirement of Hermiticity for the resulting eigenvalue problem [105] led, however, to a rather complex formalism, which was applied only to semi-empirical-model Hamiltonians at the time and might be worthy of reexamination in view of present-day computational capabilities. The idea was later systematically developed at the MR level by exploiting similarity-transformed (and thus non-Hermitian) Hamiltonians. Nowadays, three methods carrying different labels, yet conceptually closely related, exploit this avenue: the so-called symmetry-adapted cluster method followed by configuration interaction (SAC/SAC-CI) method [106, 107], and equations-of-motion (EOM) [108] and linear response (LR CC or CCLR) [109] methods combined with the CC *Ansatz* for the wave function [110–121] (see also [14, 31]; cf., however, Section II.F.3). The most recent version, referred to as similarity-transformed EOM-CC (STEOM-CC), as well as its perturbative analogue, STEOM-PT [122], exploit also the Fock space MR CC *Ansatz* for the second similarity transformation. The method was applied to very large systems, such as free base of porphin [123]. The implicit assumption of these methods is that the dynamical correlation is similar in the chosen CS reference and the related OS state, which may or may not be valid, depending on the circumstances. The approach exploiting the idea of the so-called *intermediate* or *dressed Hamiltonians* (cf., e.g., [51, 106, 124–127]) may also be delegated to this category.

Much attention is lately being paid to spin adaptation, or at least partial spin adaptation, of various CC methods (cf., e.g., [14, 29, 30, 44–46, 52, 128–139]), as will be discussed in Sections II and IV, as well as to other important aspects of the theory, such as a better description of cusps via the geminal CC and CC-R12 approach [140, 141], the account of relativistic effects [142], or an exploration of the topography of the potential-energy surfaces [143, 144].

All these developments leave little doubt about the usefulness of CC methodology in molecular electronic structure calculations. Although we cannot do justice to all aspects of this nowadays vast field of endeavor, we shall try to provide a glimpse of the status quo in most important directions pertaining to quantum chemistry, as outlined in the next section.

B. Scope of the Review

During the past three decades, CC ideas have been extensively pursued. However interesting and useful these developments may be in their own right, many of them lie beyond the scope of this article. Today, the exponential cluster *Ansatz* for the wave function is being exploited well beyond the confines of quantum chemistry, being relevant to both many-fermion as well as many-boson systems. We only mention in passing its exploitation in nuclear physics [81, 145–147] (where, in fact, the idea originated [3, 4]), in condensed-matter physics [148–150], in theories of a homogeneous electron gas [151–153], and in single-mode bosonic field theory [154, 155], as well as in relativistic quantum field theory [156, 157]. A general overview of these and other endeavors was given by Bishop [158] and, in a very general form, by Bishop and Kümmel [159]. For the same reasons, we also refrain from reporting on very interesting, yet rather specialized, work concerning the exploitation of CC theory in studies of nonconservation of parity in heavy alkaline atoms [160] or, in fact, atomic hyperfine structure and relativistic effects in atomic systems [161] or diatomics [142, 162], since our objective is to explore those aspects of CC theory that pertain to nonrelativistic molecular *ab initio* electronic structure calculations.

Our task is further facilitated by the fact that during the past few years several rather extensive reviews on various aspects of the CC theory and its application to quantum chemical calculations have been published [14, 28–32]. To avoid unnecessary duplication, we will often refer to these reviews, whose authors naturally focus on contributions by their own laboratories. Even though we shall do our best to provide an overall viewpoint and a critical assessment of the present-day coupled-cluster methodology and its application to molecular electronic structure calculations, we cannot entirely escape a subjective bias, particularly when we present examples of actual numerical results.

The main focus of the reviews just mentioned can be summarized as follows:

Earlier reviews by the senior coauthor [14, 29] were devoted to methodological aspects of SR and MR CC approaches, relying on both diagrammatic [14] and algebraic [29] techniques. Brief attention was also given to the calculation of stationary properties via linear response theory [14] and to numerical aspects of solving CC equations. The review [29] concentrated on MR approaches.

The objective of a tutorial by Bartlett and Stanton [28] is to guide the users of commercially available software (in particular, ACES II [15], GAUSSIAN [19], and related packages) in its practical and intelligent exploitation while avoiding possible pitfalls. Following a brief exposé of post-Hartree–Fock methodology and basis sets, the authors discuss the computation of potential-energy curves, vibrational, photoelectron, and electronic spectra, first- and second-order static properties, and NMR shifts, using numerous examples for illustration and comparing the performance of

SCF, CI, MBPT, and CC methods. In a more recent review [31], Bartlett first discusses the basic concepts of CC theory and outlines the derivation of explicit SR CC equations at the spin-orbital level, relying on projection operator and diagrammatic techniques and their factorization. After analyzing the MBPT structure of the CC approach and the ways for an approximate handling of higher-than-pair clusters, he discusses open-shell applications based on either the UHF or ROHF references, as well as the CC approach to properties and excited states via the EOM-CCSD method. All methodological developments are illustrated by actual numerical results obtained with an ACES II [15] set of codes.

The review by Lee and Scuseria [30] from the same year concentrates primarily on the performance of the CCSD(T) method in selected chemical applications, namely, equilibrium geometries, vibrational frequencies, heats of formation, and static molecular properties. In its theoretical part, it discusses various approximate ways of accounting for three-body clusters, as well as the computational aspects of performing highly accurate calculations using large basis sets, including vectorization. Special attention is given to the problem of symmetry adaptation—particularly spin adaptation in OS applications—and to the so-called T_1-diagnostics [135, 163], based on empirical observations allowing the assessment of the importance of nondynamical correlation effects and thus of the appropriateness of the SR CC approaches. The review also briefly addresses the computation of analytical gradients, response theory, and EOM–CCSD approaches to excitation energies.

The monograph *Recent Advances in Coupled Cluster Methods* [32] contains 10 papers reviewing diverse developments in CC theory and its applications. In addition to the already mentioned contributions by Adamowicz and Malrieu [126] on intermediate Hamiltonian-based approaches, by Kaldor [142] on relativistic CC, by Stanton and Gauss [144] on analytic second derivatives for CCSD and CCSD(T), by Noga et al. [141] on explicitly correlated CC–R12 [18] methods, by Urban et al. [133] on spin-adapted CC approaches to high-spin OS systems, and by Li and Paldus [44] on UGA–CC, the monograph contains articles on SR CC and MBPT theories of excitation energies by Head–Gordon and Lee [164], on SS MR–CC approaches by Mahapatra et al. [165], on analytic CC response approaches using multideterminantal model space by Pal and Vaval [166], and, finally, on SS MR–CC version of coupled electron-pair approximation (CEPA) by Szalay [167]. We will often refer to these papers to avoid duplication.

Since the standard SR CC theory has been derived in a number of different ways, providing diverse viewpoints and insights, we shall keep our presentation to a minimum, highlighting only the conceptually most important aspects and referring to the literature for details. We shall likewise handle applications to CS systems, wherein a number of standard codes exist and which are now carried out routinely. Also, the treatment of genuine MR CC approaches will be kept to a minimum in view of the existing vast theoretical literature, but, unfortunately, very limited practical exploitation. The reasons for this state of affairs will be briefly addressed in Section II.D. We shall thus focus our attention on SS-type approaches to open-shell systems and their various spin-adapted or partially spin-adapted versions, trying to interrelate them and point out their advantages and shortcomings. In Section II.E, we also address the most important closely related methods, particularly the MBPT and the so-called quadratic CI (QCI) [168]. Sections II.F and G will be devoted to CC approaches to properties and to computational aspects, respectively.

Applications will be presented in Section III, covering, first, energetic properties and, subsequently, other derived properties, including static moments. Dynamic properties, whose handling is still in its infancy, will be mentioned only in passing. Finally, in Section IV we shall try to summarize the status quo of both CC theory and its various applications and point out expected future developments in this field.

II. METHODOLOGY

A. Basic Notation: Second vs. First Quantization

We first establish basic notation that will be employed throughout this paper. We shall primarily employ the time-independent second-quantization formalism and its diagrammatic representation (for more details, see [14, 78, 169, 170]), as well as the ubiquitous wave function (i.e., the first-quantization) formalism if convenient. For example, the one-electron spin orbital $\phi_A(\mathbf{x})$, $\mathbf{x} \equiv (\mathbf{r}, s)$, with $\mathbf{r}$ and s the position and spin coordinates, respectively, will also be simply represented by the ket $|A\rangle$, the two notations being related by

$$\phi_A(\mathbf{x}) = \langle \mathbf{x}|A\rangle. \tag{2.1}$$

We shall consider a standard *ab initio* atomic or molecular model Hamiltonian H, involving at most two-body interactions and defined on a finite-dimensional Fock space that is spanned by $2n$ spin orbitals $|I\rangle, |J\rangle, |K\rangle$, etc., which we write in the form

$$H = Z + V = z_I^J e_J^I + \frac{1}{2} v_{IK}^{JL} e_{JL}^{IK}, \tag{2.2}$$

invoking the Einstein summation convention over repeated indices. Further,

$$z_I^J = \langle I|z|J\rangle \quad \text{and} \quad v_{IK}^{JL} = \langle IK|v|JL\rangle \tag{2.3}$$

are one- and two-electron integrals (in Dirac notation), and $e_{JLN\cdots}^{IKM\cdots}$ are the so-called *replacement operators* (sometimes called excitation operators) [29], [174–177], whose second quantized form is

$$e_{JLN\cdots}^{IKM\cdots} = X_I^\dagger X_K^\dagger X_M^\dagger \cdots X_N X_L X_J, \tag{2.4}$$

where $X_I^\dagger, X_J^\dagger, \ldots (X_I, X_J, \ldots)$ are the creation (annihilation) operators associated, respectively, with the spin orbitals $|I\rangle, |J\rangle, \ldots$, satisfying the fermionic anticommutation relations

$$\{X_I, X_J\} = \{X_I^\dagger, X_J^\dagger\} = 0, \quad \{X_I^\dagger, X_J\} = \langle I|J\rangle = \delta_{IJ} \equiv \delta_J^I, \tag{2.5a}$$

where $\{M, N\} = MN + NM$ designates the anticommutator of M and N and $M^\dagger$ the Hermitian conjugate of M, as well as the annihilation property

$$X_I|0\rangle = 0 \quad \text{or} \quad \langle 0|X_I^\dagger = 0, \tag{2.5b}$$

when acting on the physical vacuum $|0\rangle$, $\langle 0|0\rangle = 1$.

We note that the one-body replacement operators e_J^I represent U($2n$) generators (cf. [174–177]) satisfying the standard unitary group commutation relations

$$[e_J^I, e_L^K] = e_L^I \delta_J^K - e_J^K \delta_L^I, \tag{2.6a}$$

and the Hermitian property

$$(e_J^I)^\dagger = e_I^J. \tag{2.6b}$$

The higher-than-one-body replacement operators could also be defined recursively as

$$e_{JL}^{IK} = e_J^I e_L^K - e_L^I \delta_J^K, \tag{2.7a}$$

$$e_{JLN}^{IKM} = e_{JL}^{IK} e_N^M - e_{NL}^{IK} \delta_J^M - e_{JN}^{IK} \delta_L^M, \text{ etc.}, \tag{2.7b}$$

which may be easily verified using the definition (2.4) and the anticommutation relations (2.5).

The algebra of these (spin-orbital) replacement operators and of their spin-free counterparts (see later) is summarized in Appendix A. The operators form a very convenient basis for the universal enveloping algebra of $U(2n)$ [or $U(n)$] and are at the heart of algebraic approaches to the many-fermion problem, since the rules for the evaluation of their commutators are equivalent to the time-independent Wick theorem (see [171–174]) of the second-quantization formalism. Recall that the latter expresses all the relevant quantities (i.e., both wave functions and operators) via the creation and annihilation operators $X_I^\dagger$ and X_I, respectively, satisfying Eqs. (2.5). Thus, any operator mean value can be expressed as a vacuum mean value of a string of these operators $\langle 0|M|0\rangle$, $M = M_{I_1} M_{I_2} \cdots M_{I_{2k}}$, where $M_J = X_J^\dagger$ or $M_J = X_J$, $J = I_1, \ldots, I_{2k}$, and any state as a linear combination of expressions of the type $M|0\rangle$. To exploit the vacuum annihilation property (2.5b), one defines (cf. [14, 78]) the *normal product* of the operator string M, $N[M]$, in which all the creators are permuted to the left of all the annihilators, attaching the (± 1) phase, depending on the parity of the required permutation. Consequently, the vacuum mean value of any normal product (involving a nonempty string M) vanishes (i.e., $\langle 0|N[M]|0\rangle = 0$), and the expression $N[M]|0\rangle$ also vanishes, as long as M involves at least one annihilation operator. For completeness, one defines $N[\emptyset] = 1$, with $M = \emptyset$ representing the product of an empty set of operators. Defining, further, the *contraction* of two operators as the difference of their ordinary and normal products, that is,

$$M_I M_J = M_I M_J - N[M_I M_J], \qquad (2.8)$$

one then facilitates the evaluation of the aforementioned quantities $\langle 0|M|0\rangle$ or $M|0\rangle$ by exploiting the time-independent Wick theorem [14, 78], which states that

$$M \equiv M_{I_1} \cdots M_{I_k} = N[M_{I_1} \cdots M_{I_k}] + \sum N[M_{I_1} \cdots \cdots M_{I_k}], \qquad (2.9)$$

the sum extending over all possible N-products with contractions. The latter are defined as the product of individual contractions and the N-product of uncontracted operators, again with (± 1) phase, depending on the parity of the permutation required to unscramble the operators. (In fact, this phase equals $(-1)^p$, where p is the number of intersections of the contraction symbols.) Thus, in evaluating the vacuum mean values $\langle 0|M_{I_1} \cdots M_{I_k}|0\rangle$, only fully contracted terms give a nonvanishing contribution. The quantities $M|0\rangle$ may be similarly simplified. The Wick theorem can be easily generalized to products with some operators already in the N-product form: One simply drops all the terms involving contractions of operators that appear on the left-hand side in the same N-product.

Realizing that the only nonvanishing contraction is of the type $\overset{\frown}{X_I X_J^\dagger} = 1$, since $X_I X_J^\dagger = \{X_I, X_J^\dagger\} = \delta_I^J$, we see that there are generally only a few fully contracted terms that contribute to $\langle 0|M|0\rangle$. To minimize the number of operators constituting M in Eq. (2.9), one employs the *particle–hole* ($p - h$) *formalism* by introducing a new (*Fermi*) N-electron vacuum state

$$|\Phi_0\rangle = X_{A_1}^\dagger \cdots X_{A_N}^\dagger |0\rangle \equiv |\{A_1 \cdots A_N\}\rangle \,, \tag{2.10}$$

where we emphasize the antisymmetric (Slater determinant-type) nature of this independent-particle-model (IPM) state by enclosing the spin-orbital labels in braces. To simplify our notation, we shall label the spin orbitals that are occupied in the Fermi vacuum (2.10) by letters from the beginning of the alphabet ($A, B, C, D, \ldots$), those unoccupied in $|\Phi_0\rangle$ by letters from the end of the alphabet ($R, S, T, U, \ldots$), and generic spin orbitals by letters from the middle of the alphabet ($I, J, K, L, \ldots$). It is then easy to see that, by reversing the creation and annihilation character of the operators associated with occupied spin orbitals (referred to as *hole* states), i.e., by defining new $p–h$ operators $Y_A^\dagger, Y_A$ by

$$Y_A^\dagger = X_A, \quad Y_A = X_A^\dagger \,, \tag{2.11a}$$

while preserving their character for unoccupied spin orbitals (*particle* states),

$$Y_R^\dagger = X_R^\dagger, \quad Y_R = X_R \,, \tag{2.11b}$$

we preserve the anticommutation relations (note that the sets of hole and particle labels are distinct) for the $p–h$ operators $Y_I^\dagger, Y_I$; and the vacuum annihilation condition also holds if we replace the true vacuum $|0\rangle$ by the Fermi vacuum $|\Phi_0\rangle$, i.e.,

$$Y_I|\Phi_0\rangle = 0 \quad \text{or} \quad \langle \Phi_0|Y_I^\dagger = 0 \,, \tag{2.12}$$

where I designates either a particle or a hole state. We can thus replicate the particle-only formalism by defining analogous operations, namely, the normal product of $p–h$ operators $n[M]$ and contractions

$$\overset{\frown}{M_I M_J} = M_I M_J - n[M_I M_J] \,, \tag{2.13}$$

where we now employ the lowercase letter $n[\cdots]$ for the $p–h$ n-product and place the contraction symbol above the operators in order to distinguish it from the particle-only formalism.

It should be obvious by now that the number of creation and annihilation operators that must be explicitly considered in the $p–h$ formalism is independent of the total fermion number N in Eq. (2.10) characterizing our system. (In fact, this formalism can also be employed for extended systems.) The only complication that we face when transforming the X_I operators into the Y_I operators of the p–h formalism is the fact that we have to know whether I is a particle index or a hole one. For example, when transforming the Hamiltonian of Eq. (2.2), we would have to split each summation into the two separate ones, one extending over the holes and the other over the particles. (This would result in $2^4 = 16$ distinct two-body terms.) However, the complication can be avoided if we combine both representations, leaving the general operators (such as the Hamiltonian) expressed in terms of the X operators, and derive the expressions for p–h contractions. We easily find that [14, 78, 169]

$$\overline{X_A X_B} = \overline{X_A^\dagger X_B^\dagger} = 0, \ \overline{X_A^\dagger X_B} = h(A)\delta_B^A, \ \overline{X_A X_B^\dagger} = p(A)\delta_A^B, \tag{2.14}$$

where we introduce the p–h "step functions"

$$h(A) = 1, \ h(R) = 0, \ p(A) = 0, \ p(R) = 1. \tag{2.15}$$

If needed, similar expressions are easily derived for mixed cases, involving one X and one Y operator [78, 169].

To maximize the benefit offered by the generalized Wick theorem, it is convenient to transform the relevant operators to their n-product form (indicated by the subscript N). For the Hamiltonian H, Eq. (2.2), we thus get [1, 2, 78, 169]

$$H = H_N + \langle H \rangle, \quad \langle H \rangle \equiv \langle \Phi_0 | H | \Phi_0 \rangle, \tag{2.16a}$$

$$H_N = F_N + V_N, \quad F_N = Z_N + G_N, \tag{2.16b}$$

where

$$\Xi_N = \xi_I^J n[e_J^I] = \xi_I^J n[X_I^\dagger X_J], \quad \Xi = F, Z, G; \xi = f, z, g, \tag{2.17a}$$

$$V_N = \frac{1}{2} v_{IK}^{JL} n[e_{JL}^{IK}] = \frac{1}{2} v_{IK}^{JL} n[X_I^\dagger X_K^\dagger X_L X_J], \tag{2.17b}$$

with

$$f_I^J \equiv \langle I|f|J \rangle = z_I^J + g_I^J, \tag{2.18a}$$

$$g_I^J = \langle I|g|J \rangle \equiv \tilde{v}_{IA}^{JA}, \quad \tilde{v}_{IK}^{JL} \equiv v_{IK}^{JL} - v_{IK}^{LJ}, \tag{2.18b}$$

the sums over repeated indices being implied. Note that in the expression for g_I^J, the sum extends over the hole states $|A\rangle$ only, the g operator being reminiscent of the Hartree–Fock potential. It arises from single contraction terms of the two-electron part, while the fully contracted terms of H constitute the Fermi vacuum expectation value $\langle H \rangle$. Consequently, in applying the generalized Wick theorem, no internal contractions of operators constituting H_N are required.

We shall also need excitation operators that generate all possible configurations spanning the N-particle component of the Fock space when acting on the Fermi vacuum $|\Phi_0\rangle$. These operators simply arise as products of particle–hole $U(2n)$ generators e_A^R, and since particle and hole labels belong to distinct sets, we have

$$\left[e_A^R, e_B^S \right] = 0 \quad \text{and} \quad e_{ABC\cdots}^{RST\cdots} = e_A^R e_B^S e_C^T \cdots . \tag{2.19}$$

The k-excited configurations will then be denoted as

$$\left| \begin{matrix} R_1 \cdots R_k \\ A_1 \cdots A_k \end{matrix} \right\rangle = e_{A_1}^{R_1} \cdots e_{A_k}^{R_k} |\Phi_0\rangle \equiv G_j^{(k)} |\Phi_0\rangle \equiv |\Phi_j^{(k)}\rangle , \tag{2.20}$$

where, in the last two equations, we introduce a shorthand notation for the excitation operator $G_j^{(k)}$ and the k-excited configuration $|\Phi_j^{(k)}\rangle$, the superscript (k) indicating the excitation order and the subscript j the spin-orbital index set $j \equiv \{R_1, \ldots, R_k; A_1, \ldots, A_k\}$. A general k-body excitation operator Γ_k is then given by a linear combination:

$$\Gamma_k = (k!)^{-2} \tilde{\gamma}_{R_1 \cdots R_k}^{A_1 \cdots A_k} e_{A_1 \cdots A_k}^{R_1 \cdots R_k} \equiv \gamma_j^{(k)} G_j^{(k)} . \tag{2.21}$$

Finally, we recall a diagrammatic representation that not only facilitates the actual exploitation of Wick's theorem, but also provides a useful insight into the structure of the theory. For this purpose, one represents the individual creation $(X_I^\dagger)$ and annihilation (X_I) operators by oriented lines that, respectively, leave and enter various vertices, whose form indicates the operator they represent, as well as the associated scalar factor (e.g., $f_J^I, v_{JL}^{IK}, \gamma_{RS}^{AB}$, etc.). An example of diagrams representing the one- and two-body components of H or H_N and of a general excitation operator Γ_k of Eq. (2.21) is shown in Figure 1. The operators represented by a single vertex (top line) are referred to as *Hugenholtz diagrams*, while those indicating the k-body structure of the operators (bottom line) are *Goldstone diagrams*. With Hugenholtz vertices are associated antisymmetrized matrix elements (e.g., $\tilde{v}$ or $\tilde{\gamma}$), while with Goldstone vertices, one usually associates nonantisymmetrized quantities. However, most convenient is a combined representation

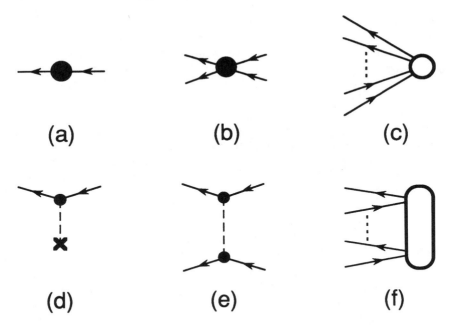

(a)   **(b)**   **(c)**

(d)   **(e)**   **(f)**

Figure 1. Hugenholtz (a–c) and Goldstone (d–f) (unlabeled) vertices representing one-body (a,d) and two-body (b,e) components of the Hamiltonian, and n-body excitation or cluster operators (c,f).

(sometimes referred to as *Brandow diagrams*), in which each Hugenholtz diagram is represented by one Goldstone version, since, in this way, each diagram uniquely determines both the algebraic expression it represents and the correct phase. (Clearly, the phase is undetermined in Hugenholtz representation). Since Hugenholtz diagrams result by collapsing the Goldstone vertices (represented by dashed lines or ovals) into a single point-like vertex, they are fewer in number. Thus, the combined (Brandow) representation inherits the advantages of both Goldstone and Hugenholtz representations: the smallest possible number of diagrams and a unique representation for the corresponding algebraic expressions [14, 78, 169].

The application of the (generalized) Wick theorem is then achieved by representing each contraction by joining the corresponding oriented lines. It is easily seen [14, 78, 169] that, in doing so, the line orientation must be preserved, lest the contribution vanish. Moreover, Eq. (2.14) implies that the *hole lines* (those carrying the hole label) are oriented from right to left, while *particle lines* extend in the opposite direction, assuming that the vertices are arranged horizontally in the same order as the corresponding operators in the evaluated expression. Each internal line then implies a summation over the hole or particle labels. Clearly, in evaluating the expectation values $\langle\Phi_0|M|\Phi_0\rangle$, only fully contracted diagrams (called *vacuum diagrams*) can contribute, while in evaluating expressions of the type $M|\Phi_0\rangle$, the uncontracted lines can only extend to the left. Following a few simple rules [14, 169], we associate with each distinct *resulting diagram* its weight (given by the reciprocal of the number of automorphisms of the bare diagram or skeleton resulting by stripping the diagram of summation labels), the phase $(-1)^{h+\ell}$ (h being the number of internal hole lines and ℓ the number of closed loops of oriented lines in the Goldstone or Brandow representation), and the product of scalar factors associated with the vertices and easily obtain the corresponding explicit algebraic form of the desired expression. Depending on the circumstances, further simplifying rules may be added, as in MBPT. (See later.) For a more detailed explanation of the diagrammatic technique, see [1, 2, 14, 78, 169, 170].

B. Many-Body Perturbation Theory (MBPT): Origins of Coupled-Cluster (CC) *Ansatz*

The idea of the exponential *Ansatz* was first employed in statistical mechanics in computing the partition function of a nonideal gas with pairwise interactions between the molecules, initially in its classical form [68, 178, 179] and later in its quantum analogue [69, 179, 180]. For many-fermion systems interacting via two-body forces, the exponential structure of the wave function in terms of the so-called *connected* terms (represented by connected MBPT diagrams; see shortly) was first clearly established by Hubbard [62]. Purely formally, we can understand this result by realizing that the wave function of a system involving two noninteracting subsystems A and B with the Hamiltonian $H = H_A + H_B$ will be given by the product of wave functions Φ_A and Φ_B characterizing the subsystems, (i.e., $\Phi_{A+B} = \Phi_A\Phi_B$), while the energy is additive, being the sum of the energies of the subsystems ($E_{A+B} = E_A + E_B$). Labeling the subsystems by suitable additive parameters a and b defining the energy (say, pair cluster amplitudes defined relative to Brueckner orbitals; see subsequently), we thus require that

$$\Phi(a + b) = \Phi(a)\Phi(b). \tag{2.22}$$

It is easy to show that any solution of this functional equation must have an exponential character, so that $\Phi(x) = e^{kx}$, with k a constant.

In view of the fact that the exponential *Ansatz* for the wave function has its origin in the MBPT and, consequently, that there exists a close relationship between the CC and MBPT approaches, we first recall the basic facts concerning the latter. The pioneering papers [61–63] establishing the structure of the MBPT employed a time-dependent formalism of quantum field theory and relied on the adiabatic theorem of Gell-Mann and Low [181]. However, when we are interested in stationary solutions of the Schrödinger equation, it is natural—and simpler—to rely on a time-independent formalism [78], as summarized earlier.

In quantum mechanical time-independent perturbation theory (PT) (see, e.g., [78, 169, 182]) we strive to find the eigenvalues k_j and the eigenvectors $|\Psi_j\rangle$ of the full (perturbed) problem

$$K|\Psi_j\rangle \equiv (K_0 + W)|\Psi_j\rangle = k_j|\Psi_j\rangle \qquad (2.23)$$

by exploiting the known corresponding quantities for an unperturbed-problem

$$K_0|\Phi_j\rangle = \kappa_j|\Phi_j\rangle, \quad \langle\Phi_j|\Phi_k\rangle = \delta_{jk}. \qquad (2.24)$$

Assuming for the sake of simplicity the nondegeneracy ($\kappa_j \neq \kappa_k$ if $j \neq k$) and using the so-called intermediate normalization for the state of interest $|\Psi_i\rangle$, viz.,

$$\langle\Phi_i|\Psi_i\rangle = 1, \qquad (2.25)$$

Eq. (2.23) implies the asymmetric energy formula

$$k_i = \kappa_i + \langle\Phi_i|W|\Psi_i\rangle. \qquad (2.26)$$

The desired PT expressions for $|\Psi_i\rangle$ and k_i then take the form [14, 78, 169, 182]

$$|\Psi_i\rangle = \sum_{n=0}^{\infty} [R_i(\lambda - k_i + W)]^n|\Phi_i\rangle, \qquad (2.27a)$$

$$k_i = \kappa_i + \sum_{n=0}^{n} \langle\Phi_i|W[R_i(\lambda - k_i + W)]^n|\Phi_i\rangle, \qquad (2.27b)$$

where

$$R_i \equiv R_i(\lambda) = \sum_{j(\neq i)} \frac{|\Phi_j\rangle\langle\Phi_j|}{\lambda - \kappa_j} \qquad (2.28)$$

is the resolvent operator and $\lambda(\neq \kappa_j)$ is an arbitrary scalar.

Now, setting $\lambda = k_i$ yields the Brillouin–Wigner (BW) PT, while the choice $\lambda = \kappa_i$ gives the Rayleigh–Schrödinger (RS) PT that is the basis of the MBPT. While the BWPT does not give-size extensive energies (when applied to the case $K = H$; see, however, [183–186]), the RSPT energy contributions can be arranged into size-extensive terms when grouped by their order in W. Thus, designating the nth-order contributions by the superscript (n) (and dropping the subscript i for simplicity), we write

$$k_i \equiv k = \sum_{n=0}^{\infty} k^{(n)}, \quad |\Psi_i\rangle \equiv |\Psi\rangle = \sum_{n=0}^{\infty} |\Psi^{(n)}\rangle \qquad (2.29a)$$

and substitute into Eq. (2.27b). Collecting terms of the same order in W, we find that the nth-order contribution has the general form

$$k^{(n)} = \langle W(RW)^{n-1} \rangle + \mathcal{R}^{(n)}, \qquad (2.29b)$$

where the first term on the right-hand side is referred to as the *principal term*, while the second term, the so-called *renormalization term*, involves products of the lower order terms, e.g.,

$$
\begin{aligned}
\mathcal{R}^{(0)} &= \mathcal{R}^{(1)} = \mathcal{R}^{(2)} = 0, \\
\mathcal{R}^{(3)} &= -\langle W \rangle \langle WR^2 W \rangle, \\
\mathcal{R}^{(4)} &= -\langle W \rangle [\langle WR(RW)^2 \rangle + \langle (WR)^2 RW \rangle] + \langle W \rangle^2 \langle WR^3 W \rangle \\
&\quad - \langle WRW \rangle \langle WR^2 W \rangle, \text{ etc.}
\end{aligned}
\qquad (2.30)
$$

Here, R designates the RS resolvent; that is,

$$R \equiv R_i^{(\text{RS})} = \sum_{j(\neq i)} \frac{|\Phi_j\rangle \langle \Phi_j|}{\kappa_i - \kappa_j}. \qquad (2.31)$$

Note also that we use the shorthand notation for the reference-state mean values,

$$\langle X \rangle \equiv \langle \Phi_i | X | \Phi_i \rangle, \qquad (2.32)$$

and that the expressions for $|\Psi^{(n)}\rangle$ are obtained by removing the leftmost bra and the first perturbation operator W in the expressions for $k^{(n)}$. [Cf. Eqs. (2.27a) and (2.27b).] The renormalization terms of any order can be directly generated by the so-called *bracketing technique* (cf. [14, 78, 169]), and their total number is $\{(2n - 2)! / [n!(n - 1)!] - 1\}$.

Applying the RSPT formalism to our *ab initio* model Hamiltonian H, Eq. (2.2), in its normal product form given by Eqs. (2.16), we set

$$K = H - \langle H \rangle = H_N, \tag{2.33a}$$

$$K_0 = H_0 - \langle H_0 \rangle, \quad H_0 = Z + U, \tag{2.33b}$$

where U designates some suitably chosen one-body potential approximating the two-body part V of H, so that

$$W = K - K_0 = V - U - \langle V - U \rangle \equiv W_N. \tag{2.33c}$$

Designating the eigenvalues of H and H_0 by E_i and ε_i, respectively, we find that the eigenvalues of K and K_0, Eqs. (2.23) and (2.24), become

$$k_j \equiv \Delta E_j = E_j - \langle H \rangle \quad \text{and} \quad \kappa_j \equiv \Delta \varepsilon_j = \varepsilon_j - \varepsilon_i. \tag{2.34}$$

The n-product form of the perturbed and unperturbed operators, Eq. (2.33), greatly simplifies the RSPT series consisting of Eqs. (2.29) and (2.30), since $k^{(0)} = \kappa_i = 0$ and $k^{(1)} = \langle W \rangle = 0$ for any choice of U, so that the renormalization terms will appear for the first time in the fourth order, i.e.,

$$\begin{aligned}
\Delta E^{(0)} &= \Delta E^{(1)} = 0, \\
\Delta E^{(2)} &= \langle WRW \rangle, \\
\Delta E^{(3)} &= \langle WRWRW \rangle, \\
\Delta E^{(4)} &= \langle W(RW)^3 \rangle - \langle WRW \rangle \langle WR^2W \rangle, \text{ etc.,}
\end{aligned} \tag{2.35}$$

where we drop the subscript i for simplicity and where the denominator in Eq. (2.31) is given by $-\kappa_j = -\Delta \varepsilon_j$.

The unperturbed eigenvalue problem for H_0, Eq. (2.33b) [or for K_0, Eq. (2.24)] is separable by definition and thus reduces to a one-electron problem,

$$(z + u)|I\rangle = \omega_I |I\rangle, \quad \langle I|J \rangle = \delta_J^I, \tag{2.36}$$

which in turn supplies us with a suitable one-electron basis $\{|I\rangle\}$. The unperturbed eigenstates are N-spin-orbital configurations given by

$$|\Phi_j\rangle = |\{I_1 I_2 \cdots I_N\}\rangle = X_{I_1}^\dagger X_{I_2}^\dagger \cdots X_{I_N}^\dagger |0\rangle, \tag{2.37}$$

and corresponding eigenvalues are given by an appropriate sum of spin-orbital energies ω_{I_k}. Choosing $|\Phi_0\rangle$, Eq. (2.10), as our reference (i.e., setting $i = 0$ and using the particle–hole formalism labeling of spin orbitals), we can represent the remaining $|\Phi_j\rangle$, $j \neq 0$, relative to $|\Phi_0\rangle$ as in Eq. (2.20) by identifying the configuration label j with the index set $\{R_1, \ldots, R_k;$

$A_1, \ldots, A_k\}$. The corresponding eigenvalues of $K_0 = H_0 - \langle H_0 \rangle$ are then given by

$$\kappa_0 = 0, \ \kappa_j = \sum_{i=1}^{k}(\omega_{R_i} - \omega_{A_i}), \quad j \equiv \{R_1, \ldots, R_k; A_1, \ldots, A_k\} \neq \emptyset, \quad (2.38)$$

so that the resolvent denominators $\kappa_0 - \kappa_j = -\kappa_j = -\Delta\varepsilon_j$ are given by the difference of hole and particle orbital energies defining a given excited configuration $|\Phi_j\rangle$ [Eq. (2.20)].

In view of the structure of the n-product form of our Hamiltonian, Eqs. (2.16), the most natural choice for U is to set $U = G$, so that $H_0 = Z + G = F$, and

$$K_0 = F_N, \quad \text{and} \quad W \equiv W_N = V_N, \quad (2.39)$$

which will be the case when we choose the Hartree–Fock (HF) reference. In this case, $\langle H \rangle = z_A^A + \frac{1}{2}g_A^A$ is the HF energy, so that $\Delta E_i \equiv \Delta E = \Sigma_{n=0}^{\infty}\Delta E^{(n)}$ gives directly the correlation energy. Note, however, that our formulation eliminates the first-order PT contribution $\Delta E^{(1)} = \langle W \rangle$, regardless of the choice of U or, equivalently, of the spin-orbital basis employed.

The structure of the resulting RSPT expansion becomes particularly transparent when we employ a diagrammatic representation. (See, e.g., [14, 78, 169, 170].) Using the HF reference, so that $W = V_N$, we have to evaluate expressions of the type $\langle\Phi_0|V_N R^k V_N R^\ell \cdots V_N|\Phi_0\rangle$ and $R^k V_N R^\ell \cdots V_N|\Phi_0\rangle$, where

$$R^k = \sum_{j(\neq 0)} \frac{|\Phi_j\rangle\langle\Phi_j|}{(-\Delta\varepsilon_j)^k}. \quad (2.40)$$

We thus replace each operator in these expressions by a corresponding diagram, construct all possible distinct *resulting diagrams* by interconnecting the oriented lines, and evaluate the diagrams by the following simple rules referred to earlier [14, 78, 169]. In the current case, we represent each V_N operator by a v-vertex (Figure 1b or 1e), placing them side by side as the V_N operators in the evaluated expression, and realize that we can account for the operators R^k ($k \geq 1$) of Eq. (2.40) that are interspersed between the V_N operators [cf. Eqs. (2.35)], by requiring that at least one oriented line cross a vertical line separating each neighboring pair of v-vertices. [In this way, we avoid the vacuum projector $|\Phi_0\rangle\langle\Phi_0|$ that is excluded from the sum in Eq. (2.40).] With each such vertical line (which we represent in the diagram by a dashed vertical line with the index k only when $k > 1$), we then associate the denominator factor $(-\Delta\varepsilon_j)^{-k}$.

In evaluating vacuum mean values $\langle WR^k W \cdots W \rangle$, only the so-called *vacuum diagrams*, representing the fully contracted terms and thus having only internal lines interconnecting distinct vertices, can contribute, while the *wave function–type diagrams*, representing the terms of the type $R^k W \cdots W |\Phi_0\rangle$, have free (i.e., uncontracted) oriented lines extending to the left. (Their number defines the excitation order of such a contribution.)

Turning our attention to the energy contributions $\Delta E^{(n)}$, we see that in the second and the third orders only connected* diagrams can arise (recall that $W = V_N$ is in the *n*-product form), whose Hugenholtz form is shown in Figure 2. Starting with the fourth order, disconnected diagrams issuing from the principal term $\langle W(RW)^3 \rangle$ can also arise. (Cf. Figures 3a and 3b; Figure 3c represents the renormalization term $\mathcal{R}^{(4)}$.) Generally, any diagram containing a disconnected vacuum part is called an *unlinked* diagram. Clearly, for energy diagrams, the terms *linked* and *connected* are synonymous, since only vacuum diagrams are involved. This is not the case, however, for wave function diagrams, where linked disconnected diagrams (no disconnected component is of a vacuum type—cf. Figure 4a) must be distinguished from unlinked ones (Figure 4b).

The essence of the MBPT *linked-cluster theorem* is the fact that, in both the energy and wave function contributions, all unlinked terms (which are not size extensive) cancel out in each order. To see how this cancellation works in the simplest case, consider the contribution from unlinked diagrams in Figures 3a and 3b. The numerators (given by the product of

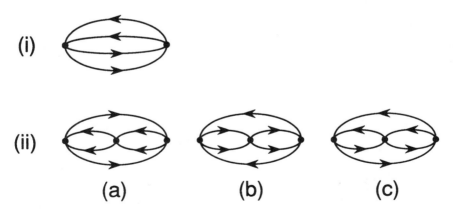

Figure 2. Hugenholtz (unlabeled) diagrams representing the second- (i) and third- (ii) order MBPT energy contributions.

---

*The terms *connected* and *disconnected* have their standard graph-theoretical meaning.

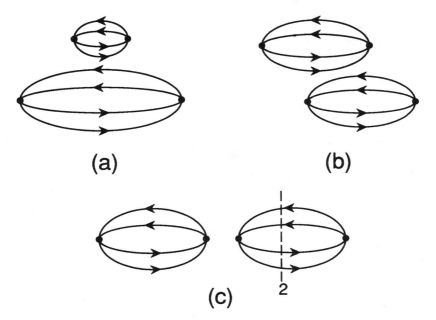

Figure 3. Hugenholtz (unlabeled) diagrams representing the fourth-order MBPT unlinked (or disconnected) energy contributions (a,b) that are canceled by the renormalization term $\langle WRW\rangle\langle WR^2W\rangle$ represented by diagram (c).

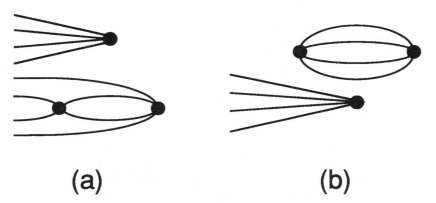

Figure 4. Example of the third-order linked (a) and unlinked (b) (nonoriented and unlabeled) MBPT wave function diagrams.

four two-electron v-integrals) being the same in each case, we consider only the denominators $(-\Delta \varepsilon_j^a)$. Designating the denominators associated with the top and the bottom disconnected component of each diagram by A and B, respectively, we find, for the overall contribution,

$$\frac{1}{B} \cdot \frac{1}{A+B} \cdot \frac{1}{B} + \frac{1}{A} \cdot \frac{1}{A+B} \cdot \frac{1}{B} = \frac{1}{AB^2} \, . \tag{2.41}$$

This is precisely (up to the sign) the contribution from the renormalization term (Figure 3c). A similar cancellation occurs for wave function diagrams, leaving only linked contributions (e.g., Figure 4a), while the unlinked ones exactly cancel out [169, 187]. We can thus express the linked-cluster theorem by writing

$$\Delta E = \sum_{n=0}^{\infty} \langle \Phi_0 | W | \Psi^{(n)} \rangle = \sum_{n=0}^{\infty} \langle W(RW)^n \rangle_L \, , \tag{2.42a}$$

$$|\Psi\rangle = \sum_{n=0}^{\infty} |\Psi^{(n)}\rangle = \sum_{n=0}^{\infty} \{(RW)^n |\Phi_0\rangle\}_L \, , \tag{2.42b}$$

where the subscript L indicates that only linked contributions or diagrams are to be retained. We thence obtain the nth-order MBPT energy contribution $\Delta E^{(n)}$ by constructing all possible distinct linked diagrams involving n interaction vertices and transcribing them into the algebraic form following a few simple rules. (See, e.g., [78, 169, 170] for details.) The corresponding wave function diagrams or expressions for $|\Psi^{(n)}\rangle$ may be obtained by removing the leftmost vertex in the $\Delta E^{(n+1)}$ diagrams and keeping all possible distinct ones.

A word of caution is in order here. By performing independent summations over the hole and particle labels carried by oriented lines, we also automatically include the so-called *exclusion-principle-violating* (EPV) terms that correspond to nonphysical configurations $|\Phi_j\rangle$ in Eq. (2.28), (2.31), or (2.40). In fact, these configurations vanish, since they involve two or more identical spin orbitals. However, the contributions from the corresponding $\Delta E^{(n)}$ or $|\Psi^{(n)}\rangle$ diagrams do not vanish, but can be shown to be compensated for by the corresponding unlinked terms. Thus, considering the principal term, we see that it is irrelevant whether we include the EPV terms or disregard them (by placing appropriate restrictions on the spin-orbital summation labels). However, since the unlinked terms arising from the principal term are used to cancel out the renormalization terms, in which the $p-h$ labels are summed independently in each disconnected part, we *must* keep the EPV terms in linked contributions in Eq. (2.42).

While both the energy and wave function MBPT series of Eq. (2.42) involve only linked terms, the former ones are always connected, which is not the case for $|\Psi^{(n)}\rangle$. (Cf. Figure 4a.) To reveal a deeper structure of the latter, we introduce the *cluster operator*

$$T|\Phi_0\rangle = \sum_{n=1}^{\infty} T^{(n)}|\Phi_0\rangle, \qquad (2.43)$$

which generates only connected wave function contributions

$$T^{(n)}|\Phi_0\rangle = \{(RW)^n|\Phi_0\rangle\}_C, \qquad (2.44)$$

as indicated by the subscript C. Since all the linked terms involving r disconnected components can be generated as the rth power of T, with each distinct contribution appearing $r!$ times, we see that the general structure of the exact wave function $|\Psi\rangle$ has the exponential character in terms of the connected cluster operator:

$$|\Psi\rangle = e^T|\Phi_0\rangle. \qquad (2.45)$$

This result is referred to as the *connected-cluster theorem* (and is often confused with the linked-cluster theorem stated earlier). Equivalently, we can write the so-called wave operator $\mathcal{W}$ transforming a given IPM reference into the exact wave function

$$|\Psi\rangle = \mathcal{W}|\Phi_0\rangle \qquad (2.46)$$

as the exponential of the cluster operator T, i.e.,

$$\mathcal{W} = e^T, \qquad (2.47)$$

which is the basis of the SR CC approaches.

Since all linked wave function diagrams, whether connected or not, have at least one pair of p–h free (i.e., uncontracted) oriented lines extending to the left, we can also classify them by the excitation order k given by the number of such p–h pairs. Designating all k-times excited contributions by $C_k|\Phi_0\rangle$, we can write

$$|\Psi\rangle = \sum_{k=0}^{N} C_k|\Phi_0\rangle, \quad C_k = (k!)^{-2}\tilde{c}_{R_1\cdots R_k}^{A_1\cdots A_k}e_{A_1}^{R_1}\cdots e_{A_k}^{R_k} = c_j^{(k)}G_j^{(k)}, \qquad (2.48)$$

where $C_0 = 1$ in view of the chosen intermediate normalization given by Eq. (2.25). Similarly, we can classify the connected-cluster components by their excitation order, writing

$$T = \sum_{k=1}^{N} T_k, \quad T_k = (k!)^{-2} \tilde{t}_{R_1 \cdots R_k}^{A_1 \cdots A_k} e_{A_1}^{R_1} \cdots e_{A_k}^{R_k} = t_j^{(k)} G_j^{(k)}. \tag{2.49}$$

In the rightmost expressions for C_k and T_k in Eqs. (2.48) and (2.49), we use the notation introduced in Eq. (2.20). (Note that in this case the implied summation extends only over the spin-orbital index set j.)

Using Eq. (2.45), we readily find the relationship between the k-body excitation operators C_k and the corresponding cluster operators T_j, namely [70, 78, 169],

$$C_k = T_k + \sum_{\mathcal{P}_k} \prod_{j=1}^{p} (n_j!)^{-1} T_j^{n_j}, \tag{2.50}$$

with the sum extending over all nontrivial partitions $\mathcal{P}_k$ of k, $k = \sum_{j=1}^{p} j n_j$, $0 \leq n_j \leq k$, $1 \leq p < k$. Thus,

$$C_1 = T_1,$$
$$C_2 = T_2 + \frac{1}{2} T_1^2,$$
$$C_3 = T_3 + T_1 T_2 + \frac{1}{3!} T_1^3, \tag{2.51}$$
$$C_4 = T_4 + \frac{1}{2} T_2^2 + T_1 T_3 + \frac{1}{2} T_1^2 T_2 + \frac{1}{4!} T_1^4, \text{ etc.}$$

The first term on the right-hand side of Eq. (2.50) or (2.51) represents the *connected* k-times excited component, while the remaining terms, involving at least two T_j operators of lower excitation order, represent the *disconnected* part of C_k. It is precisely this structure that provides a clue to the size-extensive formalism in view of the desired exponential character of the wave function in terms of the cluster operator T of Eq. (2.45), which automatically satisfies the separability property given by Eq. (2.22).

C. Single-Reference Coupled-Cluster (SR CC) Approaches

1. General Spin-Orbital Formalism

To gain insight into the basic structure of CC theory and the techniques enabling us to obtain explicit CC equations, we first consider the simplest— yet the most important—case, in which the cluster operator T is approximated by its pair cluster component T_2; that is,

$$T \approx T_2. \tag{2.52}$$

We shall exploit essentially the wave function (or first-quantization) formalism to obtain basic, energy-independent CCD equations, whose detailed explicit form in terms of one and two electron integrals and pair cluster amplitudes may be derived by using either the same formalism or second-quantization-based diagrammatic or algebraic techniques. This should illustrate the close interrelationship of these techniques, including their relative merits and potential synergy.

We start with the well-known SR CI approach, wherein we expand the exact wave function (within the finite-dimensional N-electron space implied by the choice of our basis set) in terms of N-electron configurations given by Eq. (2.20), as in Eq. (2.48), and consider the coefficients $c_{R_1 \cdots R_k}^{A_1 \cdots A_k}$, or $c_j^{(k)}$, at

$$|\Phi_j^{(k)}\rangle \equiv \left| \begin{matrix} R_1 \cdots R_k \\ A_1 \cdots A_k \end{matrix} \right\rangle, \quad j \equiv \{R_1, \ldots, R_k; A_1, \ldots, A_k\}, \quad \text{as variational para-}$$

meters. Considering, for simplicity's sake, only configurations excited an even number of times, i.e.,

$$|\Psi\rangle = (1 + C_2 + C_4 + \cdots)|\Phi_0\rangle, \tag{2.53}$$

we use either the variation principle or directly the time-independent Schrödinger equation [in the n-product form; cf. Eqs. (2.16)]

$$H_N(1 + C_2 + C_4 + \cdots)|\Phi_0\rangle = \Delta E(1 + C_2 + \cdots)|\Phi_0\rangle \tag{2.54}$$

to generate the chain of CI equations by projecting onto the configuration states spanning the trial wave function, Eq. (2.53), obtaining

$$\langle \Phi_0 | H_N C_2 | \Phi_0 \rangle = \Delta E, \tag{2.55a}$$

$$\langle \Phi_j^{(2)} | H_N(1 + C_2 + C_4) | \Phi_0 \rangle = (\Delta E) c_j^{(2)}, \tag{2.55b}$$

$$\langle \Phi_j^{(4)} | H_N(C_2 + C_4 + C_6) | \Phi_0 \rangle = (\Delta E) c_j^{(4)}, \text{ etc.}, \tag{2.55c}$$

or, in matrix form,

$$\mathbf{H}\mathbf{C} = (\Delta E)\mathbf{C}, \tag{2.56a}$$

where

$$(\mathbf{H})_{jk} \equiv H_{jk} = \langle \Phi_j | H_N | \Phi_k \rangle \quad \text{and} \quad \mathbf{C}^T = (1 \, \mathbf{C}_2^T \, \mathbf{C}_4^T \cdots), (\mathbf{C}_i)_j = c_j^{(i)}, \tag{2.56b}$$

are CI matrix elements and corresponding coefficients, with the superscript T indicating the transpose.

In all practical applications, the full CI (FCI) chain of equations (2.55) has to be truncated at some excitation level, yielding CID by setting $C_4 = C_6 = \cdots = 0$, CIDQ by setting $C_6 = C_8 = \cdots = 0$, etc. Such a truncated CI is never size extensive, the effect being particularly strong for truncation at the CID level. In fact, it can be shown that for extended systems, a CI wave function truncated at any finite excitation level has zero overlap with the exact wave function. This deficiency can be overcome by employing a size-extensive cluster *Ansatz* for the exact wave function, Eq. (2.45), permitting a much more physical truncation. For example, by truncating at the pair cluster level, Eq. (2.52), yielding the CCD method, we retain the major part of quadruples, which comes from their disconnected component $C_4 \approx \frac{1}{2}T_2^2$, since, symbolically, $T_4 \ll \frac{1}{2}T_2^2$. (Recall that we set $T_1 = 0$, which corresponds to the use of Brueckner orbitals.) Thus, using Eq. (2.51) with $T_1 = T_3 = T_4 = \cdots = 0$, the chain of equations (2.55) decouples at the double excitation level, yielding, in addition to Eq. (2.55a) for the energy,

$$\Delta E = \langle \Phi_0 | H_N T_2 | \Phi_0 \rangle \,, \tag{2.57a}$$

a precursor of the CCD equation (2.55b), namely,

$$\langle \Phi_j^{(2)} | H_N (1 + T_2 + \frac{1}{2}T_2^2) | \Phi_0 \rangle = (\Delta E) t_j^{(2)} \,, \tag{2.57b}$$

where $j \equiv \{R, S; A, B\}$.

Now, the essence of the CC approach is to cancel out all unlinked terms, so that a size-extensive formalism results. To see this cancellation most easily, we express the term involving disconnected quadruple excitations $\frac{1}{2}\langle \Phi_j^{(2)} | H_N T_2^2 | \Phi_0 \rangle$ in terms of commutators, using the simple relationship

$$[[A, B], B] = AB^2 - 2BAB + B^2 A \,. \tag{2.58}$$

Thus,

$$\begin{aligned}
\frac{1}{2}\langle \Phi_j^{(2)} | H_N T_2^2 | \Phi_0 \rangle &= \frac{1}{2}\langle \Phi_j^{(2)} | [[H_N, T_2], T_2] | \Phi_0 \rangle \\
&+ \langle \Phi_j^{(2)} | T_2 H_N T_2 | \Phi_0 \rangle - \frac{1}{2}\langle \Phi_j^{(2)} | T_2^2 H_N | \Phi_0 \rangle \,.
\end{aligned} \tag{2.59}$$

Since, obviously,

$$\langle \Phi_j^{(2)} | T_2^2 = \langle \Phi_0 | e_R^A e_S^B T_2^2 = 0 \,, \tag{2.60}$$

and

$$\langle \Phi_j^{(2)} | T_2 H_N T_2 | \Phi_0 \rangle = \sum_k \langle \Phi_j^{(2)} | T_2 | \Phi_k \rangle \langle \Phi_k | H_N T_2 | \Phi_0 \rangle$$
$$= \langle \Phi_j^{(2)} | T_2 | \Phi_0 \rangle \langle \Phi_0 | H_N T_2 | \Phi_0 \rangle \quad (2.61)$$
$$= t_j^{(2)} (\Delta E) ,$$

Eq. (2.59) becomes

$$\frac{1}{2} \langle \Phi_j^{(2)} | H_N T_2^2 | \Phi_0 \rangle = \frac{1}{2} \langle \Phi_j^{(2)} | [[H_N, T_2], T_2] | \Phi_0 \rangle + (\Delta E) t_j^{(2)} , \quad (2.62)$$

and our precursor CCD equation (2.57b) takes the energy-independent form

$$\langle \Phi_j^{(2)} | H_N | \Phi_0 \rangle + \langle \Phi_j^{(2)} | H_N T_2 | \Phi_0 \rangle + \frac{1}{2} \langle \Phi_j^{(2)} | [[H_N, T_2], T_2] | \Phi_0 \rangle = 0 . \quad (2.63)$$

Expressing T_2 in terms of two-body excitation operators [cf. Eqs. (2.20) and (2.21)], i.e.,

$$T_2 = (2!)^{-2} \tilde{t}_{RS}^{AB} e_A^R e_B^S \equiv t_j^{(2)} G_j^{(2)} , \quad (2.64)$$

we obtain a quadratic system of algebraic equations in CCD amplitudes $t_j \equiv t_j^{(2)}$ (the summation convention is implied), viz.,

$$a_i + b_{ij} t_j + c_{ijk} t_j t_k = 0, \quad (i = 1, 2, \ldots, M) \quad (2.65)$$

where

$$a_i = \langle \Phi_i^{(2)} | H_N | \Phi_0 \rangle , \quad (2.66a)$$

$$b_{ij} = \langle \Phi_i^{(2)} | H_N | \Phi_j^{(2)} \rangle , \quad (2.66b)$$

$$c_{ijk} = \frac{1}{2} \langle \Phi_i^{(2)} | \left[[H_N, G_j^{(2)}], G_k^{(2)} \right] | \Phi_0 \rangle , \quad (2.66c)$$

and M designates the number of biexcited configurations. These coefficients may be evaluated using Slater rules [75, 76] or, more efficiently, diagrammatic [1, 2, 14, 169] or algebraic [29, 173] (cf. Appendix A) techniques. Note that $a_i = H_{i0}$ and $b_{ij} = H_{ij}$ are in fact CID matrix elements, constituting the CID matrix

$$\mathbf{H}^{(\mathrm{CID})} = \begin{bmatrix} 0 & \mathbf{a}^\dagger \\ \mathbf{a} & \mathbf{b} \end{bmatrix}. \qquad (2.67)$$

However, since the CCD equations can be multiplied by any nonzero constant, and it is often convenient to define a set of suitably renormalized cluster amplitudes $\tau_j = k(j)t_j$, the matrix $\mathbf{b}$ appearing in CCD equations is usually no longer Hermitian (or symmetric). (Cf., e.g., [1, 2, 5, 14, 85, 128, 129, 169].)

In the general case, Eq. (2.45), we can achieve the same result by using the operator identity

$$e^{-B} A e^{B} = \sum_{n=0}^{\infty} (n!)^{-1} [\cdots [[A, \underbrace{B], B], \dots, B]}_{n}, \qquad (2.68)$$

so that, starting with the Schrödinger equation in its n-product form, namely,

$$H_N e^{T} |\Phi_0\rangle = \Delta E e^{T} |\Phi_0\rangle, \qquad (2.69)$$

and acting from the left with the inverse of the wave operator $\mathcal{W}^{-1} = e^{-T}$, we obtain

$$e^{-T} H_N e^{T} |\Phi_0\rangle = \Delta E |\Phi_0\rangle. \qquad (2.70)$$

Projecting now onto $|\Phi_0\rangle$, we get the energy expression

$$\Delta E = \langle e^{-T} H_N e^{T} \rangle = \langle [H_N, T_1 + T_2 + \frac{1}{2} T_1^2] \rangle = \langle H_N (T_1 + T_2 + \frac{1}{2} T_1^2) \rangle, \quad (2.71)$$

and projecting onto the excited configurations $|\Phi_i\rangle$, $i \neq 0$, we get the desired CC equations

$$\langle \Phi_i | e^{-T} H_N e^{T} |\Phi_0\rangle = 0. \qquad (2.72)$$

Using the expansion given in Eq. (2.68) and the fact that $\langle \Phi_i | = \langle \Phi_0 | G_i^\dagger$, we can write general SR CC equations in terms of the reference-state mean values as

$$\sum_{n=0}^{\infty} (n!)^{-1} \langle G_i^\dagger [\cdots [[H_N, T], T], \dots, T] \rangle = 0. \qquad (2.73)$$

This summation is in fact always finite, since our Hamiltonian involves at most two-body potentials. Thus, considering a truncation of T at the r-body level, i.e.,

$$T = \sum_{j=1}^{r} T_j, \tag{2.74}$$

we can express the standard CC equations (2.73) in the form

$$\langle G_i^{(k)\dagger} H_N \rangle + \sum_{s=1}^{4} \sum_{\substack{1 \leq j_1, \ldots, j_s \leq r \\ (k+s-2 \leq j_1 + \cdots + j_s \leq k+2)}} (s!)^{-1} \langle G_i^{(k)\dagger} [\cdots [H_N, T_{j_1}], \ldots, T_{j_s}] \rangle = 0, \tag{2.75}$$

where $k = 1, \ldots, r$. Moreover, the absolute-value term $\langle G_i^{(k)\dagger} H_N \rangle$ vanishes for $k > 2$ and also for $k = 1$ if HF spin orbitals are employed.

Using the shorthand notation

$$\Lambda_i^{(k)}(j_1, \ldots, j_s) = \langle G_i^{(k)\dagger} [\cdots [H_N, T_{j_1}], \ldots, T_{j_s}] \rangle, \tag{2.76}$$

for the general term of Eq. (2.75), we see that the basic CCSD equations $(r = 2)$ involve the following terms (dropping the index i for simplicity):

$$\langle G^{(1)\dagger} F_N \rangle + \Lambda^{(1)}(1) + \Lambda^{(1)}(2)$$
$$+ \frac{1}{2} \Lambda^{(1)}(1,1) + \Lambda^{(1)}(1,2) + \frac{1}{6} \Lambda^{(1)}(1,1,1) = 0, \tag{2.77a}$$

$$\langle G^{(2)\dagger} H_N \rangle + \Lambda^{(2)}(1) + \Lambda^{(2)}(2)$$
$$+ \frac{1}{2} \Lambda^{(2)}(1,1) + \Lambda^{(2)}(1,2) + \frac{1}{2} \Lambda^{(2)}(2,2) \tag{2.77b}$$
$$+ \frac{1}{6} \Lambda^{(2)}(1,1,1) + \frac{1}{2} \Lambda^{(2)}(1,1,2) + \frac{1}{24} \Lambda^{(2)}(1,1,1,1) = 0.$$

The detailed explicit structure of CC equations is most transparent and easy to derive when we use a diagrammatic representation. (See also the next section.) Separating the connected components (indicated by the subscript C), we can express the left-hand side of Eq. (2.69) as [1, 2, 14, 169]

$$H_N e^T |\Phi_0\rangle = (H_N e^T)_C e^T |\Phi_0\rangle = e^T (H_N e^T)_C |\Phi_0\rangle, \tag{2.78}$$

since $e^T |\Phi_0\rangle = (1 + T + \frac{1}{2} T^2 + \cdots) |\Phi_0\rangle$ is represented by diagrams having none, one, two, three, etc., disconnected t-vertices. Thus, $e^T |\Phi_0\rangle$ can always be recovered after a finite number (at most four) of its t-vertices have been contracted with f- or v-vertices, the latter representing H_N. (Again, EPV terms must be included to achieve factorization; for a graphical illustration of this result, see [14, 169].) Thus, Eq. (2.69) becomes

$$(H_N e^T)_C |\Phi_0\rangle = \Delta E |\Phi_0\rangle , \qquad (2.79)$$

which can be viewed as the connected-cluster form of the Schrödinger equation (2.54). Projecting onto $|\Phi_0\rangle$, we recover the energy expression of Eq. (2.71),

$$\Delta E = \langle \Phi_0 | (H_N e^T)_C |\Phi_0\rangle = \langle H_N e^T \rangle_C , \qquad (2.80)$$

and projecting onto $|\Phi_i\rangle \equiv |\Phi_i^{(k)}\rangle$, we recover the CC equations (2.72), viz.,

$$\langle \Phi_i | (H_N e^T)_C |\Phi_0\rangle = 0 . \qquad (2.81)$$

Since H_N entails at most two-body potentials, the corresponding vertices have at most four fermion lines, and thus any connected term can involve at most four connected T_i components, so that

$$(H_N e^T)_C |\Phi_0\rangle = \sum_{s=0}^{4} (s!)^{-1} (H_N T^s)_C |\Phi_0\rangle . \qquad (2.82)$$

Hence, projecting onto $|\Phi_i^{(k)}\rangle$, only diagrams involving k hole and k particle free lines extending to the left survive—namely, the terms

$$\Lambda_i^{(k)}(j_1, \ldots, j_s) = \langle G_i^{(k)\dagger} H_N T_{j_1} \cdots T_{j_s} \rangle_C \qquad (s < 4) \qquad (2.83)$$

that arise in the general CC equations (2.75).

In order to appreciate the interrelationship between the diagrammatic and algebraic representations, we realize that no nonvanishing contractions are possible in $T_j H_N |\Phi_0\rangle$, since T_j involves only the p–h creation operators $Y_I^\dagger$ (i.e., either $X_R^\dagger$ or X_A), so that all fermion lines issuing from any t-vertex must extend to the left. Clearly, all such terms will also arise when one applies the generalized Wick theorem to $H_N T_j |\Phi_0\rangle$ (as the uncontracted terms), in addition to the terms involving at least one contraction between the v- (or f-) and t-vertices. Thus, subtracting $T_j H_N |\Phi_0\rangle$ from $H_N T_j |\Phi_0\rangle$ will leave only contracted (and thus connected) terms, so that

$$[H_N, T_j] |\Phi_0\rangle = (H_N T_j)_C |\Phi_0\rangle . \qquad (2.84)$$

If some fermion lines issuing from H_N vertices are left uncontracted in $(H_N T_j)_C |\Phi_0\rangle$, the process can be repeated with another T_k, until no free lines issuing from f- and v-vertices are left. This will clearly occur if four t-vertices are involved, thereby explaining the connection between the "algebraic" and "diagrammatic" forms for $\Lambda_i^{(k)}(j_1, \ldots, j_s)$'s, Eqs. (2.76) and (2.83), respectively.

Using the diagrammatic representation, we can easily obtain the explicit CC equations by constructing all connected diagrams involving one f- or v-vertex and up to four t-vertices. The vacuum-type diagrams will give the expression for the energy (Figure 5), while those having k particle and k hole lines extending to the left (representing $\langle \Phi_0 | G_i^{(k)\dagger} = \langle \Phi_i^{(k)}|$) contribute to the subset of CC equations associated with k-times excited subspace. For the simplest CCD case, these are shown in Figure 6. Since no denominators are now involved, the rules for evaluating CC diagrams are particularly simple. (See, e.g., [1, 2, 14, 27, 128, 169].)

We can, of course, reach the same result in a purely algebraic manner, evaluating the required commutators by exploiting the rules summarized in Appendix A. (For an explicit example (CCD), see [173].) While the diagrammatic technique has the advantage of being very transparent and direct, the algebraic way is more suitable for automation. As we shall see later, the algebraic technique is particularly suitable when one considers MR generalizations of SR CC approaches, particularly when their spin-adapted form is required. (Cf. [45].)

The diagrammatic representation makes it especially easy to establish the relationship with the MBPT. Using HF orbitals, so that the one-body part of H_N is diagonal (i.e., $\langle i|f|j\rangle = \omega_i \delta_{ij}$), we find that the diagrams involving f-vertices (Figure 6b) contribute only to the diagonal linear term $b_{ii} t_i$ (there is no implied summation), in the amount $t_i \Delta \varepsilon_i = t_i \sum_{j=1}^k (\omega_{R_j} - \omega_{A_j})$, where we

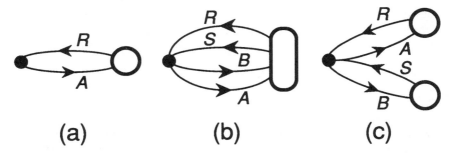

Figure 5.   Hugenholtz diagrams representing the CC energy, Eq. (2.71) or Eq. (2.80).

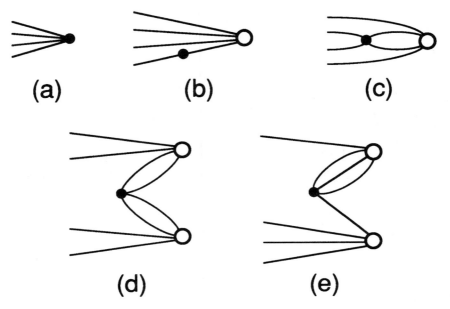

Figure 6. Nonoriented (and unlabeled) Hugenholtz diagrams characterizing the CCD method.

use the notation of Eq. (2.38). We can thus solve the CC equations (2.65) iteratively by moving this contribution to the other side of the equation and dividing by $-\Delta\varepsilon_i$, obtaining

$$t_i^{[n+1]} = (-\Delta\varepsilon_i)^{-1} \left[a_i + b'_{ij} t_j^{[n]} + c_{ijk} t_j^{[n]} t_k^{[n]} \right], \qquad (2.85)$$

where $b'_{ij} = b_{ij}$, $(i \neq j)$; $b'_{ii} = b_{ii} - \Delta\varepsilon_i$; and the superscript $[n]$ indicates the iterative order. Initially setting $t_j^{[0]} = 0$, we get $t_j^{[1]} = a_i/(-\Delta\varepsilon_i)$, so that, already in the first iteration at the CCD level, we get

$$\Delta E^{(2)} = \frac{1}{4} \tilde{v}_i a_i/(-\Delta\varepsilon_i) = \frac{1}{4} \tilde{v}_{AB}^{RS} \tilde{v}_{RS}^{AB}/(\omega_A + \omega_B - \omega_R - \omega_S), \qquad (2.86)$$

i.e., the second-order MBPT result.

This result is easy to interpret diagrammatically: $t_j^{[1]}$ is represented by the diagram of Figure 6a, regarded now as the first-order wave function diagram with the associated denominator rule, so that, when evaluating the corresponding energy by substituting into the diagram of Figure 5b, we get the second-order result of Figure 2(i). In the next iteration, already the linear terms (again at the CCD level) yield all the third-order MBPT diagrams of Figure 2(ii), while the nonlinear terms contribute in the fourth order. We see that we can generate the entire MBPT series in this way. Clearly, when T is truncated at a certain excitation level, only a certain subset of all MBPT diagrams will be generated. For example, when we disregard quadratic terms in CCD (the so-called linear CCD, or L-CCD, method), the iteration of Eq. (2.85) gives us all MBPT diagrams having doubly excited configurations as their intermediates (i.e., all ladder and ring diagrams and their combinations representing the so-called DMBPT(∞) [188] method, or CEPA(0) method; for more details, see, e.g., [14, 169, 189, 190]). Thus, solving L-CCD equations, we automatically generate and simultaneously evaluate all these diagrams to infinite order.

In general, a CC method at a certain level of truncation will automatically generate and evaluate the corresponding subset of MBPT diagrams to infinite order. The MBPT content of various CC methods was explored in detail, and we refer the reader to [191, 192].

2. Coupled-Cluster Method with Singles and Doubles (CCSD)

The MBPT structure of connected-cluster components T_i and their contribution to the energy via the T_2 (and, to a very small extent, via T_1) clusters [Eq. (2.71)] implies their relevance, at least in a semiquantitative sense, assuming the convergent behavior of the perturbation expansions involved. Obviously, the most important contribution to both the wave function and the energy arises from T_2 clusters, contributing in the first order to the wave function and in the second order to the energy. (Cf. Figure 7.) Provided that the HF reference represents a good zero-order approximation (i.e., that there is no quasi-degeneracy with higher excited configurations), which is usually the case for most closed-shell ground states near the equilibrium geometry, the contribution of doubles amounts to

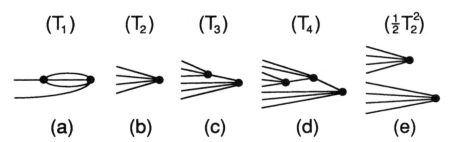

(T_1) (T_2) (T_3) (T_4) $(\frac{1}{2}T_2^2)$

(a)   (b)   (c)   (d)   (e)

Figure 7.   Schematic representation of the lowest order MBPT wave function diagrams contributing to various connected ($T_i, i = 1, 2, 3, 4$) and disconnected (T_2^2) cluster components.

approximately 94–96% of the correlation energy (as estimated from, say, CID).

The next important contribution comes from quadruples, amounting to about 2–4% of the correlation energy, while singles and triples contribute still less ($\sim$ 1–2%). It is precisely for this reason that our initial derivation in Section II.B was carried out using the assumption of Eq. (2.53). In fact, the contribution of singles is completely eliminated when one uses Brueckner orbitals and is usually rather small when HF reference is employed. Note that the main contribution of singles comes from their interaction via doubles, rather than directly through the energy term involving $\frac{1}{2}T_1^2$ [cf. Eq. (2.71) and the diagram in Figure 5c], which is usually negligible. Nonetheless, since the number of singles is very small relative to that of higher excited clusters, it is a common practice to account fully for their contribution (the standard CCSD method). This is particularly important when non-HF (spin) orbitals are employed (say, various localized orbitals), in which case the F_N operator is nondiagonal and the $\langle I|f|J \rangle$ terms with $I \neq J$ must be accounted for as well. Moreover, the main contribution of singles can be accomplished via a suitable factorization of CC equations (cf. Section II.G), namely, by treating T_2 and $\frac{1}{2}T_1^2$ terms jointly as an intermediate quantity $(T_2 + \frac{1}{2}T_1^2)$.

Just like the singles (assuming the HF reference), the triples and quadruples contribute to the wave function for the first time in the second order of PT and thus to the energy in the fourth order. (Cf. Figure 7.) However, while the primary contribution of triples (and, of course, singles) arises from their connected component T_3 (the disconnected triples of the T_1T_2 type contribute to the wave function for the first time in the third order) [5], the main contribution of quadruples comes from the disconnected term $\frac{1}{2}T_2^2$ (the connected component T_4 contributing in the next order of PT). Since the connected T and disconnected Q clusters begin to contribute in the same order of PT, the importance of their overall contribution is thus roughly proportional to their number.

The preceding qualitative considerations imply that the CCD method (originally referred to as CPMET [72]) provides the bulk of the correlation effects, being essentially equivalent to CIDQ [5], and thus represents a basic approximation of CC theory. In fact, employed with Brueckner orbitals, when T_1 clusters exactly vanish, the CCD method is equivalent to the CCSD approximation. Consequently, when one uses arbitrary orbitals, and in view of the existence of a small number of singles, it is the CCSD method that is the basic 'workhorse' of CC approaches to the many-electron correlation problem.

It must be emphasized that the aforementioned estimates of contributions by various clusters are quite approximate, since these contributions are generally not additive, the nonadditivity increasing with the quasi-degeneracy of the reference configuration. (See, e.g., [193].) Although the CCD and CCSD methods often give amazingly good results even in highly quasi-degenerate situations (cf., e.g., [190, 194–196]), they will eventually break down, particularly for large, metallic-like systems. This behavior was explored in considerable detail using a semiempirical π-electron model of cyclic polyenes C_NH_N, where such a breakdown occurs for cycles having more than about 30 carbon centers. (In this case, the lowest energy CCD $\equiv$ CCSD solution can be shown to bifurcate into the two complex solutions.) Not only does the importance of quadruples dramatically increase in such cases, but in view of the high symmetry of these model systems, there is an increasing number of connected T_4 clusters that have *no* disconnected counterpart [193]. For reasonably small systems, however, the CCSD method provides excellent results that can be further improved by a noniterative account of higher-than-pair clusters. (See Section II.C.4.) As a rule of thumb, for small systems (4–20 electrons), the CCD error is about 0.9–1.0 mhartree per electron pair, and that of CCSD is about 0.6–0.8 mhartree per electron pair. An approximate account of triples (Section II.C.4.2) will further lower the error to less than 0.1 mhartree per electron pair. However, this accuracy will be substantially compromised once we consider stretched geometries. Again, as a rule of thumb, these errors roughly double with a 50% stretch of a single bond and increase by a factor of four to six for a 100% stretch (i.e., for $R = 2R_e$), while the perturbative estimates of higher order cluster contributions break down completely. (See Section II.C.4.2.) It is thus important to consider ways that will enable us to handle OS situations as well.

In view of the basic importance of the CCSD approximation, and for future reference, we give the relevant diagrams (in Hugenholtz form) in Figures 8(i)–(iii). A CCD subset of these diagrams (Figure 8(ii)) easily follows from the nonoriented skeletons of Figure 6. Relying on Figure 8, we see that it is straightforward to write down the explicit algebraic equations. (See, e.g., [5, 7, 31, 129, 169].)

3. *Spin Adaptation*

So far we have ignored the fact that nonrelativistic *ab initio* model Hamiltonians that are employed in most molecular electronic structure calculations are *spin independent*. Employing pure spin orbitals (i.e., those that do not involve linear combinations of up and down spin states $|\alpha\rangle$ and $|\beta\rangle$, respectively), and designating the corresponding orbital part by lower case letters of the Latin alphabet, i.e.,

$$|I\rangle \equiv |i\sigma\rangle = |i\rangle \otimes |\sigma\rangle \equiv |i\rangle|\sigma\rangle, \qquad \sigma = \alpha, \beta, \tag{2.87}$$

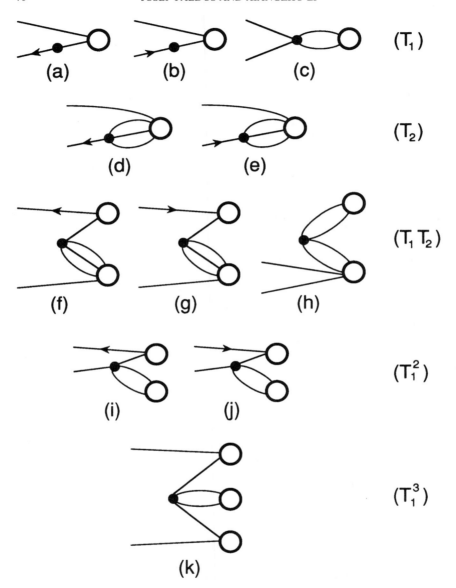

Figure 8(i). Unlabeled Hugenholtz diagrams representing CCSD equations for singles, Eq. (2.77a). The cluster components from which these terms originate are indicated on the right-hand side. Only those arrows that uniquely determine the orientation of the remaining hole and particle lines are shown explicitly.

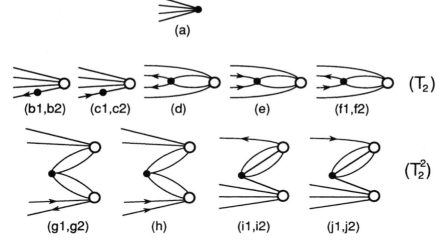

Figure 8(ii). Unlabeled Hugenholtz diagrams for the CCD subset of CCSD equations. [See also caption to Figure 8(i).] Diagrams labeled (x1,x2) correspond to the two nonequivalent labeled diagrams.

we find that a general *ab initio* model Hamiltonian takes the form

$$H = z_i^j E_j^i + \frac{1}{2} v_{ik}^{j\ell} E_{j\ell}^{ik}, \tag{2.88}$$

where $z_i^j = \langle i|z|j\rangle$ and $v_{ik}^{j\ell} = \langle ik|v|j\ell\rangle$ are the standard one- and two-electron integrals over the orthonormal molecular orbitals and E_j^i and $E_{j\ell}^{ik}$ are the *orbital replacement operators* (see Appendix A), the one-body replacement operators E_j^i being the generators of the orbital unitary group U(n) forming the basis of UGA [174–177]. In view of this spin independence, the model Hamiltonian H commutes with the operators of total spin, i.e.,

$$[H, \hat{S}_z] = [H, \hat{S}^2] = 0, \tag{2.89}$$

or, equivalently, with the generators of the spin group U(2) or SU(2). (See, e.g., Appendix B of [176].) If properly exploited, this symmetry can substantially simplify the problem while guaranteeing that the resulting wave functions have proper invariance properties, since the total spin generally represents a very good quantum number.

The main purpose of spin adaptation is thus twofold: to warrant the spin purity of the correlated wave function and to factorize the problem by eliminating spin variables, thus reducing its dimensionality. Although the

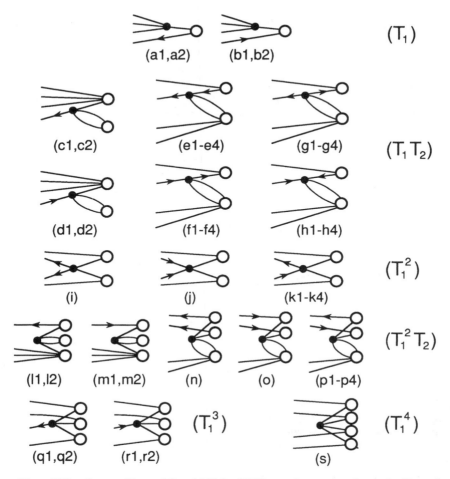

Figure 8(iii). Same as Figures 8(i) and 8(ii) for CCSD equations projected onto doubles and originating from the terms involving T_1 clusters.

latter aspect significantly reduces the "size" of the problem (i.e., the storage requirements, as well as the number of arithmetic operations required), it usually leads to a more complex formalism that is more difficult to implement (e.g., vectorization of the code, etc.). Nonetheless, properly formulated, spin adaptation will lead to substantial computational savings. For example, extensive computations on benzene dimers and DNA base pairs [33–35] that were carried out with MOLPRO 94 [21] could never have been accomplished with the GAUSSIAN 92 [19] package, which employs spin-orbital formalism and, although very fast, requires too much memory (roughly, by a factor of three).

Obviously, the basic requirement for a spin-adapted CC formalism is the use of a pure spin eigenstate for the reference $|\Phi_0\rangle$ and the formulation of the required algorithms in terms of orbitals rather than spin orbitals. In using pure spin orbitals, Eq. (2.87), the $\hat{S}_z$ symmetry is taken care of automatically. This is also the case for the $\hat{S}^2$ symmetry when one considers high spin states with $M_S = S$. However, already here the problem of proper spin adaptation is far from being trivial, even in a consideration of simple OS systems, as will be seen subsequently.

For a CS ground state, we thus employ a pure-singlet, single anti-symmetrized product (Slater determinant) wave function as a reference, usually a restricted Hartree–Fock (RHF) solution. This is fine as long as we are interested in near-equilibrium geometries or in processes that do not involve OS fragments. However, in most dissociation or reaction processes, the breaking, or a significant stretching, of even a single chemical bond will involve unpaired electrons, and the RHF wave function will be unable to properly describe the dissociation products or reaction intermediates, as is well known. In such situations, one often relies on broken spin-symmetry unrestricted HF (UHF) solutions. In common parlance, by UHF, one usually understands the so-called different-orbitals-for-different-spins (DODS) HF solution that exists even for CS systems whenever the RHF solution is triplet (or nonsinglet) unstable [197–199]. Although the DODS-type UHF solutions are most common and useful, there exist other types of broken-symmetry (spin, axial spin, and space or time reversal; see [198, 199]) solutions that deserve to be explored at the *ab initio* level, as will be pointed out in Section IV.

Since the UHF-type solution, in contrast to the RHF one, is often capable of describing the desired physical dissociation channel (sometimes referred to as the *size consistency* requirement), it is often employed in actual applications, both in computing potential-energy surfaces (PESs) or potential-energy curves (PECs) and in handling OS systems in general. When using the UHF reference, one employs the standard SR spin-orbital CC formalism, as outlined in the preceding sections.

While, in using the DODS-type reference the spin contamination of a given state may often be minor, and the bulk of it may sometimes be avoided when computing the final CC energy (by projecting onto a pure spin reference), there are situations when spin contamination becomes important. (For a thorough discussion and examples of UHF-based PECs for simple systems, see [28].) Moreover, a UHF solution does not always correctly dissociate, even when one allows a more general symmetry breaking, leading to the so-called general HF (GHF) solutions.

(See, e.g., [199].) Particularly in OS systems, when the restricted OS HF (ROHF) solution shows a singular behavior and the DODS UHF solution gives a smooth PEC, one may find that it dissociates into a wrong channel [200].

Thus, while spin (or other) broken-symmetry solutions used as a reference may often provide a reasonable and very affordable approximation, particularly when OS systems or subsystems are present, we must not forget that these are makeshift measures that are prone to failure [200–204] and, in fact, break down completely in certain circumstances (e.g., low spin states).

We can thus briefly summarize the pros and cons of spin-adapted vs. UHF-based CC approaches as follows:

(i)   A spin-adapted formalism yields wave functions that are free of any spin contamination, regardless of the spin multiplicity involved, but usually requires an MR-type OS theory to handle dissociation involving OS fragments. In contrast, UHF-based methods allow an approximate description at the SR level of the theory, but produce spin-contaminated wave functions, and the quality of the resulting potential-energy surfaces or curves can be very poor in the region of intermediate nuclear separations.

(ii) Spin-adapted approaches require fewer cluster amplitudes (by a factor of about three at the doubly excited level), but their formulation and encoding are more demanding than for the SR spin-orbital methods.

(iii) Certain properties may be very sensitive to spin contamination. (See, e.g., Section III.B.4.)

(iv) Broken-spin UHF solutions (namely, DODS-type solutions used currently in CC applications) may exist only for a limited range of geometries, being identical with the RHF solution elsewhere. Consequently, the resulting PESs are not smooth. Although in many instances the lowest energy DODS UHF solution is size consistent (i.e., dissociates properly), this is not always the case [200]. Moreover, in OS situations, the multiplicity of possible UHF solutions may also cause problems [198–200].

The ROHF solutions, required by spin-adapted approaches, may also yield nonsmooth PECs (in fact, displaying a "hysteresis" [200, 205, 206]). This is, however, to a large extent rectified already at the CCSD level. In these cases, assuming that a suitable UHF (or, in general, GHF) broken-symmetry solution properly describes the examined dissociation channel, the externally corrected CCSD approach (see Section II.D.3.3), based on such a solution, would deserve a closer examination.

(v) The UHF-based approaches cannot be used for a very important case of low spin states (e.g., the singlet excited states of molecules with CS ground state).

We now briefly address the techniques of spin adaptation that are suitable for CC approaches. They fall basically into two classes, depending on whether or not we require the excited-state manifold to be orthogonal or not. We hasten to stress that, in contrast to variational (e.g., CI) methods, the nonorthogonality (in fact, even the linear dependence [5]) of excited configurations is of much less concern in CC approaches and in many situations causes no problems. However, the so-called orthogonally spin-adapted (OSA) versions of CC methods not only have their theoretical appeal, but are essential if we wish to interrelate them with corresponding CI approaches. These OSA versions automatically eliminate linear de-pendencies, thus minimizing the number of required cluster amplitudes, and can also be shown to minimize the number of nonlinear terms that characterize CC approaches [207]. Both of these features may be important in large-scale applications when ill-conditioned nonlinear algebraic systems might arise and cause convergence problems for standard algorithms used in solving CC equations. Moreover, the OSA formalism is essential for the formulation and understanding of certain approximate theories (such as the so-called ACPQ, ACC(S)D, or CC(S)D' methods [190, 196, 208–210]) and the formulation of MR SS approaches for OS systems [45, 46].

A nonorthogonal spin adaptation pertaining to CS ground states can be very simply achieved by employing Goldstone diagrams and by assigning a factor of 2 to each closed loop of oriented lines [corresponding to the up and down orientation of spin functions, just as in the MBPT (see, e.g., [78])], while replacing the spin-orbital labels by the orbital ones [1, 2, 5]. In the CC(S)D case, this procedure can be shown to correspond to a use of VB-type biexcited configurations [Eq. (A.21) in Appendix A.] It may also be easily generalized to OS cases, as long as the SR formalism is employed. (Cf. [105].)

For an OSA, several ways are opened. These may be again basically of two types, depending on whether we exploit diagrammatic or algebraic techniques, even though a combination of both may sometimes be ad-vantageous. The diagrammatic approach is based on graphical methods of spin algebras [39, 211–214]. In the CC context, this approach was first employed in [128], and its different versions were discussed in [129]. The essence of the technique is to choose a suitable coupling scheme and to represent the OSA configurations in terms of products of Clebsch–Gordan coefficients [129, 214, 215]. The latter are then represented by the angular momentum diagrams, and the orbital diagrams of CC theory are converted into the corresponding spin graphs [129, 214]. Evaluating these graphs by means of the rules of graphical methods of spin algebras, one obtains suitable numerical factors that multiply the contributions from the associated orbital diagrams of the CC theory. Thus, these numerical spin adaptation factors replace the simple 2^ℓ factors of nonorthogonal spin adaptation described earlier. (Recall that ℓ designates the number of closed loops.) For the OSA version of the CCD and CCSD methods, see [128] and [129]. Also, note that this technique is very general and can be used in other CC approaches, including the MR CC [93] and LR CC [216] methods, as well as in CI approaches [214, 217].

The algebraic method relies on spin-free replacement operators. (See Appendix A.) In simpler situations, it can be employed directly (cf. [173]), while for more general OS situations, it is preferable to exploit the Clifford algebra UGA (CAUGA) formalism [45, 46], as will be discussed in Section II.D.3.2. The advantage of this technique is its generality, as well as the possibility of automating it [45].

Let us finally mention that in OS cases (currently restricted to high spin), various authors rely on a partial spin adaptation [131–135, 137–139]. The goal of these approaches is to eliminate the bulk of spin contamination that may be present in spin-nonadapted UHF-based methods, while keeping the formalism as simple as possible in order to permit the design of efficient computer codes. In view of a recent interest in these techniques and their diverse formulations by various authors, it is worthwhile to discuss them in greater detail and to try to interrelate them. The latter may be achieved by comparing such techniques with the UGA-based CCSD (UGA CCSD) methods [45, 46, 218], which represent a general tool for a proper spin adaptation of states of arbitrary multiplicity that is applicable both to the standard CS CC theory and to various OS MR versions.

The UGA CCSD method is outlined in Section II.D.3.2. At this stage, let us only emphasize that it represents a fully spin-adapted theory in the sense that (i) the cluster *Ansatz* $e^T|\Phi_0\rangle$ produces a pure spin eigenfunction and (ii) the cluster operator T is uniquely defined for all possible spin quantum numbers. In the closed-shell case ($S = M_S = 0$), this *Ansatz* reduces to the standard OSA one [128, 129]. For the high-spin case and spin-independent properties, when one can employ an $M_S = S$ spin component that can be described by a single Slater determinant, the UGA CC *Ansatz* may be approximated via partially spin-adapted *Ansätze*.

Generally speaking, a partially spin-adapted CC *Ansatz* requires that only the reference and the linear component, i.e., $(1 + T)|\Phi_0\rangle$, be strictly spin-adapted, but not the nonlinear components $(n!)^{-1}T^n|\Phi_0\rangle$, $n > 1$. This partial spin adaptation is usually achieved by exploiting only some subset of UGA CC excitation operators spanning T. In this way, we can interpret the partially spin-adapted *Ansätze* that are exploited by Knowles, Hampel, and Werner (KHW) [134] and by Neogrády, Urban, and Hubač (NUH)

[132, 133] (each admitting a different coupling and orthogonalization scheme), as will be explained in Section II.D.3.2.

We note that the concept of a partially spin-adapted *Ansatz* was first employed by Janssen and Schaefer [131]. However, their cluster operator was defined in terms of a small set of excitation operators that do not span even the first interacting space, as was pointed out by KHW. A very recently proposed approach by Szalay and Gauss also employs an essentially partially spin-adapted *Ansatz*, supplemented by certain additional require-ments (the so called spin-restricted CC scheme, in which the wave function is not rigorously spin adapted, but produces the correct spin eigenvalue within the truncated excitation space) [139].

The very original and extensively exploited [30] approach by Jayatilaka, Lee, et al. [135, 137, 138] that uses symmetric (or, rather, rotated) spin functions $\sigma^\pm = (\alpha + \beta)/\sqrt{2}$ instead of pure up and down spin functions α and β, respectively, is neither spin adapted nor partially spin adapted in the sense defined previously (i.e., fully spin adapted at the linear level) and thus cannot be described as an approximation to the fully spin-adapted UGA CC approach. In contrast to the latter, which defines the cluster operators T_k [or corresponding excitation operators $G_j^{(k)}$; see Eq. (2.49)] in terms of the UGA generators, the approach by Jayatilaka and Lee employs the "spin-flip" analogues of U(n) generators (distinguished by a dot over the E), namely,

$$\dot{E}_j^i = X_{i\alpha}^\dagger X_{j\beta} + X_{i\beta}^\dagger X_{j\alpha}. \tag{2.90}$$

Of course, these are not U(n) generators, but can be expressed as a difference of U($2n$) generators, viz.,

$$\dot{E}_j^i = X_{i\sigma^+}^\dagger X_{j\sigma^+} - X_{i\sigma^-}^\dagger X_{j\sigma^-} = e_{j\sigma^+}^{i\sigma^+} - e_{j\sigma^-}^{i\sigma^-}. \tag{2.91}$$

Note that the spin-orbital group U($2n$) is now defined via two rotated spin functions, σ^+ and σ^-, and thus, the generators $e_{j\sigma^+}^{i\sigma^+}$ and $e_{j\sigma^-}^{i\sigma^-}$ preserve the S_x, rather than S_z, symmetry.

At this point, recall that within the spin group U(2) with generators $\mathcal{E}_\nu^\mu = X_{i\mu}^\dagger X_{i\nu}$ (the sum over i is implied), the sum of the weight generators $\mathcal{E}_\alpha^\alpha + \mathcal{E}_\beta^\beta = N$ is a scalar operator defining the particle number, while $\mathcal{E}_\alpha^\alpha - \mathcal{E}_\beta^\beta = S_z$, $\mathcal{E}_\beta^\alpha = S_+$, and $\mathcal{S}_\alpha^\beta = S_-$ are generators of the (spin) angular momentum group SU(2). The three operators S_+, S_z, and S_- form a vector operator that may change the spin by $\Delta S = 0, \pm 1$ when acting on a pure spin eigenstate. (Cf. [176, 219].)

Thus, the E-dot operators are not U(n) generators, but instead behave as SU(2) vector operators. This is the case regardless of whether the rotated or unrotated spin functions are employed. In fact, such operators can be very useful in handling spin-dependent operators, but not in the correlation problem for spin-independent Hamiltonians. From the UGA viewpoint, E-dot operators represent a special case of general tensor operators (see, e.g., [220]) that are very useful in general many-electron problems. (See [221].) These aspects, however, go beyond the scope of this review.

Let us indicate that the use of E-dot operators leads to the nonconservation of total spin. Consider, for example, the T_1-operators involving $\dot{E}_a^r$, whose action on a three-electron, pure-doublet reference gives

$$
\begin{aligned}
\dot{E}_a^r |\{a\alpha\, a\beta\, i\sigma^+\}\rangle &= |\{a\alpha\, r\alpha\, i\sigma^+\}\rangle + |\{r\beta\, a\beta\, i\sigma^+\}\rangle \\
&= (|\{a\alpha\, r\alpha\, i\alpha\}\rangle + |\{a\alpha\, r\alpha\, i\beta\}\rangle \\
&\quad + |\{r\beta\, a\beta\, i\alpha\}\rangle + |\{r\beta\, a\beta\, i\beta\}\rangle)/\sqrt{2} .
\end{aligned} \tag{2.92}
$$

Clearly, the $M_S = +\frac{3}{2}$ and $M_S = -\frac{3}{2}$ terms $|\{a\alpha\, r\alpha\, i\alpha\}\rangle$ and $|\{r\beta\, a\beta\, i\beta\}\rangle$, respectively, belong to a quadruplet or a yet-higher spin multiplet. Similarly, the action of a two-body operator $\dot{E}_a^r \dot{E}_b^s$ on a five-electron doublet reference $|\{a\alpha\, a\beta\, b\alpha\, b\beta\, i\sigma^+\}\rangle$ will produce terms, like $|\{a\alpha\, r\alpha\, b\alpha\, s\alpha\, i\alpha\}\rangle$ and $|\{a\beta\, r\beta\, b\beta\, s\beta\, i\beta\}\rangle$, with $M_S = \frac{5}{2}$ and $M_S = -\frac{5}{2}$, respectively, indicating the spin contamination from sextuplet or higher spin states.

Since various methods rely on the interacting space, it is worth pointing out that the spin-adapted (first-order) interacting space is smaller than the spin-orbital (or determinantal) interacting space. Note also that, as long as $|\Phi_0\rangle$ is an eigenfunction of $\hat{S}^2$, it is only the wave function $|\Psi\rangle$, and not the energy, that will be spin contaminated. (Cf. [135, 222–224].) We should also mention that very recently the symmetric spin-orbital approach of Jayatilaka and Lee was implemented in the Brueckner orbital-based CCD (B-CCD) method by Crawford et al. [138].

4. Beyond CCSD

As has been pointed out, the CCSD method gives excellent results, as long as the basic assumption of the nondegeneracy of the reference state is fulfilled. In fact, it often provides surprisingly good results even in highly quasi-degenerate, or almost degenerate, situations, as was illustrated on simple model systems involving four hydrogen atoms arranged in different geometrical configurations [92, 190, 196, 225]. By varying the geometry of these models, we can continuously vary the amount of quasi-degeneracy from a nondegenerate situation (linear-configuration, or separated H_2 systems) to a complete degeneracy (a square configuration). For this very reason, these models, together with a larger H_8 model [226], have been extensively studied in the past [89, 92, 93, 95, 98, 190, 196, 210, 225, 227–233]. For example, in the simplest H4 case [190, 196], which can be regarded as modeling the breaking of a single bond, the difference between the exact full CI (FCI) and CCSD energies amounts to 0.118 mhartree in the nondegenerate limit, to only -2.159 mhartree in an almost fully degenerate situation at the minimum-basis-set (MBS) level [190], and to 0.702 and 4.700 mhartree, respectively, at the double-zeta-plus-polarization (DZP) level [196].

Nonetheless, in order to achieve $1 \, \text{kcal/mol} \approx 1.6$ mhartree "chemical accuracy" it is generally necessary to go beyond the CCSD level, especially when one considers stretched geometries. Within the SR CC theory, this goal can be achieved only by accounting for—in one way or another— higher-than-pair clusters, in particular for connected triples (T) and, eventually, quadruples (Q). In this section, we present an overview of the main undertakings in that of direction.

4.1. Standard Full Approaches (CCSDT, etc.). The most obvious and appropriate way of accounting for T_3 and T_4 clusters is to truncate the full chain of CC equations at the correspondingly higher level ($T_i = 0$ for $i \geq 4$ or $i \geq 5$), leading to the standard CCSDT and CCSDTQ methods. The role of T_3 clusters was examined already in the first *ab initio* pilot study of CC methodology [5] on the MBS model of BH_3 (using Slater AOs). The reason for this choice was the fact that, from among the systems studied at that time via CI methods, this model showed the largest relative contribution from triply excited configurations. However, in contrast to quadruples, it is the connected-cluster component T_3 that plays the major role, as suggested by a simple PT argument. (See earlier and [5].) With the use of suitably truncated CCSDT equations (an approach referred to as extended CPMET, or ECPMET-D$'$) that accounted fully for the T_3 clusters at the linear level, while neglecting the higher order nonlinear terms, the contribution of connected triples was well represented, the result differing from FCI by only $1 \, \mu$hartree. This study also discussed the problem of a simple spin adaptation at the triply excited level, where we encounter for the first time the linear dependencies of the orbital t_3-amplitudes (from $3! = 6$ amplitudes, only 5 are linearly independent), and showed two equivalent procedures for overcoming this problem [5], either by adding an extra constraining equation for each distinct orbital set or by using such an equation to eliminate one of the dependent orbital t_3-amplitudes. The latter solution leads to a rather "asymmetric" formalism, while the former solution considers a larger set of equations (and amplitudes) than is necessary, thus clearly pointing to the advantages of the OSA formalism [207, 234, 235].

Following the general-purpose implementation of the CCSD method [6, 7], which enabled its wider exploitation and testing, the design of similar codes for the full CCSDT [100–102] and even CCSDTQ [103] (cf. also

[96]) methods was accomplished. These methods indeed provide highly accurate results (cf. Table 1 of [31]), as long as they can be carried out. Unfortunately, their exploitation is still very limited, since they are computationally too demanding. Indeed, while the CCSD method scales as $n_o^2 n_u^4 \sim n^6$, where n_o and n_u designate, respectively, the number of occupied and unoccupied (or virtual) orbitals $(n = n_o + n_u)$, the CCSDT and CCSDTQ methods scale as $n_o^3 n_u^5 \sim n^8$ and $n_o^4 n_u^6 \sim n^{10}$, respectively. Nonetheless, at least for small systems, these methods provide very useful benchmarks.

4.2. Perturbative Approaches [CCSD(T), etc.]. In view of the limited applicability of the CCSDT and CCSDTQ methods, it is worthwhile to look for approximate ways of accounting for higher-than-pair clusters. Here again, the MBPT provides very useful guidance.

Clearly, the main limiting factor for the standard approaches is the rapid growth of the number of t_3- and t_4-amplitudes with the increasing orbital number $n = n_o + n_u$, particularly due to n_u. Thus, even when we simplify the full CCSDT equations by neglecting higher order terms, as was done, for example, in the aforementioned ECPMET method [5, 235], we are still faced with the problem of n^8 scaling.

Recall that the CCSD equations (2.77) result from a decoupling of the full CC chain, Eq. (2.75), by setting $T_i = 0$ for $i \geq 3$. This implies that the equations projected onto the singly excited manifold, Eq. (2.77a), are obtained by neglecting the term

$$\widetilde{\Lambda}^{(1)} \equiv \Lambda^{(1)}(3) = \langle G_i^{(1)\dagger} H_N T_3 \rangle , \qquad (2.93a)$$

and those projected onto the doubly excited manifold, Eq. (2.77b), are obtained by neglecting the terms

$$\begin{aligned} \widetilde{\Lambda}^{(2)} &\equiv \Lambda^{(2)}(3) + \Lambda^{(2)}(4) + \Lambda^{(2)}(1,3) \\ &= \langle G_i^{(2)\dagger} H_N (T_3 + T_4 + T_1 T_3) \rangle_C . \end{aligned} \qquad (2.93b)$$

Evidently, were we given the exact t_3- and t_4-amplitudes, as obtained, for example, by the cluster analysis of the FCI wave function, then evaluating the terms of Eq. (2.93) and adding them to the absolute and linear (in t_1-amplitudes) terms of our CCSD equations, we would recover the FCI result as a CCSD solution. We shall see later that this fact is exploited by the so-called *externally corrected* CCSD approaches [47–50], in which the t_3- and t_4-amplitudes are obtained independently from some external source, such as the UHF [194, 208, 210], VB [47, 236], CAS-SCF or CAS-FCI

[48, 49, 237, 238], or MR-CISD [50, 239, 240] wave functions. It is also possible to use the MBPT estimate of these contributions. This is, in fact, most effectively done in conjunction with the CCSD method, so that the approaches we are now going to outline could also be referred to as *internally corrected* CCSD approaches.

For the sake of simplicity, we again initially concentrate on pair clusters (CCD method), using the HF reference, while trying to account for T_3 clusters. With these assumptions, the neglected term $\widetilde{\Lambda}^{(2)}$ in Eq. (2.93b) becomes

$$\widetilde{\Lambda} = \langle G_i^{(2)\dagger} H_N T_3 \rangle_C = \langle \Phi_i^{(2)} | V_N T_3 | \Phi_0 \rangle .$$ (2.94)

Utilizing the MBPT expansion for $T_3|\Phi_0\rangle$, Eqs. (2.43) and (2.44), we can write

$$T_3|\Phi_0\rangle = \sum_{n=2}^{\infty} \{(RV_N)^n|\Phi_0\rangle\}_{C_3} = \{R^{(3)} V_N \sum_{n=1}^{\infty} (RV_N)^n|\Phi_0\rangle\}_C ,$$ (2.95)

where we used the fact that $W = V_N$ and the $n = 0$ and $n = 1$ contributions to $T_3|\Phi_0\rangle$ vanish, and where we keep only connected triexcited terms (represented by connected diagrams with three particle and three hole lines extending to the left), as indicated by the subscript C_3 or by the fact that the leftmost resolvent $R^{(3)}$ will involve only projection operators onto the triexcited manifold. (Cf. [215].)

Thus, the first contribution to $T_3|\Phi_0\rangle$ arises from the second-order diagrams, whose skeleton is shown in Figure 7c. This approximation can be significantly improved by replacing the right-hand-side v-vertex (which can be viewed as the first-order approximation to $T_2|\Phi_0\rangle$; see Figure 7b) by the t_2-vertex, i.e., by considering diagrams of the type shown in Figure 9a. In that way, we automatically account for an infinite class of MBPT diagrams that contribute to $T_3|\Phi_0\rangle$, although not for all such diagrams. (See, e.g., Figure 9b.) Accordingly, we get the following approximation for $\widetilde{\Lambda}$ in Eq. (2.94):

$$
\begin{aligned}
\widetilde{\Lambda} &= \langle \Phi_i^{(2)} | V_N R^{(3)} V_N \sum_{n=1}^{\infty} (RV_N)^n|\Phi_0\rangle_C \approx \langle \Phi_i^{(2)} | V_N R^{(3)} V_N T_2 | \Phi_0 \rangle_C \\
&= \sum_j \langle \Phi_i^{(2)} | V_N R^{(3)} V_N | \Phi_j^{(2)} \rangle \langle \Phi_j^{(2)} | T_2 | \Phi_0 \rangle \\
&= \sum_j {}^{(2)} W_{ij}^{(3)} t_j \equiv \widetilde{\Lambda}' .
\end{aligned}
$$ (2.96)

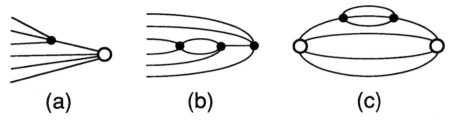

(a)   **(b)**   **(c)**

Figure 9.   Schematic representation of the terms constituting $T_3|\Phi_0\rangle$ that are (a) and are not (b) accounted for in the CCSD[T] or CCSD(T) approximation $\Delta E^{(T)}$, Eq. (2.100), represented by diagram (c).

Here,

$$W_{ij}^{(3)} = \sum_k {}^{(3)} \langle \Phi_i^{(2)}|V_N \frac{|\Phi_k^{(3)}\rangle\langle\Phi_k^{(3)}|}{-\Delta\varepsilon_k} V_N|\Phi_j^{(2)}\rangle = \sum_k {}^{(3)} (-\Delta\varepsilon_k)^{-1} H_{ik}H_{jk}^* ,$$

(2.97)

where $H_{ik} = \langle \Phi_i^{(2)}|V_N|\Phi_k^{(3)}\rangle$ and the superscript on the summation symbol designates the excitation order of the configurations that are involved.

We can now incorporate the $\widetilde{\Lambda}'$ term of Eq. (2.96) into the CCD equations (2.65) by modifying the linear term, obtaining

$$a_i + \widetilde{b}_{ij}t_j + c_{ijk}t_jt_k = 0$$

(2.98)

where now

$$\widetilde{b}_{ij} = b_{ij} + W_{ij}^{(3)}.$$

(2.99)

Of course, we could also add the $\widetilde{\Lambda}'$ term to the absolute term, evaluating it in each iteration with current $t_j^{[n]}$ values, which may be advantageous in direct algorithms. This is clearly equivalent to the approximation in which we consider only the T_2 clusters and those diagonal T_3 terms in the CCSDT equations projected onto the triexcited manifold, which involve the one-body F_N operator (i.e., the orbital energies) and produce the MBPT denominators. This approach is referred to as the *iterative account of triples*.

Note that the same procedure will account for the $\Lambda^{(2)}(3)$ term in the CCSD method (or, in fact, OSA-CCSD [129]), as well as for the $\Lambda^{(1)}(3)$ term when we replace the bra states in $W_{ij}^{(3)}$, as given by Eq. (2.97), by singly excited configurations. (Cf. also the optimized inner projection technique [215].) In the iterative scheme, one can also account for the $\Lambda^{(2)}(1,3)$ term in this way. When we evaluate $W_{ij}^{(3)}$ terms a priori, we obtain formally the same equations as for standard CCSD, Eq. (2.98), so that there is no need to store t_3-amplitudes. (This is the so-called *noniterative account of triples*.) Clearly, both the iterative and noniterative algorithms will give the same solution (assuming that they converge).

As the structure of the just-outlined formalism implies, there is considerable freedom in its actual exploitation. This explains numerous approximate methods accounting for triples.

The iterative versions are known as various CCSDT-n procedures [11, 100, 241–243]. The foregoing account of T_3 clusters corresponds to CCSDT-1 [241] or CCSDT-1a [242], which are still n^7 procedures (for each iteration). When the $\Lambda^{(2)}(1,3)$ term is also accounted for, the method is referred to as CCSDT-1b [17, 242].

Although the noniterative account of triples is clearly preferable, there is an even simpler way of estimating their contribution, first suggested by Urban et al. [242] and, independently, by Raghavachari [244]. These approaches are referred to as CCSD + T(CCSD), or CCSD[T] for short, and as CCD + ST(CCD) or CCD[ST], respectively. As the acronyms suggest, the CCSD[T] scheme uses the converged t_2-amplitudes $t_j \equiv t_j^{(\mathrm{CCSD})}$ and evaluates the triple contribution via the expression

$$\Delta E^{(\mathrm{T})} = \sum_{ij} t_i W_{ij}^{(3)} t_j, \tag{2.100}$$

based on the fourth-order-like diagram of Figure 9c. This contribution is then simply added to the CCSD energy. The CCD[ST] procedure is based on the CCD method and handles the singles in an analogous manner. Thus, the triple contribution of Eq. (2.100) is evaluated with CCD amplitudes $t_j \equiv t_j^{(\mathrm{CCD})}$, and the singles contribution is

$$\Delta E^{(\mathrm{S})} = \sum_{i,j} t_i^{(\mathrm{CCD})} W_{ij}^{(1)} t_j^{(\mathrm{CCD})}, \tag{2.101}$$

where, now,

$$W_{ij}^{(1)} = \sum_{k} {}^{(1)}(-\Delta\varepsilon_k)^{-1} \langle \Phi_i^{(2)} | V_N | \Phi_k^{(1)} \rangle \langle \Phi_k^{(1)} | V_N | \Phi_j^{(2)} \rangle. \tag{2.102}$$

Both of these approaches provide the correlation energy that is correct through the fourth order of MBPT (as may be easily verified by replacing t_2-vertices by v-vertices). Moreover, just like CCSDT-1, they account for a significant part of the fifth and higher order contributions. If one also accounts in a similar way for singles–triples interaction terms, i.e., for $\Lambda^{(1)}(3)$, one arrives at the so-called CCSD(T) method [245], which includes an additional fifth-order term that seems to systematically improve the

CCSD[T] or CCD[ST] scheme. The CCSD(T) is nowadays probably the most often-used SR CC method, yielding very accurate and reliable results. The only known cases where CCSDT-1 or CCSDT give substantially better results are the PECs for Be_2 [246] and F_2 [28], and for vibrational frequencies of O_3 [31]. A recent review by Lee and Scuseria [30] is almost exclusively devoted to this approach, demonstrating that it attains "chemical accuracy" as long as the basis sets include f- and g-type functions.

Let us, finally, briefly mention analogous methods that account for quadruples [the $\Lambda^{(2)}(4)$ term in Eq. (2.93b)]. These may be handled either as perturbative corrections to CCSDT, resulting in a CCSDT + Q(CCSDT) $\equiv$ CC5SDT[Q] procedure [247], which is again hardly affordable for larger systems, or as a CCSD-based $TQ^*(CCSD)$ and CCSD + $TQ^*(CCSD)$ $\equiv$ CC5SD[TQ*] procedure, which also requires an n^8 step [246]. For more details, see the original papers and [31]. Clearly, all these methods are computationally very demanding and, in contrast to CCSD(T), have so far enjoyed only a very limited use.

4.3. Other Approximate Approaches (ACCSD, ACPQ, etc.). As alluded to in the preceding section, another way of accounting for T_3 and T_4 clusters at the SR CC level of the theory (as well as in MR CC SS approaches; see Section II.D.3.3) is via the externally corrected CCSD, when we use some external source of non-CC origin—either implicitly or explicitly—to approximate the contribution of these clusters. For this purpose, we must exploit a wave function that, in contrast to the RHF or ROHF reference, can properly describe the studied dissociation channel and yet is inexpensive to generate and cluster analyze. In this way, an excellent performance of the SR CCSD method can be extended to highly stretched geometries. Several options have been exploited so far [47–50]. The most recently investigated approach, referred to as the reduced multireference (RMR) CCSD method [50, 239, 240], which is based on an MR-CISD wave function of a relatively small dimension, or its truncated version, seems to be particularly promising.

A prototype of these approaches was introduced by Paldus et al. [194, 208, 210] and was based on projected UHF- (PUHF)-type wave functions. It was motivated by the fact that the standard CC(S)D method breaks down completely [194] when applied to sufficiently large Pariser–Parr–Pople (PPP) models of cyclic polyenes $C_N H_N$ having a nondegenerate ground state ($N = 4\nu + 2$, $\nu = 1, 2, 3, \ldots$) or even for small cycles when approaching the fully correlated limit (i.e., when the resonance integral $\beta \to 0$), while the UHF solution (of the DODS type) yields the exact energy in this limit. (Concerning the high degeneracy arising in this limit, see [210, 248].) This led to the formulation of the so-called CCDQ' method, which exploits information about the quadruply excited cluster components provided by the singlet PUHF wave function. (Note that this wave function does not contain singlet components excited an odd number of times and thus cannot provide any information concerning the triexcited clusters.)

An analysis of CCDQ′ equations showed that in cases when the PUHF wave function provides the exact pair cluster components (which is the case in the fully correlated limit of cyclic polyenes [248]), the contribution of T_4 clusters, as given by $\Lambda^{(2)}(4)$ in Eq. (2.93b), cancels certain nonlinear terms of the standard CCD theory (namely, the diagrams of the type shown in Figure 8(ii) (h) and (j1, j2) that are separable over one or two hole lines and that produce important EPV terms). In this case, the CCDQ′ method reduces to the so-called approximate coupled pair (ACP) or approximate CCD (ACCD) approach, each of which has been independently put forward earlier [190, 196, 207, 209, 234]. The precise account of this cancellation led to the so-called approximate-coupled-pair-theory-with-quadruples (ACPQ) method (which differs from ACCD by a numerical factor of 9 multiplying the contributions from diagram (h) of Figure 8(ii) for triplet-coupled pp-hh t_2-amplitudes [208]). The ACPQ method (like ACCD or ACP-45 [190, 196, 207, 234]) gives excellent results over the whole range of the coupling constant and has proved to perform well in other quasi-degenerate situations [209], as well as at the MR SU level [225]. The assumption for its good performance is the requirement that the PUHF wave function provide a good approximation of T_2 clusters and that the T_3 amplitudes be negligible, which is not always the case. (Cf. [210, 229].) Nonetheless, the ACC(S)D methods avoid having to consider the computationally most demanding nonlinear terms of the standard CCSD method, while providing superior results to CCSD methods in all known cases [190, 194–196, 209, 229, 231, 233, 249, 250]. (See also [251].) The method was also successfully extended to include triples perturbationally (ACPTQ or ACCD + ST (ACCD) [231, 249]). Of course, the perturbative account of triple corrections breaks down when we are far away from the equilibrium, just as with the usual CCSD(T) or CCSD[T] approaches.

The CCDQ′ equations were recently extended to include singles (CCSDQ′), and the method was applied to the MBS and DZP H_4 and H_8 model systems [210], thus elucidating the performance of the simpler ACCSD method. In all cases studied [210], it was found that the CCSDQ′ approach properly corrects the CCSD energy for the PUHF T_4 clusters, while ACCSD or ACPQ approaches do so only in cases when PUHF provides a good estimate of both T_2 and T_4 clusters. This shortcoming, however, does not diminish the usefulness of the ACCSD approaches, which are computationally simpler than CCSD and provide very good results even for demanding *ab initio* models [250–252].

The inability of the DODS UHF description to provide information about T_3 clusters prompted the search for other suitable sources. One obvious option is to use a simple valence-bond (VB)-type wave function, which already correctly dissociates in its simplest perfect-pairing approximation. The approach, implemented for the semiempirical PPP-type Hamiltonians by using the UGA-VB method and codes [253, 254], provided convincing evidence for the usefulness of externally corrected CC approaches. It would certainly be of interest to explore this avenue at the *ab initio* level, although the well-known nonorthogonality problems of the VB approach present a definite challenge. For that reason, it is easier not to leave the realm of the MO theories.

The next obvious choice is to use the CAS SCF or CAS FCI wave functions [48, 49, 237, 238] of modest dimensionality. This is achieved by using a small active space involving primarily the orbitals that permit a correct dissociation of the system. However, that way, we can generate only connected three- and four-body cluster amplitudes, whose orbital labels are restricted to the active space orbitals; in other words, we generate only the active space clusters T_3 and T_4. Although this is already very helpful [49], it is desirable to have more extensive information concerning the T_3 and T_4 clusters. This could be achieved by enlarging the active space, although doing so inevitably leads to demanding CAS SCF or CAS FCI computations. It is thus worthwhile to explore more affordable schemes relying on other CI-type wave functions.

One possible avenue [238] is to subdivide the extended set of active orbitals into the two disjoint subsets of *internal* and *external active orbitals* in such a way that the internal subset warrants a dissociation into the proper channel, while the external subset can account for the remaining (mostly dynamical) correlation effects. As the external source of T_3 and T_4 clusters, one then employs the wave function obtained by performing CAS FCI with the internal active orbitals, followed by the second-order CI (SOCI), using the external active orbitals. This results in an affordable scheme, based on CI-type wave functions of modest dimension, which provides almost-as-good t_3- and t_4-amplitudes as are generated by the CAS FCI or CAS SCF scheme, using a large active space (AS) spanned by both internal and external active orbitals. We found [49] that even with rather modest ASs, we obtain a significant improvement of CCSD energies, which, for all tested benchmark systems, are invariably closer to the exact FCI values than the energies that are associated with wave functions used as an external source. Of course, the difference between the latter and the externally corrected CCSD energies decreases with increasing dimension of the AS, since both energies will coincide when all orbitals are treated as active orbitals, i.e., when the CAS FCI becomes FCI. (This is, in fact, a very good test of the correctness of the required codes, although in the OS case only an approximate equality can be achieved in view of the additional approximations made in the CCSD approach; see Section II.D.3.3). With the use, however, of modest-sized ASs, the externally corrected CCSD provides a very significant improvement over both the CAS FCI/SOCI and standard CCSD energies, while being computationally inexpensive (usually a factor of two over the standard CCSD) and reliable, thus substantially extending the usefulness of the standard CCSD.

Concerning more technical aspects, we only mention that it is of little relevance whether we employ the leading CAS SCF configuration or the true RHF or ROHF wave function as a reference, as both yield very similar results. Furthermore, proper treatment of the $\Lambda^{(2)}(1,3)$, i.e., $T_1 T_3$, term requires that we use the external t_3-amplitudes to correct the linear coefficients b_{ij} in Eq. (2.65) that are associated with the unknown t_1-amplitudes. In direct approaches we have to evaluate these terms in each iteration, and we thus refer to such procedures as the iterative account of $T_1 T_3$. Note that, in contrast, the $\Lambda^{(1)}(3)$, $\Lambda^{(2)}(3)$, and $\Lambda^{(2)}(4)$ terms involving the external t_3- and t_4-amplitudes are evaluated prior to solving the externally corrected CCSD equations and are used to correct the absolute terms a_i in Eq. (2.65) once and for all. It turns out that the same can be done with the $T_1 T_3$ correcting term when we use the approximate external T_1 amplitudes (that are obtained as a by-product of the cluster analysis of the external-source wave function). Although the error due to this noniterative handling of the $T_1 T_3$ term increases with the increasing distortion of the system, it seldom exceeds a few μhartree.

The applications of this CAS FCI/SOCI-based approach to several benchmark systems [49, 238], involving a stretching of one or more single bonds or of a single multiple bond, unequivocally indicate the usefulness of the procedure. Nonetheless, it is important to continue the search for inexpensive and reliable external sources for the higher-than-pair clusters. Our most recent work [50, 239, 240] in this direction uses suitable MR CI wave functions, resulting in the so-called reduced MR (RMR) CCSD method, which seem to represent by far the best choice, particularly when handling the stretching of triply bonded systems (e.g., N_2 or CN). Since this method (as well as, in fact, the CAS FCI/SOCI-based approaches when applied to OS systems) must be classified as an MR SS method, it is described in more detail in Section II.D.3.3.

D. Multireference Coupled-Cluster (MR CC) Approaches

The SR CC approach is clearly inadequate to describe general OS states. For example, when we examine a singlet state whose IPM description involves two singly occupied orbitals (i.e., a singlet state arising by a single excitation from a CS ground state), we have to consider both spin-orbital configurations, $|\{\cdots a\alpha\, r\beta\cdots\}\rangle$ and $|\{\cdots a\beta\, r\alpha\cdots\}\rangle$, on an equal footing. At the spin-orbital level, this situation already requires a two-reference theory [89, 230] (unlike the corresponding triplet state, wherein one can choose a single-determinantal high-spin reference), even though at the spin-adapted level we still can deal with a single, yet multideterminantal, reference. (See Section II.D.3.2.) However, when three singly occupied spin-orbitals are involved (as in doublet excited states of simple radicals), then even in the spin-adapted theory, we have to deal with two homologous configurations (or with three at the spin-orbital level).

The situation becomes even more complex when we consider the dissociation involving multiple bonds, in which one may have to handle up to six (or even eight) open-shell orbitals. In all such cases, a genuine MR approach is called for if a bona fide description is required. As already noted, a generalization of the SR CC *Ansatz* to these situations is not unique. Since, however, all MR CC approaches rely on the effective Hamiltonian formalism, we first briefly outline this concept.

1. Effective Hamiltonian Formalism

In all molecular-orbital (MO)-based MR approaches, the one-electron orbital space is partitioned by subdividing the spin orbitals into the three disjoint subsets of core (c); valence, or active (a); and virtual, or external (e) spin orbitals (or orbitals), involving m_c, m_a, and m_e spin orbitals, respectively, so that $m_c + m_a + m_e = m = 2n$. The IPM N-electron configurations $|\Phi_i\rangle$ $(i = 1, \ldots, M)$ in which all core spin-orbitals are occupied, while the remaining $(N - m_c)$ electrons are distributed in all possible ways over the m_a $(m_a > N - m_c)$ active spin-orbitals, $M = \binom{m_a}{N-m_c}$, span the model space $\mathcal{M}_0, \mathcal{M}_0 = \mathrm{Span}\{|\Phi_i\rangle : i = 1, \ldots, M\}$. With this model space, we associate a Hermitian projection operator $P : \mathcal{H}_N \rightarrow \mathcal{M}_0$ such that

$$P = \sum_{i=1}^{M} |\Phi_i\rangle\langle\Phi_i|, \qquad P^2 = P = P^\dagger, \qquad (2.103)$$

as well as the Hermitian projection operator $Q = 1 - P$, $Q : \mathcal{H}_N \rightarrow \mathcal{M}_0^\perp$, where $\mathcal{H}_N$ designates the N-electron component of the Fock space $\mathcal{F}$ and $\mathcal{M}_0^\perp$ is the orthogonal complement of $\mathcal{M}_0$ in $\mathcal{H}_N$.

In order to define an MR analogue of the wave operator $\mathcal{W}$ transforming the chosen IPM reference $|\Phi_0\rangle$ into the exact wave function $|\Psi\rangle$ in the SR theory, we have to specify (at least implicitly) a *target space* $\mathcal{M}$ that is spanned by M exact eigenstates $|\Psi_i\rangle$, $\mathcal{M} = \mathrm{Span}\{|\Psi_i\rangle : i = 1, \ldots, M\}$. The necessary condition for $|\Psi_i\rangle \in \mathcal{M}$ is that

$$P|\Psi_i\rangle =: |\widetilde{\Phi}_i\rangle \in \mathcal{M}_0 \qquad \text{and} \qquad \mathrm{Span}\{|\widetilde{\Phi}_i\rangle : i = 1, \ldots, M\} = \mathcal{M}_0 . \quad (2.104)$$

Obviously, in an ideal situation, $|\Phi_i\rangle$ or $|\widetilde{\Phi}_i\rangle$ will represent a good approximation to $|\Psi_i\rangle$, so that when $|\Psi_i\rangle$ is expanded in terms of the IPM configurations spanning $\mathcal{H}_N$, the $|\widetilde{\Phi}_i\rangle$, representing a suitable linear combination of $|\Phi_i\rangle$'s, will have the largest weight. In principle, however, $\mathcal{M}$ can be any M-dimensional subspace of $\mathcal{H}_N$ satisfying Eq. (2.104). This ambiguity is, of course, the main source of a possible high multiplicity of MR CC solutions. (See later.) At this stage, however, all that we need is the assumption of the existence of a subspace $\mathcal{M} \subset \mathcal{H}_N$ having the properties expressed in Eq. (2.104), which always holds despite the fact that $\mathcal{M}$ is far from being unique.

For a chosen target space $\mathcal{M}$, we then define a nonsingular map $\widetilde{U} : \mathcal{M}_0 \to \mathcal{M}$ by

$$\widetilde{U}|\widetilde{\Phi}_i\rangle = |\Psi_i\rangle, \tag{2.105}$$

and extend the map to the entire space $\mathcal{H}_N$, $U = \widetilde{U} \uparrow \mathcal{H}_N$, by requiring that $U = \widetilde{U}$ on $\mathcal{M}_0$ and $U = 0$ on $\mathcal{M}_0^\perp$, so that

$$U|\widetilde{\Phi}_i\rangle = |\Psi_i\rangle \tag{2.106}$$

and

$$UQ = U(1 - P) = 0, \quad \text{or} \quad U = UP. \tag{2.107}$$

Since

$$PU|\widetilde{\Phi}_i\rangle = P|\Psi_i\rangle = P^2|\Psi_i\rangle = P|\widetilde{\Phi}_i\rangle, \tag{2.108}$$

and U and P vanish on $\mathcal{M}_0^\perp$, we also require that

$$PU = P, \tag{2.109}$$

which can be shown to be equivalent to the requirement of the *intermediate* (or *Bloch*) *normalization*

$$\langle \Phi_i | \widetilde{\Psi}_j \rangle = \delta_{ij}, \tag{2.110}$$

where $|\widetilde{\Psi}_j\rangle = U|\Phi_j\rangle$, $(j = 1, \ldots, M)$. Note that in contrast to P, U is not Hermitian, although it is clearly idempotent; i.e.,

$$U^2 = (UP)^2 = U(PU)P = UP^2 = UP = U. \tag{2.111}$$

We can now introduce the concept of an *effective*, or *model*, *Hamiltonian* $H^{(\text{eff})}$, acting within $\mathcal{M}_0$, yet having the same eigenvalues as the exact Hamiltonian H defined on $\mathcal{H}_N$. Of course, this remarkable property of $H^{(\text{eff})}$

is possible only due to the fact that we restricted our attention to a finite subset of M eigenvalues of H. Indeed, starting with the exact Schrödinger equation

$$H|\Psi_i\rangle = E_i|\Psi_i\rangle, \qquad (2.112)$$

and projecting onto $\mathcal{M}_0$ with P, i.e.,

$$PH|\Psi_i\rangle = PHU|\widetilde{\Phi}_i\rangle = E_i P|\Psi_i\rangle = E_i|\widetilde{\Phi}_i\rangle, \qquad (2.113)$$

we can write

$$H^{(\text{eff})}|\widetilde{\Phi}_i\rangle = E_i|\widetilde{\Phi}_i\rangle, \qquad (2.114)$$

where

$$H^{(\text{eff})} = PHU = PHUP. \qquad (2.115)$$

Clearly,

$$QH^{(\text{eff})}P = QPHU = 0, \qquad (2.116)$$

so that the action of $H^{(\text{eff})}$ on $\mathcal{M}_0$ does not take us out of $\mathcal{M}_0$.

It thus remains to determine the MR wave operator U in order to compute M exact eigenvalues of H. Since, by Eq. (2.111), U is idempotent, it acts as the identity on $\mathcal{M}$; that is, $U|\Psi_i\rangle = |\Psi_i\rangle$ [cf. Eq. (2.106)], so that its action on the Schrödinger equation (2.112) gives

$$\begin{aligned} UH|\Psi_i\rangle &= UHU|\widetilde{\Phi}_i\rangle \\ &= E_i U|\Psi_i\rangle = E_i|\Psi_i\rangle = H|\Psi_i\rangle = HU|\widetilde{\Phi}_i\rangle, \end{aligned} \qquad (2.117)$$

and thus,

$$UHU = HU \qquad (2.118a)$$

holds in $\mathcal{H}_N$, because U vanishes on $\mathcal{M}_0^\perp$. Equation (2.118a) represents the simplest form of the *Bloch equation* determining the wave operator U. It is often expressed in various alternative forms [14, 29, 39], a particularly useful one being in terms of the effective Hamiltonian (2.115), namely,

$$HU = UH^{(\text{eff})}, \qquad (2.118b)$$

which easily follows from Eq. (2.118a), since $UP = U$, so that $UHU = UPHUP = UH^{(\text{eff})}$. Clearly, the Bloch equation can be exploited with or without the CC *Ansatz*. (For interesting applications of the latter kind, see [255].)

Note that in the SR case, when $\mathcal{M}_0 = \text{Span}\{|\Phi_0\rangle\}$ and $\mathcal{M}_0^\perp = \text{Span}\{\Phi_j^{(k)} : k = 1, \ldots, N; j = 1, \ldots, M^{(k)}\}$ (here $M^{(k)}$ designates the dimension of the k-times excited manifold), the standard CC *Ansatz* reads

$$U = \mathcal{W}P_0 = e^T P_0, \qquad P_0 = |\Phi_0\rangle\langle\Phi_0|, \qquad (2.119)$$

so that acting with Eq. (2.118a) on $\mathcal{M}_0$ gives

$$e^T P_0 H e^T |\Phi_0\rangle = H e^T |\Phi_0\rangle, \qquad (2.120)$$

or, equivalently (acting with e^{-T} from the left),

$$P_0 H e^T |\Phi_0\rangle = e^{-T} H e^T |\Phi_0\rangle. \qquad (2.121)$$

Since dim $\mathcal{M}_0 = 1$, we have [cf. Eq. (2.71)]

$$E_0 = \langle H^{(\text{eff})} \rangle \equiv \langle \Phi_0 | H^{(\text{eff})} | \Phi_0 \rangle = \langle H e^T \rangle = \langle e^{-T} H e^T \rangle, \qquad (2.122)$$

and projecting onto $\mathcal{M}_0^\perp$ gives the CC equations (2.72):

$$\langle \Phi_j^{(k)} | e^{-T} H e^T | \Phi_0 \rangle = 0. \qquad (2.123)$$

As already mentioned, several avenues are opened to us when we generalize the SR CC *Ansatz* for the wave operator U. Out of various attempts developed, we now briefly outline two viable alternatives for the genuine MR CC *Ansatz*.

2. *Proper MR CC Approaches*

In generalizing the SR CC *Ansatz* for the wave operator to the MR case, it is important to require that the relevant cluster operator T (or S) be uniquely defined, which in turn implies that we must uniquely characterize all possible excitations out of the model space $\mathcal{M}_0$. We have basically two options:

(i) We can strive to define a single, yet sufficiently general and unique, cluster operator, yielding a size-extensive cluster expansion for the wave operator U, when acting on any model space configuration [38, 39]. It turns out [43] that, in general, this requirement can be satisfied only when, together with the model space $\mathcal{M}_0$, we also consider model spaces $\mathcal{M}_0^{(k)}$ for ionized species, all the way to the system having m_c electrons occupying core spin orbitals. We thus must consider a sequence of model spaces

$$\mathcal{M}_0^{(0)} = \text{Span}\{|\Phi\rangle\},$$

$$\mathcal{M}_0^{(k)} = \text{Span}\{X_{I_1}^\dagger \cdots X_{I_k}^\dagger |\Phi\rangle : I_1 < \cdots < I_k = m_c + 1, \ldots, m_c + m_a\},$$

$$k = 1, \ldots, (N - m_c), \tag{2.124}$$

where the I_j's label active spin orbitals. (All spin orbitals are assumed to be arranged in some fixed order and numbered consecutively, starting with core spin orbitals; to label active spin orbitals, we employ the letters from the middle of the Latin alphabet, $I, J, K, \ldots$, while $A, B, C, \ldots$ and $R, S, T, \ldots$ label core or active, and virtual or active, spin orbitals, respectively, unless otherwise mentioned.) Only in this way can the aforementioned requirement of uniqueness be fulfilled. For this reason, the resulting cluster *Ansatz* and the CC method are referred to as the *valence-universal* (VU) or, in view of the fact that, in addition to the N-electron component $\mathcal{H}_N$ of the Fock space $\mathcal{F}$, we must also consider the components $\mathcal{H}_K$, $K = m_c, m_c + 1, \ldots, (N - 1)$ of $\mathcal{F}$, as the *Fock space* CC approach [37–39, 41–43, 82, 83].

(ii) If we insist on staying within $\mathcal{H}_N$, we must admit different cluster operators for different model space references $|\Phi_i\rangle$, thus representing U as a linear combination of separate CC *Ansätze* for each $|\Phi_i\rangle \in \mathcal{M}_0$. This approach, due to Jeziorski and Monkhorst [40], is referred to as the *state-universal* (SU), or *Hilbert space*, CC method.

We next briefly point out the main features and form of these approaches.

2.1. Valence-Universal CC Methods. A valence-universal operator A represents a family of linear operators A_k, with each A_k acting in the k-electron component $\mathcal{H}_k$ of $\mathcal{F}$ for $k = m_c, m_c + 1, \ldots, m_c + m_a$. It can be proved [43] that for any family $\{A_k\}$ of such linear operators, there exists a unique valence-universal operator A such that its restriction to $\mathcal{M}_0^{(k)}$, $A \downarrow \mathcal{M}_0^{(k)}$, gives the action of A_k in $\mathcal{M}_0^{(k)}$. Such a valence-universal operator A is an element of the so-called *model algebra* $\mathcal{A}$ whose basis consists of all replacement (or excitation) operators of the type $e_{AB\cdots}^{RS\cdots}$, including the identity e. (Recall that $A, B, \ldots$ now label either core or active spin orbitals and $R, S, \ldots$ either virtual or active spin orbitals.) In particular, for a valence-universal wave operator U, we require that $U \downarrow \mathcal{M}_0^{(k)} =: U_k = UP_k$, where P_k is a projector onto $\mathcal{M}_0^{(k)}$. The uniqueness requirement implies that $U = 1 + \widehat{\Omega}$, where

$$\widehat{\Omega} = \omega_R^A e_A^R + (2!)^{-2} \omega_{RS}^{AB} e_{AB}^{RS} + \cdots \tag{2.125}$$

is an element of the *external subalgebra* $\mathcal{A}_{\text{ext}}$ of $\mathcal{A}$, $\mathcal{A}_{\text{ext}} \subset \mathcal{A}$ (as indicated by a caret over Ω) that excludes replacement operators having *only* active labels (i.e., those belonging to the *internal subalgebra* $\mathcal{A}_{int}$ of $\mathcal{A}$, $\mathcal{A}_{int} \subset \mathcal{A}$, $\mathcal{A} = \mathcal{A}_{int} \oplus \mathcal{A}_{ext}$). Note that $\widehat{\Omega}$, sometimes called the *correlation operator*, does include the so-called *spectator amplitudes* (e.g., ω_{RI}^{AI}), as long as A and R are not both active labels. The spectator amplitudes are essential for achieving valence universality. They are a consequence of the fact that in active orbitals the electrons can be either annihilated or created. This fact also implies that the active labels can figure either as particle or hole labels, giving rise to the so-called *folded diagrams*. (Cf. [29, 39].)

Another implication of the foregoing fact is that the cluster *Ansatz* for U must be in the normal product form, as was first pointed out by Lindgren [38]. This can be best formulated [43] by writing

$$\Omega = \mathbf{e}^S = e + S + \tfrac{1}{2}S * S + \cdots, \qquad S \in \mathcal{A}_{ext}, \qquad (2.126)$$

where e designates the identity element of the algebra and the asterisk indicates the normal product in $\mathcal{A}_{ext}$, which can be simply defined as

$$
\begin{aligned}
e_A^R * e_B^S &= e_{AB}^{RS}, \\
e_{AB}^{RS} * e_C^T &= e_{ABC}^{RST}, \quad \text{etc.}
\end{aligned}
\qquad (2.127)
$$

Using the valence-universal representation of the Hamiltonian, and relying on Bloch's equation, one derives the VU CC equations that determine the cluster amplitudes S_R^A, S_{RS}^{AB}, etc. (cf., e.g., [29]), which in turn yield the effective Hamiltonian H^{eff}, whose diagonalization finally gives the desired energies. Although, together with the systems studied, one also has to consider all their k-times ionized species, $k = 1, 2, \ldots, N - m_c + 1$, it must be emphasized that the VU CC equations have a nice general structure based on the so-called *valence rank* concept. (An operator from $\mathcal{A}$ is of valence rank r if it involves r *valence* annihilators.) The fact that the highest valence rank component appears linearly in a given U_k permits a decoupling and thus a successive calculation of higher and higher valence rank components [256]. However, when S is truncated at the pair cluster level, a consideration of at least selective three-body amplitudes of the spectator type is called for [95, 228].

As far as we know, there is no general-purpose implementation of the VU CC method, although numerous applications [142, 257–259] and pilot studies [95, 228, 260] have been carried out. A two-reference spin-orbital version designed to handle excited states of closed-shell systems is also available [89, 230]. The main obstacle in an actual exploitation of this approach is the intruder state problem [261] and multiplicity of solutions [262], although the complexity of the formalism also plays an important role. A great advantage of the VU CC approach is the fact that it can be rather easily symmetry adapted, and the diagrammatic techniques can be employed with little modification with the core state $|\Phi\rangle$ used as a Fermi vacuum. The fact that, along with the system studied, one simultaneously obtains information concerning various ions is of limited practical value, since the ionized and neutral species will usually require the use of different basis sets.

The intruder state problem, usually associated with MR approaches, is in fact encountered already at the SR level if one of the excited configurations $|\Phi_j^{(k)}\rangle$ (or, rather, some linear combination of such configurations) becomes degenerate (or quasi-degenerate) with the reference $|\Phi_0\rangle$. In such cases, the **b** matrix of Eq. (2.67) [see also Eq. (2.65)] becomes singular or nearly singular, causing the linear approximation to fail completely and the CCSD result to deteriorate, indicating the importance of higher-than-pair clusters. Nonetheless, in many such nearly degenerate situations, the CCSD still works extremely well, as already noted earlier.

Just as the indicated quasi-degeneracy in the case of an SR CC approach calls for an MR description, by essentially adopting the quasi-degenerate excited configuration(s) as reference(s) that now span a higher-than-one-dimensional reference space $\mathcal{M}_0$, one can in principle avoid the intruder state problem by similarly enlarging $\mathcal{M}_0$ of the MR CC approach by incorporating those excited configurations, whose energies fall within the interval of energies defined by the reference configurations $|\Phi_i\rangle \in \mathcal{M}_0$. However, in order to warrant the size extensivity of the outlined MR CC method, it is essential to use a complete model space $\mathcal{M}_0$. Consequently, together with the excited configurations that become quasi-degenerate with the manifold of reference configurations, one also has to include numerous other configurations that result by enlarging the subset of active orbitals. This extension further increases the probability of additional intruders appearing, not to mention the increase in the complexity and size of the resulting MR CC problem. One only has to look at the potential-energy curves of some diatomics to realize the severity of the conundrum.

These considerations clearly indicate that, for an actual exploitation of MR CC methods, it is essential to use as small a model space as possible. Consequently, much attention has been devoted to the possibility of using *truncated*, or *incomplete*, model spaces, while preserving size extensivity [37]. In fact, most VU CC applications carried out in the past were based on truncated model spaces, often different ones for different geometries [257, 263–267].

The breakdown of size extensivity when one relies on an incomplete model space was shown to be related to the use of the intermediate normalization for the wave operator given by Eq. (2.109), and several remedies were suggested regarding how to rectify this problem. (See [37].)

However, since giving up the intermediate normalization further complicates the formalism, these various theories enjoyed little actual exploitation.

The problem of multiplicity of solutions of MR CC equations, which
in some cases may be beneficial because it permits the description of several states (belonging to different target spaces $\mathcal{M}$), can be particularly
bothersome in VU CC approaches, since this multiplicity drastically
increases in view of the hierarchical structure of the approach. Consequently, the multiplicity at a given valence rank level will propagate into
the higher levels (cf. [95, 228, 268]), resulting in numerous solutions with
different genealogies.

2.2. State-Universal CC Methods. The MR CC formalism that does not
involve the assumption of valence universality and thus relies solely on the
N-electron component $\mathcal{H}_N$ of the Fock space $\mathcal{F}$ (hence the term *Hilbert-
space approach*) requires a separate cluster *Ansatz* for each reference
configuration $|\Phi_j\rangle \in \mathcal{M}_0$. Designating the orthogonal projector onto $|\Phi_j\rangle$
by p_j, $p_j = |\Phi_j\rangle\langle\Phi_j|$, we thus generalize the SR *Ansatz* of Eq. (2.45) or
Eq. (2.119) for the wave operator as

$$U = \sum_{j=1}^{M} e^{T(j)} p_j, \tag{2.128}$$

where $T(j)$ now designates a connected-cluster operator* associated with the
reference $|\Phi_j\rangle \in \mathcal{M}_0$. Thus, as in the SR case, we express each $T(j)$ in terms
of the excitation operators and corresponding amplitudes; that is,

$$T(j) = \sum_{k} T_k(j) = \sum_{i,k} t_i^{(k)}(j)\, G_i^{(k)}(j), \tag{2.129}$$

where the superscript indicates the excitation level relative to $|\Phi_j\rangle$, while the
subscript enumerates distinct excitation operators or the corresponding
configurations $|\Phi_{j;i}^{(k)}\rangle = G_i^{(k)}(j)|\Phi_j\rangle$. (Concerning ambiguities in the definition
of the excitation level when one uses spin adaptation, see Section II.D.3.2;
we drop the superscripts when they are not essential.)

The exact eigenstates $|\Psi_i\rangle \in \mathcal{M}$ $(i = 1, \ldots, M)$ will thus be given by

$$|\Psi_i\rangle = \sum_{j=1}^{M} C_{ij}\, e^{T(j)} |\Phi_j\rangle, \tag{2.130}$$

---

*Note that we gain use T rather than S to designate cluster operators, as in the SR formalism.
This notation should help to avoid a possible confusion with the VU formalism, and it is also
consistent with current usage.

where the coefficients C_{ij} are components of the eigenvectors of the effective Hamiltonian matrix $\mathbf{H}^{(\text{eff})}$, with the eigenvalues providing the corresponding exact energies E_i in Eq. (2.112). The matrix representative $\mathbf{H}^{(\text{eff})}$ of the effective Hamiltonian is easily calculated once the cluster operators $T(j)$ are known:

$$H_{ij}^{(\text{eff})} \equiv \langle \Phi_i | H^{(\text{eff})} | \Phi_j \rangle = \langle \Phi_i | He^{T(j)} | \Phi_j \rangle. \tag{2.131}$$

The cluster amplitudes defining $T(j)$ operators are in turn determined by solving Bloch equations. Using the form given in Eq. (2.118b) and letting it act on $|\Phi_i\rangle$, we find that

$$He^{T(i)} |\Phi_i\rangle = \sum_j e^{T(j)} p_j H^{(\text{eff})} |\Phi_i\rangle = \sum_j e^{T(j)} |\Phi_j\rangle H_{ji}^{(\text{eff})}. \tag{2.132}$$

We again premultiply with $\exp(-T(i))$ and project onto $|\Phi_{i;\ell}^{(k)}\rangle \equiv |\Phi_{i;\ell}\rangle$, obtaining the SU CC equations

$$\langle \Phi_i | G_\ell^\dagger(i) e^{-T(i)} He^{T(i)} |\Phi_i\rangle = \sum_{j(\neq i)} \langle \Phi_i | G_\ell^\dagger(i) e^{-T(i)} e^{T(j)} |\Phi_j\rangle H_{ji}^{(\text{eff})}, \tag{2.133}$$

since $|\Phi_{i;\ell}\rangle$ is orthogonal to $|\Phi_i\rangle$.

Note that the left-hand side, called the direct term, has the same form for each reference configuration as in the SR case and may thus be similarly handled using the commutator expansion given by Eq. (2.68). The right-hand side, representing the coupling between the references, can be similarly handled by expanding the product of exponentials using the well-known relation of Lie group theory,

$$e^A e^B = e^{A+B+\frac{1}{2}[A,B]+\cdots} = 1 + (A+B) + \frac{1}{2}\{(A+B)^2 + [A,B]\} + \cdots. \tag{2.134}$$

The explicit form of these equations at the CCSD level for the case of two active orbitals of different symmetry in the OSA form (i.e., the simplest two-reference case) was given in [87, 93], and its computational implementation and applications were described in [90–92, 94]. The same level of theory at the spin-orbital level (i.e., a four-reference case involving both singlet and triplet states) was also implemented and tested [88, 89, 230] at approximately the same time.

The pilot studies showed that the CCSD version of this SU CC theory works extremely well as long as no intruders are present, particularly when both references are degenerate or nearly degenerate. Once the intruders set in, multiple solutions can be obtained, providing a reasonable description of both the reference and the intruder state. However, when the reference configurations are no longer degenerate, SR CCSD gives superior results. Some of these results are presented in Section III.

In view of the rapidly increasing complexity of the MR formalism and the difficulties arising due to the problems associated with the intruder state(s), one cannot expect a widespread utilization of this approach or even its general-purpose implementation in the near future, unless some new developments take place or the available computational capabilities increase substantially. For this reason, most of the present effort concentrates on the state-selective, or state-specific, approaches to OS systems. Although these provide a much more efficient and reliable description for most important cases of interest, one must keep in mind that a proper description of several quasi-degenerate states of the same symmetry or of states involving several leading OSA configurations having similar weights will require a genuine MR approach of one kind or another.

3. State-Selective, or State-Specific (SS), Approaches

We finally address essentially SR approaches that, however, exploit certain features of the general MR theory. Thus, all such approaches consider one state at a time, but invariably rely on the (spin) orbital partitioning into the core, active, and virtual orbitals, sometimes even further refining this division. Moreover, the spin-symmetry-adapted state-specific (SS) approaches generally employ a single, yet multideterminantal, reference (with fixed weights). We also include here the externally corrected approaches (see Section II.C.4.3) based on MR CI wave functions used as an external source.

3.1. SS MR CC Method of Adamowicz et al. The basic step in this approach [96–99] is to partition the SR CC cluster operator into the so-called internal and external parts, i.e.,

$$T = T^{(\text{int})} + T^{(\text{ext})} \,, \tag{2.135}$$

in such a way that $\exp(T^{(\text{int})})$ produces a wave function within a multideterminantal model space when acting on a suitably chosen formal reference configuration $|\Phi_0\rangle$, while $\exp(T^{(\text{ext})})$ then generates, in principle, the exact wave function by acting on $\exp(T^{(\text{int})})|\Phi_0\rangle$. We can write that

$$|\Phi\rangle = e^T|\Phi_0\rangle = e^{T^{(\text{ext})}} e^{T^{(\text{int})}}|\Phi_0\rangle = e^{T^{(\text{ext})}}|\Phi_0^{(\text{int})}\rangle \,, \tag{2.136}$$

since the internal and external cluster operators commute,

$$\left[T^{(\mathrm{int})}, T^{(\mathrm{ext})} \right] = 0 \,, \tag{2.137}$$

thanks to the fact that $T^{(\mathrm{int})}$ and $T^{(\mathrm{ext})}$ are defined in terms of the hole–particle operators relative to a single determinantal reference $|\Phi_0\rangle$.

To define the partitioning given by Eq. (2.135), the spin-orbitals are subdivided into the core, active, and virtual ones as usual, and the active ones are further subdivided into the so-called active-hole and active-particle ones. As the name suggests, the Fermi vacuum consists of core and active-hole spin-orbitals, defining the reference $|\Phi_0\rangle$. The internal cluster operator $T^{(\mathrm{int})}$ is then defined in terms of the replacement operators involving only active spin-orbitals (i.e., $T^{(\mathrm{int})} \in \mathcal{A}_{\mathrm{int}}$; see Section II.D.2.1), the relevant configurations thus spanning the active space $\mathcal{M}_0$, while the external cluster operator $T^{(\mathrm{ext})}$ involves the replacement operators from $\mathcal{A}_{\mathrm{ext}}$ of the type $e_{AB\cdots}^{RS\cdots}$, where $A, B, \ldots$ are either the core or active-hole labels and $R, S, \ldots$ are the virtual or active-particle labels. Hence, $T^{(\mathrm{ext})}$ excludes excitations involving only active orbitals, even though when $T^{(\mathrm{int})}$ excitations are truncated at a certain prescribed excitation level, the authors allow the higher level internal excitations to be shifted to $T^{(\mathrm{ext})}$. This shift does not violate the commutativity requirement, since all p–h operators are defined with respect to a fixed Fermi vacuum, given by the partitioning of active orbitals into the active-hole and active-particle ones.

The required CC equations and the energy expression are the same as in the SR CC theory (truncated at an appropriate excitation level). In the actual exploitation of this approach, the expansion of $T^{(\mathrm{int})}$ was truncated to a double excitation level, while that of $T^{(\mathrm{ext})}$ included all-semi-internal and all-external singles and doubles, as well as those semi-internal triples and quadruples that are external singles and doubles relative to those defining $T^{(\mathrm{int})}$.

Clearly, the approach just outlined represents an approximate version of the SR CCSDTQ method, using the MR model-space concept to achieve a physically meaningful and computationally manageable truncation of triples and quadruples. Even so, the rapid growth of the number of triples and quadruples with the increasing basis size limits the capabilities of this approach quite severely. The authors thus considered the so-called SS MR CCSD(T) method, in which external triples are restricted to those having at least one pair of active hole–particle labels, and the SS MR CCSD(TQ) method, including quadruples with at least two hole–particle pair labels [126].

The approach seems to be able to handle the breaking of single and double bonds, as well as some quasi-degenerate excited states. In order to handle more general situations, the authors recently proposed the employment of a linear CI-type *Ansatz* for the internal part [126].

3.2.   Unitary Group Approach (UGA)–Based CCSD Methods. The unitary group approach (UGA) to the many-electron correlation problem employs the so-called boson calculus [220] and the representation theory of unitary group Lie algebras in order to exploit the spin symmetry of finite-dimensional *ab initio* models. It relies on the spin independence of the standard *n*-orbital *ab initio* model Hamiltonian H of Eq. (2.88), which, in view of Eq. (A.18) in the appendix is expressible in terms of the generators E_j^i of the orbital unitary group U(n). [See Eq. (A.16).] Consequently, by exploiting the group chain (cf. [174–177])

$$U(2n) \supset U(n) \otimes U(2), \qquad (2.138)$$

the eigenvalue problem for H may be factorized into pure-spin subproblems that are characterized by two-column irreducible representations $\lambda \equiv [2^a \, 1^b \, 0^c]$ of U(n), where $a = \frac{1}{2}N - S$, $b = 2S$, and $c = n - a - b$, with N the electron number and S the total spin quantum number. The modeling of the electronic structure for each S can then be achieved within the carrier space of λ by relying on a suitably chosen effective N-electron basis set. The exact solution of the model requires, of course, the use of the full carrier space. The well-documented techniques that are based on the Gel'fand–Tsetlin canonical representations [174–177] and are utilized in large-scale CI calculations [269] cannot, however, be easily adapted to CC methodology, which requires a different approach based on the so-called Clifford algebra UGA (CAUGA) [174, 270] or bonded tableaux [271].

Thanks to its versatility, the UGA or CAUGA formalism can be just as beneficial when applied to CC theory as it is in CI or VB approaches. This should be apparent already from the fact that the so-called algebraic approach (see Appendix A) is actually based on the enveloping algebra of U(n) or U($2n$). Let us point out at least three obvious applications.

In the SR CS case, we can use this approach to derive an OSA version of CC equations. In this case, we obtain the standard OSA formalism [128], as shown in [173] for the case of CCD equations. In other words, choosing a suitable coupling scheme based on CAUGA (such as the *pp–hh* scheme for doubles—see Appendix A; note that Gel'fand states are not a good choice here [218]), we arrive at an OSA formalism that is equivalent to the one derived via graphical methods of spin algebras and that will yield the same energies as the spin-orbital formalism. This approach could be particularly beneficial when one proceeds beyond the CCSD level, since it will automatically eliminate the linear dependencies arising at the spin-orbital level. (Cf. Section II.C.4.1.)

The second prospect for CAUGA formalism in CC theory is the extension of an SR approach to OS systems, for which a zero-order wave function is fully determined by the spin symmetry. In such cases, we have a single, yet multiconfigurational, reference that is equivalent to a linear combination of Slater determinants with fixed weights. Since all these determinants have the same orbital occupancy, they can be represented by a single spin-free reference. This is the case for all high-spin OS states, as well as for low-spin OS singlet states arising via the coupling of two singly occupied MOs. In these cases, we can choose a suitable vector from the carrier space of the relevant irreducible representation λ of the orbital unitary group $U(n)$ as a spin-free reference $|\Phi_0\rangle$ and formulate the cluster *Ansatz* in terms of $U(n)$ generators. Thus, formally, we again have an SR *Ansatz*

$$|\Psi\rangle = e^T|\Phi_0\rangle \, , \tag{2.139}$$

although $|\Phi_0\rangle$ and $|\Psi\rangle$ now represent spin-free kets that generally correspond to multideterminantal states. When $|\Phi_0\rangle$ belongs to the carrier space of λ, so will any of the states that result from it through the action of $U(n)$ generators. Moreover, just as in the standard SR case the CC theory is invariant with respect to any unitary transformation of occupied and, independently, of unoccupied (spin) orbitals, we also require that in the present case the same invariance property hold for independent transformations of core, active, and virtual orbitals. In other words, we require the cluster operator T to be expressed in terms of irreducible tensor operators that are adapted to the group chain

$$U(n) \supset U(n_c) \otimes U(n_e) \otimes U(n_a) \tag{2.140}$$

(and, if desirable, to a refined chain involving other subgroups of $U(n)$, e.g., $U(n_c) \otimes U(n_e)$; see [46]).

It should be emphasized that in spite of the formal resemblance of the two *Ansätze* of Eqs. (2.45) and (2.139), the spin-free version cannot be obtained by a spin adaptation of the standard spin-orbital cluster *Ansatz*, as is the case for CS states. In this sense, the UGA CC method for OS systems that is based on the *Ansatz* of Eq. (2.139) represents an independent, new approach that cannot be interpreted as a spin-adapted version of some existing spin-orbital version of the CC theory.

Finally, in the general OS case, a zero-order wave function is not uniquely determined by the spin symmetry alone and has a genuine MR character. Nonetheless, even in this case, the UGA formalism can be very useful, as is outlined in [45] for the SU (or Hilbert-space; see Section II.D.2.2) MR CC theory. Although such a method remains to be implemented (a two-reference version would be required, for example, to handle doublet excited states of radicals), UGA CC was recently employed in the design of externally corrected SS MR methods—notably, the so-called reduced MR (RMR) CCSD method—that will be described in the next section. In the current section, we pay attention to those OS cases for which the formally SR *Ansatz* of Eq. (2.139) is applicable.

The requirement that a zero-order wave function be determined solely by the spin symmetry necessitates the use of a minimal active space. Thus, for the high-spin states, characterized by the total spin S, we take $n_a = 2S$, and for the OS singlet states mentioned earlier, $n_a = 2$. A general high-spin state of spin S has $(2S + 1)$ spin components involving the same spatial part, but different spin functions. For example, in the triplet case, either the highest spin ($|(\text{core}) i\alpha j\alpha|, M_S = 1$) or the lowest spin ($|(\text{core}) i\beta j\beta|, M_S = -1$) determinants, or the $M_S = 0$ linear combination of the two determinants $|(\text{core}) ij(\alpha\beta + \beta\alpha)|$, can be employed as a reference $|\Phi_0\rangle$. [Recall the orbital labeling convention employed for the core $(a, b, c, \ldots)$, active $(i, j, k, \ldots)$, and external $(r, s, t, \ldots)$ orbitals.] In contrast to most existing partially spin-adapted methods, which can only employ a single determinantal $M_S = \pm 1$ component as a reference, the UGA CC treats all spin components on an equal footing. It is thus worth emphasizing that the UGA CC *Ansatz* of Eq. (2.139) has two important beneficial properties:

(i) $e^T|\Phi_0\rangle$ is a pure-spin eigenstate, and
(ii) the wave operator $\mathcal{W} = e^T$ can be applied to any one of the $(2S + 1)$ spin components $|M_S| \leq S$, yielding identical results (and, of course, preserving M_S).

Usually, one pays attention only to the first property. However, the second property is also desirable: A theory should treat any spin component of a high-spin state in an equitable and uniform way. This is not only an esthetic requirement, but a very useful one if we wish to handle spin-orbital interactions or study spin-dependent properties (cf. [219]) later on.

The preceding properties of the UGA CC cluster expansion are attained by representing the cluster operator T as a linear combination of a suitable set $\{G_I\}$ of excitation operators G_I [cf. Eq. (2.49)]; that is,

$$T = \sum_I t_I G_I, \tag{2.141}$$

where t_I represents the unknown cluster amplitudes (the subscript I labels the index set that uniquely characterizes G_I and t_I), much as in the standard approaches. [Cf. Eq. (2.49).] Here, however, we require that the excitation operators represent the irreducible $U(n)$ tensor operators adapted to the chain (2.140). The chosen set $\{G_I\}$ then defines the level of the theory: When it is restricted to single and double excitations, we speak of UGA CCSD.

Now, in the CS case, the singles and doubles are always represented by one- and two-body operators, respectively. Hence, truncating T at the two-body operator level is the same as truncation at the SD excitation level or, in fact, the first-order interacting space level, all three truncation schemes yielding the same result regardless of whether we employ the spin-orbital or the spatial-orbital (i.e., spin-free) formalism. A very different situation is encountered in handling OS systems. For example, a single core-virtual excitation can arise either through the direct one-body process or indirectly via a two-body core-active and active-virtual transition. Similarly, doubly excited configurations can arise through two-, three-, or even four-body processes. Thus, the one- and two-body excitation operators generate only a proper subspace, namely, the so-called (*first order*) *interacting space* (*is*), of the *full* (*f*) SD excited space. This suggests two distinct truncation schemes leading, respectively, to the CCSD(is) and CCSD(f) methods. Experience shows that the former scheme provides almost as good results as the full one, while being computationally much less demanding [272]. Consequently, the UGA CCSD(is) method is the one generally employed and, unless confusion could arise, is simply referred to as UGA CCSD.

The set of SD excitation operators that are pertinent in the CS case consists of one- and two-body operators of the type [173]

$$T(\text{cs})= \{E_a^r, (E_a^r E_b^s \pm E_a^s E_b^r)\}. \qquad (2.142)$$

(Unnormalized operators are used for simplicity.) In OS cases, this set must be extended with operators involving active orbitals. For high-spin states, the relevant additional operators that generate the interacting space are of the following types [45, 46]: For single excitations, we need

$$T_{1A} = \{E_a^i, E_i^r\} \qquad \text{and} \qquad T_{1B} = \{E_a^i E_i^r\}, \qquad (2.143)$$

and for double excitations,

$$T_{2A} = \{(E_a^r E_b^i \pm E_b^r E_a^i), (E_a^r E_i^s \pm E_a^s E_i^r), E_a^j E_j^r\}, \qquad (2.144a)$$

$$T_{2B} = \{(E_a^i E_b^j - E_b^i E_a^j), (E_i^r E_j^s - E_i^s E_j^r)\}. \qquad (2.144b)$$

Thus, the interacting space cluster operator for the high-spin (HS) case is given by a linear combination of the excitation operators belonging to the set

$$T(\text{HS})= T(\text{CS})\cup T_{1A} \cup T_{1B} \cup T_{2A} \cup T_{2B}, \qquad (2.145)$$

with the implied sums ranging over all core, active, or virtual orbitals.

In the case of an open-shell singlet (OSS) state, a minor modification is required. Since only two active orbitals are involved (say, i and j), and they transform symmetrically (i.e., both occur in the same row of the Weyl or bonded tableau), T_{1B} is replaced by

$$T_{1C} = \{(E_a^i E_i^r - E_a^j E_j^r)\}, \tag{2.146a}$$

and T_{2B} is replaced by

$$T_{2C} = \{(E_a^i E_b^j + E_b^i E_a^j), (E_i^r E_j^s + E_i^s E_j^r)\}, \tag{2.146b}$$

where i and j are fixed and only a and r are running. Thus, for the OSS state, we have

$$T(\text{oss}) = T(\text{cs}) \cup T_{1A} \cup T_{1C} \cup T_{2A} \cup T_{2C}. \tag{2.147}$$

The full-SD excited space requires additional operators found in [45, 46] and higher-than-SD operators (of the triexcited kind) found in [46] and [273]. (Note that the sets of operators given in these references are different, but equivalent, both yielding orthonormal states.)

Once the relevant cluster operator is specified, the expression for the energy and the CC equations for cluster amplitudes are given by expressions analogous to those that were obtained via the algebraic approach in the SR case. [See Eqs. (2.71) and (2.72), respectively.] However, since the excitation operators no longer commute (recall that active orbital labels now occur as both subscripts and superscripts), the commutator expansion for a similarity-transformed Hamiltonian $e^{-T}He^T$ may involve up to eight commutator terms, resulting in octic algebraic equations of the type of Eqs. (2.65) and (2.66). Nonetheless, as we shall see shortly, the high-degree terms can be safely neglected. The actual derivation of working equations, even when truncated at the quadratic level, can be very laborious in view of various types of excitation operators that arise, particularly in the full-SD case. Since the algebraic operations involved are rather routine, they can be conveniently automated. (Cf., e.g., [45, 131].) The automation can, in fact, be complete in the sense that even the final FORTRAN code is so generated [45]. (Cf. Section II.G.3.)

The current implementation of the UGA CC method for HS doublets and triplets and low-spin OSSs has been thoroughly tested at both CCSD(is) and CCSD(f) levels [44, 45, 200, 206, 232, 272, 274–278], yielding excellent results. The restriction to the first-order interacting space is perfectly adequate in most cases, and the larger effect of higher-than-two-body biexcited clusters is felt only at stretched geometries. (At the DZP level, the difference in the CCSD(is) and CCSD(f) energies amounts to 0.1–0.3 mhartree at the equilibrium geometry and to a few mhartree at 100% stretched geometries.) At the same time, CCSD(is) represents generally a very substantial saving in the computational effort, since it requires about 2.5 times fewer cluster amplitudes than CCSD(f) for doublet states and about $(S + \frac{1}{2})(S + 2)$-times fewer for general HS states. Moreover, the effect of higher-than-two-body amplitudes can be efficiently accounted for perturbatively (by the CCSD{f} method [273, 279]). This is also the case for a perturbative account of triples [273, 279], representing OS generalizations of the CCSD(T) and CCSD[T] procedures, which, however, can be employed only near the equilibrium geometries, as in the SR case. Various applications of UGA CCSD to potential-energy surfaces [44, 45, 200, 206, 232, 239, 272], vertical excitation energies [232, 274, 276], ionization potentials [44, 275], harmonic force fields [278], and dipole moments and polarizabilities [44, 277] show the viability and usefulness of this approach and will be compared with the results obtained by other methods in Section III.

Finally, let us point out the relationship of UGA CC methods to other CC approaches to OS systems, some of which were already addressed in Section II.C.3. Most of these approaches are limited to the HS case, ranging from spin-nonadapted methods using either the UHF or the ROHF reference to various partially spin-adapted CCSD methods. The OSS case was also addressed by Balková et al. [89], relying on an SU MRCC theory that uses a two-determinantal reference space and spin-orbital formalism, while enforcing the correct spin symmetry at the zero-order (model space) level, thus ensuring a partial spin adaptation. The conditions under which this approach performs satisfactorily were outlined in [276]. A comparison with EOM CCSD methods will be addressed in Sections II.F.4 and III.D.

In comparing UGA CCSD with other approaches, it is important to keep in mind that this is a fully spin-adapted theory in the sense that the cluster *Ansatz* of Eq. (2.142) yields a pure eigenstate of the total spin, and all spin components are uniformly treated. In the HS case, to which most of the existing approaches are limited, one can consider only its single-determinantal highest S_z component, assuming that we are subsequently interested only in spin-independent properties. In such cases, the UGA CC *Ansatz* can be reduced to a partially spin-adapted one by introducing suitable approximations.

As already pointed out in Section II.C.3, by a partially spin-adapted *Ansatz*, we understand a CC *Ansatz* in which the reference and linear terms, i.e., $(1 + T)|\Phi_0\rangle$, are fully spin-adapted, while the quadratic and higher order terms are not. This weaker condition can be satisfied by employing a proper subset of UGA excitation operators that is obtained by eliminating those operators whose action on $|\Phi_0\rangle$ gives a vanishing result (but that still contribute through the nonlinear terms). Generally, we can split the fully spin-adapted cluster operator $T^{(FSA)}$ of UGA into the partially spin-adapted part $T^{(PSA)}$ and the auxiliary part $T^{(AUX)}$, where

$$T^{(FSA)} = T^{(PSA)} + T^{(AUX)}, \tag{2.148}$$

and

$$T^{(FSA)}|\Phi_0\rangle = T^{(PSA)}|\Phi_0\rangle, \qquad T^{(AUX)}|\Phi_0\rangle = 0. \tag{2.149}$$

Obviously, the whole set $\mathcal{T}(\text{cs})$ appears in $T^{(PSA)}$. Moreover, when one uses the highest $M_S = S$ spin component $|\Phi_0\rangle$ as a reference (e.g., $|\Phi_0\rangle = |\{(a\alpha\,a\beta\cdots)\,i\alpha\}\rangle$), the action of $E_a^i \in \mathcal{T}_{1A}$ [see Eq. (2.143)] is

$$E_a^i|\Phi_0\rangle = (e_{a\alpha}^{i\alpha} + e_{a\beta}^{i\beta})|\Phi_0\rangle = e_{a\beta}^{i\beta}|\Phi_0\rangle. \tag{2.150}$$

Consequently, $T^{(PSA)}$ will involve only $e_{a\beta}^{i\beta}$, and $e_{a\alpha}^{i\alpha}$ will be delegated to $T^{(AUX)}$. Similarly, for $e_{i\alpha}^{r\alpha}$ and $e_{i\beta}^{r\beta}$,

$$\mathcal{T}_{1A}^{(PSA)} = \{e_{a\beta}^{i\beta}, e_{i\alpha}^{r\alpha}\} \qquad \text{and} \qquad \mathcal{T}_{1B}^{(PSA)} = \{e_{a\beta}^{i\beta}\,e_{i\alpha}^{r\alpha}\}, \tag{2.151a}$$

and by the same argument,

$$\mathcal{T}_{2A}^{(PSA)} = \{(E_a^r e_{b\beta}^{i\beta} \pm E_b^r e_{a\beta}^{i\beta}), (E_a^r e_{i\alpha}^{s\alpha} \pm E_a^s e_{i\alpha}^{r\alpha}), e_{a\beta}^{i\beta}\,e_{i\alpha}^{r\alpha}\}, \tag{2.151b}$$

$$\mathcal{T}_{2B}^{(PSA)} = \{(e_{a\beta}^{i\beta} e_{b\beta}^{j\beta} - e_{b\beta}^{i\beta} e_{a\beta}^{j\beta}), (e_{i\alpha}^{r\alpha} e_{j\alpha}^{s\alpha} - e_{i\alpha}^{s\alpha} e_{j\alpha}^{r\alpha})\}. \tag{2.151c}$$

Hence,

$$\mathcal{T}^{(PSA)}(\text{HS}) = \mathcal{T}(\text{cs}) \cup \mathcal{T}_{1A}^{PSA} \cup \mathcal{T}_{1B}^{(PSA)} \cup \mathcal{T}_{2A}^{(PSA)} \cup \mathcal{T}_{2B}^{(PSA)}. \tag{2.152}$$

This *Ansatz* is equivalent to the partially spin-adapted *Ansatz* by KHW [134] and NUH [132, 133]. (Some operators may differ, depending on the orthogonalization scheme used.) It is worth noting that the originally proposed partially spin-adapted *Ansatz* by Janssen and Schaefer [131] spans only a proper subspace of the interacting space [134]. A recently proposed approach by Szalay and Gauss [139] also relies essentially on a partially spin-adapted *Ansatz*, which is, however, supplemented by additional constraints on the cluster amplitudes that are derived by spin projection.

Since $T^{(PSA)}$ involves only genuine excitations, while excluding de- and pseudoexcitations, the resulting formalism affords important simplifications, namely, the truncation of the commutator expansion [cf. Eqs. (2.68) and

(2.73)] at the fourth-order terms (rather than the eighth order for the general theory), at the cost of the loss of invariance with respect to an arbitrary rotation of the spin axis. In contrast, the fully spin-adapted CC equations terminate only after the eighth-power terms. Since this fact is often quoted as a severe shortcoming of the fully spin-adapted MR CC theory, it deserves at least a brief comment.

First of all, it is important to realize that all higher-than-quartic terms involve at least one operator that is absent from a partially spin-adapted CC *Ansatz*. For example, every higher-than-quartic term must involve at least one pair of one-body excitation operators E_a^i and E_i^r. These operators may be contracted so that they act as a single excitation $(a \to r)$, and the products of corresponding cluster amplitudes appear in higher-than-quartic terms. In terms of spin orbitals, in lieu of each generator E_a^i and E_i^r, we have a pair of spin orbital generators $e_{a\alpha}^{i\alpha}$, $e_{a\beta}^{i\beta}$ and $e_{i\alpha}^{r\alpha}$, $e_{i\beta}^{r\beta}$, respectively, and the generators associated with different spins cannot give nonvanishing contractions. We thus have, for example,

$$
\begin{aligned}
\langle \Phi_j | [\cdots [[[H, E_a^i], E_i^r], O_1], \ldots, O_p] | \Phi_0 \rangle \\
= \langle \Phi_j | [\cdots [[[H, e_{a\alpha}^{i\alpha}], e_{i\alpha}^{r\alpha}], O_1], \ldots, O_p] | \Phi_0 \rangle \qquad (2.153) \\
+ \langle \Phi_j | [\cdots [[[H, e_{a\beta}^{i\beta}], e_{i\beta}^{r\beta}], O_1], \ldots, O_p] | \Phi_0 \rangle .
\end{aligned}
$$

However, the generators $e_{a\alpha}^{i\alpha}$ and $e_{i\beta}^{r\beta}$ are in $T^{(\mathrm{AUX})}$, but absent from $T^{(\mathrm{PSA})}$. We can therefore conclude that the adoption of a partially spin-adapted *Ansatz* represents a more severe approximation than does the neglect of higher-than-quartic terms in the fully spin-adapted approach. Indeed, in the former case a certain class of excitation operators is excluded a priori, while in the latter approach these operators are accounted for up to the fourth order. Thus, in practice, we can safely neglect higher-than-quartic terms in the fully spin-adapted approach. We shall see later that, in fact, even quadratic approximation provides excellent results.

Another interesting scheme relying on rotated spin functions $\sigma^\pm = (\alpha \pm \beta)/\sqrt{2}$, proposed by Jayatilaka and Lee [135, 137, 138], has already been discussed in Section II.C.3. The most recently proposed spin adaptation scheme by Nooijen and Bartlett [52] also exploits UGA formalism, but employs a spin-orbital formulation ultimately based on a closed-shell reference. The entire formulation is restricted a priori to the first-order interacting space, the final equations being given in terms of the matrix elements of a similarity-transformed Hamiltonian. Although the authors do appreciate the usefulness of having the same wave operator for all the components of a given multiplet (the second property of the UGA cluster *Ansatz* pointed out earlier), they nonetheless employ the high-spin component when they consider high-spin triplet states. Clearly, a different form of the wave operator then results for the $M_S = 0$ component. Moreover, while the OSS reference can be generated from a CS configuration with a single $U(n)$ generator, this cannot be done for $S \neq 0$ states. The claim by the authors [52] that their "spin-orbital based cluster coefficients satisfy completely identical symmetry relations as in the closed shell case" is certainly true for (core, core) → (virtual, virtual) excitations (regardless of whether one uses spin-orbital or spin-adapted cluster amplitudes), but we cannot see how it can hold for semi-internal excitations. It will certainly be interesting to compare the results with those obtained from UGA CC approaches when the formalism is implemented computationally.

3.3. Externally Corrected CCSD Methods. In Section II.C.4.3, we outlined the basic idea behind the so-called externally corrected CC approaches, in which information about the higher-than-pair cluster amplitudes is imported from some independent source. In MR approaches, the higher-than-pair clusters are accounted for by considering single and double excitations from several reference configurations that span the model space $\mathcal{M}_0$. This aspect is most transparent in the variational MR CI approach and less obvious in the nonlinear, nonvariational CC theories involving an exponential cluster *Ansatz* for the wave function.

Recall (Section II.C.4.3) that in the SR CC approach, the energy is fully determined by singles and doubles: $\Delta E = f(T_1, T_2)$ [Eq. (2.71)]. However, in order to determine these SD amplitudes exactly, one has to solve the full chain of CC equations, which are equivalent to FCI, since doubles are coupled with quadruples, quadruples to hexuples, etc. We also know that the direct solution of CC equations for higher-than-pair cluster amplitudes is very inefficient. Since the pair cluster equations involve at most quadruples, the basic idea of the externally corrected CCSD is to solve for T_1 and T_2, while approximating T_3 and T_4 with the help of some external, non-CC source. The basic *Ansatz* of such an externally corrected CCSD can thus be written as

$$|\Psi\rangle = e^{T_1 + T_2 + T_3^{(0)} + T_4^{(0)}} |\Phi_0\rangle, \qquad (2.154)$$

where T_1 and T_2 are to be determined via CCSD-type equations and $T_3^{(0)}$ and $T_4^{(0)}$ represent some a priori fixed approximation of actual T_3 and T_4 clusters. The terms $\widetilde{\Lambda}^{(1)}$ and $\widetilde{\Lambda}^{(2)}$ in Eqs. (2.93), coupling the SD projected pair cluster equations with the rest of the CC chain via the T_3 and T_4 clusters, have the form (we use now the commutator representation that is more convenient in OS cases handled via UGA CC)

$$\widetilde{\Lambda}^{(1)} \equiv \Lambda^{(1)}(3) = \langle \Phi_i^{(1)} | [H, T_3^{(0)}] | \Phi_0 \rangle , \qquad (2.155a)$$

$$\widetilde{\Lambda}^{(2)} \equiv \Lambda^{(2)}(3) + \Lambda^{(2)}(4) + \Lambda^{(2)}(1,3)$$
$$= \langle \Phi_i^{(2)} | [H, T_3^{(0)}] + [H, T_4^{(0)}] \qquad (2.155b)$$
$$+ \tfrac{1}{2}([[H, T_1], T_3^{(0)}] + [[H, T_3^{(0)}], T_1]) | \Phi_0 \rangle .$$

The last two terms, involving T_1 and $T_3^{(0)}$, are identical if T_1 and $T_3^{(0)}$ commute.

After evaluating $\widetilde{\Lambda}^{(1)}$ and $\widetilde{\Lambda}^{(2)}$ by using the available $T_3^{(0)}$ and $T_4^{(0)}$ amplitudes, we employ the $\Lambda^{(1)}(3)$ and $\Lambda^{(2)}(k), k = 3, 4$ terms to correct the absolute term a_i in the system of CCSD equations (2.65), while the $\Lambda^{(2)}(1,3)$ term will correct the linear b_{ij} coefficients at singles in equations projected onto doubles. Thus, the externally corrected CCSD equations have the same form as the standard ones. [Cf. Eq. (2.65) with, possibly, cubic and quartic terms in t_1-amplitudes.]

As pointed out in Section II.C.4, the idea of externally corrected CCSD is completely general and does not depend on the source of the $T_3^{(0)}$ and $T_4^{(0)}$ amplitudes. In fact, we also used Eqs. (2.93) or Eqs. (2.155) in deriving perturbatively corrected CCSD equations (i.e., CCSD(T) or CCSD[T]), which were referred to for this reason as internally corrected, since they relied solely on the CCSD amplitudes (Section II.C.4.2). It is also irrelevant which version of the CCSD method (spin orbital, spin adapted, or partially spin adapted) we employ, even though the coupling between T_1, T_2 and $T_3^{(0)}, T_4^{(0)}$ clusters may be different in each case. The explicit spin-orbital form of the correcting terms was given in [47]. Since the cluster analysis is, in general, performed at the spin-orbital level, and, together with the evaluation of corrections, Eqs. (2.155) represent a small fraction of the required computational effort, one can calculate even the spin-adapted corrections in this way by representing the pertinent states as a linear combination of Slater determinants.

Although the externally corrected CCSD is an open-ended approach that can exploit any source of higher-than-pair clusters, we now concentrate on a very promising version relying on MR CISD-type wave functions of modest dimension, referred to as reduced MR (RMR) CCSD [50, 239, 240]. In most problems involving the breaking or the formation of chemical bonds, the zero-order description requires a set of configurations with different orbital occupancies, spanning a suitable model space $\mathcal{M}_0$. (Cf. Section II.D.1.) Any such MR *Ansatz* may be expressed in a single reference form,

$$|\Psi\rangle = e^{T_1 + T_2 + \Delta} |\Phi_0\rangle , \qquad (2.156)$$

where one chooses a suitable $|\Phi_0\rangle$ (in our case, the spin-adapted configuration having the largest weight). Here, Δ represents a subset of all higher-than-pair cluster operators that arise by SD excitations from references $|\Phi_j\rangle \in \mathcal{M}_0$ which are different from $|\Phi_0\rangle$. Thus, the explicit form of Δ, which may involve even higher-than-four-body clusters, depends on the choice of the model space $\mathcal{M}_0$.

The *Ansatz* represented by Eq. (2.156) can be exploited in two ways: (i) One can use the structure of Δ to select a suitable subset of higher-than-pair cluster amplitudes and determine them, together with T_1 and T_2, by using the corresponding set of CC equations. (That is, one can determine all the cluster amplitudes within the CC framework.) (ii) One can also solve for T_1, T_2 and Δ clusters independently, obtaining the latter from some non-CC source, and only subsequently use the CC theory to solve for the T_1 and T_2 clusters, relying on the *Ansatz* given by Eq. (2.154). The SS MR CC approach due to Adamowicz et al. (Section II.D.3.1) follows the first strategy, while the second alternative is the basis of the externally corrected CC methods—in particular, RMR CCSD.

The advantages of the latter approach are quite obvious: Even when the higher-than-pair cluster manifold is further truncated, its dimensionality, as well as the complexity of the relevant CC equations, grows rapidly with the increasing size and quasi-degeneracy of the problem. These problems would be further magnified should it be necessary to include selected six-body or even higher clusters contained in Δ to warrant proper dissociation (as is the case, for example, for N_2). In contrast, when Δ is determined via an affordable non-CC approach, then only three- and four-body components of Δ are required to be known *explicitly*, while all the higher excited cluster components will be *implicitly* accounted for (as long as they are accounted for by the external source employed, of course), since they are intimately related to the three- and four-body clusters included in Δ.

The connected clusters are present already at the linear level of the CC theory. If we thus approximate the SU CC *Ansatz* of Eq. (2.130) by its linear version,

$$|\Psi_i\rangle = \sum_j C_{ij}(1 + T(j))|\Phi_j\rangle, \qquad (2.157)$$

no connected clusters are lost. In this way, we can estimate higher-than-pair clusters in the *Ansatz* given by Eq. (2.156) that will be responsible for the main nondynamical correlation effects. The best solution, in the variational sense, for the linear *Ansatz* wave function of Eq. (2.157) is clearly provided by the MR CI method. The latter, however, introduces unlinked terms, which we can largely remove via the cluster analysis.

Thus, the RMR CCSD approach involves the following three steps: First, we compute the MR CISD wave function $|\Psi_0^{MR-CI}\rangle$, as defined by the chosen orbital partitioning for the MR CC *Ansatz*, for the energetically lowest state of a given symmetry. Such a wave function can be written in the SR CI form by choosing a proper leading configuration $|\Phi_0\rangle$ for the state of interest

$$|\Psi_0^{MR-CI}\rangle = |\Phi_0\rangle + \sum_{i,j} C_i^{(j)}|\Phi_i^{(j)}\rangle, \qquad (2.158)$$

where (j) is the excitation order relative to $|\Phi_0\rangle$. In the second step, we extract the connected-cluster components $T_i^{(0)}$ from the $C_i^{(j)}$ coefficients by carrying out the cluster analysis of Eq. (2.158), writing

$$|\Psi_0^{MR-CI}\rangle = e^{(T_1^{(0)}+T_2^{(0)}+T_3^{(0)}+T_4^{(0)}+\cdots)}|\Phi_0\rangle, \qquad (2.159)$$

and using the relationships given in Eq. (2.51). In this way, we get the full set of one- and two-body clusters, as well as subsets of three-, four-, five-body, etc., clusters, depending on the choice of the active space. Generally, the higher-than-pair clusters will represent only a very small subset of all such clusters, assuming that the chosen active space is small. In the third and final step, we employ the *Ansatz* of Eq. (2.154) and determine the new values of the T_1 and T_2 clusters from the externally corrected CCSD equations. Since the higher-than-four-body clusters do not directly couple with T_1 and T_2, they can be ignored in the second and the third steps.

Briefly, RMR CCSD involves two subproblems: MR CISD and (externally corrected) SR CCSD. The MR CI wave function provides an approximation of Δ, whose three- and four-body components enable us to properly truncate the SR CC theory at the SD level. This does not imply that RMR CCSD neglects higher-than-four-body clusters, since they are accounted for implicitly via MR CI. Indeed, if one uses a proper reference space, the MR CISD wave function will contain sextuple or higher excitations relative to the leading configuration $|\Phi_0\rangle$, and their effect will be implicitly included in the three- and four-body cluster amplitudes extracted from this wave function, allowing more precise decoupling of the CCSD equations and thus the calculation of reliable one- and two-body clusters determining the energy.

From the viewpoint of the externally corrected CC methods, RMR CCSD can be viewed as MR CISD-corrected SR CCSD. However, RMR CCSD is effectively an MR approach that is uniquely defined by the choice of the reference space. Moreover, since RMR CCSD relies on the "single reference" $|\Phi_0\rangle$, which is the leading configuration in the lowest energy MR

CISD wave function, a different reference configuration may be required for different geometries. This reference configuration may thus be different from the reference employed by standard SR CCSD or even that employed in other versions of externally corrected CCSD.

In order to better understand why RMR CCSD represents an improvement over both MR CISD and SR CCSD, we must examine the intrinsic deficiencies of the latter two methods. While MR CISD is known to provide a very good description of nondynamical correlation, SR CCSD does the same for dynamical correlation. Thus, whereas MR CISD will always provide a qualitatively correct wave function, it will yield good energies only when one employs a sufficiently large reference space. With a small reference space, the results may not be very good, since the excited state manifold resulting from SD excitations of a small number of references will involve only a small subset of higher-than-doubly excited configurations. Although these latter configurations contribute very little individually, their collective effect is significant in view of their large numbers, as we have seen in the SR case. (Cf. Section II.C.2.)

Thus, a poor description of dynamical correlation via MR CISD can be improved only by enlarging the reference space, which is not cost effective. In contrast, SR CCSD neglects all higher-than-pair connected clusters, but automatically fully accounts for their disconnected part and thus efficiently describes dynamical correlation. The only drawback of SR CCSD is the lack of important higher-than-two-body connected clusters, leading to an unsatisfactory description of nondynamical correlation. Since RMR CCSD combines both of these approaches, almost all higher-than-pair clusters are taken into consideration: The essential clusters, both connected (originating from MR CISD) and disconnected (originating in the SR CCSD *Ansatz*), are directly accounted for, while the less important ones are taken care of by the disconnected clusters of the SR CCSD *Ansatz*. One can thus expect the RMR CCSD method to provide a balanced description of both dynamical and nondynamical correlation effects.

Another great advantage of RMR CCSD is its affordability. Recall that the computational cost (per iteration) of SR CCSD, CCSDT, and CCSDTQ scales as n^6, n^8 and n^{10}, respectively. A well-designed MR CCSD method should scale as $M_{ref} n^6$, where M_{ref} is the number of references employed. This is precisely the case for RMR CCSD. To understand this fact, it is important to realize that the three- and four-body connected clusters that enter RMR CCSD are those appearing in the MR CISD expansion given by Eq. (2.158). These in turn represent a very small subset of all possible such amplitudes. For example, for the minimal reference space spanned by two CS-type configurations $|\Phi_0\rangle$ and $|\Phi_1\rangle$, the latter resulting by a double excitation (HOMO,HOMO) $\rightarrow$ (LUMO, LUMO) from $|\Phi_0\rangle$, the MR CISD configuration set consists of a union of SD excitations from both $|\Phi_0\rangle$ and $|\Phi_1\rangle$, the latter representing T and Q excitations relative to $|\Phi_0\rangle$. Thus, the total number of cluster amplitudes that must be handled is roughly twice the number of SR CCSD amplitudes. The three- and four-body amplitudes depend on at most four nonactive (i.e., core or virtual) orbital labels, the remaining ones being restricted to the active orbital labels (in the foregoing example, HOMO and LUMO). The example also shows that the minimal two-reference space cannot account for higher-than-four-body clusters, and the $T_3^{(0)}$ and $T_4^{(0)}$ clusters that enter the externally corrected CCSD are SD excitations from the second reference $|\Phi_1\rangle$.

The total cost of an RMR CCSD calculation is the sum of the costs required by the three steps involved. The first step, MR CISD, scales as $M_{ref}\,n^6$, and the last step, the solution of externally corrected CCSD, is also an n^6 procedure, as is the standard SR CCSD, since the coupling terms $\tilde{\Lambda}^{(i)}$ between T_1, T_2 and $T_3^{(0)}, T_4^{(0)}$ are computed only once. The time required by the second step—i.e., the computation of $T_3^{(0)}$ and $T_4^{(0)}$ from the MR CISD wave function and the subsequent evaluation of the correction terms $\tilde{\Lambda}^{(i)}$—is negligible in comparison with that needed for the other two steps. In general, it amounts to less than the cost of a single step in the $M_{ref}\,n^6$ procedure, since only three- and four-body cluster amplitudes need to be extracted, while the MR CISD wave function may involve higher-than-quadruply excited configurations relative to $|\Phi_0\rangle$. The latter enter only implicitly and are ignored in cluster analysis. Further, the noniterative evaluation of $T_1 T_3$ and T_4 corrections to double equations scales at most as n^6, in view of a very limited number of four-body cluster amplitudes. In fact, the most time-consuming evaluation is the account of $T_1 T_3$ correction. However, even this calculation scales as n^6 (when no intermediates are used), and this correction is most efficiently taken care of noniteratively by approximating T_1 with $T_1^{(0)}$, the latter a by-product of the cluster analysis. (For details, see [49].) We can thus conclude that the total cost of the RMR CCSD method scales as $M_{ref}\,n^6$. Another important advantage of this method is its low memory demand. Since the three- and four-body connected-cluster amplitudes are used only to correct CCSD equations, with the correction terms evaluated just once, these amplitudes can be stored on a disk.

RMR CCSD methods have been implemented and tested [50, 239, 240] within the framework of UGA CC, providing most encouraging results. (Cf. Section III.)

E. Related Approaches

We now consider two other types of approaches that are closely related to the CC methodology and are often used in the electronic structure calculations, namely, finite-order perturbation theory and the so-called quadratic CI (QCI) methods. Approaches, based on either CC or QCI formulation, that exploit the concept of Brueckner orbitals (making the one-body clusters vanish) are addressed in Section II.G. Likewise, approaches exploiting the CC idea in conjunction with the EOM, CI, and Green functions (polarization propagator), and linear-response approaches are dealt with in Section II. F in view of their importance in handling various other aspects of the molecular electronic structure than the total energy.

1.   Many-Body Perturbation Theory (MBPT)

The close relationship between MBPT and CC theory in the SR case should by now be obvious. (Cf. Sections II.B and C.) On the one hand, MBPT furnishes a conspicuous and explicit source of the CC *Ansatz*, while, on the other hand, CC formalism permits a straightforward recursive generation of MBPT diagrams and, by virtue of solving the (in practice necessarily truncated) CC equations, an implicit generation of important classes of these diagrams and their actual evaluation and summation to infinite order. This property is very important for several reasons. First, the computational demand for the evaluation of the finite-order PT contributions rapidly increases with increasing order of PT and is extremely difficult beyond the fourth order. Second, the MBPT series may converge poorly or even diverge (or display an erratic behavior in lower orders), even in cases for which the RHF reference dominates the wave function and truncated CCSD equations can be readily solved [194, 195, 227, 280–283]. Third, the CC codes can be easily modified to generate the first few orders of the MBPT series. Finally, the infinite-order "tails" of certain classes of diagrams, as supplied by the CC theory, are often essential for an accurate account of correlation effects. (Cf. [227].)

Nonetheless, the second-order PT contribution, MBPT(2) or MP(2), is very easy to generate and often provides a rather good description of, or at least an excellent starting point for, a higher order theory. For larger systems, this may be the only available post-Hartree–Fock method that can be used. (Cf., e.g., [284].) Often, even qualitative information about the size or direction of correlation corrections can be worthwhile. For nondegenerate closed-shell systems, MBPT(2) usually accounts for 80–90%, or even more, of the correlation energy, and the next third-order correction usually does not significantly change this result. (The next significant improvement is due to the fourth-order correction.) This performance deteriorates rapidly once we leave the neighborhood of the equilibrium geometry.

Unfortunately, as the last remark already indicates, neither the simple structure and relationship nor the reliability carries over in a consideration of OS systems. For this reason, the spin-orbital version based on the UHF reference is often used. Although in some cases it provides very reasonable results, even at the second-order level, it cannot be relied on. This fact is well documented in the literature [28, 201, 285–290] and need not be repeated here. We only wish to point out the close relationship between the CC and MBPT approaches, which enables us to obtain the low-order MBPT results by using only slightly modified CC codes.

Genuine MR MBPT is again based on the idea of an effective Hamiltonian, and there exists a vast literature on the subject. (Cf., e.g., [39].) Here, instead of computing the matrix representative of an effective Hamiltonian by using the MR CC cluster amplitudes, one determines the matrix order by order, using perturbation theory. This approach is often characterized as a "perturb first, diagonalize later" technique [124, 291] and, obviously, suffers from the same intruder state problems as the corresponding MR CC approaches. For that reason, one often employs the reverse technique, "diagonalize first and perturb later". This approach may also be characterized as a state-selective (SS) one, since, by carrying out CI within the model space, one generates the correct "zero-order" states as linear combinations of a quasi-degenerate manifold of configurations spanning the model space. The subsequent PT then accounts for dynamical correlation in these individual zero-order states. Of course, the possible complexity of the zero-order states makes it extremely difficult to handle higher orders of PT, so that these approaches are a priori limited to the second or, at most, the third order, as is the case of a highly successful CASPT2 (complete-active-space second-order PT) method [292]. (See also [290, 293].) Obviously, such approaches can only be successful if the zero-order wave function already represents a good approximation for a given state, so that the missing part of the dynamical correlation, arising from a large set of inactive orbitals, can be accounted for by the low-order PT. In this way, one can handle even relatively large systems [294].

The complexity of the genuine MR MBPT approaches is usually bypassed (cf., however, [233, 295, 296]) by employing a (DODS-type) UHF-based SR MBPT, as was already pointed out. The potential of this approach is well known, as are its pitfalls. (Cf., e.g., [201, 288–290].) Particularly when the spin contamination is appreciable, the UHF-based PT converges poorly and gives unsatisfactory results. More importantly, however, just like the corresponding CC approaches, it cannot handle the low-spin OS states. Since the OSS states play an important role both in molecular spectroscopy and photochemistry, we wish to point out how UGA CCSD approaches can be used to generate low-order PT results for these states [297], enabling one to treat much larger systems than the full CCSD method.

Let us remark that until recently, the only viable alternative to handle correlation in the OSS states was ROHF-based CI or CI-related spin-adapted PT [298, 299]. At the CC level, these states have been handled by both the VU and SU MR CC methods [14, 29], the latter in the form of the so-called two-determinant (TD) CCSD [89, 230, 300], or via EOM-based approaches. (See Section II.F.4.) Since all these methods were implemented in their spin-orbital form, often only partial spin adaptation was possible by imposing the constraints on spin-orbital amplitudes [89, 230, 300]. However, a full CCSD treatment cannot be easily applied to large molecular systems, for which we are presently limited to the low-order PT approaches. We thus briefly describe recently implemented second- and third-order PT for the low-spin OSSs, which complements the often-used low-order MBPT (specifically, MP2) approaches utilized for HS states.

In the OSS case, the spin-free (ROHF) reference $|\Phi_0\rangle$ is equivalent to the standard two-determinantal wave function

$$\Phi_0 = (|(\text{core}) \, i\alpha \, j\beta| - |(\text{core}) \, i\beta \, j\alpha|)/\sqrt{2}, \qquad (2.160)$$

where (core) represents the doubly occupied singlet part. The set of excitation operators (in fact, its interacting space subset) $\{G_I\}$ of UGA CCSD can then be used to generate the first-order wave function

$$|\Psi^{(1)}\rangle = \left(1 + \sum_I a_I^{(1)} G_I\right)|\Phi_0\rangle =: |\Phi_0\rangle + \sum_I a_I^{(1)}|\Phi_I\rangle, \qquad (2.161)$$

where

$$a_I^{(1)} = -V_{I0}/\Delta_I = -\langle\Phi_I|V|\Phi_0\rangle/(E_I - E_0), \qquad (2.162)$$

in which V is the perturbation, $H = H_0 + V$, and $E_K = \langle\Phi_K|H_0|\Phi_K\rangle$ represents the unperturbed Hamiltonian energies. Using the Wigner formula, we get the following simple expressions for the first-, second-, and third-order energies:

$$
\begin{aligned}
E^{(1)} &= V_0 = \langle\Phi_0|V|\Phi_0\rangle, \\
E^{(2)} &= \sum_I a_I^{(1)} V_{0I} = -\sum_I |V_{I0}|^2/\Delta_I, \\
E^{(3)} &= \sum_I a_I^{(1)} \sum_J a_J^{(1)} \langle\Phi_I|V - V_0|\Phi_J\rangle.
\end{aligned}
\qquad (2.163)
$$

For the Møller–Plesset version, we choose the diagonal Fock matrix

$$H_0 = \sum_\mu F_\mu^\mu E_\mu^\mu, \qquad (2.164)$$

where

$$F_a^a \equiv \varepsilon_a = f_a^a + \frac{1}{2}\left(v_{aa}^{ii} + v_{aa}^{jj}\right),$$
$$F_i^i \equiv \varepsilon_i = f_i^i + 2v_{ii}^{jj},$$
$$F_j^j \equiv \varepsilon_j = f_j^j + 2v_{ii}^{jj}, \qquad\qquad (2.165)$$
$$F_r^r \equiv f_r^r,$$

with

$$f_\mu^\nu = z_\mu^\nu + \sum_a \left(2v_{a\mu}^{a\nu} - v_{aa}^{\mu\nu}\right) + \left(v_{i\mu}^{i\nu} - v_{ii}^{\mu\nu}\right) + \left(v_{j\mu}^{j\nu} - v_{jj}^{\mu\nu}\right). \qquad (2.166)$$

Note that we again use the Hamiltonian of Eq. (2.88) and the orbital labeling convention introduced earlier. Then, the denominators Δ_I are given by the orbital energy differences $(\varepsilon_\mu - \varepsilon_\nu)$ or $(\varepsilon_\mu + \varepsilon_\nu - \varepsilon_\sigma - \varepsilon_\tau)$ for singly $(\mu \to \nu)$ and doubly $(\mu, \nu \to \sigma, \tau)$ excited configurations $|\Phi_I\rangle$, respectively, and the corresponding numerators by simple expressions in terms of two-electron integrals. (Cf. Table 1 of [297].) In fact, the V_{I0} matrix elements are identical with the absolute CCSD terms $(V_{I0} = a_I)$, and the $\bar{V}_{IJ} = \langle \Phi_I | V - V_0 | \Phi_J \rangle$ matrix elements needed for the third-order energy are the standard normal product CI matrix elements

$$\bar{V}_{II} = (H_N)_{II} - \Delta_I, \qquad \bar{V}_{IJ} = (H_N)_{IJ} \quad (I \neq J), \qquad (2.167)$$

where

$$(H_N)_{IJ} = \langle \Phi_0 | G_I^\dagger H_N G_J | \Phi_0 \rangle, \qquad\qquad (2.168)$$

which can be evaluated with a slightly modified routine computing CCSD b_{IJ} terms. In a similar way, one can formulate Epstein–Nesbet UGA-based PT [297]. To appreciate the performance of this approximation, the reader should consult Section III.D.

2. Quadratic Configuration Interaction (QCI)

As we have seen in Section II.C, the principal role of the exponential cluster *Ansatz* is to eliminate unlinked terms that plague the truncated CI expansions. Thus, at the double excitation level, instead of simply neglecting higher-than-pair excitation operators [i.e., $C_i = 0$ for $i > 2$; e.g., C_4 in Eq. (2.55b)], the truncation is done at the connected-cluster operator level [i.e., $T_i = 0$ for $i > 2$; e.g., Eq. (2.57b)], so that C_4 is approximated by its disconnected component $\frac{1}{2}T_2^2$ (and $\frac{1}{2}T_1^2 T_2$, etc., at the full CCSD level). This procedure results in the elimination of unlinked terms [cf. Eqs. (2.57b), (2.62), and (2.63)] and thus of the explicit energy dependence, fully restoring the size extensivity of the formalism.

Starting again from the CI chain of equations [e.g., Eqs. (2.55)], as in Section II.C.1, one can achieve the same goal by approximating C_4 not by $\frac{1}{2}T_2^2$, but by $\frac{1}{2}C_2^2$, i.e., by relying on the approximation $T_2 \approx C_2$. (Note that at the CCD level, we have exactly $T_2 = C_2$.) This is the basis of the so-called quadratic CI (QCI) approaches [168]. Of course, since at the CCD level we have $C_2 = T_2$, QCID is identical with CCD, so that the differences first appear at the SD level. Hence, QCI exploits the fact that $C_2 = T_2 + \frac{1}{2}T_1^2$, so that the energy expression, Eq. (2.71), becomes

$$\Delta E = \langle H_N C_2 \rangle \equiv \langle \Phi_0 | H_N C_2 | \Phi_0 \rangle, \qquad (2.169)$$

assuming that $|\Phi_0\rangle$ is a canonical HF reference, in which case $\langle \Phi_0 | H_N T_1 | \Phi_0 \rangle = 0$ thanks to the Brillouin theorem. To eliminate unlinked terms, C_3 is approximated by $C_1 C_2$ in CI equations projected onto singles, and C_4 is approximated by $\frac{1}{2}C_2^2$ in equations projected onto doubles, as in the QCID $\equiv$ CCD case. In this way, all the terms needed to cancel the unlinked contributions are present, so that the resulting QCISD is size extensive. However, since the energy is given by Eq.(2.169), so that the theory is restricted to the canonical HF reference (which also helps to keep the $C_1 = T_1$ amplitudes small), the same idea cannot be extended beyond the SD level, as was originally claimed in [168]. (Cf. [192, 301, 302].) Of course, a perturbative account of triples—for example, by QCISD(T)—will be again fully size extensive.

From the viewpoint of CC theory [192, 301], QCISD represents an approximation to CCSD in which the following terms are kept: (i) the diagram (b) of Figure 5 for the energy, (ii) the diagrams (a)–(h) of Figure 8(i) for singly projected equations, and (iii) all CCD diagrams of Figure 8(ii), but only the diagrams (a1, a2) and (b1, b2) of Figure 8(iii). This results in only a moderate (~ 10–20%) computational saving, since the neglected terms invariably involve monoexcited t_1 amplitudes, whose number is very small, and, moreover, since most of the neglected terms can be efficiently taken care of by a suitable factorization [103]. (Cf. Section II.G.2.)

From the viewpoint of MBPT, both CCSD and QCISD account for all the third-order terms, and in the fourth-order they miss only the diagrams involving triples as the intermediate states. [Cf. Figure 9(c).] Since the latter may be added perturbatively, both CCSD(T) and QCISD(T) account for all the fourth-order MBPT terms. A detailed MBPT analysis of both CC and QCI methods was given by several authors [191, 303–306]. (For up to the eighth-order analysis, see [306].) Of course, the account of high-order terms is not necessarily a guarantee of a superior performance, since some terms may be more important than others. (This is, for example, clearly indicated by the importance of the infinite-order sums of certain classes of diagrams, as is revealed by comparing the CC and the finite-order MBPT results; see, e.g., [227].) Nonetheless, the conclusions implied by the MBPT analysis [305, 306] do seem to be borne out by experience.

Indeed, for relatively small CS systems involving only single bonds, both CCSD and QCISD give practically identical results. This is not surprising, since QCID is identical with CCD, so that as long as the T_1 clusters are small, both approaches must give very similar results when the canonical HF reference is employed–and they do. However, it is important to realize that two problems may arise in the QCISD and QCISD(T) methods. The first has to do with the effect of T_1 clusters. It is well known that CCSD is rather insensitive to the type of the MOs employed, thanks to the presence of the $e^{T_1}|\Phi_0\rangle$ term permitting orbital rotations. In contrast, the absence of T_1^2, T_1^3, etc., terms in QCISD does not allow for a proper account of orbital relaxation effects. Second, as pointed out by He and Cremer [305, 306], QCISD(T) is likely to overestimate the role of T_3. Hence, as soon as the quasi-degeneracy sets in—particularly when triples and singles play a more significant role—CCSD is clearly superior to QCISD, being consistently built on the CC *Ansatz*. This superiority is especially manifested when triples are simultaneously accounted for perturbatively [305, 306], in which case QCISD(T) overestimates their importance.

Such behavior was clearly demonstrated recently in several studies [307–311]. Watts et al. [307] found that for BeO, the QCISD and CCSD polarizabilities and infrared intensities differ by one and three orders of magnitude, respectively. Likewise, Böhme and Frenking [308], who investigated the dissociation process $CuCH_3(^1A_1) \rightarrow Cu(^2S) + CH_3(^2A_1)$, found large discrepancies between the CC and QCI results, the latter being very unreliable and strongly basis-set dependent. The computed CCSD and CCSD(T) dissociation energies D_e were, respectively, 45.4 and 48.0 kcal/mol when the effective core potential (ECP) was used for Cu, and 46.0 and 49.7 kcal/mol with an all-electron basis set. These results compare well with the modified coupled-pair functional (MCPF) result [312] of 48.4 kcal/mol. (Experimentally [313], it is known only that $D_e \leq 60$ kcal/mol.) In contrast, QCISD and QCISD(T) results were, respectively, 55.4 and 6.0 kcal/mol with the ECP basis set and 42.7 and 29.4 kcal/mol with the all-electron basis set [308].

Obviously, the primary cause of the failure of the QCISD method stems from the fact that it is restricted by its design to canonical HF orbitals, so that systems in which HOMO–LUMO single excitations play a significant role (while the back-donated state plays a minor role) are likely to perform poorly or even fail. This was amply demonstrated by Hrušák et al. [310], who studied the CuH and CuF molecules. (See also [311] for CuH.) Particularly for CuF, the QCISD(T) PEC displays a double minimum, thus being completely unrealistic (while QCISD gives a reasonable-looking, even though too deep, PEC), and both QCISD and QCISD(T) give unrealistic moment functions and, thus, dipole moments: $\mu^{\text{CCSD}} = 5.23$ D and $\mu^{\text{CCSD(T)}} = 5.21$ D, while $\mu^{\text{QCISD}} = 2.40$ D and $\mu^{\text{QCISD(T)}} = 14.36$ D [310]. (Note that $\mu^{\text{exp}} = 5.7(7)$D.) In fact, dispensing altogether with singles, one gets $\mu^{\text{CCD}} = 6.37$ D. The authors [310] attribute this poor performance of QCI to the lack of orbital relaxation, which is especially important for calculations of properties.

It is worth noting that the QCISD method was also used to calculate excitation energies within the framework of the linear-response theory. Head-Gordon et al. [314] showed that the LR-QCISD excitation energies may differ greatly from the LR-CCSD ones. This defect was attributed to the lack of the $T_1 T_2$ term in the equations projected onto doubles. The authors concluded that LR-QCISD is not to be recommended if an accurate description of the excited states is desired.

We can thus say that, in view of the fact that both QCI and CC methods scale with the same power of the basis-set size and therefore do not differ significantly in the required computational effort [315]—not to mention the solid rooting of the coupled-cluster methods in MBPT (cf. Section II.B), implying their hierarchical structure and generality—it is safer to employ the latter. Indeed, in the absence of quasi-degeneracy, CCSD(T) seems to be an optimal choice in every respect [30].

F. Coupled-Cluster Approaches to Properties

The computation of the average value of any observable $\widehat{O}$, other than the energy, in a state $|\Psi\rangle$, given by the expectation value

$$\langle \widehat{O} \rangle = \langle \Psi | \widehat{O} | \Psi \rangle / \langle \Psi | \Psi \rangle, \tag{2.170}$$

is not straightforward within the CC formalism, as it is in variational approaches that rely on the linear *Ansatz* for the wave function, but is, in fact, highly nontrivial in view of the exponential *Ansatz*. Since the disconnected terms in the numerator on the right hand side of Eq. (2.170) that do not involve the operator $\widehat{O}$ cancel the normalization integral in the denominator [2], the expectation value $\langle \widehat{O} \rangle$ can again be expressed only in terms of connected (or linked) contributions as

$$\langle \widehat{O} \rangle = \langle \Phi_0 | e^{T^\dagger} \widehat{O} e^T | \Phi \rangle_C = \langle e^{T^\dagger} \widehat{O} e^T \rangle_C. \tag{2.171}$$

Nonetheless, the expansion (2.171) does not terminate and involves arbitrarily high powers of the cluster amplitudes, even when T is truncated at, say, the pair cluster level. This is precisely the reason that one employs the inverse of the wave operator $\mathcal{W}^{-1} = e^{-T}$ or, equivalently, the similarity-transformed Hamiltonian $\bar{H} = e^{-T}He^{T}$, which involves only a *finite* number of contractions of H with T, rather than its Hermitian conjugate $\mathcal{W}^{\dagger} = e^{T^{\dagger}}$, as one must when evaluating the expectation values of operators other than the energy. Of course, this is also the reason for the nonvariational character of CC methods.

Although a direct evaluation of the property expectation values $\langle \widehat{O} \rangle$ in Eq. (2.171) may seem intractable and unappealing in view of the necessity to truncate a nonterminating expansion, leading essentially to similar problems encountered in finite-order MBPTs (which probably contributed to their being characterized as "prohibitively expensive in practice" [316]), these difficulties can be overcome as shown in Section II.F.1. Nonetheless, it is worthwhile to look for alternative, more general, ways of computing CC properties. Such approaches are usually based on the Hellmann–Feynman theorem [317] or its generalizations, as well as on linear (or higher order) response theory (e.g., [109, 318, 319]). These techniques are, in fact, often employed even with variational approaches (e.g., [320]). The whole plethora of different methods for various purposes, which are based on the same principles, but couched in diverse disguises, has been developed in the meantime. This is understandable in view of the well-known interrelations between the underlying concepts—e.g., between the linear response of quantum systems and retarded Green functions (specifically, the retarded polarization propagators that arise in dealing with hole–particle excitation processes), as given by Kubo formulas, or the relationship between nonlinear response functions and a general evolution operator that characterizes field-theoretical techniques, or even relationships implied by dispersion relations. It is well beyond the scope and purpose of this review to explore these various connections. We thus restrict ourselves, as in the rest of the review, to those developments that have been successfully exploited computationally in the context of CC methodology, and particularly those that have not been reviewed elsewhere.

As was already indicated, one of the key principles invariably relied upon is that of the Hellmann–Feynman theorem [317]. Adding the property operator $\widehat{O}$ to the standard Hamiltonian H as a perturbation, i.e.,

$$H(\lambda) = H + \lambda\widehat{O}, \qquad (2.172)$$

and designating the eigenvalues and eigenstates of $H(\lambda)$ by $E(\lambda)$ and $|\Psi(\lambda)\rangle$, respectively, we obtain the desired result by differentiating the expectation value

$$E(\lambda) = \langle H(\lambda) \rangle_\lambda \equiv \langle \Psi(\lambda) | H(\lambda) | \Psi(\lambda) \rangle / \langle \Psi(\lambda) | \Psi(\lambda) \rangle \tag{2.173}$$

with respect to the coupling parameter λ:

$$\frac{\partial E(\lambda)}{\partial \lambda} = \left\langle \frac{\partial H(\lambda)}{\partial \lambda} \right\rangle_\lambda . \tag{2.174}$$

Relying on this theorem and on Eq. (2.174), and writing $|\Psi\rangle \equiv |\Psi(0)\rangle$, we see that we can calculate the expectation value as

$$\langle \widehat{O} \rangle = \langle \Psi | \widehat{O} | \Psi \rangle / \langle \Psi | \Psi \rangle = \left. \frac{\partial E(\lambda)}{\partial \lambda} \right|_{\lambda=0} , \tag{2.175}$$

and similarly for higher order properties. (Cf., e.g., [14, 216, 320].) For this purpose, it is again convenient to use the normal product form of the Hamiltonian with respect to the reference state $|\Phi_0\rangle$, so that

$$\Delta E(\lambda) \equiv \langle \Psi(\lambda) | H_N(\lambda) | \Psi(\lambda) \rangle = E(\lambda) - \langle \Phi_0 | H(\lambda) | \Phi_0 \rangle$$
$$= \sum_{n=0}^{\infty} \lambda^n \Delta E^{(n)} \tag{2.176}$$

and

$$\Delta E^{(k)} = (k!)^{-1} \left. \frac{\partial^k \Delta E(\lambda)}{\partial \lambda^k} \right|_{\lambda=0} . \tag{2.177}$$

Then,

$$\Delta E^{(0)} = E(0) - \langle \Phi_0 | H | \Phi_0 \rangle = \Delta E \tag{2.178}$$

gives the correlation energy (assuming that $|\Phi_0\rangle$ is chosen to be the RHF reference),

$$\Delta E^{(1)} = \left. \frac{\partial \Delta E(\lambda)}{\partial \lambda} \right|_{\lambda=0} = \left. \frac{\partial E(\lambda)}{\partial \lambda} \right|_{\lambda=0} - \langle \Phi_0 | \widehat{O} | \Phi_0 \rangle \tag{2.179}$$

gives the desired property

$$\langle \widehat{O} \rangle = \Delta E^{(1)} + \langle \Phi_0 | \widehat{O} | \Phi_0 \rangle , \tag{2.180}$$

and $\Delta E^{(k)}$, $k > 1$, gives the higher order properties. For example [216, 321], when $\widehat{O}$ is the dipole moment operator, $\Delta E^{(k)}$ describes the (hyper)polar-izabilities $a, b,$ and g for $k = 2, 3,$ and 4, respectively.

These results immediately suggest the simplest possible approach to calculating properties, namely, the so-called finite-field (FF) method, in which one simply adds the operator characterizing the desired property to the Hamiltonian (in fact, $\widehat{O}$ is usually a one-particle operator) and computes the energy for a few small values of λ, subsequently obtaining $\langle \widehat{O} \rangle$ via numerical differentiation. (Cf. Section II.F.2.)

A more general approach, applicable to both static and dynamic properties, relies on response theory [109, 318, 319]. In the CC context, this approach was first outlined by Monkhorst [110] (cf. also [14, 114, 216, 321]) about two decades ago. Although other versions of this formalism (based primarily on variational formulations of CC theory by Arponen [322]) have been developed and exploited, the original approach [110] was first used for the direct determination of excitation energies [114], and only recently was its full, explicit formulation at the OSA level utilized for the computation of static properties [14, 216, 321]. We address these various developments in Section II.F.3.

1. Direct Calculation of Expectation Values

In spite of the infinite nature of the expansion for CC expectation values [Eq. (2.171)], it was used to develop a relatively inexpensive technique, providing satisfactory results at least in specific cases. The expansion was first employed in calculations of nuclear form factors [323]; more recently, it was used in the computation of static one-electron properties [324–326]. As shown by Noga and Urban [324], the key to a successful truncation is to rely on the MBPT structure of the CC method employed and to retain all the second-order contributions to the wave function in both the bra and the ket in Eq. (2.171). These two authors correspondingly designate their approach as WF(2), so that for CC methods up to the CCSD, they define

$$\langle \widehat{O} \rangle_{\mathrm{WF}(2)} = \langle \Psi^{\mathrm{WF}(2)} | \widehat{O} | \Psi^{\mathrm{WF}(2)} \rangle_C , \qquad (2.181)$$

where

$$|\Psi^{\mathrm{WF}(2)}\rangle = (1 + T_1 + T_2 + T_3 + \tfrac{1}{2} T_2^2) |\Phi_0\rangle . \qquad (2.182)$$

Adding also the terms involving $T_1 T_2$ clusters [325], as well as the third-order corrections to T_1 and T_2 arising via T_3 clusters, Noga and Urban formulated a WF(3)-CCSD(T) method [326] that seems to provide as good results as the method based on analytical derivatives of the CCSD(T) energy.

A related approach to the evaluation of property expectation values by Pal [327] and Ghose et al. [328] uses a slightly different truncation scheme based on linear, quadratic, and cubic truncation for the exponential CC wave function. In this context, we note that a very similar truncation scheme was employed earlier by Geertsen and Oddershede [115] (see also [251]) in their CC-based polarization propagator approximation (CCPPA). Being faced with the same problem of handling an $e^{T^\dagger}$ operator when they employed the CC(S)D wave function as a reference, while extending their second-order PPA (SOPPA) to CCPPA (requiring them to evaluate quantities of the type $\langle e^{T^\dagger}[X, E^{rs\cdots}_{ab\cdots}]e^T\rangle$, with X equal to either H, V, F or $E^{ab\cdots}_{rs\cdots}$), these authors resolved to truncate the exponential expansion in the expressions required by CCPPA in such a way that one obtains the MBPT-based SOPPA expressions when replacing the t-amplitudes by their second-order approximation (i.e., when replacing t-vertices by v-vertices, in diagrammatic terms).

It is worth mentioning that the extension of SOPPA to CCPPA led to a considerable improvement in the computed excitation energies and transition moments, particularly for low-lying excited states [251], as well as to a much faster convergence. The use of the CC wave function as a reference proved to be generally beneficial in directly computing excitation, ionization, or electron attachment energies and related properties. (Cf. [31, 105, 106].) The expectation value for the energy, $[\hat{O} = H$, Eq. (2.171)] referred to as XCC (expectation value CC) *Ansatz* (cf. [329, 330]), or UCC (unitary CC) *Ansatz* when one employs the skew-Hermitian form of the cluster operator $(T - T^\dagger)$ [330, 331], was also employed in designing the noniterative schemes for the perturbative account of higher-than-pair cluster components [31].

2. Finite-Field Approaches

The most straightforward and often-employed method of computing response properties that can be used conjointly with any method capable of providing reasonably accurate energies for a molecule immersed in a weak field of interest is the so-called finite-field (FF) or finite-perturbation (or even finite-point-charges) method. One simply employs the Hamiltonian of Eq. (2.172) and computes $E(\lambda_i)$ for a few values of $\lambda_i, i = 1, 2, \ldots, r$, that are required for a numerical differentiation yielding the desired $\Delta E^{(k)}$. This approach needs only a small modification of the existing codes, supplying the necessary integrals involving the computed property $\hat{O}$. Since, in most cases of interest, $\hat{O}$ is a one-body operator, one has to modify only the one-electron part of H. In computing various response properties, as defined by the multipole expansion of the energy in terms of the field components and its gradients, it is sometimes beneficial (especially for small, highly symmetrical systems) to use strategically placed point charges rather than uniform fields. (Cf. [113, 332–334].)

We have to keep in mind that the Hellmann–Feynman theorem, Eq. (2.174), holds only for exact eigenstates or energies (or, when generalized, for wave functions with variationally optimized parameters), so that we only obtain approximate expectation values $\langle \hat{O} \rangle$ when employing approximate $E(\lambda)$. In other words, the expectation value $\langle \hat{O} \rangle$ computed as a numerical derivative of the (necessarily) approximate energy, say, $E^{CCSD}(\lambda)$, does not equal the true mean value $\langle \Psi^{CCSD} | \hat{O} | \Psi^{CCSD} \rangle / \langle \Psi^{CCSD} | \Psi^{CCSD} \rangle$, the difference being given by the so-called non–Hellmann–Feynman terms. (See [113, 324, 335].) Numerical experience shows, however, that these terms are negligible when one employs CC wave functions in which T_1 clusters are properly accounted for. (This is, for example, the main reason for the reported failure of QCI [307–310]; cf. Section II.E.2.) A proper treatment of T_1 clusters also allows for an automatic accommodation of orbital polarization or relaxation effects, as implied by Thouless' theorem. [Indeed, CC properties are practically independent of the kind of molecular orbitals (i.e., relaxed vs. fixed) employed.]

The main weak point of the FF approach stems from the contradictory demands imposed by the numerical differentiation: To obtain accurate derivatives (especially those of higher order that are required for hyper-polarizabilities), we have to use very small fields (i.e., small values of the coupling constant λ). This in turn requires that we compute the energies $E(\lambda)$ to a very high precision, even though a convenient (symmetric) choice of the fields seems to help. The problem is felt particularly when one computes property functions for nonequilibrium geometries [216, 321].

Another disadvantage of the FF approach is the necessity to perform a number of energy calculations for different λ values, which will be especially demanding if highly correlated methods are used if we are in the region of geometries where the method used for the energy evaluation converges poorly, or if large systems are considered. Moreover, each property requires a new set of such computations. Nonetheless, this method is used often, and was successfully exploited in the past, to yield a wealth of information about various properties. (Cf., e.g., [31, 333].) Recently, it was also exploited in conjunction with the two-reference OSA SU CCSD method [94] and yielded very good results for both the ground and the excited states, regardless of whether the relaxed or nonrelaxed orbitals were employed, in both degenerate and nondegenerate regions. Most importantly, the method was able to correct a qualitatively wrong behavior of broken-symmetry SCF and SR CCSD results.

3. Linear-Response (LR) Approaches

There are several ways in which linear- or general-response theory can be (and has been) employed in the context of CC theories. (Cf. [30, 31, 336]; for applications within variational approaches, see, e.g., [119, 336, 337].) All these approaches are invariably based on the Hellmann–Feynman theorem or any of its generalizations. Here we shall briefly outline the original approach by Monkhorst [110], both for the sake of simplicity and for the advantages it offers in computing high-order properties when it is properly implemented [216, 321], even though an equivalent formalism can be derived from more general formulations. In particular, one can exploit a variational formulation of CC theory [322] relying on the energy functional $\langle \widetilde{\Psi} | H | \Psi \rangle$, which uses the standard CC *Ansatz* in the ket, but a more general one in the bra, namely,

$$\langle \widetilde{\Psi} | = \langle \Phi_0 | e^{\Sigma} e^{-T} , \qquad (2.183)$$

defining the so-called extended CC method (ECCM). When only the linear approximation is used for e^{Σ}, i.e., $e^{\Sigma} \approx 1 + \Sigma \equiv 1 + \Lambda$, one obtains the so-called normal CCM (NCCM) [149, 322]. When Λ is determined variationally, the latter version is equivalent to Monkhorst's approach. The parameters defining Λ (which, in contrast to T, represents both connected and disconnected—although always linked—de-excitation operators) may also be regarded as Lagrange multipliers [338–340]. Closely related are techniques for computing molecular gradients and/or Hessians, enabling the optimization of molecular geometries and calculation of force fields and harmonic vibrational frequencies [338–342]. Although in this case one considers the response to changes in the molecular geometry, the relevant codes are easily adapted to compute other response properties as well.

Considering the CS (or the spin-orbital) case, we employ the standard cluster *Ansatz* for the eigenstates $|\Psi(\lambda)\rangle$ of the perturbed Hamiltonian $H(\lambda)$, Eq. (2.172), to obtain

$$|\Psi(\lambda)\rangle = e^{T(\lambda)} |\Phi_0\rangle , \qquad (2.184)$$

and expand $T(\lambda)$ in powers of the coupling constant λ, resulting in

$$T(\lambda) = \sum_{n=0}^{\infty} \lambda^n T^{(n)} \equiv T + U(\lambda) , \qquad (2.185)$$

with $T^{(0)} = T(0) \equiv T$ being the cluster operator defining the unperturbed wave function. Since we assume a single-determinantal nondegenerate reference $|\Phi_0\rangle$, all $T^{(n)}$ components commute, and the right-hand side of Eq. (2.184) is well defined. Substituting the latter equation into the Schrödinger equation and employing the connected-cluster theorem, we get [just as in

Section II.C.1; cf. Eq. (2.79)]

$$\{(H_N + \lambda \widehat{O}_N)e^{T(\lambda)}|\Phi_0\rangle\}_C = \Delta E(\lambda)|\Phi_0\rangle, \qquad (2.186)$$

with ΔE given by Eq. (2.176). Considering the first-order terms in Eq. (2.186) and designating the first-order component $T^{(1)}$ of $T(\lambda)$, or $U(\lambda)$, simply by U, i.e.,

$$T^{(1)} \equiv U = u_j G_j, \qquad (2.187)$$

we get

$$(H_N e^T U|\Phi_0\rangle)_C + (\widehat{O}_N e^T|\Phi_0\rangle)_C = \Delta E^{(1)}|\Phi_0\rangle. \qquad (2.188)$$

In view of the Hellmann–Feynman theorem, the desired expectation value $\langle \widehat{O} \rangle$ of Eq. (2.180) is given by $\Delta E^{(1)}$, which results by projecting Eq. (2.188) onto $|\Phi_0\rangle$,

$$\Delta E^{(1)} = \langle (H_N U + \widehat{O}_N)e^T \rangle_C. \qquad (2.189)$$

Assuming the unperturbed cluster operator T to be known, it remains to find U of Eq. (2.187), whose components must satisfy a linear system of equations obtained by projecting Eq. (2.188) onto the excited-state manifold $\{G_j|\Phi_0\rangle\}$; that is,

$$A_i + B_{ij}u_j = 0, \qquad (2.190)$$

where

$$A_i = \langle G_i^\dagger \widehat{O}_N e^T \rangle_C \quad \text{and} \quad B_{ij} = \langle G_i^\dagger H_N e^T G_j \rangle_C, \qquad (2.191)$$

or, in matrix form,

$$\mathbf{A} + \mathbf{B} \cdot \mathbf{U} = \mathbf{0}, \qquad (2.192)$$

in which $\mathbf{A}$ and $\mathbf{U}$ are column vectors with components A_i and u_i, respectively, and $\mathbf{B} = [B_{ij}]$. Once we solve for $\mathbf{U}$, we have

$$\Delta E^{(1)} = \mathbf{C}^t \cdot \mathbf{U} + D, \qquad (2.193)$$

where

$$C_i = \langle H_N e^T G_i \rangle_C \qquad (2.194)$$

are the components of the column vector $\mathbf{C}$ (with the superscript t indicating the transpose) and

$$D = \langle \widehat{O}_N e^T \rangle_C. \tag{2.195}$$

For the most-often-arising case of a one-body operator $\widehat{O}_N$, the explicit expression for $\Delta E^{(1)}$ is very simple and follows immediately from the diagrams of Figure 10. (Note that diagrams (a)–(c) represent $\mathbf{C}' \cdot \mathbf{U}$, and diagram (d) gives $D = \langle \widehat{O}_N T_1 \rangle$.) Likewise, it is straightforward to generate the explicit form of the linear system represented by Eq. (2.190) or Eq. (2.192). At the CCSD level, when $T^{(1)} \equiv U = U_1 + U_2$ and $U_i = u_j^{(i)} G_j^{(i)}$, the orthogonally spin-adapted form of these linear-response (LR) CCSD equations may be found in [216, 321]. It is important to note that only the absolute terms A_i of Eq. (2.191) and the constant D of Eq. (2.195) depend on the property $\widehat{O}$, while the coefficients B_{ij} in Eq. (2.191) and C_i in Eq. (2.194) depend only on the unperturbed $T \equiv T^{(0)}$ amplitudes that are determined a priori by solving standard CCSD equations.

It is not difficult to see that we get a similar system of algebraic equations for higher order properties given by $\Delta E^{(n)}$, $n > 1$, as well. Thus, generally,

$$\mathbf{A}(\mathbf{T}, \mathbf{U}^{(1)}, \ldots, \mathbf{U}^{(n-1)}) + \mathbf{B}(\mathbf{T}) \cdot \mathbf{U}^{(n)} = \mathbf{0}, \tag{2.196}$$

where we emphasize the recursive nature of these equations by indicating the dependence of matrices $\mathbf{A}$ and $\mathbf{B}$ on the lower order amplitudes. The explicit, detailed form of these equations (for the OSA version at the SD level) shows that this recursive nature allows the design of general-purpose codes that can, in principle, handle properties of an arbitrarily high order [216, 321]. In fact, these codes were exploited to numerically estimate the radii of convergence of the energy expansions in electrostatic fields that define permanent multipole moments, polarizabilities, etc. For this purpose, very accurate $\Delta E^{(n)}$ values were generated up to the 20th order [343].

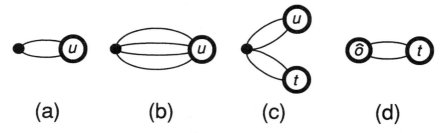

Figure 10.   Nonoriented and unlabeled Hugenholtz diagrams yielding the expectation value $\langle\Psi|\widehat{O}|\Psi\rangle/\langle\Psi|\Psi\rangle$ for a one-body operator $\widehat{O}$ as given by the LR CC method. The $T_1^{(1)} \equiv U$ and $T_1^{(0)} \equiv T$ amplitudes are labeled by u and t, respectively, while the $\hat{o}$ vertex represents the one-body operator $\widehat{O}$.

Obviously, one can employ different strategies for solving Eq. (2.192). In particular, one can exploit Wigner's $(2n + 1)$ rule and compute $\Delta E^{(2n+1)}$ using only the $T^{(k)}$ clusters with $k \leq n$. Since, for the first-order property $\Delta E^{(1)}$, this requires only the zero-order clusters $T \equiv T^{(0)}$, there is, in fact, no need to solve Eq. (2.192) for $U \equiv U^{(1)}$, which we can write formally as $U = B^{-1}A$. In view of Eq. (2.193), we can instead solve the equation

$$B^t \cdot \Lambda = C, \qquad (2.197)$$

for the auxiliary vector Λ and compute $\Delta E^{(1)}$ as

$$\Delta E^{(1)} = \Lambda^t \cdot A + D, \qquad (2.198)$$

since $\Lambda^t \cdot A = [(B^t)^{-1}C]^t A = C^t B^{-1} A = C^t \cdot U$. The advantage of this procedure lies in the "universality" of Λ, since B and C—and thus Λ—are property independent. Hence, once we know Λ and evaluate A and D for a given $\widehat{O}$, we can obtain $\langle \widehat{O} \rangle$ as a simple scalar product. The technique, often referred to as the Z-vector procedure [344] and echoing the well-known method of Dalgarno and Lewis [345, 346], is very much exploited in analytic approaches to energy gradients. (Cf., e.g., [320, 338–342].) It can be extended to higher order properties, assuming that the unknown coefficients appear linearly and the coefficient matrix is independent of the perturbation. However, since in the general case the absolute terms A involve, in principle, several lower order components $U^{(k)}$, $k < n$, it is advantageous to exploit the recursive nature of the linear system of Eq. (2.196) and compute higher order properties successively. The aforementioned high-order results [343] were indeed generated in this way.

Note that the auxiliary vector Λ represents, in fact, the Λ-operator of the NCCM. This follows from an insightful MBPT analysis of the CC response function by Nooijen and Snijders [316], who carried out a kind of "retro derivation" of the CC LR formalism, starting from the MBPT expansion for $\langle \widehat{O} \rangle$ rather than from the perturbed Schrödinger equation and collecting the subgraphs defining various cluster components and then interrelating them in the spirit of a derivation of Dyson-like equations. This approach also enabled Nooijen and Snijders to directly derive Eq. (2.197) for the Λ amplitudes for one- and two-body density matrices. An equivalent result can be obtained by a formal "transposition" of the standard formalism given by Eqs. (2.192) and (2.193). (See [374] for the OSA CCSD case.)

4. Equations-of-Motion (EOM)-Based Approaches

All the relevant information concerning the stationary bound states of a molecular system, as described by a spin-free ab initio model Hamiltonian given by Eq. (2.88), is contained in the time-independent Schrödinger equation (2.112). Yet, in exploiting various approximation schemes (as we must in order to arrive at whatsoever a solution) and employing suitable *Ansätze* for the wave function, it is often beneficial to rely on alternative, though equivalent, formulations, be they of a PT nature, including field-theoretical techniques, or be they based on the variation principle. Especially when one considers "two-state" properties, such as excitation energies and corresponding transition probabilities, it may be advantageous to employ the so-called *direct* approaches, rather than the standard ones, treating each state per se. Thus, just as it is desirable to rely on the Bloch equation in handling quasi-degenerate states, it may be likewise useful to exploit the CC *Ansatz* in conjunction with other methods. We have already seen that combining the CC approach with MBPT enabled us to account effectively for higher-than-pair clusters via a CCSD(T) method or some similar scheme (cf. Section II.C.4.2) or with the linear-response theory as in the preceding section. For the same reason, the CC wave function was also successfully exploited in SOPPA, leading to the CCPPA method. (Cf. [115, 251] and Section II.F.1.) To some extent, this is also true of QCI (cf. Section II.E.2) or dressed Hamiltonian methods [126, 348]. Clearly, the CC method or wave function can be useful in combination with other "derived" approaches, be they oriented towards the "direct" calculation of energy differences or other properties, particularly those involving two distinct states, such as electric or magnetic dipole or electric quadrupole oscillator strengths.

As a paradigm of such combined approaches, we consider nowadays often-used methods of computing the excitation, ionization, or electron attachment energies that rely on the CC ground-state description, together with the linear *Ansatz* for the relevant "excitation" operator, while exploiting the equations-of-motion formalism (the EOM-CC methods discussed in [31, 112, 113, 115–118, 122, 123]) or simply the CI-type procedure (the SAC/SAC-CI methods examined in [106, 107]). In fact, a precursor of these approaches, using the same *Ansatz* for the wave function, was already formulated two decades ago [105]. We shall briefly discuss these developments and relationships (including the Green-function-based approaches set forth in [115, 251, 349]) after outlining the basic underlying ideas involved.

The basic equations of the EOM formalism, being aimed at the direct determination of energy differences (excitation, ionization, etc., energies), are most easily obtained by considering an appropriate difference between the Schrödinger equations [e.g., Eq. (2.112)] for the initial ($|\Psi_0\rangle$) and final ($|\Psi_i\rangle$) states. For this purpose, we define an excitation operator $R_i^\dagger$ that promotes $|\Psi_0\rangle$ to $|\Psi_i\rangle$ (note the difference from the replacement operators that are sometimes referred to by the same name), via the equation

$$|\Psi_i\rangle = R_i^\dagger|\Psi_0\rangle\,, \qquad (2.199)$$

and that enables us to write the Schrödinger equation for the ith state in the form

$$H R_i^\dagger|\Psi_0\rangle = E_i R_i^\dagger|\Psi_0\rangle\,. \qquad (2.200)$$

Subtracting, thus, the same equation for $|\Psi_0\rangle$, i.e., $H|\Psi_0\rangle = E_0|\Psi_0\rangle$, on which we first act with $R_i^\dagger$, we obtain the basic EOM

$$\left[H, R_i^\dagger\right]|\Psi_0\rangle = \omega_i R_i^\dagger|\Psi_0\rangle\,, \qquad (2.201)$$

where

$$\omega_i = E_i - E_0 \qquad (2.202)$$

is the desired energy difference. Since $E_i - E_0 = (E_i - \langle H\rangle) - (E_0 - \langle H\rangle)$, where $\langle H\rangle = \langle\Phi_0|H|\Phi_0\rangle$ for a suitable IPM reference $|\Phi_0\rangle$, we can assume the Hamiltonian in Eq. (2.201) to be in the normal product form (relative to $|\Phi_0\rangle$).

Obviously, the excitation operator is far from being unique, since we can write, for example,

$$R_i^\dagger = |\Psi_i\rangle\langle\Psi_0| + \sum_{j(\neq i)}\sum_{k(\neq 0)} C_{jk}|\Psi_j\rangle\langle\Psi_k|\,, \qquad (2.203)$$

where the C_{jk} are arbitrary. The arbitrariness in turn endows this approach with great flexibility. One thus often requires that R_i annihilate $|\Psi_0\rangle$, i.e.,

$$R_i|\Psi_0\rangle = 0\,, \qquad (2.204)$$

or, equivalently,

$$[H, R_i]|\Psi_0\rangle = 0\,. \qquad (2.205)$$

Designating the first variation of R_i by δR_i, one then easily derives the standard form of EOM, viz.,

$$\langle\Psi_0|[\delta R_i, [H, R_i^\dagger]]|\Psi_0\rangle = \omega_i\langle\Psi_0|[\delta R_i, R_i^\dagger]|\Psi_0\rangle. \qquad (2.206)$$

Or, in a more symmetric form, one replaces the left-hand side of this equation by [108]

$$\langle\Psi_0|[\delta R_i, H, R_i^\dagger]|\Psi_0\rangle := \tfrac{1}{2}\langle\Psi_0|[\delta R_i, [H, R_i^\dagger]]$$
$$+ [[\delta R_i, H], R_i^\dagger]|\Psi_0\rangle, \tag{2.207}$$

relying on Jacobi's identity.

In view of the arbitrariness of $R_i^\dagger \equiv R^\dagger$ (we drop the subscript i for simplicity's sake), one can make various choices of *Ansätze* for $R^\dagger$, even when employing finite-dimensional ab initio model spaces. Thus, approximating $|\Psi_0\rangle$ by the Hartree–Fock reference $|\Phi_0\rangle$ and choosing

$$R^\dagger = C_R^A e_A^R \quad \text{or} \quad R^\dagger = c_r^a E_a^r \tag{2.208}$$

(with summations over hole and particle labels implied), one easily arrives at the CIS (or Tamm–Dancoff) approximation. Since this approximation assumes $|\Phi_0\rangle$ to be the exact ground state, the CIS method provides a reasonable description of excitation energies only when the HF wave function represents a good approximation of the ground state. To remedy the shortcomings of CIS, one thus employs a better approximation $|\widetilde\Psi_0\rangle$ for $|\Psi_0\rangle$ that involves excited configurations $e_A^R|\Phi_0\rangle$, $e_{AB}^{RS}|\Phi_0\rangle$, etc., while including the "de-excitations" in $R^\dagger$, i.e.,

$$R^\dagger = Y_R^A e_A^R - Z_A^R e_R^A, \tag{2.209}$$

and requiring that $R|\widetilde\Psi_0\rangle = 0$. This leads to various random phase approximation (RPA) schemes, depending on the truncation used for $|\widetilde\Psi_0\rangle$.

Now, in the context of CC methods, we can employ the CC wave function $|\Psi_0\rangle = e^T|\Phi_0\rangle$ for the ground state, representing an excellent approximation. Assuming, further, that the bulk of the correlation effects is the same in the ground and excited states, one can employ the linear *Ansatz* for $R^\dagger$, viz.,

$$R^\dagger = R_R^A e_A^R + R_{RS}^{AB} e_{AB}^{RS} + \cdots = r_I G_I, \tag{2.210}$$

truncated at a suitable (usually monoexcited) level. Consequently, $R^\dagger$ and T commute, i.e.,

$$[R^\dagger, T] = 0, \tag{2.211}$$

although $R|\Psi_0\rangle$ does not vanish. Nonetheless, using the CC wave function $|\Psi_0\rangle = e^T|\Phi_0\rangle$ in Eq. (2.201) and premultiplying with e^{-T}, the basic relationship of the EOM-CC formalism takes the form

$$[\bar H_N, R^\dagger]|\Phi_0\rangle = \omega R^\dagger|\Phi_0\rangle, \tag{2.212}$$

where

$$\bar{H}_N = e^{-T} H_N e^T \tag{2.213}$$

is the similarity-transformed Hamiltonian in normal product form.

We can obtain an equivalent formulation directly from the Schrödinger equation, as in the standard CC formalism. Indeed, premultiplying $H_N |\Psi_i\rangle = \Delta E_i |\Psi_i\rangle$ with e^{-T}, while using Eq. (2.199), we find that

$$\bar{H}_N R_i^\dagger |\Phi_0\rangle = \Delta E_i R_i^\dagger |\Phi_0\rangle. \tag{2.214}$$

Next, separating the connected and disconnected parts on the left-hand side (the latter being given by $R_i^\dagger \bar{H}_N |\Phi_0\rangle$, since no contraction can link $\bar{H}_N$ and $R_i^\dagger$), results in

$$\begin{aligned}
\bar{H}_N R_i^\dagger |\Phi_0\rangle &= R_i^\dagger \bar{H}_N |\Phi_0\rangle + (\bar{H}_N R_i^\dagger)_C |\Phi_0\rangle \\
&= \Delta E_0 R_i^\dagger |\Phi_0\rangle + (\bar{H}_N R_i^\dagger)_C |\Phi_0\rangle,
\end{aligned} \tag{2.215}$$

and we have

$$(\bar{H}_N R^\dagger)_C |\Phi_0\rangle = \omega R^\dagger |\Phi_0\rangle, \tag{2.216}$$

where, again, we have dropped the subscript i in the last equation and used the fact that $\omega_i \equiv \omega = E_i - E_0 = (E_i - \langle H \rangle) - (E_0 - \langle H \rangle) = \Delta E_i - \Delta E_0$. Because the left-hand-side of Eq. (2.212) clearly involves only connected terms, Eqs. (2.212) and (2.216) are equivalent, the latter stemming directly from the Schrödinger equation or CI formalism (hence, the designation SAC-CI). Either equation, when written in matrix form, represents a non-Hermitian eigenvalue problem

$$\bar{\mathbf{H}} \mathbf{R} = \omega \mathbf{R}, \tag{2.217}$$

where $\mathbf{R}$ is the matrix of right-hand-side eigenvectors, whose entries are the amplitudes in Eq. (2.210). Since nowadays there are efficient algorithms and codes for finding both right- and left-hand-side eigenvectors of a non-Hermitian matrix, the non-Hermiticity of this formalism does not represent a major obstacle. Moreover, the non-Hermiticity being only minor, at least when the problem is correctly formulated, no appearance of complex roots was reported so far. The left-hand-side eigenvectors can, in fact, be related with the Λ operator of the analytical-gradient theory or linear-response theory. (See Section II.F.3.) Nonetheless, in computing higher order properties, the necessity to handle both the left- and right-hand-side eigenvectors can become rather awkward.

To avoid the non-Hermiticity problems, the original formalism [105] exploiting the *Ansatz*

$$|\Psi_i\rangle = R_i^\dagger e^T |\Phi_0\rangle \qquad (2.218)$$

(note that $R^\dagger$ was designated by W in [105]) relied on the energy expectation value difference

$$\omega_i = \langle \Psi_i | H_N | \Psi_i \rangle / \langle \Psi_i | \Psi_i \rangle - \langle \Psi_0 | H_N | \Psi_0 \rangle / \langle \Psi_0 | \Psi_0 \rangle , \qquad (2.219)$$

which corresponds to a premultiplication of the Schrödinger equation with $e^{T^\dagger}$ rather than e^{-T}. It can be shown that even in this case, the expression for $\omega_i \equiv \omega$ involves only connected terms, since (cf. [105])

$$\langle \Psi_0 | H_N | \Psi_0 \rangle = \langle \Psi_0 | H_N | \Psi_0 \rangle_C \langle \Psi_0 | \Psi_0 \rangle,$$
$$\langle \Psi_i | \Psi_i \rangle = \langle \Psi_i | \Psi_i \rangle_C \langle \Psi_0 | \Psi_0 \rangle, \qquad (2.220a)$$

and

$$\langle \Psi_i | H_N | \Psi_i \rangle = \langle \Psi_0 | H_N | \Psi_0 \rangle_C \langle \Psi_i | \Psi_i \rangle$$
$$+ \langle \Psi_i | H_N | \Psi_i \rangle_C \langle \Psi_0 | \Psi_0 \rangle, \qquad (2.220b)$$

so that

$$\omega_i = \langle \Psi_i | H_N | \Psi_i \rangle_C / \langle \Psi_i | \Psi_i \rangle_C . \qquad (2.221)$$

Thus, representing $R_i^\dagger \equiv R^\dagger$ in the form of Eq. (2.210), we have

$$\omega = \sum_{I,J} H_{IJ} r_I^* r_J / \sum_{I,J} S_{IJ} r_I^* r_J, \qquad (2.222)$$

leading to a Hermitian eigenvalue problem

$$(\mathbf{H} - \omega \mathbf{S})\mathbf{r} = \mathbf{0} . \qquad (2.223)$$

One can, of course, choose $G_I | \Phi_0 \rangle$ to be orthogonally spin-adapted, in which case $\mathbf{S} = \mathbf{I}$, the identity matrix.

The diagrammatic form of the matrix elements H_{IJ} and S_{IJ} can be found in [105]; the second reference [105] also containing examples of computations of excitation energies, ionization potentials, and electron affinities for several PPP model systems. Of course, the Hermitization leads to a more complex formalism and the absence of a natural truncation that characterized the similarity-transformed Hamiltonian. Nonetheless, in view of present-day computing facilities, this version of the formalism—particularly if it is orthogonally spin-adapted—would be worth reexamining.

Let us also note that when we set $T = 0$ in *Ansatz* (2.218) (i.e., if we employ $|\Phi_0\rangle$ as $|\Psi_0\rangle$), the preceding formalism reduces to a simple CI (involving the excitations defining $R_i^\dagger$). Thus, we can view this approach as a kind of "renormalized" CI in which the higher excitations are represented by disconnected clusters of the type $R^\dagger T_k$. If we are interested only in low-lying excited (or ionized) states, it is sufficient to represent $R^\dagger$ by singly excited (one-body) terms. Likewise, it is easy to relate the diagrammatic form of this formalism to the corresponding Green-function-based approaches [115, 251, 349], as outlined in [105].

Let us finally comment on the extensively exploited EOM-CC [31, 112, 113, 115–118] and SAC/SAC-CI [106, 107] methods, both based on *Ansatz* (2.199), (2.210), or (2.218) and using non-Hermitian similarity-transformed Hamiltonians. The latter approach, in addition to emphasizing the (spin-)symmetry adaptation of the cluster *Ansatz* employed, seems to differ from the former in other important details that, however, are not easy to precisely discern in view of the sparseness of the available information concerning the algorithms employed (e.g., the configuration selection scheme, particularly for the disconnected terms, the spin adaptation of the latter, etc.), but also in view of certain more fundamental differences. For example, $|\Psi_g^{SAC}\rangle$, as defined by Eq. (25) of [107], clearly employs the intermediate normalization. This, however, seems to contradict Eqs. (30)–(32) of the same paper, since Eq. (32), namely, $\langle \Psi_g^{SAC}|\Phi_K\rangle = 0$, with $|\Phi_K\rangle$ defined by Eqs. (30) and (31), can hold only if $|\Psi_g^{SAC}\rangle$ is normalized to 1. It would certainly be worthwhile to carry out a detailed comparison, both at the fundamental and at the numerical level, of these often-used approaches.

Finally, we mention the most recent formulation of the EOM-CC approach that employs two successive similarity transformations of the Hamiltonian and is thus referred to as the similarity-transformed EOM-CC (STEOM-CC) method [122]. The first similarity transformation is the same one as in the standard EOM-CC, while the second relies on the VU or Fock-space CC *Ansatz*. The method seems to provide results that are comparable in accuracy to those obtained with MR CISD, CASPT2, EOM-CC, or SAC-CI approaches, but it requires a considerably reduced computational effort [122, 123]. Nonetheless, the method is again restricted to predominantly singly excited states out of the closed-shell ground state and is sensitive to the choice of active orbitals, which in turn is expected to cause complications in generating the entire PESs for the excited states [122].

G. Computational Aspects

In this section, we briefly address the technical side of finding numerical solutions of CC equations and the creation of suitable codes for this purpose. We focus primarily on those facets that are specific to CC approaches and only allude to those that pertain to all ab initio post–Hartree–Fock methods. In particular, we underscore the incipient capabilities of an automated code generation that will undoubtedly permeate future developments.

1. Orbital Choice

Any ab initio model is defined by the choice of the atomic orbital (AO) basis set. The essential desideratum for a suitable basis is its flexibility in order to span that "region" of the one-electron space that is essential for the computed property. Although no universal prescription or procedure exists for constructing or selecting the AO basis, there is nowadays enough practical experience in this regard, as well as a stockpile of widely used basis sets, that facilitate the task. As a general rule, the so-called *radial* correlation is taken care of by doubling (DZ), tripling (TZ), etc., the minimum basis set (MBS), while the *angular* correlation requires the introduction of polarization (P or p) functions having a higher angular momentum quantum number than those of the MBS AOs (e.g., p and d functions for H, d, f, etc., orbitals for the first-row elements, etc.), leading to DZP, TZ2P, etc., basis sets.

It would be ideal to possess a sequence of basis sets converging to the basis-set limit, sometimes referred to as the complete basis-set (CBS) limit. Such a capability is to some extent realized by the so-called hierarchical basis sets that are nowadays more and more often employed in high-level correlated calculations. Using such basis sets, one sometimes carries out the extrapolation to the CBS limit (cf. [350–352]), assuming that the method employed and the system considered are not computationally overly demanding. On the other hand, when one is trying to obtain optimal results for a given system with the available computing facility, it is important to properly balance the complexity of the model employed, as determined by the size of the AO basis set, and the sophistication of the method used to perform the computation.

The hierarchical basis sets used in high-level approaches are usually built from a large number of Gaussians and rely on correlated atomic calculations. Basis sets that have proved to be particularly useful are the atomic natural orbital (ANO) bases [353, 354] and, especially, the correlation-consistent (cc) basis sets of Dunning and collaborators [36], designated as cc-pVxZ [correlation-consistent polarized valence double $(x = D)$, triple $(x = T)$, quadruple $(x = Q)$, or quintuple $(x = 5)$ zeta bases].

For special purposes (e.g., computing electron affinities), the latter are supplemented with small-exponent, diffuse functions that are localized predominantly in the outer regions, or with large-exponent functions enabling a better representation of the wave function in the inner valence and core regions. These basis sets are labeled, respectively, aug-cc-pVxZ (augmented cc-pVxZ, $x =$ D,T,Q,5 bases) [355] and cc-pCVxZ (cc polarized core valence xZ bases) [356]. Special bases for calculating properties (e.g., POL 1 of Sadlej [357]) are also very useful.

Since the AO basis-set problem is common to all ab initio approaches, it has been addressed in numerous monographs and even textbooks and need not be reviewed here. (For a concise description, see, e.g., [28, 286].) Let us only mention that at a DZP level, we already obtain results that can be meaningfully compared with experiment, although larger basis sets (including f functions) are required to guarantee $\sim$1 kcal/mol "chemical accuracy". (Cf. [30, 31].)

Once the AO basis is chosen, we must generate a suitable molecular orbital (MO) basis and transform the one- and two-electron integrals accordingly. Here, one most often relies on various SCF procedures, generating RHF, ROHF, QRHF, UHF (of DODS type), or other IPM orbitals, in terms of which the MO formalism employed, including the Hamiltonian of Eq. (2.2) or Eq. (2.88), is defined. While some methods are a priori restricted to canonical SCF orbitals (e.g., QCI), the general CC methods are invariant with respect to an independent unitary transformation of occupied and virtual MOs, thus permitting the use of localized orbitals if desirable. For OS systems, in addition to the RHF or UHF MOs, it may be useful to employ MC SCF or CAS SCF orbitals.

A particularly desirable choice of MOs is represented by the Brueckner, or maximum overlap, orbitals (BOs), in which case $T_1 = 0$, thus reducing CCSD to CCD. Unfortunately, to generate the "exact" BOs for a given ab initio model, one has to know the exact (i.e., the FCI) wave function (except for highly symmetric MBS models when the MOs are fully determined by the symmetry of the system). However, one can employ approximate BOs along the lines originally suggested by Dykstra [358] and later exploited in the so-called Brueckner doubles (BD) methods [359–366], in which case the orbitals are simultaneously varied with the cluster amplitudes by supplementing the CC equations with constraints requiring the maximum overlap property at a chosen level of truncation (i.e., doubles).

The problem of an optimal choice of MOs in various correlated methods, including CC approaches, is far from trivial. It influences not only the quality of the results, since in practical applications we invariably have to employ truncated CC schemes, but also the convergence behavior of the numerical algorithms used. Particularly in OS situations, one often makes use of MOs for related ionized species or states of another multiplicity (QRHF MOs). The role of the possible choices of MOs has very recently been studied by Jankowski et al. [367] for standard SR as well as MR (of both the VU and SU kind) CC methods using a simple MBS H_4 model [190], in which case any symmetry-adapted MOs are given by two parameters. The authors investigated the full range of these parameters for various geometries that displayed different degrees of quasi-degeneracy, thus covering all the standard choices of MOs (HF, Brueckner, natural, g-Hartree, and various MCSCF orbitals), as well as HF orbitals for ions and even Kohn–Sham orbitals, in addition to very exotic MOs corresponding to the possible range of defining parameters. Most of the "standard" orbitals correspond to a relatively small-parameter region and provide very similar CCSD and CCSDT energies, while CCD energies show larger differences, particularly in the case of weak quasi-degeneracy. For MR theories, the authors also introduced the so-called maximum-proximity orbitals (MPOs), which optimize the "proximity" (defined via a suitable norm) of the model and target spaces in MR theories [367, 368], thus playing the role of Brueckner orbitals for the MR case.

2. Basic Algorithms and Numerical Techniques

Except in the case of a linear-response approach to properties, the CC amplitudes are invariably given as a solution of some system of nonlinear algebraic equations. Thus, in general, one has to rely on iterative algorithms to find these solutions, starting with some initial guess for the amplitude vector $\mathbf{t}^{[0]}$ and using the nonlinear CC equations to generate suitable corrections until the convergence, each iteration providing a new amplitude vector $\mathbf{t}^{[n]}$. To accelerate the convergence, it is advisable to regard a new amplitude vector $\mathbf{t}^{[n+1]}$ as a correction to the preceding vector $\mathbf{t}^{[n]}$, so that at any stage in the iteration process, the desired amplitudes are given by the linear combination of the initial guess $\mathbf{t}^{[0]}$ and all the subsequent correction vectors $\mathbf{t}^{[m]}$, i.e.,

$$\mathbf{t} = \sum_{k=0}^{m} \lambda_k \mathbf{t}^{[k]} , \tag{2.224}$$

with the coefficients λ_k given by a small, m-dimensional system of linear equations that is easily solved using Gaussian elimination. This type of algorithm is modelled on a similar technique used in direct CI methods [369] and includes the so-called *reduced linear-equation method* of Purvis and Bartlett [370] and its modifications [99], as well as the adaptation of Pulay's direct-inversion-in-the-iterative-subspace (DIIS) method [371] to CC equations by Scuseria et al. [372]. The reduced linear-equation algorithm may also be viewed as a preconditioned conjugate gradient method [373]. Since these techniques require a storage of m **t**-vectors, the iteration number m should be kept as small as possible. Thus, in slowly converging cases, when a considerable number of iterations may be required (e.g., in quasi-degenerate situations for highly stretched geometries), one employs the technique of microiterations: After a preset maximum number of iterations $m^{(\text{max})}$ has been reached, the current amplitude vector **t** for $m = m^{(\text{max})}$ is used as a new initial-guess vector $\mathbf{t}^{[0]}$, and the next microiteration cycle is started. The limit $m^{(\text{max})}$ is usually chosen to be between 8 and 10.

The nonlinear terms of CC equations involve products of a molecular integral and two or more cluster amplitudes. For efficient handling of such terms, it is essential to introduce suitable intermediate quantities. This is a well-known technique that has been exploited for some time. For example, when transforming two-electron integrals from the AO basis to the MO basis, we have to evaluate products of an integral with four LCAO coefficients. The direct transformation of all four indices requires an n^8 procedure. Introducing suitable intermediates enables one to transform the indices one at a time, leading to an n^5 procedure. (Cf. [374].)

A similar technique may be exploited when one deals with nonlinear terms of CC equations by defining the intermediates as sums over products of an integral (or another intermediate) with a cluster amplitude. For example, the evaluation of quadratic terms

$$A_i = \sum_{j,k} c_{ijk} t_j t_k \tag{2.225}$$

proceeds in two steps: First we compute the intermediates

$$X_{ij} = \sum_k c_{ijk} t_k , \tag{2.226}$$

and then we obtain the final result via another scalar product

$$A_i = \sum_j X_{ij} t_j . \tag{2.227}$$

Thus, the direct handling of the T_2^2 terms in CC(S)D equations represents an n^8 procedure, while the stepwise computation via intermediates, Eqs. (2.225)–(2.227), requires only an n^6 algorithm. Of course, the foregoing outline is only a schematic one, since each molecular integral and each cluster amplitude (at the CCSD level) depends on up to four orbital indices.

In general, each method calls for different intermediates. In addition to the just-outlined technique of introducing intermediates, which is sometimes referred to as a quasi-linearization, it is advantageous to gather different types of cluster components and handle them as an intermediate quantity. For example, in the CCSD method, an obvious saving is achieved by treating the clusters T_2 and $\frac{1}{2}T_1^2$ jointly. This type of clustering is directly suggested by the structure of the corresponding diagrams (cf. [103]): Subgraphs having the same external lines and joined with an identical subgraph may be conveniently handled as intermediates. A similar grouping of connected or disconnected cluster amplitudes (or both), or of their products with molecular integrals, is possible even when one uses the OSA formalism. (Cf. [129].) Although the diagrammatic formulation helps one to spot obvious intermediates, hardware characteristics (e.g., vector vs. parallel computers) may also play an important role. (Cf. [375].)

3.   Automated Generation of Computer Codes

The CC *Ansätze* for the wave function—particularly their MR or OS versions—may lead to a rather formidable collection of formulas or equations. The explicit form of many a general formulation has never been worked out, not to mention its computer implementation or the testing of its performance. Even in relatively simple CS or SR cases, one can find papers with a long list of formulas or equations that are extensive in scope, yet simple and highly repetitive in their structure. Needless to say, the complexity and diversity of these expressions significantly increases when one handles OS systems. Consequently, the actual implementation and testing of various new ideas and propositions can be very time consuming, although it is unavoidable if we are to find their merits or weaknesses. Moreover, a manual derivation and implementation of complex formulas can be very tedious and thus error prone, as has occasionally been witnessed in years past.

An obvious way to avoid these impediments in handling conceptually simple, yet technically complex, formalism is to make use of ever-advancing computer technology. By a *computer automation* of a given formalism, we shall understand the derivation of its fully explicit form via symbolic manipulation, as well as its subsequent encoding, resulting in a reasonably efficient source code. There can be little doubt that such an approach will become a norm in the future. Already today, moderately complex problems can often be handled via symbolic manipulation software, such as MAPLE [376], and the MAPLE output can be automatically converted into FORTRAN code. In the present context, such an automation was accomplished [45] (cf. also [44]) for the UGA CCSD methods described in Section II.D.3.2 and will be briefly addressed next.

The basic structure of UGA CCSD is simple and straightforward, and the pertinent equations have the same general form as the SR spin-orbital or

CS CCSD ones. The dissimilarity from standard CCSD thus stems from the details of its structure: UGA CCSD equations comprise a much greater variety of distinct terms, and the OS excitation operators of the method do not generally commute. The relevant commutators are best handled algebraically, since the diagrammatic approach, even if feasible, is more difficult to encode. Nonetheless, most UGA CCSD terms have the same structure as the corresponding terms of the OSA CS CCSD. This is the case, for example, for all-external doubles [i.e., (core,core) to (virtual,virtual) excitations], whose number ($\sim n_c^2 n_e^2$) far exceeds that of semi-internal doubles ($\sim n_c n_e^2$ and $n_c^2 n_e$), which involve active orbitals. Thus, the T_2^2 terms involving all-external double excitations projected onto the doubles of the same type are identical to the corresponding terms in CS CCSD. Being most numerous, these terms require most of the computational effort. In high-spin cases, even the semi-internal excitations may be partitioned into the partially spin-adapted (PSA) and other (AUX) types, spanning $T^{(PSA)}$ and $T^{(AUX)}$, respectively. [See Eq. (2.148).] The terms involving $T^{(PSA)}$ may again be handled diagrammatically if desired, while the remaining ones are better handled algebraically. Hence, the excitations involving active orbitals are few in number, but lead to a variety of distinct terms, which complicates their treatment.

The required algebraic manipulations can be carried out either directly in terms of UGA generators or indirectly by exploiting the Slater rules. In the former case, we have to evaluate matrix elements of multiple products of unitary group generators in a spin-adapted basis generated by UGA CC excitation operators. (Cf. Section II.D.3.2.) In the latter case, we apply Slater rules after representing the spin-adapted configurations by linear combinations of Slater determinants and expand U(n) generators in terms of spin-orbital SU($2n$) ones.

In our own implementation [44, 45], we relied directly on the (CA)UGA formalism. In spite of simplifications afforded by the spin-adapted nature of this formalism, numerous expressions involving all possible types of excitation operators (for both the interacting space and full SD manifolds) and all the terms arising in UGA CC equations (at the quadratic approximation level) had to be accounted for. Although such individual matrix elements are easily calculated, the entire derivation is very time consuming and error prone, thus calling for an automated handling in view of the repetitive nature of the task at hand. We have thus designed a suite of codes automating this procedure. (See [44] and [45] for more details.)

In general, the crux of such an automation is to build in sufficient safety measures that guarantee its completeness in the sense that all the required terms are properly accounted for, without any omissions or duplications. The derivation of final expressions is carried out via a symbolic manipulation, and the output is written in a format that can be easily translated into FORTRAN code. As already mentioned, each term in CC equations is given by the product of a numerical coefficient, a molecular integral, and one or more cluster amplitudes. For example, in the quadratic term $c_{ijk}t_j t_k$, the coefficient c_{ijk} is determined by the product of a numerical coefficient (given by the types of the excitation operators that are associated with t_j and t_k amplitudes), say, γ_{ijk}, and a two-electron integral. The integral and the amplitudes involved depend on the orbital labels and are thus symbolic. For given excitation types i, j, and k, the integral labels are uniquely determined by the orbital labels defining i, j, and k. Thus, only γ_{ijk} remains to be computed, and this quantity is given by the U(n) matrix elements $\langle \Phi_0 | E_n^m E_q^p \cdots | \Phi_0 \rangle$ involving strings of U(n) generators that are characteristic of a given type of the term. Hence, the value of γ_{ijk} can be calculated for a special case of that type.

The correctness of the resulting suite of codes was checked in several ways. When the codes are applied to the standard CS CCSD case, we simply recover the well-known result. In OS situations, the correctness can be further checked by exploiting different, yet equivalent, sets of excitation operators (e.g., normalized vs. unnormalized formalism, choice of different orthogonalization schemes, etc.; cf. [108, 109]), which must yield the same CCSD energies. This type of verification is easy to carry out using the automated implementation, while it would be very time consuming if carried out manually (thus requiring a new code for each modification of excitation operators). Clearly, this feature of automation is highly desirable in implementing various versions of the theory or in making modifications to the theory.

III. APPLICATIONS

The CC method in its various forms—particularly its SR version—is nowadays one of the most often-exploited approaches to the many-electron correlation problem. During the past several decades, it was used to compute miscellaneous properties for a wide range of systems. Many of these applications, having primarily to do with computed properties, relied on standard software packages and were thus of a routine nature from a methodological viewpoint. Hence, these applications deserve to be reviewed from the standpoint of the properties and phenomena that have been studied.

Undoubtedly, the state-of-the-art technique that balances accuracy and computational effort is the SR-based CCSD(T) method. It can reliably handle many ground-state properties, such as geometries, fundamental frequencies, heats of formation, dissociation energies, etc. The overall performance of CCSD(T) has been thoroughly reviewed by Lee and Scuseria [30], who summarized a vast number of applications in order to estimate the average errors of the CCSD(T) method for certain types of basis sets. They concluded that the CCSD(T) method, when employed with *spdf-* or *spdfg-* quality basis sets, is capable of attaining "chemical accuracy". This is a fair statement that pretty well sums up the status quo of the SR-based CC methods. Consequently, our review focuses primarily on open-shell, or quasi-degenerate, systems.

Recently, the results of an extensive test of the performance of the four widely used, size-extensive, electronic structure methods [MP2, MP4, CCSD, and CCSD(T)] relying on hierarchical, diffuse function-augmented, correlation-consistent basis sets (cf. Section II.G.1) was published by Feller and Peterson [352]. The authors examined over 200 molecular systems for which the reliable experimental data for atomization energies, EAs, IPs, proton affinities, geometries, and vibrational frequencies are available. The study clearly demonstrates an excellent performance of CC methods, especially CCSD(T). A particularly satisfying feature of these methods is the "uniform convergence" of both the maximum errors and standard deviations as the basis set is expanded to approach the CBS limit, in contrast to rather erratic behavior on the part of finite-order MBPT approaches. (Cf. also [283].) This convergence pattern may be partially due to the fact that the systems which were explored involved both CS and OS species, the latter handled using the UHF reference. It would certainly be interesting to see separate data for these two classes of systems. (For such data concerning equilibrium geometries and involving 19 small CS molecules, see [351].)

Before turning our attention to specific applications of CC methods, let us make a few general remarks concerning the evaluation of the performance of various theoretical approaches. From a conceptual viewpoint, the most satisfactory and reliable test of any method is to compare its results with the exact solution for a given model, as provided by FCI. Even though such a comparison may be feasible only for relatively small model systems, our experience indicates that an excellent performance of a given method with a relatively small basis set (for which the FCI or nearly FCI result can be generated) warrants a reliable and satisfactory performance with a realistic basis set as well. (See, however, a peculiar behavior of the finite-order MBPT methods [281–283].)

Of course, the ultimate goal is to compare the theory's predictions with experimental results and to predict reliable data for properties that are difficult to investigate experimentally. Here, however, a word of caution is in order. There are very few instances in which the computed quantity directly corresponds to the measured one. Particularly when highly accurate computations are carried out, one must pay attention to this fact. Only too often do we witness such predictions being compared with "experiment" without paying due attention to the approximations made on the theoretical side or to the secondary nature of the available data and the analysis that extracted them from the primary data on the experimental side. For example, careful attention should be paid to rovibrational effects (cf. Section III.E), zero-point energies (ZPEs), or even the differences between the true equilibrium and zero-point (or vibrationally averaged) geometries. On the theoretical side, attempts are made to reach the "true experimental values" while using frozen core approximation or when ignoring core-valence effects, not to mention nonadiabatic and relativistic corrections. Thus, one should be always aware of how good an agreement it is reasonable to expect at a given level of theory and with the nature of the available experimental data.

Careful attention must be paid especially to the level of theory employed when one is predicting unknown molecular properties. Here, one can sometimes account for small discrepancies by empirical calibration of computed results based on related systems for which sufficiently accurate experimental information is available. In this way, very precise and useful information can be generated even for relatively large systems. (See, e.g., the computation of accurate equilibrium geometries and of corresponding rotational constants for cyanopolyines and their ions [377].) In cases where only derived properties are experimentally accessible (say, scattering cross sections or infrared spectra), even a simple scaling of theoretical correlation energies can help to produce the desired PESs, allowing a better interpretation of experimental results. (Cf., e.g., [378].)

A. Correlation Energies

The correlation energy is defined as the difference between the (nonrelativistic) exact and Hartree–Fock energies and, for ab initio models, is given by the difference between the FCI (or FCC) and R(O)HF energies. Configurations that have a large weight in the FCI wave function and that are thus required for a physically correct zero-order description contribute primarily to nondynamical correlation, while those with small individual contributions, but plentiful in number, thus having a significant cumulative effect, are responsible for dynamical correlation.

The CC methods proved to be extremely effective in treating dynamical correlation. This efficacy stems from the exponential cluster *Ansatz*, which represents higher order excitations by products of the lower order ones. The result is a combination of efficient handling of dynamical correlation with minimization of computational effort. For example, at the SD level, CCSD and CISD involve the same number of unknowns, yet the former approach is much more powerful, since higher-than-doubles are more or less accounted for by the products of singles and doubles. The most important quadruples are thus represented by products of doubles, providing a very good approximation in nondegenerate situations. Consequently, for nondegenerate states, the CCSD energy should be close to the CISDQ energy.

The efficiency of CCSD-type methods relative to SR CI approaches is illustrated in Table 1 for the two open-shell states (triplet and open-shell singlet) of the DZP model of methylene, for which we can easily carry out FCI. The table compares the percentage of the exact correlation energy (given by the difference of the FCI and ROHF energies) that is recovered by interacting space UGA-CCSD(is) (UGA-CCSD for short) and various SR CI methods, as well as the number of unknown amplitudes or coefficients required by each method. We see that CCSD, though nonvariational, provides as accurate energies as does limited CI of much higher dimensionality. This power of the CC theory is the main reason for its popularity in computing energetic quantities of chemical interest that are sensitive to the effects of electron correlation. A large number of systems of a single-reference character can be treated with high accuracy already at the CCSD level of theory, as has been documented in several earlier reviews [9, 10, 31, 32].

For states with a multireference character, the accurate computation of energetic quantities represents a much greater challenge, requiring an equitable handling of dynamical and nondynamical correlation effects. Currently, such states are most often treated via the MR CISD method. However, when used with a small reference space, MR CISD is incapable of properly describing dynamical correlation. To compensate for this deficiency, one has to employ a much larger reference space than is required for an appropriate zero-order description, which in turn leads to problems associated with too large an active space. Since the exponential CC *Ansatz*

TABLE 1
Performance of UGA CCSD(is) vs. CI methods

System	Basis	Method	Dimension [a]	E_c (%) [b]
CH_2 ($^1B_1$)	DZP	UGA-CCSD	883	98.58
		CISD	2,018	96.16
		CISDT	18,618	98.21
CH_2 ($^3B_1$)	DZP	UGA-CCSD	899	98.14
		CISD	2,930	96.34
		CISDT	29,390	98.44

[a] Fully spin- and space symmetry-adapted configurations are used throughout.

[b] Percentage of correlation energy recovered by a given method relative to FCI.

handles dynamical correlation very effectively, the most compact and cost-effective approach should be realized by some suitable version of MR CCSD: The exponential *Ansatz* would eliminate the large-active-space problem of the MR CISD method, while an appropriate zero-order reference space would ensure an unbiased account of dynamical and nondynamical correlation.

The capability of the MR CCSD types of methods to properly describe highly sensitive energetic quantities is illustrated in Table 2, where the SU and RMR CCSD results for the singlet-triplet separation in CH_2 are compared with those obtained by means of other highly correlated methods. From the theoretical standpoint, the primary challenge in this case is to achieve an accurate and balanced description of correlation effects in two states of different spin multiplicity. From a large number of existing theoretical studies of this system, we selected only a few representative ones for Table 2. Currently, the best experimental result is believed to be $T_0 = 8.992 \pm 0.014$ kcal/mol. When corrected for the zero-point energy, relativistic effects, and (diagonal) Born–Oppenheimer correction, the "experimental" T_e becomes 9.37 kcal/mol. Without any electron correlation, the HF limit gives 24.8 kcal/mol. A proper zero-order description requires a two-electron, two-orbital (HOMO and LUMO) (2,2) active space, resulting in a two-dimensional reference space for the $^1A_1$ state and a

TABLE 2
Singlet–triplet separation $T_e^{\mathrm{nr,BO}}$ in CH_2

Method	Basis	$E(^3B_1)$ (a.u.)	$E(^1A_1)$ (a.u.)	$T_e^{\mathrm{nr,BO}}$ (kcal/mol)
FCI [380]	$4s2p1d; 2s1p$	−39.046260	−39.027183	11.97
(UGA) SR CCSD	$4s2p1d; 2s1p$	−39.044064	−39.023639	12.82
CCSDT [379]	$4s2p1d; 2s1p$	−39.046243	−39.026976	12.09
(2,2)-SU CCSD [274]	$4s2p1d; 2s1p$	−39.044064	−39.024914	12.02
(2,2)-RMR CCSD [50]	$4s2p1d; 2s1p$	−39.044064	−39.024826	12.07
(2,2)-CISD [380]	$4s2p1d; 2s1p$	−39.041602	−39.022156	12.20
(6,6)-CASSCF SOCI [380]	$4s2p1d; 2s1p$	−39.044872	−39.025804	11.97
(UGA) SR CCSD [274]	$5s4p3d2f1g; 3s2p1d$	−39.082499	−39.065804	10.48
(2,2)-SU CCSD [274]	$5s4p3d2f; 3s2p1d$	−39.081028	−39.065742	9.59
	$5s4p3d2f1g; 3s2p1d$	−39.082499	−39.067393	9.48
(2,2)-RMR CCSD [381]	$5s4p3d2f; 3s2p1d$	−39.081028	−39.065361	9.83
	$5s4p3d2f1g; 3s2p1d$	−39.082499	−39.066998	9.73
(6,6)-CASSCF SOCI [380]	$5s4p3d2f1g; 4s3p2d$	−39.084972	−39.070250	9.23
experiment [274]				9.37[a]

[a] Experimental value corrected for the zero-point vibrational energy, (diagonal) Born–Oppenheimer correction, and relativistic effects.

one-dimensional reference space for the $^3B_1$ state. Such a two-configurational MCSCF wave function produces a gap of 11 kcal/mol. [382]

To assess the performance of various methods with respect to a given problem, prior to applying them to a computationally demanding model characterized by a good-quality basis set, it is expedient, if at all feasible, to examine less demanding, yet realistic, models for which we can compute the exact FCI energies or properties. In this way, a definite comparison that is not encumbered by any uncertainties can be made. Using a DZP quality model of CH_2 for that purpose, we find that the UGA CCSD error amounts to 0.85 kcal/mol. In fact, a very similar result is obtained with UHF CCSD and spin-nonadapted ROHF CCSD in this case [379], so that when we rely on the standard SR CC formalism, we have to use CCSDT in order to get within 0.12 kcal/mol of the FCI results.

Turning next to the simplest possible MR CC theory that uses the minimal (2,2) active space leads us to a two-reference (2R) description for the $^1A_1$ state, while still only an SR formalism for the $^3B_1$ state. For the $^1A_1$ state, we thus employed the 2R SU CCSD method, relying on the effective Hamiltonian formalism and Bloch equations (cf. Section II.D.2.2), as well as the 2R RMR CCSD method that combines 2R CISD and SR CCSD by extracting the connected higher-than-pair clusters from the 2R CISD wave function and using them in the externally corrected SR CCSD to more reliably account for nondynamical correlation. Both approaches perform very well in this case: The errors in the singlet–triplet separation when one uses 2R SU CCSD and 2R RMR CCSD are only 0.05 and 0.1 kcal/mol, respectively. We can thus hope that with a sufficiently large basis set, these methods will provide a reliable result that can be compared with experiment. This is indeed the case, as the (2,2) SU CCSD and (2,2) RMR CCSD results indicate (Cf. Table 2.) The 2R-SU CCSD result was later confirmed by Balková and Bartlett [383], who used their two-determinant (TD) version of MR CCSD.

When we compare (2,2)-CISD with (2,2)-SU CCSD or (2,2)-RMR CCSD for the multireference state $^1A_1$, we see that the latter CC methods recover a significantly larger part of the dynamical correlation energy (by about 2.7 mhartree). This is easy to understand when we realize that the (2,2)-CISD wave function contains only SD excitations from two references, so that a large number of excitations are not at all accounted for, while the MR CC methods employing an identical reference space recover a larger portion of the dynamical correlation, thanks to the product terms involving lower excitations. Thus, although the minimal two-reference space is sufficiently large to handle nondynamical correlation, one would have to employ a much larger active space when using MR CISD in order to recover more dynamical correlation. Indeed, we find that when we use the six-electron, six-orbital (6,6) active space, the CAS SCF-SOCI recovers about the same amount of the correlation energy as (2,2)-SU CCSD or (2,2)-RMR CCSD and yields an excellent result for the singlet-triplet splitting, using either a DZP basis (in which case we compare with FCI) or a large ANO basis set (when we can compare with experiment). Nonetheless, the MR CC approaches seem to be most efficient and affordable in recovering both dynamical and nondynamical correlation.

In Table 3, we illustrate the performance of representative CC methods by examining the correlation energies for the ground state of a water molecule, representing a typical single-reference state, for the first excited state ($^1A_1$) of CH_2, which manifests some multireference character, and for the X^1A_g state of regular octagonal H_8, which has a strong two-reference character (the two most important configurations have identical weights in the FCI wave function). All methods perform very well for the nondegenerate ground state of H_2O. Already at the SR-CCSD level, we

TABLE 3

Correlation energies E_c for typical nondegenerate (H_2O), quasi-degenerate (CH_2), and degenerate (H_8) states

Method	E_c (mhartree)	$E_c/E_c^{FCI}(\%)$ [a]
H_2O, X^1A_1 state, DZ basis		
FCI [384]	148.03	100.00
SR CCSD	146.24	98.79
SR CCSD(T)	147.45	99.61
SR CCSDT [103]	147.59	99.70
SR CCSDTQ [103]	148.01	99.99
(2,2)-RMR CCSD [239]	146.49	98.96
(4,4)-RMR CCSD [239]	147.00	99.30
CH_2, first excited $^1A_1$ state, DZP basis		
FCI [380]	140.89	100.00
SR CCSD	137.34	97.48
SR CCSD(T)	140.02	99.38
SR CCSDT [379]	140.68	99.85
(2,2)-RMR CCSD [50]	138.53	98.32
(2,2)-SU CCSD [274]	138.62	98.39
H_8, regular octagonal geometry, X^1A_g state, DZ basis		
FCI	161.06	100.00
SR CCSD	152.31	94.57
SR CCSD(T)	159.44	98.91
SR CCSDT [230]	169.84	105.45
(2,2)-RMR CCSD [50]	159.60	99.09
(2,2)-SU CCSD [230]	163.97	101.81

[a] Percentage of correlation energy relative to the exact FCI value for a given model.

recover 98.79% of the correlation energy. To account for the remaining correlation energy, either we can employ the SR CCSDT, SR CCSDTQ, etc., hierarchy, which is computationally very demanding, or we can use MR CC with a larger and larger reference space. Both approaches can eventually reach the exact FCI (or FCC) result. We note that the perturbative CCSD(T) performs very well, too, recovering 99.61% of the correlation energy. However, one of the weaknesses of perturbatively corrected methods is that there is no well-defined hierarchy of approaches that would enable their systematic refinement towards FCI.

For the $^1A_1$ state of CH_2, the SR CCSD still recovers 97.48% of the correlation energy, only 1.3% less than for a typical single-reference state. However, this small fraction of the correlation energy has a nonnegligible effect on the computed singlet–triplet separation, as has already been discussed.

The largest differences in the performance of various CC methods are found in the case of the degenerate H_8 ground state, as might be expected. With SR CCSD, we recover less than 95% of the FCI correlation energy. The SR CCSDT, on the other hand, overshoots by about 5%, while the (2,2)-SU CCSD overestimate is much smaller ($\sim$2%). The best results are obtained with (2,2)-RMR CCSD and SR CCSD(T), both yielding about 99% of the correlation energy. We shall see, however, that the latter method fails in genuine OS situations that are encountered when we are far away from the equilibrium geometry.

B.   Potential-Energy Surfaces (PESs) and Related Properties

1.   Equilibrium Geometries

By and large, correlation effects do not unduly influence the determination of equilibrium geometries, so that even Hartree–Fock geometries are usually in good agreement with experiment. For example, for closed-shell systems, one often obtains bond distances to within ± 0.01 Å already at the SCF level [385]. When the correlation effects are properly incorporated via CC methods employing large basis sets, equilibrium bond lengths can be determined to within ± 0.005 Å. The performance of the CCSD(T) method in computations of equilibrium geometries has been thoroughly reviewed by Lee and Scuseria [30]. Excluding molecules for which experimental data are believed to be unreliable, these authors find that at the CCSD(T)/*spdfg* quality level of theory, the average errors for single (X-H), double, and triple bonds are 0.001, 0.002, and 0.0026 Å, respectively, and conclude that "the accuracy of the CCSD(T) method has reached the point where in many instances theory is more reliable than experiment in determining polyatomic molecular geometries" [p. 74]. With basis sets of *spdf* quality, the average error can be larger by about 0.001–0.002 Å for single bonds and 0.002–0.003 Å for multiple bonds. In general, theoretical bond lengths are slightly longer than experimental ones.

Very recent studies [351, 352] reconfirm this result. In their examination of 19 small CS molecules, Helgaker et al. [351] find the mean absolute deviation of cc-pVQZ CCSD(T) bond lengths from experiment to be 0.0022 Å, which is comparable to experimental results given their uncertainties. The maximum error, amounting to 0.012 Å (for the HN bond in HNO), enabled these researchers to conclude that the experimental value must be incorrect. A slightly larger deviation found by Feller and Peterson [352] is partially due to a much larger set they examined, but also due to the inclusion of many OS systems for which the UHF reference was employed. When a yet higher accuracy is required (as in predicting the rotational constants for species of astrophysical interest), one has to rely on small empirical corrections for specific types of bonds. (Cf. [377].)

As an illustration of open-shell applications—particularly those relying on the recently introduced RMR CCSD method—we have examined [381] the ground-state bond distances in multiply bonded N_2^+ and CN radicals. (Cf. Table 4.) The methods illustrated include both SR-type UGA CCSD and MR-type RMR CCSD. In the latter case, four references (4R) are employed: the leading configuration $|\Phi_0\rangle$, two doubly excited configurations $\pi_x^2 \to \pi_x^{*2}$ and $\pi_y^2 \to \pi_y^{*2}$, and one quadruply excited configuration $\pi_x^2\pi_y^2 \to \pi_x^{*2}\pi_y^{*2}$. These configurations span the generalized-valence-bond (GVB) type of reference space, describing two π bond pairings (i.e., two configurations for every bond pairing). The results obtained with other reference spaces, for both the ground and low-lying excited states of these systems, can be found in our forthcoming paper [381]. For the sake of comparison, we also include in Table 4 the CCSD(T) results for the related N_2 and CO molecules. In all cases, a cc-pVTZ basis set [36] was employed. Such a basis set seems to give bond lengths close to the basis-set limit. For

TABLE 4
Bond distances for selected multiply bonded diatomics

Molecule	Method[a]	R_e (Å)	$R_e^{\exp}$(Å)[b]	$R_e - R_e^{\exp}$
N_2^+	UGA CCSD	1.1134	1.1164	-0.0030
	(4R)-RMR CCSD	1.1203	1.1164	0.0039
CN	UGA CCSD	1.1684	1.1718	-0.0034
	(4R)-RMR CCSD	1.1736	1.1718	0.0018
N_2 [386]	CCSD(T)	1.1042	1.0977	0.0065
CO [387]	CCSD(T)	1.1361	1.1283	0.0078

[a] All results obtained with cc-pVTZ basis set.
[b] In [388].

both OS systems, UGA CCSD slightly underestimates, while the (4R)-RMR CCSD slightly overestimates, the experimental bond length, even though the errors involved are smaller than the CCSD(T) ones for the corresponding CS systems. A general trend is that both CCSD(T) and RMR CCSD slightly overestimate bond distances. The RMR CCSD errors are in line with the average CCSD(T) error given by Lee and Scuseria [30]. The remaining residual errors seem to be due primarily to the core–valence correlation. For example, in the N_2 case, an account of core-valence correlation leads to a shortening of the bond length by 0.002 Å [389].

Generally, determining the bond angles is easier than determining the bond lengths. Often, a qualitatively correct picture is already provided by the hybridization rules. Although there are fewer actual data concerning the bond angles, in which case the changes made in improving the basis set are not always systematic (cf., e.g., [30]), it is fair to say that the errors are usually smaller than one degree, except where the PES is very shallow, as in cases involving torsional coordinates. Moreover, the available experimental data are generally vibrationally averaged, so that a direct comparison at the highest level of accuracy is usually not possible.

2. Dissociation Energies

Dissociation, or binding, energies play a very important role in chemistry. Their determination often requires highly correlated methods and large basis sets. The preeminence of CC methods in this regard stems from their size-extensive character, since the electron correlation is typically more significant in a molecule than in its fragments. Nonetheless, to attain "chemical accuracy," it is usually necessary to take into account higher-than-pair clusters, particularly triples. For example, the error that arises in using CCSD for a four-electron system that dissociates into two two-electron subsystems is due to the fact that CCSD is exact for two-electron systems but not for four-electron systems. Although the quadruple excitations will be reasonably well accounted for through the products of doubles, the triples will be lacking, since their main contribution comes from connected, rather than disconnected, clusters. Nonetheless, the CCSD result will be vastly superior to the CISD one, which accounts neither for quadruples nor for triples. When the wave function acquires a multi-reference character, even four-body connected clusters become important.

The hitherto accumulated evidence in the literature clearly indicates that the dissociation energies of diatomic molecules can be accurately estimated at the CCSD(T)/$spdfg$ level of theory. For example, the dissociation energies for F_2 and CO are predicted with 0.6-kcal/mol accuracy using a cc-pVQZ basis [390]. With an aug-cc-pV5Z basis, the dissociation energy of HF is predicted to be 141.3 kcal/mol, only 0.3 kcal/mol smaller than the experimental value (141.6 kcal/mol) [391]. Multiple-bonded systems are, of course, more difficult to describe. Table 5 shows dissociation energies for N_2, PN, and P_2, all having a similar bonding structure. At the CCSD(T)/cc-pVQZ level, the dissociation energies are still underestimated by, 3.5, 8.4, and 6.0 kcal/mol, respectively. The basis-set limit estimates for PN and P_2 are smaller by 3.4 and 1.3 kcal/mol, respectively, than the experimental values. This discrepancy is due to connected four-body clusters. Note that the triple corrections are very significant here, increasing the binding energy by about 11 kcal/mol for PN and by 9 kcal/mol for P_2.

In Table 6, we consider the binding energy of the HF dimer, calculated with various CC and related methods. With basis sets of similar quality,

TABLE 5
Dissociation energies D_e for selected diatomics

Molecule	Method	Basis Set	D_e (kcal/mol)
N_2 [390]	CCSD(T)	cc-pVQZ	221.46
	Experiment		225.10
PN [392]	CCSD	cc-pVTZ	122.00
		cc-pVQZ	129.13
		Basis-set limit	134.0
	CCSD(T)	cc-pVTZ	132.88
		cc-pVQZ	140.20
		Basis-set limit	145.2
	Experiment		148.6
P_2 [392]	CCSD	cc-pVTZ	95.40
		cc-pVQZ	102.13
		Basis-set limit	106.6
	CCSD(T)	cc-pVTZ	104.27
		cc-pVQZ	111.19
		Basis-set limit	115.9
	Experiment		117.2

TABLE 6
Binding energies D_e for the HF dimer

Method	Basis Set	D_e (kcal/mol)	Reference
ACCD	$4s3p1d; 3s1p$	4.55	[393]
ACPF	$8s6p2d; 4s1p$	4.33	[394]
CCSD	aug-cc-pVQZ	4.59	[391]
	Basis-set limit estimate	4.39	[391]
CCSD(T)	$4s3p2d1f; 3s2p1d$	4.35	[395]
	aug-cc-pVQZ	4.72	[391]
	Basis-set limit estimate	4.60	[391]
Experiment		4.63 ± 0.17	[391]

CCSD and ACCD give almost identical binding energies (4.55 vs. 4.59 kcal/mol), which are in good agreement with experiment. This is partially due to the cancellation of a basis-set error with that arising through the neglect of higher-than-pair clusters. The basis-set limit at the CCSD level is estimated to be 4.39 kcal/mol. Perturbative triple correction systematically increases the binding energy by about 0.2 kcal/mol.

On the whole, CC methods are capable of yielding very accurate binding energies. However, if we wish to attain "chemical accuracy," connected triples must be included at least perturbatively. When the wave function exhibits a strong multireference character, a suitable version of the MR CC method is called for. Since these effects are especially important for highly stretched geometries that come into play when one examines the entire PES or PEC, and less so when we are interested only in the dissociation energy given by the depth of the potential well, we now turn our attention to this more general problem.

3. Full Potential-Energy Surfaces

The generation of full PESs is one of the most challenging problems of CC theory. The bond-breaking process will generally create various kinds of open-shell subsystems that may defy the standard SR CC description. The ability of the SR CC formalism to provide a reasonable PES depends very much on the ability of the reference configuration employed, usually represented by the lowest SCF solution, to describe the dissociation process studied. The SCF solution may dissociate (i) correctly, describing the resulting fragments approximately, (ii) incorrectly, into the excited states of the fragments, or (iii) incorrectly, into the wrong fragments (for example, ions rather than neutral species, etc.) The SR CCSD then reflects the nature of the reference employed.

For most closed-shell types of systems, the UHF wave function (of DODS type) usually dissociates correctly, while the RHF one does not. In general, these are cases in which a low-spin species separates into high-spin fragments—i.e., when bond breaking creates unpaired electrons. In the simplest case of a single bond breaking, the RHF wave function consists of a 50–50 mixture of covalent and ionic species, so that it involves, at least partially, the correct dissociation products. Hence, the SR CCSD method is capable of recovering most of the nondynamical correlation effects, and the resulting PESs are generally qualitatively correct when one uses the RHF reference. However, the description of a multiple bond breaking can hardly be accomplished with the RHF reference, while the UHF CCSD method can often provide qualitatively reasonable PESs (disregarding the nonanalytic behavior in the vicinity of the triplet instability threshold). An instructive example is given for the N_2 molecule in the recent review by Bartlett [31].

A particularly challenging task is the computations of PESs for weakly bound van der Waals complexes, where the use of very large basis sets is imperative. The shortcomings of theoretical PESs (due to both basis-set limitations and higher order excitations) can be partially compensated for by scaling of computed correlation energy contributions, as suggested by Yang et al. [378] in their study of the Ar–C_2H_2 system. Since the entire PES is required here, it is not surprising that the best results were obtained with the appropriately scaled CCSD PES, even though the results based on the CCSD(T) PES were only slightly less accurate. Here, however, one deals with the interaction of two CS subsystems. Interestingly enough, the computed PESs give good results for the total differential cross section, as well as for spectroscopic parameters, but are in qualitative disagreement with experimentally or semiempirically derived potentials. In view of the ambiguities that are involved in the inversion problem, the theoretical PES is likely much closer to reality.

Open-shell systems are still more complicated. For example, in breaking a three-electron bond (i.e., a doublet state dissociates into a CS singlet and a doublet species), both the ROHF and UHF solutions dissociate correctly, as do the ROHF and UHF-based SR CCSD PESs. However, when a high-spin species dissociates into two species with lower spin (e.g., a triplet species separates into two doublets), it is not uncommon that the lowest UHF solution dissociates incorrectly, as does the UHF-based CCSD potential. This incorrect dissociation is due to the fact that the DODS-type UHF solution tends to maximize the number of unpaired electrons and thus dissociates into higher spin species. The asymmetric dissociation of methylene, CH_2 (X^3B_1) → $CH(X^2\Pi)$ + $H(^2S)$, is such an example [200]. The lowest UHF solution dissociates into the $CH(^4\Sigma^-)$ and $H(^2S)$ fragments, i.e., into the excited state of CH. As a consequence, all UHF(DODS)-based correlated methods lead to an incorrect PES in the dissociation limit. On the other hand, the ROHF wave function dissociates correctly into the $CH(X^2\Pi)$ and $H(^2S)$ species, since it tends to maximize the number of paired electrons. Unfortunately, in such cases ROHF solutions are plagued with Hartree–Fock instabilities, leading to highly nonanalytic PESs. Nonetheless, using the ROHF reference, SR CCSD methods, such as UGA CCSD, generate a correct potential. (See [200] for details.) Both of the aforementioned problems—namely, the incorrect dissociation of DODS-type UHF solutions and the symmetry breaking associated with the ROHF method—could be avoided by exploiting other types of UHF, sometimes called generalized HF (GHF), solutions. This feature deserves a closer examination. (Cf. also Sec. II.C.4.3.)

One of the major problems of the UHF-based correlated methods is the severity of spin contamination in the intermediate-coupling (bond-breaking) region. Thus, although the qualitative behavior of the UHF CCSD potentials may be correct, these potentials are not satisfactory quantitatively. The primary concern here is the fact that the region between the equilibrium and intermediate bond-breaking internuclear separations is very important in many interesting physical and chemical problems. Obviously, one way to overcome the difficulties associated with UHF- or RHF (ROHF)-based SR-CCSD is to develop suitable MR CCSD methods. Hence, in the examples that follow, we focus our attention on the performance of the RMR CCSD method that employs RHF orbitals.

Besides comparing actual total energies, the quality of a computed PES is better assessed by its parallelism with the FCI PES for the same basis set. We introduced [232] the so-called nonparallelism error (NPE) of an approximate PES, defined as follows: For a given range of geometries, the NPE is defined as the difference between the maximal and minimal signed deviations from the exact FCI PES. Clearly, NPE = 0 when the computed PES differs from the FCI one by a constant (positive or negative) shift. Of course, the NPE depends on the range of geometries considered and usually increases with widening range. Nonetheless, it provides us with a convenient simple measure of the quality of the computed PES.

The general character of various correlated methods relying on the RHF reference may be illustrated by the simplest case, involving the breaking of a single bond, as exemplified by the HF molecule. In Figure 11, we present PECs for the HF molecule, obtained with a DZ basis and lower level theories; in Figure 12, we do the same with PECs obtained with higher level theories. In both figures, a comparison with the exact FCI PES can be made. For internuclear separations R that lie between R_e and $2R_e$, most methods give a qualitatively correct PEC. Once the range of geometries is enlarged to, say, $4R_e$, all perturbation-theory-based methods (MP2 through MP4) break down. The SR CI approaches do not break down qualitatively, but quantitative agreement with the FCI is poor unless higher excitations are included. For example, the NPE of the SR CISDT PES is still 18.63 mhartree for the region $[R_e, 3R_e]$. Other lower level theories (cf. Figure 11) have even larger NPEs, and MBPT PECs are meaningless for large internuclear separations. The higher level theory PECs, shown in Figure 12, include SR CCSD, CCSD(T), and CISDTQ, as well as (2,2)-RMR CCSD [239]. The last method employs two-electron, two-orbital (σ, σ^*) active space, representing effectively a two-reference approach. The smallest NPE is obtained with SR CISDTQ, namely, 1.11 mhartree. (All NPEs are for the region $[R_e, 3R_e]$.) The SR CCSD PES is qualitatively correct, but its NPE is 9.91 mhartree. It is worth noting that SR CCSD(T) does not produce a qualitatively correct PES, due to the perturbative nature of the correction for triples, which breaks down for large R values. Finally, the improvement

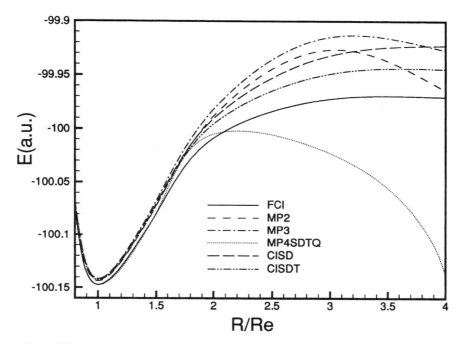

Figure 11. Potential-energy curves for the ground state of the HF molecule obtained with various "lower level" theories and a DZ basis set. An FCI PEC is given for comparison.

brought about by (2,2)-RMR CCSD relative to SR CCSD is obvious: Its NPE, at 1.73 mhartree, is almost an order of magnitude smaller than that for SR CCSD [239]. The improvement is even larger when we employ a DZP basis (0.97 mhartree for (2,2)-RMR CCSD vs. 18.55 mhartree for SR CCSD).

An excellent performance of RMR CCSD in the case of the HF molecule is not accidental. Similarly, good performance is found for F_2 and H_2O [239], as well as for the H_4 and H_8 model systems [50]. For example, with the use of a DZ basis for F_2, the NPEs of SR CCSD and (2,2)-RMR CCSD (for the region $[R_e, 3R_e]$) are 17.83 and 1.49 mhartree, respectively. For the H_8 DZ model and geometries ranging from a regular octagonal structure to a completely separated $2H_2 + H_4$ limit, the NPE for SR CCSD is 7.95 mhartree, while for (2,2)-RMR CCSD it is only 0.95 mhartree. The performance of RMR CCSD is even more remarkable in view of the fact that NPEs of SR CCSDT and (2,2)-SU CCSD are also considerably larger, amounting to 8.8 and 2.72 mhartree, respectively.

An excellent parallelism of the RMR CCSD potential with the FCI one is a consequence of the basic nature of the former method. Recall that RMR

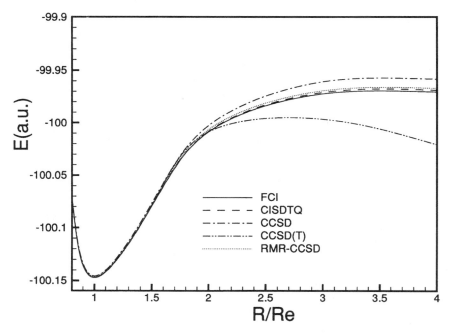

Figure 12. Potential-energy curves for the ground state of the HF molecule obtained with various "higher level" theories and a DZ basis set.

CCSD exploits three- and four-body connected clusters that are extracted from the corresponding MR CISD wave function to ensure a better decoupling of SR CCSD from the rest of the FCC chain. When the state that is under consideration is nondegenerate and well represented by a single reference, as is the case for the equilibrium geometry, the T_3 and T_4 clusters that are extracted from the MR CISD are small, so that RMR CCSD essentially reduces to SR CCSD. However, as the bond is stretched and the multireference character of the wave function increases, T_3 and T_4 can no longer be neglected, and RMR CCSD gives an important improvement over SR CCSD, yielding PESs that are almost parallel with the FCI ones.

As a final example, we consider the demanding case of the N_2 molecule, requiring sixfold excitations in the zero-order wave function in order to describe its dissociation into ground-state atoms. The six-electron, six-orbital CAS should provide an adequate description. However, the exploitation of CAS(6,6) is computationally very demanding. We thus investigated [240] various subspaces of CAS(6,6), and here we illustrate only a few cases. The "minimal" reference space for the breaking of a triple bond that is worth investigating is a GVB-type, eight-reference (8R) space, which consists of two CS-type references for each σ and π bond. Other small reference spaces (e.g., 11R or 13R) are obtained by including some additional configurations from CAS(6,6). (See [240] for details.)

In Table 7, we present some typical results for geometries ranging from R_e to $2R_e$. The complete failure of SR CISD is particularly noteworthy (NPE = 187 mhartree). In fact, even the very demanding SR CISDTQ gives very poor results (NPE = 38 mhartree). The standard SR CCSD method breaks down after about $2R_e$. (Formally, NPE = 103 mhartree, but in this case the SR CCSD potential crosses the FCI one.) Switching to MR approaches, we find that both CAS(6,6)-MR CISD and RMR CCSD perform very well, as indicated by their small NPEs of about 1 mhartree. However, it is difficult to maintain this excellent performance of RMR CCSD when smaller reference spaces are used [240]. With a 44-determinant reference space (note that CAS(6,6) has 175 spin-adapted configurations), we achieve such a performance, while smaller spaces lead to larger NPEs, particularly due to the error at $2R_e$.

Obviously, small reference spaces can be useful when only moderately stretched bonds need to be examined. In Table 8, we examine the

TABLE 7
Deviations from the FCI energy (mhartree) for a DZ model of N_2 as a function of the internuclear separation ($R_e = 2.068$ a.u.)

Method	R_e	$1.25R_e$	$1.5R_e$	$1.75R_e$	$2R_e$	NPE[a]
SR CISD	23.64	49.20	90.11	147.55	210.90	187.
SR CISDTQ	1.38	5.12	13.94	26.93	39.36	38.0
SR CCSD	8.29	19.06	33.55	17.71	−69.92	103.
CAS(6,6)-MR CISD	6.80	7.54	6.42	6.26	6.17	1.37
CAS(6,6)-RMR CCSD	1.76	2.27	2.53	2.73	2.80	1.04
44R-RMR CCSD	2.27	2.80	2.78	2.95	3.12	0.85
16R-RMR CCSD	2.98	3.95	6.20	3.41	−8.71	14.9
11R-RMR CCSD	2.59	3.96	6.44	7.14	−8.65	15.8

[a] Nonparallelism error (NPE) for the interval $[R_e, 2R_e]$. (See text for definition.)

TABLE 8
Performance of CC vs. CI methods for a DZ model of N_2 (cf. Table 7)

Method	R_e	$1.25R_e$	$1.5R_e$	NPE[a]
SR CISDTQ	1.38	5.12	13.94	12.56
SR CCSD	8.29	19.06	33.55	25.26
8R-MR CISD	13.17	18.99	21.79	8.62
8R-RMR CCSD	4.12	5.66	5.83	1.71
11R-MR CISD	9.03	12.55	16.03	7.00
11R-RMR CCSD	2.59	3.96	6.44	3.85

[a] Nonparallelism error (NPE) for the interval $[R_e, 1.5R_e]$. (See text for definition.)

performance of some of the models of Table 7 in a smaller region extending from R_e to $1.5R_e$. Even for this smaller region, the NPEs of SR CISDTQ and SR CCSD are quite significant, namely, 12.56 and 25.26 mhartree, respectively. Here it is more revealing to compare the corresponding MR CISD and RMR CCSD results, both using the same small reference space. For example, with the GVB-type 8R space, the 8R-RMR CCSD NPE is only one-fifth the 8R-MR CISD NPE.

In essence with large reference spaces, RMR CCSD and MR CISD yield similar results. However, when small reference spaces are used, RMR CCSD usually outperforms the corresponding MR CISD.

4. Vibrational Frequencies

We next consider, as a derived property of PESs, the computation of vibrational frequencies, which provide a rather severe test for any approach in view of their sensitivity to the deficiencies of the method employed. For small, "well-behaved" molecules, the average discrepancy between the calculated and experimental frequencies at the CCSD/DZP level of theory is slightly above 2%, i.e., roughly 50–70 cm^{-1}, although for "difficult" systems, errors exceeding 100 cm^{-1} are not unusual. The agreement can be improved either by using larger basis sets or higher level methods. As a rule, it is most beneficial to proceed simultaneously in both directions. For example, the average CCSD(T)/DZP error is about the same as the CCSD/DZP one, making it unnecessary to use a higher level theory. The use of a larger basis set is usually more beneficial. Thus, to halve the error of CCSD/DZP requires both a proper account of connected triples and an *spdf*-type basis. The CCSD(T) method with basis sets containing f functions, such as TZ2Pf, cc-pVTZ, or ANO bases of similar quality, have a reliability of about 1%, i.e., 20–30 cm^{-1}. With even larger basis sets containing g functions, the reliability of CCSD(T) is increased to ~10 cm^{-1}. With such information at one's hand, it is possible to assign the vibrational spectrum of a molecule by relying on the ab initio data. Systematic studies of the performance of the CCSD and CCSD(T) methods in this regard have been reported by Schaefer et al. [385], Bartlett [31], and Lee and Scuseria [30].

Focusing now on the open-shell applications, we have carried out a systematic study at the UGA CCSD/6-31G(d) level of theory, involving 48 distinct electronic states of nine diatomic hydrides and 18 diatomics, including ground- and excited-state doublets, triplets, and open-shell singlets [278]. Some of these states that are regarded as "difficult" rather than "well behaved," are addressed in greater detail subsequently. For diatomic hydrides at this level of theory, we find the mean absolute deviation from the experimental harmonic frequencies to be 55 cm^{-1}. For diatomics, the mean error is slightly larger, at 73 cm^{-1}. For the sake of brevity, we present in Table 9 only some selected examples for open-shell singlet states. All of these states are, in fact, excited states that require a two-determinantal zero-order wave function and thus cannot be described by conventional SR CCSD methods.

Using larger basis sets will generally improve the results. For example, with an ANO [$6s5p3d2f$] basis set for a heavy atom and a [$4s3p2d$] one for hydrogen, the errors in computed frequencies for CH($X^2\Pi$), NH($X^3\Sigma^-$), and OH($X^2\Pi$) are in the range of 13–45 cm^{-1} [44]. For a series of "difficult" diatomics, including NO($X^2\Pi$), $O_2(X^3\Sigma_g^-)$, $O_2^+(X^2\Pi_g)$, OF($X^2\Pi$), and $F_2^+(X^2\Pi_g)$, the discrepancies in harmonic frequencies are nonetheless in the range of 30–115 cm^{-1}, even at the UGA CCSD/TZ2P level of theory [44], with the largest error occurring for $F_2^+(X^2\Pi_g)$. It is worth recalling that the F_2^+ ground state is an especially "difficult" case for which the computed harmonic frequency is too small by 500–600 cm^{-1}, even at the spin-restricted MP4/TZ2P level of theory [396]. On the whole, the performance of UGA CCSD for open shells is on a par with that of the standard CCSD for closed-shell molecules.

The effect of spin contamination on computed frequencies deserves notice. It only pertains to high-spin states, since the UHF-based methods are incapable of handing low-spin cases. In making use of UHF-based MBPT, the spin contamination may cause a substantial error in computed frequencies, depending on how rapidly the contamination builds up as the bond is being stretched [397]. For example, the frequency for the $X^2\Pi$ state of NO at the UMP2/6-31G(d) level is twice as large as the experimental value; that is, it is in error by $\sim$ 2000 cm^{-1}! At the CCSD level of theory, the spin contamination is significantly reduced thanks to the exponential *Ansatz*, since full CC is exact, as is full CI, regardless the MOs employed. (Cf., e.g., [398].) Yet more effective spin annihilation is achieved when the ROHF orbitals are used. Otherwise, even at the CCSD(T) level of theory,

TABLE 9
Harmonic frequencies ω for selected open-shell singlet states[a]

System	State	ω (cm^{-1})	ω^{exp} (cm^{-1})[b]	$\omega - \omega^{\mathrm{exp}}$
BH	$A^1\Pi$	2,191	2,251	−60
	$c'^1\Delta$	2,731	2,610	121
CH$^+$	$A^1\Pi$	1,851	1,865	−14
	$B^1\Delta$	2,169	2,076	93
CO	$A^1\Pi$	1,468	1,518	−50
NO$^-$	$a^1\Delta$	1,514	1,492	22
O_2	$a^1\Delta_g$	1,577	1,484[c]	93

[a] Obtained with UGA-CCSD(is)/6-31G(d) method.
[b] In [388].
[c] $\Delta G(1/2)$ value.

the spin contamination may still be quite significant. This is illustrated in Table 10, which lists both the spin-restricted (R) and -unrestricted (U) CCSD(T) frequencies and their differences. The largest difference, 213 cm^{-1} for NO ($X^2\Pi$), represents more than 11% of the harmonic-frequency value. Clearly, in many cases the spin contamination is not very important when UHF-based CCSD is used, although in certain cases it does have a substantial effect. For the sake of comparison, we also list the UGA CCSD(is) values in the last column of the table [278]. We see that in this case the errors are more systematic, since the spin contamination effects have been eliminated. Note that these latter values are at the CCSD, rather than CCSD(T), level.

We must emphasize that although CCSD(T) is capable of attaining "chemical accuracy" in many important applications, it is not without problems. In particular, when the wave function exhibits a severe multireference character, CCSD(T) may give above-average errors, since it does not properly account for the connected four-body clusters that are usually important in such cases. Another potential problem stems from the fact that CCSD(T) has a tendency to overestimate the effect of triples. Finally, there is no well-defined hierarchy of perturbatively corrected methods that allow further improvement from CCSD(T). Usually, CCSD gives larger frequencies than the experimental ones, and the triple corrections act in the opposite direction, reducing the computed frequencies. Sometimes, however, the triple corrections are excessively large, yielding frequencies that are too small. This usually happens for states with a multireference character. For example, the two-reference character of the ozone molecule makes it difficult to correctly describe the antisymmetric stretch vibration ω_3. With a DZP basis, the CCSD frequency ω_3 is 150 cm^{-1} too large [399], while the CCSD(T) value is 113 cm^{-1} too small [244]. With

TABLE 10
Harmonic frequencies^a ω for "difficult" high spin-states: Comparison of UGA CCSD and CCSD(T) methods

System	$\omega^{exp\,b}$	$\omega^{UCCSD(T)\,c}$	$\omega^{RCCSD(T)\,c}$	Δ^d	$\omega^{UGA-CCSD}$
NO ($X^2\Pi$)	1,904	2,095 (191)	1,882 (−22)	213	1,986 (82)
O$_2$ ($X^3\Sigma_g^-$)	1,580	1,561 (−19)	1,563 (−17)	−2	1,620 (40)
O$_2^+$ ($X^2\Pi_g$)	1,905	1,890 (−15)	1,892 (−13)	−2	2,010 (105)
OF ($X^2\Pi$)	1,053	1,060 (7)	1,014 (−39)	46	1,085 (32)
F$_2^+$ ($X^2\Pi_g$)	1,073	1,071 (−2)	1,238 (165)	−168	1,188 (115)

^{a} All values in cm^{-1}. Theoretical values obtained with $5s4p2d$ basis. Deviations from experiment are given in parentheses.
^{b} In [388].
^{c} In [396].
^{d} $\Delta \equiv \omega^{UCCSD(T)} - \omega^{RCCSD(T)}$.

an ANO $[5s4p3d2f1g]$ basis, the CCSD error for ω_3 is even larger, namely, 191 cm^{-1} [400]. CCSD(T) again overcorrects by 32 cm^{-1} using an ANO $[4s3p2d1f]$ basis and by 36 cm^{-1} with an ANO $[5s4p3d2f]$ basis [400]. Most recent CCSDT/cc-pVTZ and CCSD(T)/cc-pCVTZ results also indicate that even in the CBS limit, CCSDT will overestimate ω_3 by about 50 cm^{-1} [401]. This effect is likely due to the quadruple excitations that are double excitations from the second reference*.

It is thus interesting to find out whether MR CC SS types of approaches perform as well as, or better than, CCSD(T), especially in quasi-degenerate situations. Our preliminary results [381] in this direction are very encouraging. In Table 11, we present harmonic frequencies for several states of N_2^+ and CN obtained with the RMR CCSD method using small reference spaces and cc-pVTZ basis sets. We also include, for the sake of comparison, the CCSD(T) results obtained with an ANO basis set of similar quality for the closed-shell N_2 molecule [386]. In all MR-type CCSD calculations labeled as (4R)-RMR, we employ four references obtained by associating two configurations with each bonding orbital that was not involved in the ionization process (i.e., the two π bond orbitals for $^2\Sigma$ state and the σ and π bond orbitals for $^2\Pi$ state). For the $B^2\Sigma_u$ state of N_2^+, a

TABLE 11

Harmonic frequencies^a ω for some triple-bonded systems: Comparison of RMR CCSD and CCSD(T) methods

Molecule	State	Method	Basis	ω	$\omega^{\exp b}$
N_2 [386]	$X^1\Sigma_g^+$	CCSD(T)	ANO[$4s3p2d1f$]	2,338 (-21)	2,359
		CCSD(T)	ANO[$5s4p3d2f1g$]	2,352 (-7)	2,359
N_2^+	$X^2\Sigma_g^+$	(4R)-RMR-CCSD	cc-pVTZ	2,195 (-12)	2,207
	$A^2\Pi_u$	(4R)-RMR-CCSD	cc-pVTZ	1,918 (14)	1,904
	$B^2\Sigma_u^+$	UGA-CCSD	cc-pVTZ	2,646 (226)	2,420
		(7D)-RMR-CCSD	cc-pVTZ	2,425 (5)	2,420
CN	$X^2\Sigma$	(4R)-RMR-CCSD	cc-pVTZ	2,081 (13)	2,069
	$A^2\Pi$	(4R)-RMR-CCSD	cc-pVTZ	1,839 (26)	1,813

^{a}All values in cm^{-1}. Deviations from experiment are given in parentheses.
^{b} In [388].

---

*We have recently explored the importance of double excitations from the second reference for the prediction of vibrational frequencies of ozone using our RMR CCSD method [X. Li and J. Paldus, *J. Chem. Phys.* **110**, 2844 (1999)]. In fact, it turns out that for an accurate prediction of the asymmetric stretching frequency ω_3, not only the second reference, but also the third reference, corresponding to a single HOMO to LUMO excitation, is important. Although this third reference does not contribute directly at the equilibrium geometry for symmetry reasons, it does contribute for asymmetrically stretched geometries at which HOMO and LUMO have identical symmetry, enabling the correct description of ω_3.

seven-determinant (7D) reference, equivalent to five spin-adapted configurations, was used. Except for the leading reference, the remaining
configurations are generated from the coupling of two sets of three singly
occupied orbitals, (σ, π_x, π_x^*) and (σ, π_y, π_y^*), respectively. The average
absolute error for the five states considered at the RMR-CCSD/cc-pVTZ
level of theory is $14\,\mathrm{cm}^{-1}$. This is as good a performance as that of CCSD(T)
using similar basis sets, in spite of the fact that the $B^2\Sigma_u$ state of N_2^+ is
somewhat more "difficult" than the other states, since the CCSD frequency
is off by more than $200\,\mathrm{cm}^{-1}$. It can be shown [279] that this error is not due
to an inadequacy of the basis set (different basis sets give essentially the
same result), but is due to the multireference character of the state. This is
why the RMR CCSD method is able to provide a satisfactory description.

It is certainly encouraging that even with relatively small reference spaces,
RMR CCSD is capable of yielding highly accurate results in the computation of vibrational spectra. Obviously, a more extensive study is
required to fully assess the method's general performance.

C. Ionization Potentials and Electron Affinities

The computation of accurate ionization potentials (IPs), and especially of
electron affinities (EAs), has proved to be a very demanding task. Although
there is a little difference between the two problems at the fundamental level,
the electron detachment leads to the formation of an anion that requires
more extensive (diffuse) basis sets. From the electron correlation viewpoint,
this calls for a highly correlated, yet balanced, description of both moieties
involved that possess a different number of electrons. The simplest approach
is to employ correlated methods that can handle both closed and open shells
in an unbiased manner. In this regard, CC methods offer the advantages
of being size extensive and rather insensitive to the nature of the MOs
employed. These advantages are important, since one cannot always
generate SCF orbitals for the open shells that are involved in the ionization
or electron attachment processes. An adequate description calls for at least
the CCSD level of theory, at which T_1 provides for the orbital relaxation
while T_2 takes care of the bulk of the correlation effects.

In the most often-encountered case, one moiety is a closed-shell singlet
and the other one a doublet. Here one can employ the standard CC methods
for CS and HS cases, even though in the latter case spin-adapted methods
offer some advantages. More involved are situations in which the parent
moiety is an OS system. In these instances, the ionization or electron
attachment may result in more intricate OS states, such as OSSs. Under such
circumstances, one needs an MR CC type of approach that is capable of
describing two-determinantal OSS states. For example, the ionization of the
$X^2\Pi$ doublet ground state of SH leads to the SH^+ ion in its triplet or OSS

states, of which several low-lying ones are experimentally known [402], including the triplets $X^3\Sigma$ and $A^3\Pi$ and the open-shell singlets $a^1\Delta$ and $B^1\Pi$. In Table 12, the computed IPs at the UGA CCSD level of theory, employing an ANO $[6s5p3d2f; 4s3p2d]$ basis [44], are compared with experiment. The largest error is 0.24 eV (for the $B^1\Pi$ state), the largest percentage error is about 1.9% (for the $X^3\Sigma$ state), and the average absolute error is 0.18 eV.

For states having a single-reference character, a proper account of three-body clusters, either perturbatively or iteratively, can, in general, increase the values of computed IPs or EAs by 0.1–0.2 eV and bring theoretical results closer to the experimental ones. As an illustrative example, Table 13 shows the IPs and EAs obtained by Urban et al. [133] for coinage metals. These results demonstrate not only the performance of CCSD(T)-type approaches, but also that of "partially spin-adapted" ones [132, 133]. The spin-adapted (SA) version of CCSD(T) reduces the error of SA CCSD by about 0.1 eV for IPs and by about 0.15 eV for EAs. An earlier study by Raghavachari [403] also showed that the discrepancies between the CCD + ST(CCD) EAs and experiment for the B, C, O, and F atoms are within 0.05–0.1 eV.

Conceptually, a very attractive way of handling the electron ionization and attachment processes relies on valence-universal (Fock space) MR CC theories, EOM CC methods, or closely related SAC-CI. (Cf. Sections II.D.2.1 and II.F.4.) Within the framework of EOM CC, the IPs or EAs are obtained directly by diagonalizing the similarity-transformed Hamiltonian

TABLE 12
Ionization potentials[a] for several SH → SH$^+$ transitions

Method	$X^3\Sigma$	$a^1\Delta$	$A^3\Pi$	$B^1\Pi$
UGA CCSD[b]	10.22	11.48	14.21	15.45
Experiment[c]	10.42	11.65	14.11	15.69

[a] All values in eV.
[b] Using ANO basis.
[c] In [402].

TABLE 13
Ionization potentials[a] and electron affinities[a] for Cu, Ag, and Au [133]

Method	Cu		Ag		Au	
	IP	EA	IP	EA	IP	EA
SA-CCSD	7.621	1.033	7.394	1.135	9.029	2.051
SA-CCSD(T)	7.770	1.196	7.492	1.255	9.118	2.201
Experiment	7.735	1.226	7.575	1.303	9.225	2.309

[a] All values in eV.

$e^{-T}He^{T}$ in a basis of determinants containing $N \pm 1$ electrons. Many applications [404, 405], including a larger system such as porphin [123], have yielded interesting and encouraging results. The EOM CCSD results for the IPs of porphin were shown [123] to be in qualitative agreement with the related SAC-CI. The agreement of the EOM CCSD results with the known experimental data are somewhat better. The theoretical results are about 0.3–0.7 eV smaller than the experimental values. The errors are partially due to the small basis set used. For the sake of brevity, we refer to the original references [122, 404, 405].

D. Electronic Excitation Energies and Excited-State PESs

Like the breaking of chemical bonds, electron excitation can create a variety of OS cases, often of a multireference type, providing a considerable challenge for CC theories. The CC methods that are designed to treat excited states are of two basic varieties. In the first class of approaches, the excited states are treated on an equal footing with the ground state, allowing for an entirely different nature of the electron correlation in both states, as is the case, for example, in the so-called collective-excitation processes. Such a treatment calls for an exponential *Ansatz* of one kind or another and the related choice of the reference or references that are capable of an accurate description of both dynamical and nondynamical correlation effects. The MR CC methods of either the SU or SS type fall into this category. Unfortunately, these approaches are very difficult to exploit in actual applications, and their implementations are presently limited to only a few special cases. Typical examples are OSS and HS excited states. The approaches are particularly useful in generating the entire PESs in a state-selective manner (i.e., independently of the ground-state PES). The excitation energy is then evaluated as the difference of the ground- and excited-state energies. For each state considered, the computational cost is at least as large as that for the CS ground state at the CCSD level.

In the second class of approaches, the exponential CC *Ansatz* is employed only for the ground state, while the excited-state wave function is generated via a linear, CI-type excitation operator acting on the ground-state CC wave function. Early approaches [105] to the electron excitation and ionization processes, as well as more recent EOM CCSD and LR CCSD techniques or the closely related SAC-CI approaches, are of this type. (Cf. Section II.F.4.) An implicit assumption here is that the dynamical correlation effects are very similar in the ground and the excited states, so that the exponential *Ansatz*, truncated at the pair cluster level, can account for the bulk of the dynamical correlation in all states concerned. From a technical viewpoint, these methods are much simpler than those of the first type, since the exponential wave operator—which is at the heart of the nonlinear nature of

CC approaches—is associated only with the ground state, while the additional new linear *Ansatz* for the excited state can be treated as in the CI problem. The methods are most useful for the direct calculations of vertical excitation energies, which may be viewed as a response property of the ground state, but are less suitable for the computation of the entire PESs.

In Table 14, we reproduce vertical excitation energies of ethylene, computed with the EOM CCSD method by Watts et al. [406], and compare them with CASPT2 results, as well as with experiment. In most cases, the discrepancies with experiment are less than 0.2 eV. It was shown [406] that for larger systems, some excited states may require perturbative triple corrections. On the whole, a very satisfactory agreement was found between the CASPT2 and EOM CCSD(T) results [406]. It is important to note that, as indicated by the average excitation level (AEL), all these excited states are one-electron (1e) excited states. This is an essential requirement for good performance of EOM CCSD (or LR CCSD), since the dynamical correlation effects are similar in the ground and 1e excited states, so that the basic assumption underlying the second class of methods holds.

Many other systems have been studied with EOM/LR-CCSD by Bartlett and coworkers[117, 118, 406–408], Koch et al. [409, 410], and Head-Gordon and coworkers [314], as well as by Nakatsuji et al. who used SAC/SAC-CI [106, 107]. In general, for singly excited states, a reliability of ~ 0.2 eV is feasible. However, for double (2e) or higher excitation-level excited states, the aforementioned basic assumption is no longer valid, and the error produced by EOM/LR-CCSD can be substantial. To overcome this problem, higher-than-pair clusters must be included. It is often found, however, that for 2e excited states, three-body clusters do not suffice.

TABLE 14
Vertical excitation energies ^{a} ΔE of ethylene [406]

State	AEL^b	$\Delta E^{\text{EOM-CCSD}}$	ΔE^{CASPT2}	ΔE^{exp}
$1^1 B_{3u}$	1.058	7.28	7.17	7.11
$1^1 B_{1g}$	1.055	7.94	7.85	7.80
$1^1 B_{2g}$	1.055	7.99	7.95	7.90
$1^1 B_{1u}$	1.049	7.98	8.40	8.0
$2^1 A_g$	1.055	8.45	8.40	8.28
$2^1 B_{3u}$	1.055	8.79	8.66	8.62
$1^1 A_u$	1.055	9.02	8.94	
$3^1 B_{3u}$	1.055	9.08	9.03	8.90
$1^1 B_{2u}$	1.054	9.26	9.18	9.05
$2^1 B_{1u}$	1.052	9.30	9.31	9.33

^{a} All values in eV. Theoretical values obtained with a large basis set.
^{b} Average excitation level.

To demonstrate the latter point, we consider a benchmark system CH^+. Selected results are given in Table 15; the exact FCI excitation energies [411] and AEL values are listed first, while, for all other methods, the deviations from the FCI values are presented. The selected methods include EOM/LR-CCSD, as well as various versions accounting for triples—that is, EOM-CCSD(T), CC3, and CCSDT-3 [408–410]. Since most of the states are OSS states (i.e., with two singly occupied orbitals coupled to a singlet), they can also be handled with UGA-based CCSD. These results are given in the last column of the table. As already mentioned, UGA CCSD belongs to the first class of approaches in that it uses the OSS reference as a vacuum and employs the full exponential-cluster *Ansatz* for the wave operator in order to describe the excited state.

The excitation level indicates that the excited states of CH^+ are of the 1e or 2e type. The EOM/LR-CCSD deviations from FCI are an order of magnitude larger for 2e states (such as $B^1\Delta$) than for 1e states. When triple excitations are taken into account, the large errors are greatly reduced, but still remain significantly larger than those for 1e states. At the EOM/LR-CCSD level of theory, the largest error (0.924 eV) is found for the lowest $^1\Delta$ state, which is a 2e state. Likewise, further large errors are found for other 2e states. Obviously, these 2e states are difficult to describe by the EOM/LR-CC theory using the ground-state CCSD reference, since they entail quite different dynamical correlation effects than the ground state does. The character of state-selective UGA CCSD is different in this respect: It is irrelevant whether the state considered has a 1e or 2e nature, since the method will perform well as long as the state is of a simple OSS type. In general, this method gives good results for the lowest state of a given symmetry, while for higher states of the same symmetry (which would

TABLE 15
Vertical excitation energies[a] ΔE of CH^+

State	AEL[b]	ΔE^{FCI} [411]	$\Delta\Delta E^c$ for EOM- or LR-type method				$\Delta\Delta E^c$
			CCSD [408, 409]	CCSD(T) [408]	CC3 [410]	CCSDT-3 [408]	UGA CCSD [381]
$A^1\Pi$	1.03	3.230	0.031	0.016	0.012	0.014	−0.020
$B^1\Delta$	2.00	6.964	0.924	0.272	0.318	0.315	0.001
$C^1\Sigma^+$	1.96	8.549	0.560	0.200	0.230	0.233	
$D^1\Sigma^+$	1.06	13.525	0.055	0.031	0.016	0.022	0.033
$E^1\Pi$	1.24	14.127	0.327	0.256	0.219	0.226	0.220
$F^1\Delta$	1.99	16.833	0.856	0.239	0.261	0.272	
$G^1\Sigma^+$	1.13	17.217	0.098	0.045	0.026	0.035	0.367

[a] All values in eV.

[b] Average excitation level.

[c] Deviations from FCI: $\Delta\Delta E \equiv \Delta E - \Delta E^{FCI}$.

require a genuine MR version), larger errors may occur. For example, for the lowest lying 2e state $B^1\Delta$, for which the EOM/LR-CCSD error is 0.924eV, the UGA CCSD result is excellent, deviating from FCI by only 0.001 eV.

Another interesting aspect of the degenerate $^1\Delta$ state is its rich multi-reference character, which allows for two distinct treatments exploiting either of its two components, $^1\Delta_{xy}$ and $^1\Delta_{x^2-y^2}$. The $^1\Delta_{xy}$ component represents an OSS state that is dominated by two determinants, $|(\text{core})\pi_x\pi_y(\alpha\beta - \beta\alpha)|$. The UGA CCSD result given in Table 15 is based on this component. The other component, $^1\Delta_{x^2-y^2}$, is of a two-CS-type, being dominated by the two CS-type configurations, $|(\text{core}) (\pi_x\pi_x - \pi_y\pi_y)\alpha\beta|$. Thus, the $B^1\Delta$ state can also be handled using a two-reference RMR CCSD method. This yields an excitation energy of 6.981 eV for the $B^1\Delta_{x^2-y^2}$ component—i.e., an error of 0.017eV. The minor difference between both values (0.001 vs. 0.017 eV), obtained with $^1\Delta_{xy}$ and $^1\Delta_{x^2-y^2}$ components, is due to a different excited-state manifold used in each case and is a good indication of the consistency of both types of rather diverse approaches.

It must be emphasized that the state-selective MR CC methods, such as UGA CCSD, are designed to handle only the lowest state of a given symmetry, although they often provide reasonable results even for higher states. This can be understood from the viewpoint of the MR CC theory. For example, to handle an OSS state that has the same symmetry as the CS ground state, a rigorous treatment would require that we use MR CC theory, such as the SU CC theory employing two references—Φ_0 for the CS ground state and Φ_1 for the OSS excited state, with Φ_1 being a singly excited configuration from Φ_0. Since these states are assumed to have the same symmetry, they will couple with one another. However, the coupling should be small. In the SU CC theory, it is described by the product of the effective Hamiltonian and the so-called coupling term, $\langle G_I(0)\Phi_0|e^{-T(0)}e^{T(1)}|\Phi_1\rangle$. [Cf. Eq. (2.133).] The leading term in the effective Hamiltonian is the CI matrix element $H_{10} = \langle\Phi_1|H|\Phi_0\rangle$ between Φ_0 and Φ_1, which vanishes in view of the Brillouin theorem. The largest contribution to the coupling term should originate from the pair clusters and is given by $\langle\Phi_{0,I}^{(2)}|e^{-T_2(0)}e^{T_2(1)}|\Phi_1\rangle$, where $\Phi_{0,I}^{(2)}$ is the Ith doubly excited configuration relative to Φ_0. These terms again vanish when Φ_1 is a single (or odd) excitation from Φ_0. Consequently, the two-reference MR CC problem is effectively diagonalized, so that the excited state can be handled independently of the ground state by using state-selective UGA-based OSS CCSD, even though the excited state is not the lowest state of a given symmetry. Some states in Table 15 are of this type. Of course, when T_1 is important, the coupling terms are no longer small, and the genuine version of MR CC theory must be applied.

As a by-product of UGA CCSD, we formulated UGA-based second- and third-order perturbation theories [297]. (Cf. Section II.E.1.) Their major attraction is not only their ability to handle rather large systems, but also the low-spin OSS states, for which the UHF-based PT cannot be used. For the sake of brevity, we refer to these theories simply as MPn theories. Their performance is again best assessed by applying them to benchmark systems for which we can generate the exact FCI results [297]. For example, considering vertical excitation energies of the water molecule at a DZ level, the absolute error for both MP2 and MP3 is less than 0.1 eV [297]. The best results are, of course, obtained with UGA CCSD, in which case the errors are less than 0.02 eV. In Table 16, the MP2, MP3, and UGA CCSD excitation energies of H_2O, obtained with various basis sets, are compared with experiment. The CCSD energies are within 0.2 eV of the experimental results and are bracketed by the MP2 and MP3 excitation energies. The difference between the CCSD and MP3 values never exceeds 0.05 eV.

These results also nicely illustrate the importance of an appropriate choice of a basis set. For example, a DZP basis and an ANO $[4s2p1d; 2s1p]$ basis involve the same number of AOs, yet the computed excitation energies differ by more than 1 eV. In contrast, within the ANO basis-set sequence, the behavior of the computed excitation energies is relatively stable. Even when going from the largest $[4s3p2d1f; 3s2p1d]$ to the smallest $[4s2p; 2s]$ ANO basis set, the differences do not exceed 0.1 eV, although the total energies of the states involved change significantly. In fact, already the smallest $[4s2p; 2s]$ ANO basis, which contains no polarization functions, gives rather good agreement with experiment for the lowest excited state. This suggests that, in dealing with large systems, where we have to rely on relatively small basis sets, it is advantageous to use an ANO, even if severely truncated, basis set.

TABLE 16
Vertical excitation energies[a] ΔE of H_2O

Basis[b]	$^1B_1$			$^1A_2$			$^1B_2$		
	MP2	MP3	CCSD	MP2	MP3	CCSD	MP2	MP3	CCSD
DZP	9.14	9.05	9.10	11.15	11.07	11.11	14.10	14.02	14.04
[42;2]	7.74	7.55	7.62	10.10	9.93	10.01	12.61	12.45	12.52
[421;21]	7.84	7.66	7.70	10.23	10.07	10.11	12.89	12.74	12.77
[432;32]	7.64	7.41	7.45	9.43	9.18	9.21	11.80	11.57	11.60
[4321;321]	7.73	7.51	7.54	9.54	9.28	9.30	11.86	11.62	11.64
Experiment	7.49 [413], 7.4 [412]			9.1 [412]					

[a] All values in eV. Theoretical results obtained with UGA MPn and UGA CCSD(is) methods [297].

[b] Numbers in brackets specify the ANO basis employed (e.g., [42;2] implies $4s2p$ for O and $2s$ for H).

As has been alluded to, an important asset of UGA CCSD is its capability to selectively generate PESs for the states of interest, particularly OSS states. As an illustrative example, the PEC for the $a^1\Delta$ state of the OH$^+$ ion [44] is shown in Figure 13. Both a moderate DZP basis set, as well as a large $[6s5p3d2f;4s3p2d]$ ANO basis set were employed. The internuclear separations that were considered ranged from $\sim 0.6R_e$ to $\sim 2R_e$, in addition to a very large separation of $\sim 5R_e$. The ability of UGA CCSD to generate high-quality PESs is born out by a comparison of the UGA CCSD and FCI potentials obtained with a DZP basis. The deviation between the two potentials ranges from 1.1 to 2.8 mhartree, the maximum deviation appearing around $R = 3.5$ bohr. In terms of the "non-parallelism error," the NPE of the UGA CCSD potential is 1.7 mhartree for geometries ranging from $0.6R_e$ to $2R_e$ (or, in fact, to $5R_e$). Except for being shifted to lower energies, the potential computed with a large ANO basis shows very much the same behavior as the DZP one. From this potential, we derive the equilibrium bond distance $R_e = 1.0242$ Å and $R_0 = 1.035$ Å, the harmonic frequency $\omega_e = 3182$ cm^{-1}, and the dipole moment $\mu = 2.291$ D. Experimentally, only the effective internuclear separation in the lowest vibrational state is known and has a value of $R_0 = 1.043$ Å. The computed R_e should be

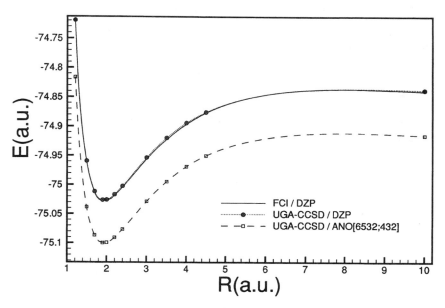

Figure 13. Potential-energy curves for the $a^1\Delta$ excited state of the OH$^+$ ion obtained with UGA CCSD(is) and FCI methods and a DZP basis, as well as with a UGA CCSD(is) method and ANO basis.

reliable to within 0.005 Å and should be likely shorter than the experimental R_e [44]. The experimental ω_e and μ are unknown, and the computed results should represent a reasonable prediction.

Finally, we must mention the possibilities offered by the Hilbert space or SU CCSD method. Only a few applications have been carried out so far, and a simple example is given in Table 17 for the energy difference between the first and second $^1A_1$ states of methylene, as obtained with two-CS-reference SU CCSD at a DZP level [93]. (Note that these are not excitation energies, since the ground state is a triplet.) Compared with FCI, the energy differences are well reproduced at the equilibrium geometry and worsen when the C–H bonds are symmetrically stretched. Note that the underlying assumption for a satisfactory performance of the SU CCSD is that, for a chosen M-dimensional model space (in our example, the dimension is 2), the zero-order wave function of the M low-lying states is well described by these M configurations. In general—especially for larger model spaces—this condition is too strong and is thus hard to be satisfied. Usually, the model space is good enough for the ground state, but not for all the excited states, which may require a yet larger or different space. This is essentially the source of the intruder state problem.

In our example, the two CS-type configurations do indeed provide a very good zero-order description for both the $1\,^1A_1$ and $2\,^1A_1$ states at the equilibrium geometry. However, the situation becomes less ideal when the bonds are stretched. For this reason, the SU CCSD theory in its present form is likely to be most useful for genuinely degenerate states, such as the states originating from a partially filled set of degenerate orbitals. Typical examples include excited states of diatomics from partially filled π and δ shells and states arising from the configuration $e^m t^n$ that are found in ligand field theory. An application of SU CCSD to excited states of the N_2 molecule was carried out by Berkovic and Kaldor [414]. Although the computed excitation energies were generally quite good, they showed that the spin-orbital formulation preserves neither spin nor spatial symmetry, causing a spin or spatial degeneracy breaking (i.e., different components of

TABLE 17
$1\,^1A_1$ and $2\,^1A_1$ energy differences[a] for methylene at three geometries [93]

Method[b]	R_e	$1.5R_e$	$2R_e$
SU CCSD	4.62	4.66	2.86
CISDTQ	4.64	4.56	2.67
FCI	4.60	4.45	2.55

[a] All values in eV.
[b] Using a DZP basis set.

the same multiplet having different energies). This is another example of the importance of using a properly spin- and/or symmetry-adapted formalism.

In short, CC theories for the description of excited states are, in general, lacking. Methods belonging to the second class of approaches, such as EOM-CC and LR-CCSD, are technically simpler, but their applicability is limited to singly excited states. Methods belonging to the first class of approaches, based on some kind of MR CC formulation, are much more accurate, but technically more difficult to execute. Currently, they can handle only a few special types of states. The most satisfactory approaches that are capable of producing PESs or PECs for a number of low-lying excited states, as well as for Rydberg states, are undoubtedly based on an MR CI formalism.

E.   Static Properties (Dipole Moments, etc.)

An accurate prediction of molecular properties—particularly higher order ones—represents a challenging problem that requires a suitable choice of a basis set and a proper account of correlation effects. The choise of a basis set is very important, since prevaling basis sets have been optimized for energy calculations and may not be suitable for predicting properties. For static properties, such as dipole moments, correlation effects may account for 10–15% of the value obtained.

In the 1970s, most quantum chemical computations were directed towards molecular energetics. By contrast, the past decade has witnessed a tremendous rise in interest in both static and dynamic response properties. The advantages and limitations of several techniques employed in both variational and nonvariational approaches—the latter including CC methods—were outlined in Section II.F. It is beyond the scope of this article to address these developments in full; rather, they deserve a specialized treatment. (See, e.g., the exhaustive review of hyperpolarizabilities by Shelton and Rice [415].) We shall concentrate here on some new possibilities in computation of response properties, offered by UGA and LR CCSD. At the same time, we wish to emphasize the importance of generating property functions that enable one to assess the rovibrational effects, as well as the computation of various transition moments that are often hardly accessible to experiment, rather than concentrating only on equilibrium properties.

The reliability of UGA CCSD in computing static properties was first confirmed for benchmark systems. For example, with the use of a DZP model, the dipole moment of the open-shell triplet $^3B_1$ and singlet $^1B_1$ states of CH_2 were computed to be 0.668 D and 0.925 D [44], respectively, the corresponding FCI values being 0.671 and 0.925 D. The computed polarizabilities are identical to the FCI ones [44].

For accurate computations that can be compared with experiment, it is important to use AO basis sets that include a sufficient number of diffuse functions, rather than high-angular-momentum AOs. Such a property-oriented basis that usually gives good results is the POL1 basis set [357]. The standard and UGA CCSD dipole moments for selected CS and OS systems obtained with this basis set are compared with experiment in Table 18. Recalling that the errors at the SCF level of theory for these systems are about 10–15%, we find an order-of-magnitude improvement which we use UGA CCSD. Except for NH, the performance of the standard CS CCSD and that of the OS UGA CCSD methods are comparable in terms of the resulting percentage error. The NH radicals being out of line with other examples illustrates how ab initio results can help in identifying insufficiently accurate experimental data. Indeed, all available ab initio results, including those of SR CI, MR CI, CEPA, and MR PT [227], give values exceeding 1.5 D. Rovibrational averaging could decrease the computed purely electronic dipole moment by $\sim$0.03 D, but that would still leave a discrepancy with experiment. We thus believe that in this case the discrepancy is due to the experiment.

The UGA CCSD method was also employed to predict static properties of the excited state [277]. This is illustrated in Table 19, listing the dipole moment of several excited states of NH, including triplet and open-shell singlet states. A comparison is also made with the corresponding CISD values. For the $a^1\Delta$ state, both CISD and CCSD results are within the experimental bounds. For the $A^3\Pi$ and $c^1\Pi$ states, neither CISD nor CCSD values are within the experimental bounds, although the CCSD results are considerably closer to them by 0.055 and 0.044 D, respectively.

Polarizabilities and hyperpolarizabilities are often crucially dependent on dynamical correlation effects. In this regard, CC theories, which are highly effective in handling dynamical correlation, are the method of choice. For

TABLE 18
Dipole moments ^{a} μ for selected CS and OS molecules

Molecule (State)	μ^b	μ^{exp}	$\mu - \mu^{\mathrm{exp}}$	Error (%)
H_2O (X^1A_1) [28]	1.852	1.833	0.019	1.1%
OH ($X^2\Pi$) [277]	1.631	1.655	-0.024	-1.4%
NH_3 (X^1A_1) [28]	1.519	1.469	0.050	3.4%
NH ($X^3\Sigma^-$) [277]	1.497	1.38 ± 0.07	0.117 ± 0.07	3.4%–13.6%
H_2S (X^1A_1) [28]	0.997	1.019	-0.022	-2.1%
SH ($X^2\Pi$) [277]	0.774	0.758	0.016	2.1%

^{a} All values in debye.

^{b} Standard CCSD for CS molecules and UGA CCSD(is) for open-shell systems, both using POL1 basis [357].

TABLE 19
Dipole moments[a] μ for excited states of NH

State	$\mu^{\text{CISD } b}$	$\mu^{\text{CCSD } b}$	μ^{exp}
$a^1\Delta$	1.499	1.473	1.49 ± 0.06
$A^3\Pi$	1.161	1.216	1.31 ± 0.03
$c^1\Pi$	1.544	1.588	1.70 ± 0.07

[a] All values in debye.
[b] Obtained with POL1 basis [277] and CISD or UGA CCSD(is) method.

example, for the three-electron Li atom, the polarizability obtained with a UGA CCSD and ANO/$spdf$ basis is 165.1 a.u., which agrees well with the experimental value of 164±3 a.u. [277] and is identical to the FCI value for the same basis. In Table 20, we present the hyperpolarizabilities for the HF molecule obtained by Kondo et al. [321] with a POL1 basis and by Hättig et al. [121] with a considerably larger t-aug-cc-pVTZ basis. Both results were obtained using response theory, the linear version in the former case and the quadratic response (QR) in the latter. The results show that both independent implementations agree fairly well with one another and experiment. It would be interesting to find out to what extent the superiority of the latter result is due to the larger basis set used and the quadratic-vs.-linear nature of the response theory. In general, correlation effects are more important for hyperpolarizabilities than for dipole moments. At the SCF level, the error in hyperpolarizabilities is almost twice as large as that for dipole moments.

As with energy, static properties depend on the nuclear geometry, and the corresponding property surfaces are important in calculations of various spectral properties. For example, the quadrupole moment functions of the HF and N_2 molecules were recently studied via linear-response CCSD theory [250, 252]. (For the dipole moment function and radiative transition probabilities of HF, see [416].) These functions were subsequently used to

TABLE 20
Dipole moments[a] μ_z, polarizability[a] α_{zz}, and hyperpolarizability[a] β_{zzz} for the HF molecule

Method	$\mu_z{}^b$	$\alpha_{zz}{}^b$	β_{zzz}
SCF/POL1 [321]	0.8116 (14.8)	4.261 (−33.4)	−14.6
SCF/t-aug-cc-pVTZ [121]	0.756 (6.9)	5.758 (−10.0)	−8.399
LR-CCSD/POL1 [321]	0.6995 (1.1)	6.521 (1.9)	−9.869
LR/QR-CCSD/t-aug-cc-pVTZ [121]	0.706 (−0.1)	6.410 (0.2)	−9.841
Experiment	0.707	6.40	

[a] All values in a.u.
[b] Percentage errors are enclosed in parentheses.

compute effective rovibrational averages and transition moments. The reliability of the approximations involved (basis set, level of CC theory, etc.) was tested by simultaneously computing the PECs and the actual vibrational levels. An excellent agreement for vibrational energies up to the 12th vibrational level for HF and the 6th for N_2 was found. However, while the computed quadrupole moments and vibrational transition quadrupole moments crucially depend on the property surface employed, the dependence on the PEC used for the rovibrational averaging is minimal [250, 252].

Due to very limited experimental data for quadrupole moments, comparison with experiment is possible only for the rotationless values. For the HF molecule, the experimental rotationless ($\nu = 0$) value is 1.75 ± 0.02 a.u., while the CCSD with Sadlej's basis [357] rovibrationaly averaged value is 1.773 a.u. (The purely electronic value is 1.716 a.u.) Thus, rovibrational averaging accounts for about 3.2% (0.057 a.u.) of the experimental value, even in the zero-point state ($\nu = J = 0$). For the N_2 molecule, the experimental rotationless value is -1.09 ± 0.06 a.u., which again compares well with the CCSD value of -1.156 a.u. Since the purely electronic value is 1.169 a.u., rovibrational averaging is in this instance responsible for a 1.1% (0.013 a.u.) increase, again improving the agreement with experiment. Note that in this case the rovibrational averaging lowers the purely electronic value, while for HF it works in the opposite direction, each time, however, improving the agreement with experiment. (For a similar study of ro-inversional effects in NH_3, see [417].) This good agreement gives one confidence that the wealth of other ab initio results [250, 252] for higher vibrational levels, quadrupole transition moments, etc., that rely on the same property functions represents reliable predictions that should prove useful in explaining future experimental results.

F. Dynamic Properties

Correlated calculations of time- or frequency-dependent dynamic properties of molecules are still in their infancy. Recent developments include CC calculations based on the response theory [121, 418] and the EOM methods [419, 420]. These CC calculations agree reasonably well with available experiments.

For dynamic properties, the role and relative significance of the electron correlation, orbital relaxation, and basis-set effects are not yet well understood. Thus, although CC methods account for dynamical correlation satisfactorily, time-dependent HF (TDHF) may be more effective in accounting for orbital relaxation effects. When the relaxation effects are more important, TDHF seems to give very good results. For example, Rozyczko et al. [420] recently found that, for the CS_2 molecule, the orbital relaxation is of such importance that, for static polarizabilities, TDHF gives better results than EOM CCSD. In this case, the orbital relaxation can be taken into account by employing finite-field CCSD. For the dispersion coefficients, the correlated results at the CCSD level of theory are usually significantly better than the TDHF ones. In general, however, more work and experience are needed to establish a reliable strategy for the exploitation of CC theories in computations of various dynamic properties.

IV. CONCLUSIONS AND FUTURE PROSPECTS

The methodological developments, computer implementations, and practical applications of various versions of CC theory that have taken place since its inception in the 1960s leave little doubt about the fruitfulness of the exponential cluster *Ansatz* for the wave function and the formalism that ensued. The standard CCSD and CCSD(T) methods, in either spin-orbital or spin-adapted form, are presently available in most quantum chemical software packages and are routinely used in diverse applications in which a highly correlated approach is called for and can be carried out. Together with MR CISD, these methods represent the most often-used high-level correlated methods. The advantage of CCSD/CCSD(T) is that it performs nearly as well as MR CISD for many problems, while being much more affordable computationally. Moreover, single-reference approaches are easy to apply, while MR CISD requires the choice of a suitable reference space that is neither unambiguous nor straightforward. Of course, the requirement of a large AO basis set, if one strives for a "chemical accuracy" of ~ 1 kcal/mol, limits the range of possible applications to relatively small systems. However, the basis-set problem is common to all correlated methods, and the result of any quantum chemical calculation is only as good as is the basis set defining the ab initio model employed. Nonetheless, already with basis sets of DZP quality, one can obtain results that can withstand comparison with experiment.

The new methodological and algorithmic developments, as well as the ever-increasing computational power and its affordability, will undoubtedly enable future applications to consider larger molecules or medium-sized ones with large basis sets. Nevertheless, we must keep in mind that, even at the experimental level, one asks very different questions when investigating small and large systems. Thus, for simple atomic systems, we can make highly precise measurements and theoretical predictions that enable us to test our knowledge of physics at the very fundamental (QED) level. For diatomics, we can study molecular force fields and electronic structure in great detail by probing the rotational and vibrational fine structure with modern sophisticated spectroscopic techniques, and, at least for two- and three-electron systems, we can verify—and even predict—the results by precise computations going beyond the Born–Oppenheimer approximation and considering relativistic effects. Even for molecules with three to six first-row atoms, one can still sometimes unscramble the details of the rotational fine structure of either vibrational or electronic spectra and make very precise calculations. However, we have to forego the high precision and finer details when we explore larger systems at both experimental and theoretical levels. Even if we could precisely determine all the anharmonicities, centrifugal distortion constants, transition moments, etc., for, say, a 10-atom molecule, the plenitude of such data would be so overwhelming as to render most of the data useless. Thus, for large systems, even qualitative results—say, the ordering of the excited states or approximate geometries of PES crossings—can be (and are) a great help in the interpretation of various experimental results.

Although greater computational power and new algorithmic developments (cf., e.g., [421]) are of much help, the nature of the problem is such that the basis-set bottleneck demands an exploration of new possibilities offered by localized orbitals and more efficient representations for virtual orbitals (e.g., along the lines suggested by Adamowicz and Bartlett [422] concerning optimized virtual orbitals), as well as a better understanding of the mathematical structure of general open-shell wave functions. Of course, the use of localized orbitals is essential in applications involving large systems. (See, e.g., [194].) For very interesting recent applications of CC theory to polymers using localized orbitals, see the work of the Erlangen group [423] and references therein.

Nonvariational methods conceal more pitfalls than do variational ones, so that users of CC methods must be aware of the dangers of a black-box approach and should equip themselves with a solid understanding of the theory. Perhaps the most serious shortcoming of the standard CCSD—and especially of CCSD(T)—is its breakdown in quasi-degenerate, or open-shell, situations. Thus, while the latter method can render excellent binding energies or heats of formation thanks to its size extensivity, it will fail for the intermediate internuclear separations, being unable to generate realistic PESs. Especially if one wishes to generate PESs for a number of low-lying excited states, CC methods can hardly compete with MR CI approaches, except those like the EOM-CC, STEOM-CC or SAC-CI methods, that use the linear CI *Ansatz* combined with the exponential CC *Ansatz* for the ground state. However, even these methods (cf. Section II.F.4) are limited to states that do not show a collective behavior. Likewise, the treatment of general OS systems is far from routine, even in the vicinity of the equilibrium geometry, and is limited to HS cases when one relies on the UHF reference.

In this regard, we can speak of a certain "convergence" of CC and CI methodologies. This "cross-fertilization" seems not only natural for high-level theories, since both approaches become equivalent in the FCC or FCI limit, but also very desirable in view of their complementarity in handling dynamical and nondynamical correlation. Thus, the computation of vertical excitation energies using a CIS-like procedure, but relying on the CCSD ground-state description, as practiced in EOM-based methods, provides very good results in many cases. On the other hand, the RMR CCSD approach exploits the information provided by MR CISD to account for nondynamical correlation in quasi-degenerate situations, while relying on essentially the SR CCSD method. We expect that this "confluence" of CI and CC methodologies holds great potential for future developments.

It is reassuring that nowadays the importance of a proper spin adaptation is enjoying more and more appreciation [52, 131–134, 139], even though the effort to develop such a formalism is still regarded as a quixotic pursuit by more pragmatic practitioners. At this time, the only fully spin-adapted approach to OS systems that has been computationally implemented and tested is the UGA-based CCSD method [44, 45]. Even this method, however, is presently restricted to handling only high-spin cases and low-spin open-shell singlets that require a single spin-free reference. Existing codes cannot presently handle those low-spin cases which involve more than two singly occupied orbitals—e.g., singly excited states of simple radicals. States of this type require a genuine MR formalism. Such an extension is formally straightforward, but it has yet to be worked out in detail and implemented.

Although not fully spin adapted, a number of algorithms have been developed that are at least partially spin adapted [131–134, 139]. The most recent development [52] in this direction that strives for full spin adaptation is also essentially based on the UGA formalism, even though it employs spin-orbital cluster amplitudes rather than spin-adapted ones, as does UGA CCSD. The authors of these algorithms express their doubt that UGA CCSD "will lead to a widely applicable computational scheme," while ignoring the fact that the method already has been successfully exploited in a number of highly nontrivial applications [44, 45, 49, 50, 200, 206, 232, 238–240, 272, 274–279, 297], all carried out on a very modest workstation, as this review also testifies. In any case, theirs is a laudable development and we look forward to its implementation and test results.

Concerning the automation of algorithms to derive and implement a given formalism, first carried out in the context of CC approaches by Janssen and Schaefer [131], we strongly believe it to be the technique of the future. More and more powerful symbolic computation software is constantly being developed, eliminating the need for the design of special-purpose codes as practiced so far. Automated derivation and encoding of formulas is error free and, most importantly, can handle various distinct cases that otherwise must be handled individually (doublets, triplets, OSS, etc.) It also enables one to quickly test various approximations and program modifications. The fact that the derivation and implementation of UGA CC explicit formulas have been automated certainly reflects the complexity of the formalism, but in no way implies its inefficiency. The only difference from a simple CS case is that one has to distinguish many more cases. Thus, instead of computing 1,000 expressions of a single type, the algorithm handles 10 times 100 expressions of slightly different types. We believe that these developments will become standard in the not-too-distant future, too.

The MR CC methods are still too complex to be efficiently implemented at this time, and any further developments that will facilitate their general-purpose use are most welcome. The recently proposed RMR CCSD [50] seems to be the most promising avenue, thanks to its conceptual simplicity, ease of implementation, and flexibility in the use of various (even incomplete) reference spaces. An extension of the current version of the one-state RMR CCSD to multiple-state RMR CCSD should be explored. We also would like to emphasize other possibilities offered by externally corrected CCSD methods. The original idea of using the UHF solution for this purpose, which has its limitations when it is applied to CS systems (cf. Section II.D.3.3), might actually prove to be useful in OS cases, where the UHF wave function will provide both T_3 and T_4 clusters. In fact, the possibilities offered by solutions other than DODS-type UHF would be worth investigating, although there are certainly cases (e.g., the dissociation of O_2) that cannot be described by any single-reference UHF (or GHF)-type wave function [199].

The great advantage of CI approaches is their ability to simultaneously provide information about a multitude of low-lying states. This is, of course, a consequence of the linear nature of this approach and the availability of efficient algorithms that are capable of extracting a number of eigenstates and corresponding eigenvalues even from matrices of enormous dimension ($\sim 10^9$). Now, in the SR CCSD formalism, which is highly nonlinear, we invariably look only for the lowest-lying state. However, the nonlinear algebraic equations possess, in principle, a number of solutions. In fact, their number, as claimed earlier [424] on the basis of the Bezout theorem, was estimated to be very large. On the other hand, the CC equations represent another formulation of CI equations and are equivalent to them (or to the original Schrödinger equation) in the FCI or FCC limit. Although a solution of CCSD equations other than the lowest one may occasionally be obtained [425] (particularly in the MR case [91, 92, 262]), there has been no systematic attempt to examine these "other" solutions. Of course, from the viewpoint of the effective Hamiltonian formalism, such solutions can be "physical" only if the reference configuration $|\Phi_0\rangle$ has significant weight in the resulting wave function. However, this is precisely the case that characterizes quasi-degenerate situations.

The main reason that solutions of CC equations other than the lowest energy one have not been explored is the well-known lack of algorithms capable of computing all the solutions of nonlinear algebraic systems. However, thanks to recent developments in the new field of *computer algebras* and in *computational algebraic geometry*, even this age-old problem is dramatically changing. The solutions of a system of polynomial equations form a geometric object called a *variety*. The corresponding algebraic concept is that of a *radical ideal*. There is a one-to-one correspondence between the two, revealing an intimate relationship between projective geometry, dimension theory, algebraic invariant theory, etc. The new algorithmic approach to algebraic geometry—in particular, the theory of Gröbner bases generated via the Buchberger algorithm [426–428]—together with the power of fast computers, enables us to find the dimension of the solution set and the exact number of solutions if they are finitely many, as well as to compute the solutions with arbitrary precision. Presently, only polynomial systems of modest dimension can be handled; nonetheless, even a study of small model problems can teach us much about the higher energy solutions.

Another technique serving the same purpose, based on the stability theorem of homotopy theory [429], seems to be even more promising [430]. The basic idea of the homotopy method is to replace the original system of nonlinear equations $\mathbf{F}(\mathbf{x}) = \mathbf{0}$ by the system

$$\mathbf{H}(\mathbf{x}, \lambda) = (1 - \lambda)\mathbf{G}(\mathbf{x}) + \lambda\mathbf{F}(\mathbf{x}) = \mathbf{0}, \tag{4.1}$$

where $\mathbf{G}(\mathbf{x}) = \mathbf{0}$ is a system that can easily be solved exactly. Clearly, solutions corresponding to $\lambda = 0$ and $\lambda = 1$ give those for $\mathbf{G}(\mathbf{x}) = \mathbf{0}$ and $\mathbf{F}(\mathbf{x}) = \mathbf{0}$, respectively. Thus, the main task is to continue the known solution of $\mathbf{H}(\mathbf{x}, 0) = \mathbf{0}$ to $\mathbf{H}(\mathbf{x}, 1) = \mathbf{0}$. This technique was very recently employed by Kowalski and Jankowski [53] to locate all CCD solutions for the MBS H4 model. These researchers found that the number of solutions is much smaller than the upper bound based on Bezout's theorem (12 rather than 64 in the spin-adapted case and 20 instead of 1,024 in the spin-orbital case). In fact, the number of solutions they found is the same as the dimension of the corresponding FCI problem. Out of these solutions, at least 5 were deemed to be physically meaningful. Kowalski and Jankowski also used the same technique to generate a complete set of real solutions of RHF equations for the same model [53]. This is indeed a very exciting development.

The preceding pages should leave little doubt that the CC formalism and its exploitation will continue to prosper in the future. Nonetheless, as we address finer and finer facets of molecular electronic structure and develop and test new high-level approaches, the following caveat should always be kept in mind: The danger that is lurking here stems from the fact that the CCD and CCSD methods already represent excellent approximations and often give unexpectedly good results, even in highly quasi-degenerate situations. Consequently, various ad hoc modifications involving cluster components that contribute in higher-than-second order to the wave function will almost always yield reasonably looking results. Often, it takes considerable time to recognize the weaknesses of alternative approaches that are supposed to improve on the existing "workhorses" of CC theory. (Cf., e.g., Section II.E.2.) It is thus essential that any such new development be based on solid theoretical grounds.

Clearly, CC theory in its standard form will continue to be routinely applied when highly correlated results are called for. Judged by its potential, its range of applicability will undoubtedly expand, not only thanks to the increasing computational capabilities, but also owing to new algorithmic and methodological developments.

APPENDIX A: ALGEBRA OF REPLACEMENT OPERATORS

The one-body spin-orbital *replacement operator*

$$e_J^I = X_I^\dagger X_J, \tag{A.1}$$

defined on an orthonormal spin-orbital set $\mathcal{L} \equiv \{|I\rangle, |J\rangle, \ldots\}$, where $\mathrm{card}(\mathcal{L}) = 2n$, replaces the spin orbital $|J\rangle$ with the spin orbital $|I\rangle$ when acting on any antisymmetrized product of spin orbitals $|\{I_1, I_2, \cdots\}\rangle$ in which $|J\rangle$ is occupied and $|I\rangle$ is unoccupied; otherwise $|\{I_1, I_2, \cdots\}\rangle$ is annihilated by e_J^I. These operators possess the Hermitian property (2.6b), while being closed with respect to the Lie (or commutator) product of Eq. (2.6a), and may thus be regarded as generators of the spin-orbital unitary group $U(2n)$. (See, e.g., [174–177].) They are not closed, however, with respect to an ordinary product, since

$$e_J^I e_L^K = X_I^\dagger X_J X_K^\dagger X_L = X_I^\dagger X_K^\dagger X_L X_J + \delta_J^K X_I^\dagger X_L$$
$$= e_{JL}^{IK} + \delta_J^K e_L^I, \tag{A.2}$$

where e^{IK}_{JL} designates a two-body replacement operator given by Eq. (2.7a). A general r-body spin-orbital replacement operator is thus defined as follows:

$$e^{I_1 \cdots I_r}_{J_1 \cdots J_r} = X^\dagger_{I_1} \cdots X^\dagger_{I_r} X_{J_r} \cdots X_{J_1} . \tag{A.3}$$

The general replacement operators can also be defined recursively by relying on the relationship (A.2) and its generalization, Eqs. (2.7). Clearly, in employing a finite-dimensional spin-orbital space $\mathcal{V}$, where $\dim \mathcal{V} = 2n \equiv m$, the higher-than-$m$-body replacement operators vanish.

The basic properties and algebraic structure of these operators, as well as of their orbital analogues, were explored in MBPT [171, 172] and in connection with the Clifford algebra UGA [270]. The usefulness of these operators in CC theory was first illustrated by the derivation of the orthogonally spin-adapted (OSA) form of CCD equations [173]. (See also [29, 174].) We only mention here that the replacement operators, together with the identity operator e that may be regarded as the zero-body operator, form a very convenient basis for the *universal enveloping algebra* of $U(2n)$. In contrast to a Lie algebra of $U(2n)$, this is an associative algebra, and the replacement operator basis $\{e, e^I_J, e^{IK}_{JL}, \ldots\}$, which is distinct from the well-known Poincaré–Birkhoff–Witt basis, representing the standard basis for universal enveloping algebras, possesses some rather remarkable properties. (Cf., e.g., [431].) In what follows, we only highlight some basic properties of these operators that are useful in the context of CC theories [29, 173].

Clearly, such operators are invariant with respect to any simultaneous permutation of upper and lower indices and change at most their phase when the subscripts (or superscripts) alone are permuted among themselves. Using the shorthand notation

$$e^{I_1 I_2 \cdots I_r}_{J_1 J_2 \cdots J_r} \equiv e^{12 \cdots r}_{12 \cdots r}, \qquad e^{K_1 \cdots K_s}_{L_1 \cdots L_s} \equiv e^{1'2' \cdots s'}_{1'2' \cdots s'}, \text{ etc.,} \tag{A.4}$$

we thus have

$$e^{12 \cdots r}_{12 \cdots r} = e^{p_1 \cdots p_r}_{p_1 \cdots p_r} = (-1)^p e^{12 \cdots r}_{p_1 p_2 \cdots p_r} , \tag{A.5}$$

where p is the parity of the permutation $P : i \to p_i$ $(i = 1, \ldots, r)$. Clearly, those replacement operators having two identical subscripts (or superscripts) vanish.

The recursive definition now takes the form

$$e^{12 \cdots k(k+1)}_{12 \cdots k(k+1)} = e^{12 \cdots k}_{12 \cdots k} e^{k+1}_{k+1} - \sum_i \delta^{k+1}_i e^{12 \cdots \cdots i \cdots \cdots k}_{12 \cdots (k+1) \cdots k} , \tag{A.6}$$

and may be interpreted as a special case of Wick's theorem with Kronecker deltas indicating the contractions. Introducing the generalized Kronecker symbols

$$\delta_{J_1 J_2 \cdots J_r}^{I_1 I_2 \cdots I_r} \equiv \delta_{12 \cdots r}^{12 \cdots r} = \delta_1^1 \delta_2^2 \cdots \delta_r^r, \tag{A.7}$$

as well as their antisymmetrized analogues [29]

$$\Delta_{12 \cdots r}^{12 \cdots r} = \sum_{P \in \mathcal{S}_r} (-1)^P \delta_{p_1 p_2 \cdots p_r}^{12 \cdots r}, \tag{A.8}$$

where the sum extends over all permutations of the symmetric group $\mathcal{S}_r$, we can generalize the Wick theorem property to arbitrary products of replacement operators,

$$\begin{aligned}
e_{12 \cdots r}^{12 \cdots r} e_{1'2' \cdots s'}^{1'2' \cdots s'} &= e_{12 \cdots r 1'2' \cdots s'}^{12 \cdots r 1'2' \cdots s'} \\
&+ \sum_{i=1}^{s} \sum_{j=1}^{r} \delta_j^{i'} e_{12 \cdots i' \cdots r 1'2' \cdots (i-1)'(i+1)' \cdots s'}^{12 \cdots j \cdots r 1'2' \cdots (i-1)'(i+1)' \cdots s'} \\
&+ \sum_{i<j=1}^{s} \sum_{k<\ell=1}^{r} \Delta_{k\ell}^{i'j'} e_{12 \cdots i' \cdots j' \cdots r 1'2' \cdots (i-1)'(i+1)' \cdots (j-1)'(j+1)' \cdots s'}^{12 \cdots k \cdots \ell \cdots r 1' \cdots (i-1)'(i+1)' \cdots (j-1)'(j+1)' \cdots s'} \quad \text{(A.9)} \\
&+ \cdots \\
&+ \sum_{k_1 < k_2 < \cdots < k_{s'} = 1}^{r} \Delta_{k_1 k_2 \cdots k_{s'}}^{1'2' \cdots s'} e_{1 \cdots 1' \cdots 2' \cdots s' \cdots r}^{1 \cdots k_1 \cdots k_2 \cdots k_{s'} \cdots r},
\end{aligned}$$

where we assume that $s < r$. Clearly, when $r < s$, the last term takes the form

$$\sum_{i_1 < i_2 < \cdots < i_r = 1}^{s} \Delta_{1\ 2 \cdots r}^{i'_1 i'_2 \cdots i'_r} e_{1' \cdots i'_1 \cdots i'_2 \cdots i'_r \cdots s'}^{1' \cdots 1 \cdots 2 \cdots r \cdots s'}, \tag{A.10}$$

and when $r = s$, we have only one maximally contracted term,

$$\Delta_{1\,2 \cdots r}^{1'2' \cdots r'} e_{1'2' \cdots r'}^{1\,2 \cdots r}.$$

We also note that by using the coset representatives for the subgroup chain $\mathcal{S}_{r+s} \supset \mathcal{S}_r \otimes \mathcal{S}_s$, we can decompose the antisymmetric Kronecker deltas of Eq. (A.8) into the sum of products of lower order deltas [29], e.g.,

$$\Delta_{1234}^{1234} = \Delta_{12}^{12}\Delta_{34}^{34} + \Delta_{34}^{12}\Delta_{12}^{34} + \Delta_{14}^{12}\Delta_{23}^{34} + \Delta_{23}^{12}\Delta_{14}^{34} - \Delta_{13}^{12}\Delta_{24}^{34} - \Delta_{24}^{12}\Delta_{13}^{34}. \tag{A.11}$$

The product rule (A.9) immediately implies that the uncontracted terms cancel when one evaluates commutators of replacement operators, so that, for example,

$$\left[e^{12}_{12}, e^{1'}_{1'}\right] = \delta^{1'}_1 e^{12}_{1'2} + \delta^{1'}_2 e^{12}_{11'} - \delta^{1}_{1'} e^{1'2}_{12} - \delta^{2}_{1'} e^{11'}_{12} \qquad (A.12)$$

and

$$\left[e^{12}_{12}, e^{1'2'}_{1'2'}\right] = \sum^{2}_{i,j=1} (-1)^{i+j+1} \left(\delta^{i'}_j e^{1\,2\,\bar{i}'}_{1'2'\bar{j}} - \delta^{i}_{j'} e^{1'\,2'\,\bar{i}}_{1\,2\,\bar{j}'}\right)$$
$$+ \Delta^{1'2'}_{12} e^{12}_{1'2'} - \Delta^{12}_{1'2'} e^{1'2'}_{12}, \qquad (A.13)$$

where $\bar{i} = 3 - i$ ($i = 1, 2$). In general, uncontracted terms cancel out, leaving only the contracted (and thus connected in a diagrammatic sense) ones. For a commutator $[A, B]$, terms involving subscripts of A and superscripts of B appear with the plus sign, while those involving superscripts of A and subscripts of B carry a minus sign.

We also note that the sum of replacement operators over all permutations of their subscripts (superscripts) vanishes, i.e.,

$$\sum_{P \in \mathcal{S}_r} e^{12 \cdots r}_{p_1 p_2 \cdots p_r} = 0, \qquad (A.14)$$

as does the expectation value in a given reference state $|\Phi_0\rangle$, i.e.,

$$\langle \Phi_0 | e^{I_1 \cdots I_r}_{J_1 \cdots J_r} | \Phi_0 \rangle \equiv \langle e^{12 \cdots r}_{12 \cdots r} \rangle = 0, \qquad (A.15)$$

unless all spin-orbital labels involved are occupied in $|\Phi_0\rangle$, in which case

$$\langle e^{12 \cdots r}_{12 \cdots r} \rangle = \Delta^{12 \cdots r}_{12 \cdots r} = (-1)^p, \qquad (A.15')$$

where p is the parity of the permutation transforming the superscripts into the subscripts. This property reflects the normal-order character of replacement operators. Note, however, that an analogous rule does not apply to products, since, e.g., $\langle e^I_J e^J_I \rangle \neq 0$ even when $|I\rangle$ is occupied and $|J\rangle$ is unoccupied in $|\Phi_0\rangle$—i.e., $\langle e^A_R e^R_A \rangle = 1$.

A remarkable feature of the replacement operator algebra is the fact that its extension to orbital formalism, along the lines of UGA, provides a spin-free generalization of Wick's theorem and enables one to obtain a spin-adapted or an OSA form of the theory. Thus, the generators of the orbital unitary group $U(n)$, viz.,

$$E_j^i = \sum_\sigma e_{j\sigma}^{i\sigma} = e_{j\alpha}^{i\alpha} + e_{j\beta}^{i\beta}, \tag{A.16}$$

where we use lowercase letters to label the orbitals, so that $|I\rangle = |i\rangle|\sigma\rangle$, the spin function $|\sigma\rangle$ being either an up or a down eigenstate of S_z, $|\sigma\rangle = |\alpha\rangle$ or $|\sigma\rangle = |\beta\rangle$, respectively (i.e., $|I\rangle$ cannot represent a general spin orbital involving a linear combination of $|\alpha\rangle$ and $|\beta\rangle$ spin functions), have the same structure constants as the spin-orbital generators

$$\left[E_j^i, E_\ell^k \right] = \delta_j^k E_\ell^i - \delta_\ell^i E_j^k. \tag{A.17}$$

The higher order orbital replacement operators may then be similarly defined by recursion relations [171, 173] as

$$\begin{aligned} E_{j\ell}^{ik} &= E_j^i E_\ell^k - \delta_j^k E_\ell^i, \\ E_{j\ell n}^{ikm} &= E_{j\ell}^{ik} E_n^m - \delta_j^m E_{n\ell}^{ik} - \delta_\ell^m E_{jn}^{ik}, \quad \text{etc.,} \end{aligned} \tag{A.18}$$

so that they possess the same algebraic structure as the spin-orbital operators. However, in contrast to the latter, the *orbital* replacement operators do not produce orthogonal states when acting on some reference configuration. They can, nonetheless, be orthogonally spin adapted by choosing a suitable coupling scheme. This is particularly important in considering a UGA-based CC theory, in which case one defines the OSA excitation operators that are adapted to the group chain of Eq. (2.138) or Eq. (2.140).

Note that, whereas in the CS case we have

$$E_{ab\cdots}^{rs\cdots} = E_a^r E_b^s \cdots, \tag{A.19}$$

where $a, b, \ldots$ and $r, s, \ldots$ label hole and particle orbitals, respectively, this is no longer true in OS situations when we split the orbitals into the core (c), active or valence (a), and external or virtual (e) subsets, so that $n = n_c + n_a + n_e$ and the active labels can appear both as sub- and superscripts. Moreover, even in the CS case, the replacement operators $E_{ab\cdots}^{rs\cdots}$ (which can now be justifiably referred to as *excitation operators*) do not generate orthogonal states, since the property (A.5) no longer holds. For example, $|\Phi_1\rangle = \frac{1}{2} E_{ab}^{rs} |\Phi_0\rangle$ and $|\Phi_2\rangle = \frac{1}{2} E_{ba}^{rs} |\Phi_0\rangle$ represent linearly independent and normalized, but nonorthogonal, states with overlap $\langle \Phi_1 | \Phi_2 \rangle = \frac{1}{4} \langle E_{ba}^{ab} \rangle = -\frac{1}{4} \langle E_a^a \rangle = -\frac{1}{2}$. (In view of the property (A.19), these states are identical to the so-called *generator states* [177].) The two states are associated with valence bond (VB)-type couplings $(ar)(bs)$ and $(as)(br)$, so that only $|\Phi_1\rangle$ represents the *ph–ph* coupled state [214]. Thus, they are neither *ph–ph* nor *pp–hh* coupled orthonormal states that are employed in CI approaches. It is, however, the latter *pp–hh* states that possess the most desirable transformation properties. (These states change at most the phase when the particle and/or hole labels are interchanged.)

Designating the six corresponding spin-orbital configurations as in [214] by

$$|G_1\rangle = e^{r\beta\,s\beta}_{a\beta\,b\beta}|\Phi_0\rangle, \qquad |G_2\rangle = e^{r\alpha\,s\alpha}_{a\alpha\,b\alpha}|\Phi_0\rangle,$$

$$|G_3\rangle = e^{r\alpha\,s\beta}_{a\alpha\,b\beta}|\Phi_0\rangle, \qquad |G_4\rangle = e^{r\beta\,s\alpha}_{a\beta\,b\alpha}|\Phi_0\rangle, \qquad \text{(A.20)}$$

$$|G_5\rangle = e^{r\beta\,s\alpha}_{a\alpha\,b\beta}|\Phi_0\rangle, \qquad |G_6\rangle = e^{r\alpha\,s\beta}_{a\beta\,b\alpha}|\Phi_0\rangle,$$

we have

$$|\Phi_1\rangle \equiv \left|{}^{\text{VB}}\Phi\!\left(\begin{array}{cc} r & s \\ a & b \end{array}\right)\right\rangle = \frac{1}{2}E^{rs}_{ab}|\Phi_0\rangle = \frac{1}{2}(|G_1\rangle + |G_2\rangle + |G_3\rangle + |G_4\rangle),$$

$$|\Phi_2\rangle \equiv \left|{}^{\text{VB}}\Phi\!\left(\begin{array}{cc} r & s \\ b & a \end{array}\right)\right\rangle = \frac{1}{2}E^{rs}_{ba}|\Phi_0\rangle = -\frac{1}{2}(|G_1\rangle + |G_2\rangle + |G_5\rangle + |G_6\rangle).$$

$$\text{(A.21)}$$

Orthogonalizing the second state to the first one, we obtain the *ph–ph* coupled states [214],

$$|^{(1)}\Phi_{ph-ph}\rangle = |\Phi_1\rangle,$$

$$|^{(3)}\Phi_{ph-ph}\rangle = 2(|\Phi_1\rangle + \tfrac{1}{2}|\Phi_2\rangle)/\sqrt{3} \qquad \text{(A.22)}$$

$$= (|G_1\rangle + |G_2\rangle - |G_3\rangle - |G_4\rangle + 2|G_5\rangle + 2|G_6\rangle)/(2\sqrt{3}),$$

while the desirable *pp–hh* coupled states [214]

$$|^{(1)}\Phi_{pp-hh}\rangle = \tfrac{1}{2}(|G_3\rangle + |G_4\rangle - |G_5\rangle - |G_6\rangle),$$

$$|^{(3)}\Phi_{pp-hh}\rangle = (2|G_1\rangle + 2|G_2\rangle + |G_3\rangle + |G_4\rangle + |G_5\rangle + |G_6\rangle)/(2\sqrt{3}),$$

$$\text{(A.23)}$$

are obtained as symmetric and antisymmetric linear combinations of $|\Phi_1\rangle$ and $|\Phi_2\rangle$, so that

$$|^{(m)}\Phi_{pp-hh}\rangle = {}^{(m)}E^{rs}_{ab}|\Phi_0\rangle \qquad (m = 1, 3), \qquad \text{(A.24)}$$

where

$$^{(m)}E_{ab}^{rs} = {}^{(m)}S_{ab}E_{ab}^{rs}, \quad {}^{(m)}S_{ab} = e + (2 - m)(ab) . \qquad (A.25)$$

Here, the symbol (ab) designates the transposition $a \leftrightarrow b$ and the superscripts (m), $m = 1, 3$, indicate the intermediate spin coupling quantum number [214]. Clearly, if either $a = b$ or $r = s$, the triplet coupled pp–hh state vanishes, while this is not the case when other coupling schemes are employed.

APPENDIX B: LIST OF ACRONYMS

ACCD + ST(ACCD)	ACCD, perturbatively corrected for singles and triples using ACCD cluster amplitudes
ACC(S)D	Approximate CC(S)D
ACP	Approximate coupled pair (functional)
ACP-45	Approximate coupled-pair theory accounting only for hole-separable pair cluster interaction diagrams (labeled D4 and D5)
ACPQ	Approximate coupled-pair theory with quadruples
ACPTQ	Approximate coupled-pair theory with triples and quadruples
AEL	Average excitation level
ANO	Atomic natural orbital
AO	Atomic orbital
AS	Active space
aug	augmented ... (basis)
B-CCD	CCD (method) with Brueckner orbitals
BD	Brueckner CCD with doubles (method)
BO	Brueckner orbital
CAS-FCI	Complete active-space FCI (method)
CAS-PT2	Complete active-space second-order perturbation theory
CAS-SCF	Complete active-space self-consistent field (method)
CAUGA	Clifford algebra UGA
CBS	Complete basis set
cc	Correlation consistent
CC	Coupled cluster(s)
CCD	Coupled clusters with doubles (method)

CCDQ'	CCD with an approximate account of quadruples
CCD[ST]	CCD with a perturbative account of singles and triples
CCLR	Coupled-cluster linear-response (theory)
CCPPA	Coupled-cluster polarization propagator approximation
CC-R12	Coupled clusters with r_{12} coordinates
CCSD	Coupled clusters with singles and doubles (method)
CCSD(f)	UGA CCSD with full singles and doubles manifold
CCSD{f}	UGA CCSD with a perturbative account of pseudo-doubles
CCSD(is)	UGA CCSD with the first-order interacting space
CCSDQ'	CCSD with an approximate account of quadruples
CCSD(T)	CCSD with a perturbative account of triples
CCSD[T]	CCSD with a perturbative account of triples; same as $CCSD + T(CCSD)$
CCSD{T}	UGA CCSD with a perturbative account of triples
CCSDT	Coupled clusters with singles, doubles, and triples (method)
CCSDT-n	Approximate CCSDT
CCSDTQ	Coupled clusters with singles, doubles, triples, and quadruples (method)
CCSDT[Q]	CCSDT accounting perturbatively for quadruples; same as $CCSDT + Q(CCSDT)$
CEPA	Coupled electron pair approximation
CI	Configuration interaction
CIS	Configuration interaction with singles
CPMET	Coupled-pair many-electron theory
CS	Closed-shell
DIIS	Direct inversion in the iterative subspace (algorithm)
DMBPT(∞)	MBPT with intermediate doubles summed to infinite order
DNA	Deoxyribonucleic acid
DODS	Different orbitals for different spins
DZ	Double zeta (basis)
DZP	Double zeta plus polarization (basis)
DZ2P	Double zeta double polarization (basis)
EA	Electron affinity

ECCM	Extended coupled-cluster method
ECPMET	Extended coupled pair many-electron theory
EOM	Equations of motion (method)
EPV	Exclusion principle violating (diagrams)
FCC	Full coupled cluster (method)
FCI	Full configuration interaction (method)
FF	Finite field (methods)
GHF	General Hartree–Fock (method)
GVB	Generalized valence bond
HF	Hartree–Fock (method)
HOMO	Highest occupied molecular orbital
HS	High spin
IP	Ionization potential
IPM	Independent particle model
KHW	Knowles, Hampel, and Werner
L-CC(S)D	Linear approximation to CC(S)D
LR	Linear response
LUMO	Lowest unoccupied molecular orbital
MBPT	Many-body perturbation theory
MBS	Minimum basis set
MC SCF	Multiconfigurational self-consistent field (method)
MO	Molecular orbital
MP	Møller–Plesset perturbation theory
MPn	nth-order Møller–Plesset perturbation theory
MPO	Maximum proximity orbital
MR	Multireference
MR-CISD	Multireference CI with singles and doubles (method)
NCCM	Normal coupled-cluster method
NPE	Nonparallelism error
NUH	Neogrády, Urban, and Hubač
OS	Open shell
OSA	Orthogonally spin adapted
OSS	Open-shell singlet (states)
p (or P)	polarized ... (basis)
PEC	Potential-energy curve
PES	Potential-energy surface
$p–h$	particle–hole (formalism)
$pp–hh$	particle, particle–hole, hole (coupling)
PPA	Polarization propagator approximation
PPP	Pariser–Parr–Pople (model or Hamiltonian)
PT	Perturbation theory

PUHF	Projected UHF
QCI	Quadratic configuration interaction
QCID	QCI with doubles ($\equiv$ CCD)
QCISD	QCI with singles and doubles
QED	Quantum electrodynamics
QR	Quadratic response
QRHF	Quasirestricted Hartree–Fock (method)
RHF	Restricted Hartree–Fock (method)
RMR	Reduced multireference
ROHF	Restricted open-shell Hartree–Fock (method)
RPA	Random phase approximation
SA	Spin adapted
SAC/SAC-CI	Symmetry-adapted cluster/configuration interaction (method)
SCF	Self-consistent field
SOCI	Second-order CI
SOPPA	Second-order polarization propagator approximation
SR	Single reference
SS	State selective or state specific
STEOM-CC	Similarity-transformed EOM-CC
SU	State universal
TD-CCSD	Two-determinant SU CCSD
TDHF	Time-dependent Hartree–Fock (method)
TZ	Triple zeta
TZ2P	Triple zeta, double polarization (basis)
UCC	Unitary coupled-cluster (*Ansatz*)
UGA	Unitary group approach
UHF	Unrestricted Hartree–Fock (method)
VB	Valence bond (method)
VDZ	Valence double zeta (basis)
VQZ	Valence quadruple zeta (basis)
VTZ	Valence triple zeta (basis)
V5Z	Valence quintuple zeta (basis)
VU	Valence universal
XCC	Expectation value CC (method)
ZPE	Zero-point energy

Acknowledgments

The continued support by the National Science and Engineering Research Council of Canada (J. P.) is gratefully acknowledged. The senior author deeply appreciates an Alexander von

Humboldt Research Award, as well as the hospitality of Professor G. H. F. Diercksen and the Max-Planck-Institute for Astrophysics in Garching during his visit. He also wishes to express his most sincere gratitude to Prof. J. Čížek for his longtime collaboration and friendship; to his past and present graduate students, postdoctoral fellows, and research associates, Drs. B. G. Adams, M. J. Boyle, R. W. H. Cho, A. E. Kondo, G. Peris, P. Piecuch, L. Pylypow, M. Saute, J. L. Stuber, M. Takahashi, H. C. Wong, and S. Zarrabian; and to the Faculty Members Visiting Quantum Theory Group, Drs. M. Bénard, J.-Q. Chen, M. D. Gould, K. Jankowski, B. Jeziorski, A. Laforgue, P. Piecuch, J. Planelles, C. R. Sarma, V. Špirko, E. O. Steinborn, and P. E. S. Wormer, who contributed in one way or another to the developments reported in this review. Special thanks are due as well to Dr. P. Hobza of the Academy of Sciences of the Czech Republic, Prof. K. Jankowski and Dr. K. Kowalski of the Nicholas Copernicus University in Toruń, Prof. B. Jeziorski of the University of Warsaw, and Profs. I. Hubač, V. Kellö, and M. Urban and Drs. J. Mášik and P. Neogrády of the Comenius University in Bratislava for stimulating discussions, advice, and preprints of their work. Last, but not least, our thanks are due to Mrs. H. A. Warren for her constant help, patience, and immaculate typesetting.

References

[1] J. Čížek, *J. Chem. Phys.* **45**, 4256 (1966).

[2] J. Čížek, *Adv. Chem. Phys.* **14**, 35 (1969).

[3] F. Coester, *Nucl. Phys.* **7**, 421 (1958).

[4] F. Coester and H. Kümmel, *Nucl. Phys.* **17**, 477 (1960).

[5] J. Paldus, J. Čížek, and I. Shavitt, *Phys. Rev. A* **5**, 50 (1972).

[6] J. A. Pople, R. Krishnan, H. B. Schlegel, and J. S. Binkley, *Intern. J. Quantum Chem. Symp.* **14**, 545 (1978).

[7] R. J. Bartlett and G. D. Purvis, III, *Intern. J. Quantum Chem.* **14**, 561 (1978).

[8] R. J. Bartlett and G. D. Purvis, III, *Phys. Scripta* **21**, 255 (1980).

[9] R. J. Bartlett, *Annu. Rev. Phys. Chem.* **32**, 359 (1981).

[10] R. J. Bartlett, C. E. Dykstra, and J. Paldus, in *Advanced Theories and Computational Approaches to the Electronic Structure of Molecules*, C. E. Dykstra, Ed., Reidel, Dordrecht, The Netherlands, 1984, pp. 127–159.

[11] M. Urban, I. Černušák, V. Kellö, and J. Noga, in *Methods in Computational Chemistry*, Vol. 1, *Electron Correlation in Atoms and Molecules*, S. Wilson, Ed., Plenum Press, New York, 1987, pp. 117–250.

[12] K. Jankowski, in *Methods in Computational Chemistry*, Vol. 1, *Electron Correlation in Atoms and Molecules*, S. Wilson, Ed., Plenum Press, New York, 1987, pp. 1–116.

[13] R. J. Bartlett, *J. Phys. Chem.* **93**, 1697 (1989).

[14] J. Paldus, in *Methods in Computational Molecular Physics*, NATO ASI Series B: Physics, Vol. 293, S. Wilson and G. H. F. Diercksen, Eds., Plenum Press, New York, 1992, pp. 99–194.

[15] ACES II, a CC and an MBPT suite of codes by J. F. Stanton, J. Gauss, J. D. Watts, W. J. Lauderdale, and R. J. Bartlett, *Intern. J. Quantum Chem. Symp.* **26**, 879 (1992), with contributions by J. Almlöf and P. R. Taylor (MOLECULE), T. Helgaker, H. J. Jensen, P. Jørgensen, J. Olsen, P. R. Taylor (ABACUS), and D. Bernholdt.

[16] CADPAC 5.0, a suite of programs (Cambridge Analytical Derivatives Package) by R. D. Amos, with contributions from I. L. Alberts, J. S. Andrews, S. M. Colwell, N. C. Handy,

D. Jayatilaka, P. J. Knowles, R. Kobayashi, N. Koga, K. E. Laidig, P. E. Maslen, C. W. Murray, J. E. Rice, J. Sanz, E. D. Simandiras, A. J. Stone, and M.-D. Su.

[17] COMENIUS (version 1989), a suite of MBPT and CC codes by J. Noga, V. Kellö, and M. Urban, Comenius University, Bratislava, Slovakia.

[18] DIRCCR12-95, a direct integral-driven CC-R12 program by J. Noga and W. Klopper (unpublished), 1995.

[19] GAUSSIAN 92, a system of codes written by M. J. Frisch, G. W. Trucks, M. Head-Gordon, P. M. V. Gill, M. W. Wong, J. B. Foresman, B. G. Johnson, H. B. Schlegel, M. A. Robb, E. S. Replogle, R. Gomperts, J. L. Andres, K. Raghavachari, J. S. Binkley, C. Gonzales, R. L. Martin, D. J. Fox, D. J. Defrees, J. Baker, J. J. P. Stewart, and J. A. Pople; Gaussian, Inc., Pittsburgh, Pennsylvania.

[20] MOLCAS-3, a system of programs by K. Andersson, M. R. A. Blomberg, M. P. Fülscher, V. Kellö, R. Lindh, P.-Å. Malmqvist, J. Noga, J. Olsen, B. O. Roos, A. J. Sadlej, P. E. M. Siegbahn, M. Urban, and P.-O. Widmark, University of Lund, Lund, Sweden, 1994; MOLCAS-4, K. Andersson, M. R. A. Blomberg, M. P. Fülscher, G. Karlström, R. Lindh, P.-Å. Malmqvist, P. Neogrády, J. Olsen, B. O. Roos, A. J. Sadlej, M. Schütz, L. Seijo, L. Serrano-Andrés, P. E. M. Siegbahn, and P.-O. Widmark, University of Lund, Lund, Sweden, 1997.

[21] MOLPRO 94, a system of programs by H.-J. Werner and P. J. Knowles, with contributions by J. Almlöf, R. D. Amos, M. Deegan, S. T. Elbert, C. Hampel, W. Meyer, K. A. Peterson, R. M. Pitzer, E.-A. Reinsch, A. J. Stone, and P. R. Taylor.

[22] OPENMOL, an expert system by E. Steiner, J. Karwowski, V. Kellö, M. Urban, and G. H. F. Diercksen, Max-Planck-Institut für Astrophysik, Garching bei München, Germany, 1996.

[23] PSI 2.0, a CC package by G. E. Scuseria, PSITECH, Inc., Athens, Georgia, 1991.

[24] TITAN, a suite of codes written by T. J. Lee, A. P. Rendell, and J. E. Rice.

[25] A. E. Kondo, X. Li, P. Piecuch, J. Planelles, and J. Paldus, a Waterloo system of CC codes based on the GAMESS [26] SCF and integral packages.

[26] GAMESS, a system of programs by M. W. Schmidt, K. K. Baldridge, J. A. Boatz, S. T. Elbert, M. S. Gordon, J. H. Jensen, S. Koseki, N. Matsunaga, K. A. Nguyen, S. J. Su, and T. L. Windus, together with M. Dupui and J. A. Montgomery, *J. Comput. Chem.* **14**, 1347 (1993).

[27] R. J. Bartlett, Ed., *Proceedings of the Workshop on CC Theory at the Interface of Atomic Physics and Quantum Chemistry, Theor. Chim. Acta* **80**, Nos. 2–6 (1991).

[28] R. J. Bartlett and J. F. Stanton, in *Reviews in Computational Chemistry*, Vol. 5, K. B. Lipkowitz and D. B. Boyd, Eds., VCH Publishers, New York, 1994, pp. 65–169.

[29] J. Paldus, in *Relativistic and Electron Correlation Effects in Molecules and Solids*, NATO ASI Series B: Physics, Vol. 318, G. L. Malli, Ed., Plenum Press, New York, 1994, pp. 207–282.

[30] T. J. Lee and G. E. Scuseria, in *Quantum Mechanical Electronic Structure Calculations with Chemical Accuracy*, S. R. Langhoff, Ed., Kluwer Academic Publishers, Dordrecht, The Netherlands, 1995, pp. 47–108.

[31] R. J. Bartlett, in *Modern Electronic Structure Theory*, Part I, D. R. Yarkony, Ed., World Scientific, Singapore, 1995, pp. 1047–1131.

[32] R. J. Bartlett, Ed., *Recent Advances in Computational Chemistry*, Vol. 3, *Recent Advances in Coupled-Cluster Methods*, World Scientific, Singapore, 1997.

[33] P. Hobza, H. L. Selzle, and E. W. Schlag, *J. Phys. Chem.* **100**, 18790 (1996).

[34] O. Bludský, J. Šponer, J. Leszczynski, V. Špirko, and P. Hobza, *J. Chem. Phys.* **105**, 11042 (1996).

[35] P. Hobza and J. Šponer, *J. Mol. Struct.* (*THEOCHEM*) **388**, 115 (1996); J. Šponer and P. Hobza, *Chem. Phys. Lett.* **267**, 263 (1997).

[36] T. H. Dunning, Jr., *J. Chem. Phys.* **90**, 1007 (1989); R. A. Kendall, T. H. Dunning, Jr., and R. J. Harrison, *J. Chem. Phys.* **96**, 6796 (1992); D. E. Woon and T. H. Dunning, Jr., *ibid.* **98**, 1358 (1993).

[37] D. Mukherjee and S. Pal, *Adv. Quantum Chem.* **20**, 292 (1989).

[38] I. Lindgren, *Intern. J. Quantum Chem. Symp.* **12**, 33 (1978).

[39] I. Lindgren and J. Morrison, *Atomic Many-Body Theory*, Springer-Verlag, Berlin, 1982.

[40] B. Jeziorski and H. J. Monkhorst, *Phys. Rev. A* **24**, 1668 (1981).

[41] I. Lindgren and D. Mukherjee, *Phys. Rep.* **151**, 93 (1987).

[42] L. Stolarczyk and H. J. Monkhorst, *Phys. Rev. A* **32**, 725,743 (1985); *ibid.* **37**, 1908, 1926 (1988); M. Barysz, H. J. Monkhorst, and L. Z. Stolarczyk, in [27], pp. 483–507.

[43] B. Jeziorski and J. Paldus, *J. Chem. Phys.* **90**, 2714 (1989).

[44] X. Li and J. Paldus, in [32], pp. 183–219.

[45] X. Li and J. Paldus, *J. Chem. Phys.* **101**, 8812 (1994).

[46] B. Jeziorski, J. Paldus, and P. Jankowski, *Intern. J. Quantum Chem.* **56**, 129 (1995).

[47] J. Paldus and J. Planelles, *Theor. Chim. Acta* **89**, 13 (1994).

[48] L. Z. Stolarczyk, *Chem. Phys. Letters* **217**, 1 (1994).

[49] X. Li, G. Peris, J. Planelles, F. Rajadell, and J. Paldus, *J. Chem. Phys.* **107**, 90 (1997).

[50] X. Li and J. Paldus, *J. Chem. Phys.* **107**, 6257 (1997).

[51] M. Nooijen, *J. Chem. Phys.* **104**, 2638 (1996).

[52] M. Nooijen and R. J. Bartlett, *J. Chem. Phys.* **104**, 2652 (1996).

[53] K. Kowalski and K. Jankowski, *Phys. Rev. Lett.* **81**, 1195 (1998); *Chem. Phys. Lett.* **290**, 180 (1998).

[54] J. Čížek and J. Paldus, *Phys. Scripta* **21**, 251 (1980).

[55] H. Kümmel, in [27], pp. 81–89.

[56] J. Čížek, in [27], pp. 91–94.

[57] W. Macke, *Z. Naturforsch.* **5a**, 192 (1950).

[58] M. Gell-Mann and K. A. Brueckner, *Phys. Rev.* **106**, 364 (1957).

[59] H. A. Bethe, *Phys. Rev.* **103**, 1353 (1956).

[60] K. A. Brueckner, *Phys. Rev.* **100**, 36 (1955).

[61] J. Goldstone, *Proc. Roy. Soc.* **A239**, 267 (1957).

[62] J. Hubbard, *Proc. Roy. Soc.* **A240**, 539 (1957); **A243**, 336 (1958); **A244**, 199 (1958).

[63] N. M. Hugenholtz, *Physica* (Utrecht) **23**, 481 (1957).

[64] K. H. Lührmann and H. Kümmel, *Nucl. Phys. A* **194**, 225 (1972).

[65] P.-O. Löwdin, *Adv. Chem. Phys.* **2**, 207 (1959); **14**, 283 (1969).

[66] H. P. Kelly, *Phys. Rev.* **131**, 684 (1963); *Adv. Chem. Phys.* **14**, 129 (1969).

[67] O. Sinanoğlu, *J. Chem. Phys.* **36**, 706 (1962); *Adv. Chem. Phys.* **6**, 315 (1964); **14**, 237 (1969).

[68] J. E. Mayer and M. G. Mayer, *Statistical Mechanics*, Chap. 13, J. Wiley & Sons, New York, 1940.

[69] B. Khan and G. E. Uhlenbeck, *Physica* (Utrecht) **5**, 399 (1938).

[70] H. Primas, in *Modern Quantum Chemistry*, O. Sinanoğlu, Ed., Academic Press, New York, 1965, pp. 45–74.

[71] V. V. Tolmachev, *The Field Theoretical Form of the Perturbation Theory Applied to Atomic and Molecular Electronic Problems*, University of Tartu, 1963 (in Russian).

[72] O. Sinanoğlu and J. Čížek, *Chem. Phys. Lett.* **1**, 337 (1967).

[73] M. R. Hoffmann and H. F. Schaefer, III, *Adv. Quantum Chem.* **18**, 207 (1986).

[74] J. Paldus and H. C. Wong, *Comput. Phys. Commun.* **6**, 1 (1973); H. C. Wong and J. Paldus, *ibid.* **6**, 9 (1973).

[75] J. Paldus and J. Čížek, in *Energy, Structure and Reactivity: Proceedings of the 1972 Boulder Summer Research Conference on Theoretical Chemistry*, D. W. Smith and W. B. McRae, Eds., J. Wiley & Sons, New York, 1973, pp. 198–212.

[76] J. Čížek and J. Paldus, *Intern. J. Quantum Chem.* **5**, 359 (1971).

[77] J. Čížek, J. Paldus, and L. Šroubková, *Intern. J. Quantum Chem.* **3**, 149 (1969).

[78] J. Paldus and J. Čížek, *Adv. Quantum Chem.* **9**, 105 (1975).

[79] W. H. Louisell, *Radiation and Noise in Quantum Electronics*, McGraw-Hill, New York, 1964. (See especially Theorem 3 of Sec. 3.2.)

[80] F. E. Harris, in *Electrons in Finite and Infinite Structures*, P. Phariseau and L. Scheire, Eds., Plenum Press, New York, 1977, pp. 274–320.

[81] H. Kümmel, K. H. Lührmann, and J. G. Zabolitzky, *Physics Reports C* **36**, 1 (1978).

[82] F. Coester, in *Lectures in Theoretical Physics*, Vol. 11B, K. T. Mahanthappa and W. E. Brittin, Eds., Gordon and Breach, New York, 1969, pp. 157–186.

[83] R. Offerman, W. Ey, and H. Kümmel, *Nucl. Phys. A* **273**, 349 (1976); R. Offerman, *ibid.* **273**, 368 (1976); W. Ey, *ibid.* **296**, 189 (1978).

[84] W. Kutzelnigg, D. Mukherjee, and S. Koch, *Chem. Phys.* **87**, 5902 (1987).

[85] J. Paldus, in *New Horizons of Quantum Chemistry*, P.-O. Löwdin and B. Pullman, Eds., Reidel, Dordrecht, The Netherlands, 1983, pp. 31–60.

[86] W. D. Laidig and R. J. Bartlett, *Chem. Phys. Lett.* **104**, 424 (1984); W. D. Laidig, P. Saxe, and R. J. Bartlett, *J. Chem. Phys.* **86**, 887 (1987).

[87] B. Jeziorski and J. Paldus, *J. Chem. Phys.* **88**, 5673 (1988).

[88] L. Meissner, K. Jankowski, and J. Wasilewski, *Intern. J. Quantum Chem.* **34**, 535 (1988); L. Meissner, S. A. Kucharski, and R. J. Bartlett, *J. Chem. Phys.* **91**, 6187 (1989).

[89] A. Balková, S. A. Kucharski, L. Meissner, and R. J. Bartlett, in [27], p. 335.

[90] J. Paldus, P. Piecuch, B. Jeziorski, and L. Pylypow, in *Recent Progress in Many Body Theories*, Vol. 3, T. L. Ainsworth, C. E. Campbell, B. E. Clements, and E. Krotscheck, Eds., Plenum Press, New York, 1992, pp. 287–303.

[91] J. Paldus, L. Pylypow, and B. Jeziorski, in *Many-Body Methods in Quantum Chemistry*, Lecture Notes in Chemistry, Vol. 52, U. Kaldor, Ed., Springer-Verlag, Berlin, 1989, pp. 151–170.

[92] J. Paldus, P. Piecuch, L. Pylypow, and B. Jeziorski, *Phys. Rev. A* **47**, 2738 (1993); P. Piecuch and J. Paldus, *ibid.* **49**, 3479 (1994).

[93] P. Piecuch and J. Paldus, *J. Chem. Phys.* **101**, 5875 (1994).

[94] P. Piecuch and J. Paldus, *J. Phys. Chem.* **99**, 15354 (1995).

[95] K. Jankowski, J. Paldus, I. Grabowski, and K. Kowalski, *J. Chem. Phys.* **97**, 7600 (1992); **101**, 1759 (1994) (E); **101**, 3085 (1994).

[96] N. Oliphant and L. Adamowicz, *J. Chem. Phys.* **94**, 1229 (1991); **96**, 3739 (1992); *Int. Rev. Phys. Chem.* **12**, 339 (1993).

[97] P. Piecuch, N. Oliphant, and L. Adamowicz, *J. Chem. Phys.* **99**, 1875 (1993).

[98] P. Piecuch and L. Adamowicz, *Chem. Phys. Lett.* **221**, 121 (1994); *J. Chem. Phys.* **100**, 5792 (1994).

[99] N. Oliphant and L. Adamowicz, *Chem. Phys. Lett.* **190**, 13 (1992); P. Piecuch and L. Adamowicz, *J. Chem. Phys.* **100**, 5857 (1994).

[100] J. Noga and R. J. Bartlett, *J. Chem. Phys.* **86**, 7041 (1987); *ibid.* **89**, 3401 (1988) (E).

[101] J. D. Watts and R. J. Bartlett, *J. Chem. Phys.* **93**, 6104 (1990).

[102] G. E. Scuseria and H. F. Schaefer, III, *Chem. Phys. Lett.* **152**, 382 (1988).

[103] S. A. Kucharski and R. J. Bartlett, *J. Chem. Phys.* **97**, 4282 (1992).

[104] F. Harris, *Intern. J. Quantum Chem. Symp.* **11**, 403 (1977).

[105] J. Paldus, J. Čížek, M. Saute, and A. Laforgue, *Phys. Rev. A* **17**, 805 (1978); M. Saute, J. Paldus, and J. Čížek, *Intern. J. Quantum Chem.* **15**, 463 (1979).

[106] H. Nakatsuji and K. Hirao, *J. Chem. Phys.* **68**, 2053 (1978); H. Nakatsuji, *Chem. Phys. Lett.* **59**, 362 (1978); **67**, 329, 334 (1979).

[107] H. Nakatsuji, in *Computational Chemistry: Reviews of Current Trends*, Vol. 2, Chap. 2, J. Leszczynski, Ed., World Scientific, Singapore, 1997, pp. 62–124.

[108] D. J. Rowe, *Nuclear Collective Motion: Models and Theory*, Chaps. 13–15, Methuen, London, 1970.

[109] A. L. Fetter and J. D. Walecka, *Quantum Theory of Many-Particle Systems*, Chap. 5, McGraw-Hill, New York, 1971.

[110] H. J. Monkhorst, *Intern. J. Quantum Chem. Symp.* **11**, 421 (1977); E. Dalgaard and H. J. Monkhorst, *Phys. Rev. A* **28**, 1217 (1983).

[111] D. Mukherjee and P. K. Mukherjee, *Chem. Phys.* **37**, 325 (1979); S. Ghosh, D. Mukherjee, and S. N. Bhattacharyay, *Mol. Phys.* **43**, 173 (1981); *idem.*, *Chem. Phys.* **72**, 961 (1982); S. Ghosh and D. Mukherjee, *Proc. Indian Acad. Sci.* **93**, 947 (1984); M. D. Prasad, S. Pal, and D. Mukherjee, *Phys. Rev. A* **31**, 1287 (1985).

[112] K. Emrich, *Nucl. Phys. A* **351**, 379 (1981).

[113] H. Sekino and R. J. Bartlett, *Intern. J. Quantum Chem. Symp.* **18**, 255 (1984); E. A. Salter, H. Sekino, and R. J. Bartlett, *J. Chem. Phys.* **87**, 502 (1987).

[114] M. Takahashi and J. Paldus, *J. Chem. Phys.* **85**, 1486 (1986).

[115] J. Geertsen and J. Oddershede, *J. Chem. Phys.* **85**, 2112 (1986).

[116] J. Geertsen, C. M. L. Rittby, and R. J. Bartlett, *Chem. Phys. Lett.* **164**, 57 (1989).

[117] D. C. Comeau and R. J. Bartlett, *Chem. Phys. Lett.* **207**, 414 (1993).

[118] J. F. Stanton and R. J. Bartlett, *J. Chem. Phys.* **98**, 7029 (1993).

[119] J. Olsen and P. Jørgensen, in *Modern Electronic Structure Theory*, Part II, D. R. Yarkony, Ed., World Scientific, Singapore, 1995, pp. 857–990, and references therein.

[120] T. B. Pedersen and H. Koch, *J. Chem. Phys.* **106**, 8059 (1997).

[121] C. Hättig, O. Christiansen, H. Koch, and P. Jørgensen, *Chem. Phys. Lett.* **269**, 428 (1997).

[122] M. Nooijen and R. J. Bartlett, *J. Chem. Phys.* **107**, 6812 (1997).

[123] M. Nooijen and R. J. Bartlett, *J. Chem. Phys.* **106**, 6441, 6449 (1997).

[124] J.-P. Malrieu, Ph. Durand, and J.-P. Daudey, *J. Phys. A: Math. Gen.* **18**, 809 (1985).

[125] L. Adamowicz, R. Caballol, J.-P. Malrieu, and J. Meller, *Chem. Phys. Lett.* **259**, 619 (1997); L. Adamowicz and J.-P. Malrieu, *J. Chem. Phys.* **105**, 9240 (1997).

[126] L. Adamowicz and J.-P. Malrieu, in [32], pp. 307–330.

[127] L. Meissner and M. Nooijen, *J. Chem. Phys.* **102**, 9604 (1995); L. Meissner, *ibid.* **108**, 9227 (1998).

[128] J. Paldus, *J. Chem. Phys.* **67**, 303 (1977).

[129] P. Piecuch and J. Paldus, *Intern. J. Quantum Chem.* **36**, 429 (1989).

[130] P. Piecuch and J. Paldus, *Theor. Chim. Acta* **83**, 69 (1992).

[131] C. L. Janssen and H. F. Schaefer, III, *Theor. Chim. Acta* **79**, 1 (1991).

[132] P. Neogrády, M. Urban, and I. Hubač, *J. Chem. Phys.* **97**, 5074 (1992); *ibid.* **100**, 3706 (1994); P. Neogrády and M. Urban, *Intern. J. Quantum Chem.* **55**, 187 (1995).

[133] M. Urban, P. Neogrády, and I. Hubač, in [32], pp. 275–306.

[134] P. J. Knowles, C. Hampel, and H.-J. Werner, *J. Chem. Phys.* **99**, 5219 (1993).

[135] D. Jayatilaka and T. J. Lee, *J. Chem. Phys.* **98**, 9734 (1993).

[136] T. D. Crawford, T. J. Lee, and H. F. Schaefer, III, *J. Chem. Phys.* **107**, 7943 (1997).

[137] D. Jayatilaka and G. S. Chandler, *Mol. Phys.* **92**, 471 (1997).

[138] T. D. Crawford, T. J. Lee, N. C. Handy, and H. F. Schaefer, III, *J. Chem. Phys.* **107**, 9980 (1997).

[139] P. G. Szalay and J. Gauss, *J. Chem. Phys.* **107**, 9028 (1997).

[140] B. Jeziorski, H. J. Monkhorst, K. Szalewicz, and J. G. Zabolitzky, *J. Chem. Phys.* **81**, 368 (1984); K. B. Wenzel, J.G. Zabolitzky, K. Szalewicz, B. Jeziorski, and H. J. Monkhorst, *ibid,* **85**, 3964 (1986); J. Noga, W. Kutzelnigg, and W. Klopper, *Chem. Phys. Lett.* **199**, 497 (1992); J. Noga and W. Kutzelnigg, *J. Chem. Phys.* **101**, 7738 (1994); H. Müller, W. Kutzelnigg, and J. Noga, *Mol. Phys.* **92**, 535 (1997).

[141] J. Noga, W. Klopper, and W. Kutzelnigg, in [32], pp. 1–48.

[142] U. Kaldor, in [32], pp. 125–153.

[143] R. J. Bartlett, in *Geometrical Derivatives of Energy Surfaces and Molecular Properties*, P. Jørgensen and J. Simons, Eds., Reidel, Dordrecht, The Netherlands, 1986, pp. 35–61.

[144] J. F. Stanton and J. Gauss, in [32], pp. 49–79.

[145] H. Kümmel, K. H. Lührmann, and J. G. Zabolitzky, *Physics Reports C* **90**, 160 (1982).

[146] H. Kümmel, in *Nucleon–Nucleon Interaction and Nuclear Many-Body Problem*, S. S. Wu and T. T. S. Kuo, Eds., World Scientific, Singapore, 1984, p. 46.

[147] R. F. Bishop, *Phys. Rev. C* **42**, 1341 (1990).

[148] J. S. Arponen, R. F. Bishop, and E. Pajanne, *Phys. Rev. A* **36**, 2519, 2539 (1987); J. Arponen, R. F. Bishop, E. Pajanne, and N. I. Robinson, *ibid.* **37**, 1065 (1988).

[149] R. F. Bishop and J. S. Arponen, *Intern. J. Quantum Chem. Symp.* **24**, 197 (1990).

[150] R. F. Bishop, J. B. Parkinson, and Y. Xian, in *Recent Progress in Many-Body Theories*, Vol. 3, T. L. Ainsworth, C. E. Campbell, B. E. Clements, and E. Krotscheck, Eds., Plenum Press, New York, 1992, pp. 117–133.

[151] R. F. Bishop and K. H. Lührmann, *Phys. Rev. B* **17**, 3757 (1978); *ibid.* **26**, 5523 (1982).

[152] J. Arponen and E. Pajanne, *J. Phys. C* **15**, 2665, 2683 (1982).

[153] K. Emrich and J. G. Zabolitzky, *Phys. Rev. B* **30**, 2049 (1984).

[154] U. Kaulfuss and M. Altenbokum, *Phys. Rev. D* **33**, 3658 (1986).

[155]   R. F. Bishop, M. C. Boscá, and M. F. Flynn, *Phys. Rev. A* **40**, 3484 (1989); J. S. Arponen and R. F. Bishop, *Phys. Rev. Lett.* **64**, 111 (1990).

[156]   C. S. Hsue, H. Kümmel, and P. Ueberholz, *Phys. Rev. D* **32**, 1435 (1985); M. Altenbokum and H. Kümmel, *ibid.* **32**, 2014 (1985).

[157]   U. Kaulfuss, *Phys. Rev. D* **32**, 1421 (1985); M. Funke, U. Kaulfuss, and H. Kümmel, *ibid.* **35**, 621 (1987).

[158]   R. F. Bishop, in [27], pp. 95–148.

[159]   R. F. Bishop and H. Kümmel, *Physics Today* **40**, 52 (1987).

[160]   S. A. Blundell, A. C. Hartley, Z. Liu, A.-M. Mårtensson-Pendrill, and J. Sapirstein, in [27], p. 257–288.

[161]   I. Lindgren, in *Recent Progress in Many-Body Theories*, Vol. 3, T. L. Ainsworth, C. E. Campbell, B. E. Clements, and E. Krotscheck, Eds., Plenum Press, New York, 1991, pp. 245–276.

[162]   E. Eliav, U. Kaldor, and B. A. Hess, *J. Chem. Phys.* **108**, 3409 (1998).

[163]   T. J. Lee, J. E. Rice, G. E. Scuseria, and H. F. Schaefer, III, *Theor. Chim. Acta* **75**, 81 (1989); T. J. Lee and P. R. Taylor, *Intern. J. Quantum Chem. Symp.* **23**, 199 (1989).

[164]   M. Head-Gordon and T. J. Lee, in [32], pp. 221–253.

[165]   U. S. Mahapatra, B. Datta, and D. Mukherjee, in [32], pp. 155–181.

[166]   S. Pal and N. Vaval, in [32], pp. 255–273.

[167]   P. G. Szalay, in [32], pp. 81–123.

[168]   J. A. Pople, M. Head-Gordon, and K. Raghavachari, *J. Chem. Phys.* **87**, 5968 (1987).

[169]   J. Paldus, *Diagrammatic Methods for Many-Fermion Systems*, Katholieke Universiteit, Nijmegen, The Netherlands, 1981.

[170]   F. E. Harris, H. J. Monkhorst, and D. L. Freeman, *Algebraic and Diagrammatic Methods in Many-Fermion Theory*, Oxford University Press, New York, 1992.

[171]   W. Kutzelnigg, *J. Chem. Phys.* **82**, 4166 (1985), and references therein.

[172]   J. Hinze and J. T. Broad, in *The Unitary Group for the Evaluation of Electronic Energy Matrix Elements, Lecture Notes in Chemistry*, Vol. 22, J. Hinze, Ed., Springer-Verlag, Berlin, 1981, pp. 332–344.

[173]   J. Paldus and B. Jeziorski, *Theor. Chim. Acta* **73**, 81 (1988).

[174]   J. Paldus, in *Contemporary Mathematics*, Vol. 160, American Mathematical Society, Providence, RI, 1994, pp. 209–236.

[175]   J. Paldus, in *The Unitary Group for the Evaluation of Electronic Energy Matrix Elements, Lecture Notes in Chemistry*, Vol. 22, J. Hinze, Ed., Springer-Verlag, Berlin, 1981, pp. 332–344.

[176]   J. Paldus, in *Theoretical Chemistry: Advances and Perspectives*, Vol. 2, D. Henderson and H. Eyring, Eds., Academic Press, New York, 1976, pp. 131–290.

[177]   F. A. Matsen and R. Pauncz, *The Unitary Group in Quantum Chemistry*, Elsevier, Amsterdam, 1986.

[178]   K. Huang, *Statistical Mechanics*, J. Wiley & Sons, New York, 1967, Chap. 14.

[179]   R. K. Pathria, *Statistical Mechanics*, Pergamon Press, Oxford, U.K., 1977, Chap. 9.

[180]   T. D. Lee and C. N. Yang, *Phys. Rev.* **117**, 12, 22, 897 (1960); F. Mohling, *Phys. Rev. A* **139**, 664 (1965).

[181]   M. Gell-Mann and F. Low, *Phys. Rev.* **84**, 350 (1951).

[182] J. Paldus, in *Atomic, Molecular, and Optical Physics Handbook*, Chap. 5, *Perturbation Theory*, G. W. F. Drake, Ed., American Institute of Physics, Woodbury, N. Y., 1996, pp. 76–87.

[183] I. Hubač and P. Neogrády, *Phys. Rev. A* **50**, 4558 (1994).

[184] I. Hubač, in *New Methods in Quantum Theory*, NATO ASI Series, C. A. Tsipis, V. S. Popov, D. R. Herschbach, and J. S. Avery, Eds., Kluwer, Dordrecht, The Netherlands, 1996, pp. 183–202; J. Mášik and I. Hubač, in *Quantum Systems in Chemistry and Physics: Trends in Methods and Applications*, R. McWeeny, J. Maruani, Y. G. Smeyers, and S. Wilson, Eds., Kluwer Academic Publishers, Dordrecht, The Netherlands, 1997; pp. 283–308; I. Hubač, J. Mášik, P. Mach, J. Urban, and P. Babinec, in *Computational Chemistry: Reviews of Current Trends,* Vol. 3, J. Leszczynski, Ed., World Scientific, Singapore, 1998; in press.

[185] J. Mášik and I. Hubač, *Coll. Czech. Chem. Commun.* **62**, 829 (1997).

[186] J. Mášik and I. Hubač, *Adv. Quantum Chem.* **31**, 75 (1998). J. Másik, I. Hubač, and P. Mach, *J. Chem. Phys.* **108**, 6571 (1998). P. Mach, J. Mášik, J. Urban, and I. Hubač, *Mol. Phys.* **94**, 173 (1998).

[187] L. M. Frantz and R. L. Mills, *Nucl. Phys.* **15**, 16 (1960).

[188] R. J. Bartlett and I. Shavitt, *Chem. Phys. Lett.* **50**, 190 (1977), **57**, 157 (1978) (E).

[189] J. Paldus, P. E. S. Wormer, F. Visser, and A. van der Avoird, *J. Chem. Phys.* **76**, 2458 (1982).

[190] K. Jankowski and J. Paldus, *Intern. J. Quantum Chem.* **18**, 1243 (1980).

[191] S. A. Kucharski and R. J. Bartlett, *Adv. Quantum Chem.* **18**, 281 (1986).

[192] J. Paldus, J. Čížek, and B. Jeziorski, *J. Chem. Phys.* **90**, 4356 (1989).

[193] J. Paldus and M. J. Boyle, *Intern. J. Quantum Chem.* **22**, 1281 (1982).

[194] J. Paldus, M. Takahashi, and R. W. H. Cho, *Phys. Rev. B* **30**, 4267 (1984); M. Takahashi and J. Paldus, *ibid.* **31**, 5121 (1985).

[195] P. Piecuch, J. Čížek, and J. Paldus, *Intern. J. Quantum Chem.* **42**, 165 (1992).

[196] J. Paldus, P. E. S. Wormer, and M. Bénard, *Collect. Czech. Chem. Commun.* **53**, 1919 (1988).

[197] J. Čížek and J. Paldus, *J. Chem. Phys.* **47**, 3976 (1967).

[198] J. Paldus, in *Self-Consistent Field: Theory and Applications*, R. Carbó and M. Klobukowski, Eds., Elsevier, Amsterdam, 1990, pp. 1–45.

[199] H. Fukutome, *Progr. Theor. Phys.* **52**, 115 (1974); *Intern. J. Quantum Chem.* **20**, 955 (1981).

[200] X. Li and J. Paldus, *J. Chem. Phys.* **103**, 6536 (1995).

[201] N. C. Handy, P. J. Knowles, and K. Somasundrum, *Theor. Chim. Acta* **68**, 87 (1985).

[202] R. H. Nobes, J. A. Pople, L. Radom, N. C. Handy, and P. J. Knowles, *Chem. Phys. Lett.* **138**, 481 (1987); R. Murphy, H. F. Schaefer, III, R. H. Nobes, L. Radom, and R. S. Pitzer, *Intern. Rev. Phys. Chem.* **5**, 229 (1986).

[203] J. D. Watts and R. J. Bartlett, *J. Chem. Phys.* **95**, 6652 (1991).

[204] G. E. Scuseria, *Chem. Phys. Lett.* **176**, 27 (1991).

[205] R. S. Mulliken, *Intern. J. Quantum Chem. Symp.* **5**, 95 (1971).

[206] X. Li and J. Paldus, *J. Chem. Phys.* **102**, 2013 (1995).

[207] B. G. Adams, K. Jankowski, and J. Paldus, *Phys. Rev. A* **24**, 2316 (1981).

[208] J. Paldus, J. Čížek, and M. Takahashi, *Phys. Rev. A* **30**, 2193 (1984).

[209] R. A. Chiles and C. E. Dykstra, *Chem. Phys. Lett.* **80**, 69 (1981); S. M. Bachrach, R. A. Chiles, and C. E. Dykstra, *J. Chem. Phys.* **75**, 2270 (1981); S.-Y. Liu, M. F. Daskalakis, and C. E. Dykstra, *ibid.* **85**, 5877 (1986); D. E. Bernholdt, S.-Y. Liu, and C. E. Dykstra, *ibid.* **85**, 5120 (1986); C. E. Dykstra, S.-Y. Liu, M. F. Daskalakis, J. P. Lucia, and M. Takahashi, *Chem. Phys. Lett.* **137**, 266 (1987); C. E. Dykstra, *Ab Initio Calculation of the Structure and Properties of Molecules*, Elsevier, Amsterdam, 1988.

[210] P. Piecuch, R. Toboła, and J. Paldus, *Phys. Rev. A* **54**, 1210 (1996).

[211] A. P. Jucys, I. B. Levinson, and V. V. Vanagas, *Mathematical Apparatus of the Theory of Angular Momentum*, Institute of Physics and Mathematics of the Academy of Sciences of the Lithuanian S.S.R., Mintis, Vilnius, 1960 (in Russian); English translations: Israel Program for Scientific Translations, Jerusalem, 1962, and Gordon and Breach, New York, 1964.

[212] E. El Baz and B. Castel, *Graphical Methods of Spin Algebras in Atomic, Nuclear and Particle Physics*, M. Dekker, New York, 1972.

[213] S. Wilson, *Electron Correlation in Molecules*, Clarendon Press, Oxford, 1984.

[214] J. Paldus, B. G. Adams, and J. Čížek, *Intern. J. Quantum Chem.* **11**, 813 (1977).

[215] P. Piecuch and J. Paldus, *Theor. Chim. Acta* **78**, 65 (1990).

[216] E. A. Kondo, P. Piecuch, and J. Paldus, *J. Chem. Phys.* **102**, 6511 (1995).

[217] J. Paldus and M. J. Boyle, *Physica Scripta* **21**, 295 (1980).

[218] X. Li and J. Paldus, *Intern. J. Quantum Chem. Symp.* **27**, 269 (1993).

[219] M. D. Gould and J. Paldus, *J. Chem. Phys.* **92**, 7394 (1990).

[220] L. C. Biedenharn and J. D. Louck, *Angular Momentum in Quantum Mechanics, Theory and Application*, and *The Racah–Wigner Algebra in Quantum Theory*, Addison-Wesley, Reading, MA, 1981.

[221] X. Li and J. Paldus, *J. Math. Chem.* **4**, 295 (1990); **13**, 273 (1993); **14**, 325 (1993).

[222] H. B. Schlegel, *J. Chem. Phys.* **84**, 4530 (1986).

[223] M. Rittby and R. J. Bartlett, *J. Chem. Phys.* **92**, 3033 (1988).

[224] J. F. Stanton, *J. Chem. Phys.* **101**, 371 (1994).

[225] P. Piecuch, R. Toboła, and J. Paldus, *Chem. Phys. Lett.* **210**, 243 (1993).

[226] K. Jankowski, L. Meissner, and J. Wasilewski, *Intern. J. Quantum Chem.* **28**, 931 (1985).

[227] S. Wilson, K. Jankowski, and J. Paldus, *Intern. J. Quantum Chem.* **23**, 1781 (1983); **28**, 525 (1985), U. Kaldor, *ibid.* **28**, 103 (1985); N. Iijima and A. Saika, *ibid.* **27**, 481 (1985); S. Zarrabian and J. Paldus, *ibid.* **38**, 761 (1990).

[228] L. Meissner, K. Jankowski, and J. Wasilewski, *Intern. J. Quantum Chem.* **34**, 535 (1988); K. Jankowski, J. Paldus, and J. Wasilewski, *J. Chem. Phys.* **95**, 3549 (1991).

[229] S. A. Kucharski, A. Balková, and R. J. Bartlett, in [27], pp. 321–334.

[230] S. A. Kucharski, A. Balková, P. G. Szalay, and R. J. Bartlett, *J. Chem. Phys.* **97**, 4289 (1992).

[231] P. Piecuch, R. Toboła, and J. Paldus, *Intern. J. Quantum Chem.* **55**, 133 (1995).

[232] X. Li and J. Paldus, *J. Chem. Phys.* **103**, 1024 (1995).

[233] J. P. Finley, R. K. Chaudhuri, and K. F. Freed, *J. Chem. Phys.* **103**, 4990 (1995).

[234] B. G. Adams, K. Jankowski, and J. Paldus, *Phys. Rev. A* **24**, 2330 (1981).

[235] B. G. Adams and J. Paldus, *Phys. Rev. A* **24**, 2302 (1981).

[236] J. Planelles, J. Paldus, and X. Li, *Theor. Chim. Acta* **89**, 33, 59 (1994).

[237] G. Peris, J. Planelles, and J. Paldus, *Intern. J. Quantum Chem.* **62**, 137 (1997).

[238] G. Peris, F. Rajadell, X. Li, J. Planelles, and J. Paldus, *Mol. Phys.*, **94**, 235 (1998).

[239] X. Li and J. Paldus, *J. Chem. Phys.* **108**, 637 (1998).

[240] X. Li and J. Paldus, *Chem. Phys. Lett.* **286**, 145 (1998).

[241] Y. S. Lee and R. J. Bartlett, *J. Chem. Phys.* **80**, 4371 (1984); Y. S. Lee, S. A. Kucharski, and R. J. Bartlett, *J. Chem. Phys.* **81**, 5906 (1984), **82**, 5761 (1985) (E).

[242] M. Urban, J. Noga, S. J. Cole, and R. J. Bartlett, *J. Chem. Phys.* **83**, 4041 (1985).

[243] J. Noga, R. J. Bartlett, and M. Urban, *Chem. Phys. Lett.* **134**, 126 (1987).

[244] K. Raghavachari, *J. Chem. Phys.* **82**, 4607 (1985).

[245] K. Raghavachari, G. W. Trucks, J. A. Pople, and M. Head-Gordon, *Chem. Phys. Lett.* **157**, 479 (1989).

[246] R. J. Bartlett, J. D. Watts, S. A. Kucharski, and J. Noga, *Chem. Phys. Lett.* **165**, 513 (1990).

[247] S. A. Kucharski and R. J. Bartlett, *Chem. Phys. Lett.* **158**, 550 (1989).

[248] P. Piecuch and J. Paldus, *Intern. J. Quantum Chem. Symp.* **25**, 9 (1991).

[249] P. Piecuch, S. Zarrabian, J. Paldus, and J. Čížek, *Phys. Rev. B* **42**, 3351 (1990); J. Paldus and P. Piecuch, *Intern. J. Quantum Chem.* **42**, 135 (1992).

[250] P. Piecuch, A. E. Kondo, V. Špirko, and J. Paldus, *J. Chem. Phys.* **104**, 4699 (1996).

[251] J. Geertsen, S. Ericksen, and J. Oddershede, *Adv. Quantum Chem.* **22**, 167 (1991).

[252] V. Špirko, P. Piecuch, A. E. Kondo, and J. Paldus, *J. Chem. Phys.* **104**, 4716 (1996).

[253] J. Paldus and X. Li, *Israel J. Chem.* **31**, 351 (1991); X. Li and J. Paldus, *J. Mol. Structure (THEOCHEM)* **229**, 249 (1991); *Intern. J. Quantum Chem.* **41**, 117 (1992); J. Paldus and X. Li, in *Group Theory in Physics, AIP Conference Proceedings*, No. 266, A. Frank, T. H. Seligman, and K. B. Wolf, Eds., American Institute of Physics, New York, 1992, pp. 159–178; in *"Pauling's Legacy: Modern Modelling of the Chemical Bond", Theoretical and Computational Chemistry Series*, Vol. 6, Chap. 17, Z. B. Maxić and W. J. Orville-Thomas, Eds., Elsevier, Amsterdam, 1999, pp. 481–501.

[254] J. Paldus and X. Li, in *Symmetries in Science VI: From the Rotation Group to Quantum Algebras*, B. Gruber, Ed., Plenum Press, New York, 1993, pp. 573–592.

[255] H. Meissner and E. O. Steinborn, *Intern. J. Quantum Chem.* **61**, 777 (1997); **63**, 257 (1997); *Phys. Rev. A* **56**, 1189 (1997); *J. Mol. Structure (THEOCHEM)*, **433**, 119 (1998).

[256] L. Meissner, *J. Chem. Phys.* **103**, 8014 (1995).

[257] U. Kaldor, in *Many-Body Methods in Quantum Chemistry, Lecture Notes in Chemistry*, Vol. 52, U. Kaldor, Ed., Springer-Verlag, Berlin, 1989, pp. 199–213.

[258] K. Jankowski and P. Malinowski, *Chem. Phys. Lett.* **205**, 471 (1993).

[259] P. Malinowski and K. Jankowski, *J. Phys. B: At. Mol. Opt. Phys.* **26**, 3035 (1993); K. Jankowski and P. Malinowski, *ibid.* **27**, 829 (1994).

[260] K. Jankowski and I. Grabowski, *Intern. J. Quantum Chem.* **55**, 205 (1995).

[261] K. Jankowski and P. Malinowski, *J. Phys. B: At. Mol. Opt. Phys.* **27**, 1287 (1994).

[262] L. Meissner, *Chem. Phys. Lett.* **255**, 244 (1996).

[263] A. Haque and U. Kaldor, *Intern. J. Quantum Chem.* **29**, 425 (1986); S. Ben-Schlomo and U. Kaldor, *J. Chem. Phys.* **89**, 956 (1988).

[264] U. Kaldor, *J. Chem. Phys.* **87**, 467 (1987).

[265] S. Pal, M. Rittby, R. J. Bartlett, D. Sinha, and D. Mukherjee, *J. Chem. Phys.* **88**, 4357 (1988).

[266] K. Jankowski and P. Malinowski, *Intern. J. Quantum Chem.* **55**, 269 (1955); **59**, 239 (1996).

[267] P. Malinowski and K. Jankowski, *Phys. Rev. A* **51**, 4583 (1995).

[268] K. Jankowski and P. Malinowski, *Intern. J. Quantum Chem.* **48**, 59 (1993).

[269] J. Paldus, in *Mathematical Frontiers in Computational Chemical Physics*, IMA Series, Vol. 15, D. G. Truhlar, Ed., Springer-Verlag, Berlin, 1988, pp. 262–299; I. Shavitt, *ibid.*, pp. 300–349.

[270] J. Paldus and C. R. Sarma, *J. Chem. Phys.* **83**, 5135 (1985); J. Paldus, M.-J. Gao, and J.-Q. Chen, *Phys. Rev. A* **35**, 3197 (1987).

[271] X. Li and Q. Zhang, *Intern. J. Quantum Chem.* **36**, 599 (1989).

[272] X. Li and J. Paldus, *J. Chem. Phys.* **102**, 8897 (1995).

[273] X. Li and J. Paldus, *Intern. J. Quantum Chem.*, **70**, 60 (1998).

[274] X. Li, P. Piecuch, and J. Paldus, *Chem. Phys. Lett.* **224**, 267 (1994); P. Piecuch, X. Li, and J. Paldus, *ibid.* **230**, 377 (1994).

[275] X. Li and J. Paldus, *Chem. Phys. Lett.* **231**, 1 (1994).

[276] X. Li and J. Paldus, *J. Chem. Phys.* **102**, 8059 (1995).

[277] J. Paldus and X. Li, *Can. J. Chem.* **74**, 918 (1996).

[278] X. Li and J. Paldus, *J. Chem. Phys.* **104**, 9555 (1996).

[279] X. Li and J. Paldus, *Mol. Phys.* **94**, 41 (1998).

[280] M. Takahashi, J. Paldus, and J. Čížek, *Intern. J. Quantum Chem.* **24**, 707 (1983).

[281] D. Cremer and Z. He, *J. Phys. Chem.* **100**, 6173 (1996).

[282] O. Christiansen, J. Olsen, P. Jørgensen, H. Koch, and P.-Å. Malmqvist, *Chem. Phys. Lett.* **261**, 369 (1996); J. Olsen, O. Christiansen, H. Koch, and P. Jørgensen, *J. Chem. Phys.* **105**, 5082 (1996).

[283] T. H. Dunning, Jr., and K. A. Peterson, *J. Chem. Phys.* **108**, 4761 (1998).

[284] H. Guo and J. Paldus, *Intern. J. Quantum Chem.* **63**, 345 (1997).

[285] R. J. Bartlett, *Annu. Rev. Phys. Chem.* **32**, 359 (1981).

[286] P. Čársky and M. Urban, *Ab Initio Calculations: Methods and Applications in Chemistry*, Lecture Notes in Chemistry, Vol. 16, Springer-Verlag, Berlin, 1980.

[287] W. J. Hehre, J. A. Pople, P. v. R. Schleyer, and L. Radom, *Ab Initio Molecular Orbital Theory*, J. Wiley & Sons, New York, 1986.

[288] A. T. Amos and G. G. Hall, *Proc. Roy. Soc.* (London) **A263**, 483 (1961).

[289] P. Čársky and I. Hubač, *Theor. Chim. Acta* **80**, 407 (1991).

[290] C. W. Murray and E. R. Davidson, *Intern. J. Quantum Chem.* **43**, 755 (1992).

[291] K. F. Freed, *Acc. Chem. Res.* **16**, 127 (1983).

[292] K. Andersson and B. O. Roos, in *Modern Electronic Structure Theory*, Part I, *Advanced Series in Physical Chemistry*, Vol. 2, D. R. Yarkony, Ed., World Scientific, Singapore, 1995, pp. 55–109.

[293] M. P. Fülscher, K. Andersson, and B. O. Roos, *J. Chem. Phys.* **96**, 9204 (1992); K. Wolinski and P. Pulay, *ibid.* **90**, 3647 (1989).

[294] B. O. Roos, K. Andersson, M. P. Fülscher, P.-Å. Malmqvist, L. Serrano-Andrés, K. Pierloot, and M. Merchán, *Adv. Chem. Phys.* **93**, 219 (1996).

[295] G. Hose and U. Kaldor, *J. Phys. B: Atom. Mol. Phys.* **12**, 3827 (1979).

[296] M. G. Shepard and K. F. Freed, *J. Chem. Phys.* **75**, 4525 (1981); J. P. Finley, R. K. Chaudhuri, and K. F. Freed, *Phys. Rev. A* **54**, 343 (1996); R. K. Chaudhuri, J. P. Finley, and K. F. Freed, *J. Chem. Phys.* **106**, 4067 (1997); R. K. Chaudhuri, A. Mudholkar, K. F. Freed, C. H. Martin, and H. Sun, *ibid.* **106**, 9252 (1997); R. K. Chaudhuri and K. F. Freed, *ibid.* **107**, 6699 (1997).

[297] X. Li and J. Paldus, *Adv. Quantum Chem.* **28**, 15 (1997).

[298] I. Hubač and P. Čársky, *Phys. Rev. A* **22**, 2392 (1980).

[299] C. W. Murray and E. R. Davidson, *Chem. Phys. Lett.* **187**, 451 (1991).

[300] A. Balková and R. J. Bartlett, *Chem. Phys. Lett.* **193**, 364 (1992).

[301] J. Paldus, J. Čížek, and B. Jeziorski, *J. Chem. Phys.* **93**, 1485 (1990).

[302] J. A. Pople, M. Head-Gordon, and K. Raghavachari, *J. Chem. Phys.* **90**, 4635 (1989); K. Raghavachari, M. Head-Gordon, and J. A. Pople, *ibid.* **93**, 1486 (1990).

[303] S. Kucharski, J. Noga, and R. J. Bartlett, *J. Chem. Phys.* **90**, 7282 (1989).

[304] K. Raghavachari, G. W. Trucks, J. A. Pople, and E. Replogle, *Chem. Phys. Lett.* **158**, 207 (1989); K. Raghavachari, J. A. Pople, E. S. Replogle, and M. Head-Gordon, *J. Phys. Chem.* **94**, 5579 (1990).

[305] Z. He and D. Cremer, *Intern. J. Quantum Chem. Symp.* **25**, 43 (1991).

[306] Z. He and D. Cremer, *Theor. Chim. Acta* **85**, 305 (1993).

[307] J. D. Watts, M. Urban, and R. J. Bartlett, *Theor. Chim. Acta* **90**, 341 (1995).

[308] M. Böhme and G. Frenking, *Chem. Phys. Lett.* **224**, 195 (1994).

[309] J. Hrušák, W. Koch, and H. Schwarz, *J. Chem. Phys.* **101**, 3898 (1994).

[310] J. Hrušák, S. Ten-no, and S. Iwata, *J. Chem. Phys.* **106**, 7185 (1997).

[311] T. J. Lee, A. P. Rendell, and P. R. Taylor, *J. Phys. Chem.* **94**, 5463 (1990).

[312] C. W. Bauschlicher, Jr., S. R. Langhoff, H. Partridge, and L. A. Barnes, *J. Chem. Phys.* **91**, 2399 (1989).

[313] P. B. Armentrout and R. Georgiadis, *Polyhedron* **7**, 1573 (1988).

[314] R. J. Rico and M. Head-Gordon, *Chem. Phys. Lett.* **213**, 224 (1993); R. J. Rico, T. J. Lee, and M. Head-Gordon, *Chem. Phys. Lett.* **218**, 139 (1994).

[315] G. E. Scuseria and H. F. Schaefer, III, *J. Chem. Phys.* **90**, 3700 (1989).

[316] M. Nooijen and J. G. Snijders, *Intern. J. Quantum Chem.* **47**, 3 (1993).

[317] H. Hellmann, *Einführung in die Quantenchemie*, Franz Deuticke, Vienna, 1937, p. 285 (in German); R. Feynman, *Phys. Rev.* **56**, 340 (1939).

[318] D. N. Zubarev, *Nonequilibrium Statistical Thermodynamics*, Consultants Bureau, New York, 1974, Chap. 3.

[319] E. K. U. Gross, E. Runge, and O. Heinonen, *Many-Particle Theory*, Adam Hilger, Bristol, 1991.

[320] P. Pulay, in *Modern Electronic Structure Theory*, Part II, D. R. Yarkony, Ed., World Scientific, Singapore, 1995, pp. 1191–1240.

[321] E. A. Kondo, P. Piecuch, and J. Paldus, *J. Chem. Phys.* **104**, 8566 (1996).

[322] J. Arponen, *Ann. Phys.* **151**, 311 (1982).

[323] M. Fink, *Nucl. Phys. A* **221**, 163 (1974).

[324] J. Noga and M. Urban, *Theor. Chim. Acta* **73**, 291 (1988).

[325] M. Urban, G. H. F. Diercksen, A. J. Sadlej, and J. Noga, *Theor. Chim. Acta* **77**, 29 (1990).

[326] M. Medved', M. Urban, and J. Noga, *Theor. Chem. Accounts*, **98**, 75 (1997).

[327] S. Pal, *Theor. Chim. Acta* **66**, 151 (1984).

[328] K. B. Ghose and S. Pal, *Phys. Rev. A* **36**, 1539 (1987); K. B. Ghose, P. G. Nair, and S. Pal *Chem. Phys. Lett.* **211**, 15 (1993); K. B. Ghose, *Intern. J. Quantum Chem.* **53**, 275 (1995).

[329] P. G. Szalay, M. Nooijen, and R. J. Bartlett, *J. Chem. Phys.* **103**, 281 (1995).

[330] R. J. Bartlett, S. A. Kucharski, J. Noga, J. D. Watts, and G. W. Trucks, in *Many-Body Methods in Quantum Chemistry, Lecture Note in Chemistry*, Vol. 52, U. Kaldor, Ed., Springer-Verlag, Berlin, 1989, pp. 125–149.

[331] W. Kutzelnigg, *J. Chem. Phys.* **77**, 3081 (1982); **80**, 822 (1984); S. Pal, *Theor. Chim. Acta* **66**, 207 (1984); K. Tanaka and H. Terashima, *Chem. Phys. Lett.* **106**, 558 (1984); M. Hoffman and J. Simons, *J. Chem. Phys.* **88**, 993 (1988).

[332] H. D. Cohen and C. C. J. Roothaan, *J. Chem. Phys.* **43**, S34 (1965); A. D. McLean and M. Yoshimine, *J. Chem. Phys.* **46**, 3682 (1967).

[333] A. J. Thakkar, *Phys. Rev. A* **40**, 1130 (1989); G. Maroulis and A. J. Thakkar, *Chem. Phys. Lett.* **156**, 87 (1989); *idem.*, *J. Chem. Phys.* **92**, 812 (1990); E. F. Archibong and A. J. Thakkar, *Chem. Phys. Lett.* **173**, 579 (1990); G. Maroulis, *J. Chem. Phys.* **101**, 4949 (1994).

[334] H. Sekino and R. J. Bartlett, *Intern. J. Quantum Chem. Symp.* **21**, 487 (1987).

[335] P. O. Nebrandt, B. O. Roos, and A. J. Sadlej, *Intern. J. Quantum Chem.* **15**, 135 (1979); T. U. Helgaker and J. Almlöf, *ibid.* **26**, 275 (1984).

[336] T. Helgaker and P. Jørgensen, *Adv. Quantum Chem.* **19**, 183 (1988).

[337] Y. Luo, H. Ågren, P. Jørgensen, and K. V. Mikkelsen, *Adv. Quantum Chem.* **26**, 165 (1995); H. Ågren, O. Vahtras, and B. Minaev, *ibid.* **27**, 71 (1996).

[338] A. C. Scheiner, G. E. Scuseria, J. E. Rice, T. J. Lee, and H. F. Schaefer, III, *J. Chem. Phys.* **87**, 5361 (1987).

[339] H. Koch, H. J. Aa. Jensen, P. Jørgensen, T. Helgaker, G. E. Scuseria, and H. F. Schaefer, III, *J. Chem. Phys.* **92**, 4924 (1990).

[340] L. Adamowicz, W. D. Laidig, and R. J. Bartlett, *Intern. J. Quantum Chem. Symp.* **18**, 245 (1984).

[341] E. A. Salter, G. W. Trucks, and R. J. Bartlett, *J. Chem. Phys.* **90**, 1752 (1988).

[342] J. Gauss, J. F. Stanton, and R. J. Bartlett, *J. Chem. Phys.* **95**, 2623, 2639 (1991).

[343] P. Piecuch and J. Paldus, *J. Math. Chem.* **21**, 51 (1997).

[344] N. C. Handy and H. F. Schaefer, III, *J. Chem. Phys.* **81**, 5031 (1984).

[345] A. Dalgarno and J. T. Lewis, *Proc. Roy. Soc.* (London) **A233**, 70 (1955); A. Dalgarno and A. L. Stewart, *ibid.* **A238**, 269, 276 (1956); **A247**, 245 (1958).

[346] L. Schiff, *Quantum Mechanics*, 3d ed., McGraw-Hill, New York, 1968.

[347] J. Stuber, *M. Math. Thesis*, University of Waterloo, Waterloo, Ontario, Canada, 1997.

[348] J.-P. Daudey, J.-L. Heully, and J.-P. Malrieu, *J. Chem. Phys.* **99**, 1240 (1990); J.-P. Malrieu, J.-P. Daudey, and R. Caballol, *ibid.* **101**, 8908 (1994); I. Nobet-Gil, J. Sáchez-Marín, J.-P. Malrieu, J.-L. Heully, and D. Maynau, *ibid.* **103**, 2576 (1995).

[349] J. Paldus and J. Čížek, *J. Chem. Phys.* **60**, 149 (1974).

[350] A. K. Wilson and T. H. Dunning, Jr., *J. Chem. Phys.* **106**, 8718 (1997).

[351] T. Helgaker, J. Gauss, P. Jørgensen, and J. Olsen, *J. Chem. Phys.* **106**, 6430 (1997).

[352] D. Feller and K. A. Peterson, *J. Chem. Phys.* **108**, 154 (1998).

[353] J. Almlöf and P. R. Taylor, *J. Chem. Phys.* **86**, 4070 (1987).

[354] P.-O. Widmark, P.-Å. Malmqvist, and B. O. Roos, *Theor. Chim. Acta* **77**, 291 (1990); P.-O. Widmark, B. J. Persson, and B. O. Roos, *ibid.* **79**, 419 (1991).

[355] D. E. Woon and T. H. Dunning, Jr., *J. Chem. Phys.* **100**, 2975 (1994).

[356] D. E. Woon and T. H. Dunning, Jr., *J. Chem. Phys.* **103**, 4572 (1995).

[357] A. Sadlej, *Coll. Czech. Chem. Commun.* **53**, 1995 (1988); *Theor. Chim. Acta*, **79**, 123 (1991).

[358] R. A. Chiles and C. E. Dykstra, *J. Chem. Phys.* **74**, 4544 (1981).

[359] L. Z. Stolarczyk and H. J. Monkhorst, *Intern. J. Quantum Chem. Symp.* **18**, 267 (1984).

[360] G. E. Scuseria and H. F. Schaefer, III, *Chem. Phys. Lett.* **142**, 354 (1987).

[361] N. C. Handy, J. A. Pople, M. Head-Gordon, K. Raghavachari, and G. W. Trucks, *Chem. Phys. Lett.* **164**, 185 (1989); K. Raghavachari, J. A. Pople, E. S. Replogle, M. Head-Gordon, and N. C. Handy, *ibid.* **167** 115 (1990).

[362] T. J. Lee, R. Kobayashi, N. C. Handy, and R. D. Amos, *J. Chem. Phys.* **96**, 8931 (1992); R. Kobayashi, H. Koch, P. Jørgensen, and T. J. Lee, *Chem. Phys. Lett.* **211**, 94 (1993).

[363] K. Hirao, in *Self-Consistent Field Theory, Theory and Applications. Studies in Physical and Theoretical Chemistry*, Vol. 70, R. Carbó and M. Klobukowski, Eds., Elsevier, Amsterdam, 1990, pp. 531–567.

[364] J. F. Stanton, J. Gauss, and R. J. Bartlett, *J. Chem. Phys.* **97**, 5554 (1992).

[365] C. Hampel, K. A. Peterson, and H.-J. Werner, *Chem. Phys. Lett.* **190**, 1 (1992); **192**, 332 (1992) (E).

[366] L. A. Barnes and R. Lindh, *Chem. Phys. Lett.* **223**, 207 (1994).

[367] K. Jankowski, K. Kowalski, K. Rubiniec, and J. Wasilewski, *Intern. J. Quantum Chem.* **67**, 205 (1998); K. Jankowski, J. Gryniaków, and K. Rubiniec, *ibid.* **67**, 221 (1998); K. Jankowski, L. Meissner, and K. Rubiniec, *ibid.* **67**, 239 (1998).

[368] K. Jankowski, K. Rubiniec, and P. Sterna, *Mol. Phys.*, **94**, 29 (1998).

[369] B. O. Roos and P. E. M. Siegbahn, in *Methods of Electronic Structure Theory, Modern Theoretical Chemistry*, Vol. 3, H. F. Schaefer III, Ed., Plenum Press, New York, 1977, pp. 277–318.

[370] G. D. Purvis and R. J. Bartlett, *J. Chem. Phys.* **75**, 1284 (1981).

[371] P. Pulay, *Chem. Phys. Lett.* **73**, 393 (1980); *J. Comp. Chem.* **3**, 556 (1982); T. P. Hamilton and P. Pulay, *J. Chem. Phys.* **84**, 5728 (1986).

[372] G. E. Scuseria, T. J. Lee, and H. F. Scheafer, III, *Chem. Phys. Lett.* **130**, 236 (1986).

[373] P. E. S. Wormer, F. Visser, and J. Paldus, *J. Comp. Phys.* **48**, 23 (1982).

[374] S. Wilson, in *Methods in Computational Chemistry*, Vol. 1, S. Wilson, Ed., Plenum Press, New York, 1987, pp. 251–309.

[375] T. J. Lee and J. E. Rice, *Chem. Phys. Lett.* **150**, 406 (1988).

[376] B. W. Char, K. O. Geddes, G. H. Gonnet, B. L. Leong, M. B. Monagan, and S. M. Watt, *First Leaves: A Tutorial Introduction to Maple V*, Springer-Verlag, Berlin, and Waterloo Maple Publishing, Waterloo, Ont., Canada, 1992.

[377] P. Botschwina, M. Horn, K. Markey, and R. Oswald, *Mol. Phys.* **92**, 381 (1997).

[378] M. Yang, M. H. Alexander, H.-J. Werner, and R. J. Bemish, *J. Chem. Phys.* **105**, 10462 (1996).

[379] J. D. Watts, J. Gauss, and R. J. Bartlett, *J. Chem. Phys.* **98**, 8718 (1993).

[380] C. W. Bauschlicher, Jr., and P. R. Taylor, *J. Chem. Phys.* **85**, 6510 (1986); C. W. Bauschlicher, Jr., S. R. Langhoff, and P. R. Taylor, *ibid.* **87**, 387 (1987).

[381] X. Li and J. Paldus, *Coll. Czech. Chem. Commun.* **63**, 1381 (1998); *idem*, unpublished results.

[382] J. H. Meadows and H. F. Schaefer, III, *J. Amer. Chem. Soc.* **98**, 4383 (1976).

[383] A. Balková and R. J. Bartlett, *J. Chem. Phys.* **102**, 7116 (1995).

[384] P. Saxe, H. F. Schaefer, III, and N. C. Handy, *Chem. Phys. Lett.* **79**, 202 (1981); R. J. Harrison and N. C. Handy, *ibid.* **95**, 386 (1983).

[385] H. F. Schaefer, III, J. R. Thomas, Y. Yamaguchi, B. J. DeLeeuw, and G. Vacek, in *Modern Electronic Structure Theory*, Part I, D. R. Yarkony, Ed., World Scientific, Singapore, 1995, pp. 3–54.

[386] T. J. Lee and J. E. Rice, *J. Chem. Phys.*, **94**, 1215 (1991).

[387] L. A. Barnes, B. Liu, and R. Lindh, *J. Chem. Phys.* **98**, 3972 (1993).

[388] K. P. Huber and G. Herzberg, *Molecular Spectra and Molecular Structure*, Vol. 4, *Constants of Diatomic Molecules*, Van Nostrand and Reinhold, New York, 1979.

[389] C. W. Bauschlicher and H. Partridge, *J. Chem. Phys.* **100**, 4329 (1994).

[390] J. M. L. Martin, *J. Chem. Phys.* **97**, 5012 (1992).

[391] K. A. Peterson and T. H. Dunning, Jr., *J. Chem. Phys.* **102**, 2032 (1995).

[392] D. E. Woon and T. H. Dunning, Jr., *J. Chem. Phys.* **101**, 8876 (1994).

[393] G. C. Hancock, D. G. Truhlar, and C. E. Dykstra, *J. Chem. Phys.* **88**, 1786 (1988).

[394] P. R. Bunker, P. Jensen, A. Karpfen, M. Kofranek, and H. Lischka, *J. Chem. Phys.* **92**, 7432 (1990).

[395] S. C. Racine and E. R. Davidson, *J. Phys. Chem.* **97**, 6367 (1993).

[396] D. J. Tozer, N. C. Handy, R. D. Amos, J. A. Pople, R. H. Nobes, Y. Xie, and H. F. Schaefer, III, *Mol. Phys.* **79**, 777 (1993).

[397] F. Jensen, *Chem. Phys. Lett.* **169**, 519 (1990).

[398] J. S. Francisco, *J. Chem. Phys.* **107**, 9039 (1997); **108**, 659 (1998).

[399] J. F. Stanton, W. N. Lipscomb, D. H. Magers, and R. J. Bartlett, *J. Chem. Phys.* **90**, 1077 (1989).

[400] T. J. Lee and G. E. Scuseria, *J. Chem. Phys.* **93**, 489 (1990).

[401] J. D. Watts and R. J. Bartlett, *J. Chem. Phys.* **108**, 2511 (1998).

[402] J. B. Milan, W. J. Buma, and C. A. de Lange, *J. Chem. Phys.* **104**, 512 (1996); S. J. Dunlavey, J. M. Dyke, N. K. Fayad, N. Jonathan, and A. Morris, *Mol. Phys.* **38**, 729 (1979).

[403] K. Raghavachari, *J. Chem. Phys.* **82**, 4142 (1985).

[404] R. Chaudhuri, D. Mukhopadhyay, and D. Mukherjee, *Chem. Phys. Lett.* **190**, 231 (1992); D. Sinha, D. Mukhopadhyay, R. Chaudhuri, and D. Mukherjee, *Chem. Phys. Lett.* **154**, 544 (1989).

[405] C. M. L. Rittby and R. J. Bartlett, *Theor. Chim. Acta* **80**, 469 (1991); M. Nooijen and R. J. Bartlett, *J. Chem. Phys.* **102**, 3629 (1995).

[406] J. D. Watts, S. R. Gwaltney, and R. J. Bartlett, *J. Chem. Phys.* **105**, 6979 (1996).

[407] J. D. Watts and R. J. Bartlett, *J. Chem. Phys.* **101**, 3073 (1994).

[408] J. D. Watts and R. J. Bartlett, *Chem. Phys. Lett.* **258**, 581 (1996).

[409] H. Koch, H. J. Aa. Jensen, P. Jørgensen, and T. Helgaker, *J. Chem. Phys.* **93**, 3345 (1990).

[410] O. Christiansen, H. Koch, and P. Jørgensen, *J. Chem. Phys.* **103**, 7429 (1995).

[411] J. Olsen, A. M. Sanchez de Meras, H. J. Aa. Jensen, and P. Jørgensen, *Chem. Phys. Lett.* **154**, 380 (1989).

[412] D. Yeager, V. McKoy, and S. A. Segal, *J. Chem. Phys.* **61**, 755 (1974).

[413] K. Watanabe and M. Zelikoff, *J. Opt. Soc. Amer.* **43**, 753 (1953).

[414] S. Berkovic and U. Kaldor, *J. Chem. Phys.* **98**, 3090 (1993).

[415] D. P. Shelton and J. E. Rice, *Chem. Revs.* **94**, 3 (1994).

[416] P. Piecuch, V. Špirko, A. E. Kondo, and J. Paldus, *Mol. Phys.* **94**, 55 (1998).

[417] P. Piecuch, V. Špirko, and J. Paldus, *Polish J. Chem.* **72**, 1635 (1998).

[418] C. Hättig, O. Christiansen, and P. Jørgensen, *J. Chem. Phys.* **107**, 10592 (1997).

[419] J. F. Stanton and R. J. Bartlett, *J. Chem. Phys.* **99**, 1752 (1993).

[420] P. B. Rozyczko, S. A. Perera, M. Nooijen, and R. J. Bartlett, *J. Chem. Phys.* **107**, 6736 (1997).

[421] A. P. Rendell, T. J. Lee, A. Komornicki, and S. Wilson, *Theor. Chim. Acta* **84**, 271 (1993); A. P. Rendell, T. J. Lee, and R. Lindh, *Chem. Phys. Lett.* **194**, 84 (1992); A. P. Rendell, M. F. Guest, and R. A. Kendall, *J. Comp. Chem.* **14**, 429 (1993).

[422] L. Adamowicz and R. J. Bartlett, *J. Chem. Phys.* **86**, 6314 (1987).

[423] Y.-J. Ye, W. Förner, and J. Ladik, *Chem. Phys.* **178**, 1 (1993); R. Knab, W. Förner, J. Čížek, and J. Ladik, *J. Mol. Struct.* (*THEOCHEM*) **366**, 11 (1996); R. Knab, W. Förner, and J. Ladik, *J. Phys.: Condens. Matter* **9**, 2043 (1997); W. Förner, R. Knab, J. Čížek, and J. Ladik, *J. Chem. Phys.* **106**, 10248 (1997).

[424] T. P. Živković, *Intern. J. Quantum Chem. Symp.* **11**, 413 (1977); T. P. Živković and H. J. Monkhorst, *J. Math. Phys.* **19**, 1007 (1978).

[425] K. Jankowski, K. Kowalski, and P. Jankowski, *Intern. J. Quantum Chem.* **50**, 353 (1994); **53**, 501 (1995); *Chem. Phys. Lett.* **222**, 608 (1994).

[426] D. Cox, J. Little, and D. O'Shea, *Ideals, Varieties, and Algorithms: An Introduction to Computational Algebraic Geometry and Commutative Algebra*, 2d ed., Springer-Verlag, New York, 1992.

[427] T. Becker and V. Weispfenning, *Gröbner Bases; A computational Approach to Commutative Algebra*, Springer-Verlag, New York, 1993.

[428] R. Fröberg, *An Introduction to Gröbner Bases*, J. Wiley & Sons, Chichester, 1997.

[429] V. Guillemin and A. Pollack, *Differential Topology*, Prentice-Hall, Englewood Cliffs, NJ, 1974.

[430] A. Morgan, *Solving Polynomial Systems Using Continuation for Engineering and Scientific Problems*, Prentice-Hall, Englewood Cliffs, NJ, 1987.

[431] M. D. Gould, J. Paldus, and G. S. Chandler, *J. Chem. Phys.* **93**, 4142 (1990).

CHAPTER 2

ON THE ELECTRONIC SPECTRA OF SMALL LINEAR POLYENES

RUTH McDIARMID

Laboratory of Chemical Physics, National Institute of Diabetes and Digestive and Kidney Diseases, National Institutes of Health, Bethesda, Maryland 20892-0510

CONTENTS

I. INTRODUCTION

An understanding of the electronic spectra of small polyenes such as ethylene, butadiene, and hexatriene was recognized as fundamental to electronic theory at its birth [1]. Yet despite repeated reinvestigations of these spectra since they were first observed more than 60 years ago [2, 3], many features of the spectra of these molecules remain unassigned, and many excited states of the molecules have not been observed.

The reasons for these failures lie in the properties of these molecules. The polyenes possess both low-lying intravalence excited states and the usual manifold of extravalence (Rydberg) excited states [4, 5]. The equilibrium geometries of the excited states of each molecule differ from that of its

Advances in Chemical Physics, Volume 110, Edited by I. Prigogine and Stuart A. Rice. ISBN 0-471-33180-5. © 1999 John Wiley & Sons, Inc.

ground state. Consequently, the electronic transitions of the molecules possess long Franck–Condon (vibrational) progressions. And because the electronic transitions of each molecule are relatively closely spaced, vibrational progressions belonging to different electronic transitions overlap and complicate even the identifications of the origins of the electronic transitions. In addition, in the small linear polyenes—ethylene, butadienes, hexatrienes, etc.—the intravalence excited states are exceedingly short lived. This means that they both give rise to broad spectral bands and, through various perturbations, broaden the spectral bands of transitions to adjacent Rydberg states. Rotational analysis, the bedrock of spectroscopic assignment, is not possible for most transitions of ethylene, butadiene, etc. These intrinsic molecular properties hinder our ability to understand the experimental spectra of the small polyenes.

Theoretical models describing these molecules have also been in development for over 60 years [1]. During this time, numerical calculations based on these models have been carried out to determine the excited state energies and geometries of polyenes [4, 6]. Like the experimental attempts to locate and identify these excited states, the numerical calculations have not been fully successful. Handicaps to accurate calculations arise from the different natures of the intra- and extravalence states, which generate difficulties in calculating the two types of states at equivalent levels of accuracy. Additional complications arise from the geometry dependence of the couplings between the two classes of states. There is also a controversy as to whether two orbitals of the same nodal structure can have independent existences [7]. Such orbitals pairs include the b_{2g} $3d\pi$ and π^* orbitals of ethylene and the a_u $3p\pi$ and π^*, and the b_g $3d\pi$ Rydberg and π, orbitals of butadiene.

Throughout the history of the investigations of these molecules, experimentalists have sought to validate their assignments by correlation with the results of empirical calculations. Similarly, theorists have sought to validate their calculations by correlation with the results of experimental investigations. There is, unfortunately, a discrepancy between the number of states that can be calculated and the number that have been observed in each molecule. These disparities are displayed in Table 1. Faced with these disparities and the ensuing mapping problem, both experimentalists and theorists have tended to "correlate" the states they find with those states obtained by the other method that are closest in energy. And, unfortunately, such "correlations" need not yield the best or the most internally consistent set of assignments. The observation of fewer states than calculated also prevents the resolution of the Rydberg-valence conjugate-state problem.

The goal of this article is to rectify the situation to the greatest extent possible. First, newer spectroscopic techniques that can be used to assign

TABLE 1

Number of possible, calculated, and observed excited states of polyenes (no./no. calculated/no. observed)

State[a]	Molecule			
	Ethylene	Butadiene	Trans-hexatriene	Cis-hexatriene
$\pi\pi^*$	2/1/1	5/2/1	> 5/2/1	> 5/2/2
$\pi 3s$	1/1/1	1/1/1	1/1/1	1/1/1
$\pi 3p$	3/3/3	3/3/2	3/3/2	3/3/2
$\pi 3d$	5/5/3	5/5/2	5/5/1	5/5/2

[a] Only the partially filled orbitals are indicated.

rotationally unresolvable spectra will be discussed. Next, these techniques will be used to maximally assign the excited states of ethylene, butadiene, and the hexatrienes. The experimental transition energies to the states so assigned will then be compared with corresponding transition energies obtained from high-quality modern quantum mechanical calculations. Reasonable experimentally based estimates of the adiabatic and vertical transition energies to the important, unobserved, A_g valence-excited state of butadiene and of its vertical transition energy in hexatriene will be made and compared with calculated vertical transition energies. Experimental and computational evidence concerning the relative energies and equilibrium geometries of the $2A_g$ and B_u excited states of butadiene will be discussed. Finally, on the basis of the previous comparisons, conclusions will be drawn concerning the strengths and weaknesses of the best current quantum mechanical calculations of excited states of ethylene, butadiene, and hexatriene, the excited-state ordering in butadiene, and the equilibrium geometries of the excited states of butadiene.

II. EXPERIMENTAL TECHNIQUES

Molecular spectroscopy was initially "the study of the absorption or emission of electromagnetic radiation by molecules"[8]. The observation of both the absorption and emission spectra of a transition facilitates the identification of its origin and the assignment of its vibrational substructure, the former by the appearance of the origin in both absorption and emission spectra, the latter by correlation of the emission substructure with the ground-state vibrational assignment of the molecule. This advantage cannot be used in assigning the excited states of ethylene, butadiene, and the hextrienes, because these states do not emit light. The other traditional technique, the analysis of the rotational structure of an absorption or emission band, is likewise precluded for ethylene, butadiene, etc., because few of the transitions of these molecules are rotationally resolved.

Other useful older spectroscopic techniques include isotopic (mass) substitution and temperature variation. Isotopic substitution does not change electronic potentials or, therefore, electronic spectra, but because vibrational frequencies are dependent on mass, the same electronic transition in different isotopomers possesses different vibrational and vibronic substructures. Isotopic substitution can thus be used both to identify electronic origins, which minimally shift on isotopic substitution, and to assign vibrational substructures [9].

Temperature variation methods include an older and a newer technique. In the older method, a molecule is heated sufficiently to populate, or cooled sufficiently to depopulate, excited vibrational levels of the ground electronic state; then the spectra of "hotter" and "colder" molecules are measured and compared. Bands enhanced at higher temperatures include both "hot" transitions (bands whose vibrational quantum numbers decrease on excitation) of symmetric modes and sequence bands (bands that originate on vibrationally excited levels of the ground electronic state and whose vibrational quantum numbers remain the same on excitation) of all vibrations. The hot bands arise from Franck–Condon-enabled vibrations, the relative intensities of which depend on their intrastate vibrational overlap integrals and, through the Boltzmann equation, on their ground-state fundamental frequencies and the temperature of the sample. The sequence bands, the relative intensities of which depend only on their Boltzmann factors, arise predominantly from the lower frequency ground-state vibrational modes. The observation of hot bands thus aids in identifying the Franck–Condon active modes. The observation of sequence bands aids in assigning upper-state vibrational frequencies. In the newer method, the molecule of interest is entrained in a carrier gas, the gas mixture is expanded, and the spectrum of the sample is observed in the supersonic jet [10]. Collisions within the jet cool the internal rotational and vibrational temperatures of the molecule unequally [10]: Under normal expansion conditions, the rotational temperature of the sample is $10°–20°$ and the vibrational temperature is at, or near, room temperature. Reducing rotational temperatures sharpens the spectral bands and thus facilitates their detection. Reducing vibrational temperatures, as in the traditional technique, depopulates vibrationally excited bands of the ground electronic state and hence serves as in the traditional method described above.

Newer spectroscopic techniques were developed subsequent to the dissemination of the laser as an experimental tool. One such technique is *resonant multiphoton spectroscopy* [11]. With this techniques an electronic excited state of an atom or a molecule is created by the simultaneous absorption of more than one photon. Thus, excited states to which transitions are forbidden by one photon absorption, but are allowed by n-photon absorption can be promoted. By observing an experimental multiphoton selection rule and comparing it with theoretical selection rules for the appropriate symmetry group of the molecule, some transitions can be partially assigned. These transitions include $g \leftrightarrow g$ transitions in molecules that possess a center of symmetry and A_2 transitions in molecules with a two- or more-fold rotation axis [12]. Note, however, that states to which one-photon allowed transitions are weak, but two-photon allowed transitions are strong, may appear to be one-photon forbidden, two-photon allowed—hence A_2. Resonant multiphoton absorption is usually detected by monitoring emitted light (fluorescence) or emitted heat (photoacoustics) as the molecule relaxes back to the ground state or by monitoring emitted ions [11]. The latter method, called *resonantly enhanced multiphoton ionization spectroscopy* (REMPI), requires the ionization of the excited state of the molecule and is thus a time-dependent spectroscopic tool, as well as a selection-rule-based tool. For REMPI to occur the excited state must exist long enough to absorb the additional photon(s) needed for ionization. For typical ionization cross sections and nsec lasers (power $\sim 10^{28}$ photons/sec), this is roughly 100 psec. Thus, transitions to very short-lived states can be distinguished from transitions to longer lived states by comparing their REMPI and absorption spectra.

Polarization-selected multiphoton resonant spectroscopy [13], another laser-based technique, is very useful in assigning some nonrotationally resolvable transitions. In this technique, the relative intensity of the signal detected when a molecule is excited with circularly polarized light is compared with that detected when the molecule is excited with linearly polarized light. The ratio of the two intensities depends on the number of resonant photons and the symmetry of the transition. For two-photon resonances, this ratio is 3/2 for all non-totally symmetric transitions and, although not predicted by group theory for totally symmetric transitions, it is often significantly less. Thus, symmetric transitions can often be identified by simply observing the intensity ratios of the polarization-selected two-photon resonant experimental spectra of the transitions [14]. For three-photon resonances, other selection rules apply.

An example of the first two techniques, the use of selection rules and excited-state lifetimes in assigning electronic spectra, is illustrated in Figure 1, in which the one-photon absorption spectrum and the two-photon and three-photon REMPI spectra of part of the spectrum of *cis*-hexatriene are compared [15]. Many of the same bands appear in all three spectra. However, the absorption spectrum shows increasing background absorption on the long-wavelength side that is absent from the REMPI spectra. This

TRANSITION ENERGY (cm⁻¹)

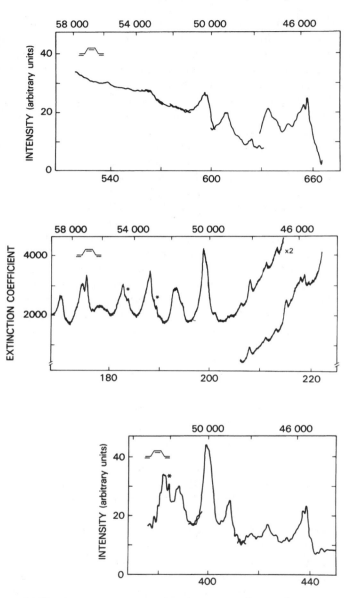

EXPERIMENTAL WAVELENGTH (nm)

Figure 1. Two- (lower) and three- (upper) photon resonantly enhanced multiphoton ionization and one- (center) photon absorption spectra of *cis*-hexatriene. The asterisks indicate impurity bands. (Reproduced from [15] by permission of the American Institute of Physics.)

difference indicates that the lifetime of the upper state of the transition underlying the long-wavelength side of Figure 1 is shorter than those of the other transitions that appear in the figure. The two-photon REMPI spectrum also contains bands around 49,000 and 52,500 cm^{-1} that are absent from the absorption spectrum. These bands arise from transitions that are one-photon forbidden, two-photon allowed in the C_{2v} symmetry of *cis*-hexatriene. An example of the third technique, the use of polarization-selected excitation in assigning electronic spectra, is given in Figure 2. Here, the REMPI spectrum of *cis*-hexatriene obtained with circularly polarized excitation is compared with the spectrum obtained with linearly polarized excitation. Six strong bands are observed in each spectrum. With circularly polarized excitation, the 390- and 400-nm (experimental wavelength) bands are greatly reduced in intensity relative to the 440-, 425-, 410-, and 381-nm bands compared with the relative intensities of these bands generated with linearly polarized excitation. Thus, in the C_{2v} symmetry of *cis*-hexatriene, the use of these three techniques suggests assignments for four of the six transitions observed in the experimental REMPI spectrum presented here. These results will be discussed further in Section V.A.

Electron energy loss spectroscopy is another technique that has been employed to locate and identify excited states [16]. Assignments based on the results of measurements made with this technique, especially in the earlier investigations, must be used with great caution. Because the transitions of ethylene, butadiene, and hexatriene have extensive and overlapping vibrational substructures, and because the resolution of

TRANSITION ENERGY (cm⁻¹)

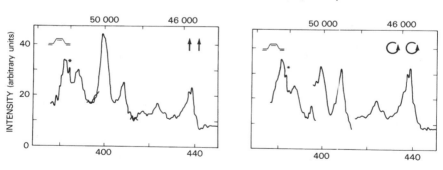

EXPERIMENTAL WAVELENGTH (nm)

Figure 2. Two-photon resonantly enhanced multiphoton ionization spectra of *cis*-hexatriene excited with circularly (right) and linearly (left) polarized light. The asterisk indicates an impurity band. (Reproduced from [15] by permission of the American Institute of Physics.)

electron energy loss spectroscopy is poor by spectroscopic criteria, many vibrational subbands have been incorrectly assigned as electronic origins either by the original or subsequent investigators. One variant of the technique, *parameter-selected electron impact spectroscopy*, is more useful in locating and identifying electronic transitions in small organic molecules. (In electron-impact spectroscopy, the experimental parameters of incident energy and scattering angle can be independently varied.) Theoretically, this variant can unambiguously distinguish singlet from triplet, and dipole allowed from dipole forbidden, transitions. It has been successfully used to distinguish vibrational substructures from electronic origins [16]. However, like ordinary electron energy loss spectroscopy, it, too, suffers from poor resolving power by even medium-resolution spectroscopic criteria. This means that the vibrational substructures of closely spaced electronic transitions can be so significantly overlapped, that the interpretation of the results is ambiguous at best. As applied, the technique is also not a purely experimental one, because it depends on assumptions about the relative intensities of different types of transitions. Parameter-selected electron impact spectroscopy is, therefore, useful in classifying electronic transitions, but must be employed with caution.

Once an electronic transition has been identified and, if possible, assigned by symmetry, the next step is to give the transition an electronic assignment. In the polyenes, one-photon allowed valence-excited transitions are usually assigned because of the distinctive appearance of their absorption bands. Rydberg transitions, by contrast, all look quite similar. However, because the ionization energies of Rydberg states of a given (s,p,d) type (i.e., their term values) are usually similar, empirical rules have been developed to assist in their assignment [17]. According to these rules, effective quantum numbers or quantum defects are calculated from the experimental transition energy via the equation $n^* = (R/(IP - \bar{\nu}))^{1/2}$, where R is the Rydberg constant (13.6 eV), IP is the ionization potential, $\bar{\nu}$ is the experimental frequency, and n^* is the effective quantum number. Note that $n^* = n - \delta$, where n is the primary quantum number and δ is the quantum defect. For unperturbed transitions, $\delta \sim 1$ for transitions to s Rydberg states, $\delta = 0.4$–0.6 for transitions to p Rydberg states, $\delta = 0$–0.2 for transitions to d Rydberg states, and $\delta \sim 0$ for transitions to f and higher Rydberg states.

III. ETHYLENE

Ground-state ethylene is a planar molecule of D_{2h} symmetry. The symmetry species of its lower excited states are given in Table 2. These states include two excited valence states plus numerous Rydberg states. As noted earlier, there is a controversy as to whether the B_{1u} $(\pi\pi^*)$ state is distinct from the

TABLE 2
Symmetries of the excited states of polyene

State[a]	Molecule			
	Ethyl.[b]	Butyl[b]	Trans-hexatriene[b]	Cis-hexatriene[b]
$\pi\pi^*$	B_{1u}	B_u	B_u	B_1
$\pi\pi^*$	A_g	A_g	A_g	A_1
$\pi3s$	B_{3u}	B_g	A_u	B_1
$\pi3p\pi$	B_{1g}	A_u	B_g	B_1
$\pi3p\sigma$	B_{2g}	A_u	B_g	A_2
$\pi3p\pi$	A_g	B_u	A_g	A_1
$\pi3d\sigma$	B_{3u}	B_g	A_u	B_1
$\pi3d\pi$	B_{1u}	A_g	B_u	B_2
$\pi3d\pi$	A_u	B_g	A_u	A_2
$\pi3d\delta$	B_{2u}	A_g	B_u	A_1
$\pi3d\delta$	B_{3u}	B_g	A_u	B_1

[a] Only the partially filled orbitals are indicated.

[b] For ethylene, the z-axis is along the C=C bonds and the x-axis is out of plane. For butadiene and trans-hexatriene, the z-axis is out of plane and the x-axis is along the carbon chain. For cis-hexatriene, the z-axis is in the molecular plane, perpendicular to the carbon chain, and the x-axis is along the carbon chain. These are not the axis orientations used in some of the numerical calculations to be discussed subsequently. The axes and symmetry labels of these calculations have been rotated to correspond with the conventions presented in this table.

B_{1u} ($\pi3d\pi$) Rydberg state of ethylene [7]. The lower energy one-photon allowed transitions of ethylene were first studied before 1940 [4]. The spectrum of this energy region, presented in Figure 3 [18] is seen to be composed of a large number of poorly resolved bands.

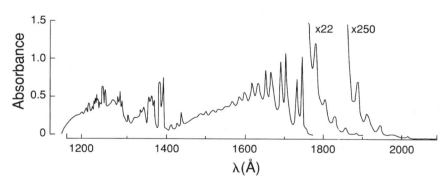

Figure 3. Survey optical spectrum of ethylene. (Reproduced from [18] by permission of the Optical Society of America.)

The lowest energy transition—the diffuse bands extending from 200 to 180 nm in the figure—has been assigned as the lowest energy valence-excited transition [4]. Because ethylene is isoelectronic with O_2, in which the O–O bond length in the lowest energy valence-excited state is much longer than in the ground electronic state, it was proposed that the C–C bond length in the lowest energy valence-excited state of ethylene is elongated [4]. It was also proposed that the CH_2 groups of ethylene are rotated 90° from each other in this excited state [4, 19]. Therefore, initially it was uncertain whether the long vibrational progression observed in the 200-nm transition was a progression in the CH_2 torsion or in the C=C stretch mode [20]. This vibration has since been demonstrated, by temperature variation and isotopic substitution techniques, to be a progression in the torsional mode [21]. The electronic origin of the state, although not directly observed, has been extrapolated to 5.5 eV in an investigation of all six ethylene isotopomers [22]. The maximum of the $\pi\pi^*$ transition, which is more than 2 eV above its origin, cannot be unambiguously located because it is overlapped by the much stronger $3s$ Rydberg $\leftarrow$ X transition. The observed irregularity in the vibrational progression of the $\pi\pi^*$ transition around 6.5 eV has not been interpreted.

The remainder of the absorption bands of ethylene shown in Figure 3 arise from Rydberg $\leftarrow$ X transitions. The intense one-photon allowed transition observed around 175 nm has long been assigned as the $3s$ Rydberg $\leftarrow$ X transition converging on the first ionization potential [23]. Initially, it was included in a Rydberg series with a quantum defect of .08 despite its experimental quantum defect of 1.0 [23]. It is no longer considered to be part of this series [24]. Although the 175-nm transition is rotationally diffuse, from a careful vibrational analysis of the spectrum, ethylene was deduced to be nonplanar in that excited state [25]. This deduction was subsequently confirmed by a rotational contour analysis of the origin band of the transition in C_2D_4 [22]. In the same investigation, the equilibrium geometry and symmetry assignment of the $3s$ Rydberg state of ethylene were also deduced [22]. Both the B_{3u} symmetry assignment of the transition and its quantum defect confirmed its $3s$ Rydberg assignment.

np Rydberg $\leftarrow$ X transitions of ethylene are optically (one-photon) forbidden. One $3p \leftarrow$ X transition is observed in the optical spectrum, presumably as a vibronically enabled false origin [23, 24, 26]. This transition has also been observed by threshhold electron impact [27] and variable-parameter electron impact [28] spectroscopies. The $np \leftarrow$ X Rydberg transitions of ethylene are two-photons allowed. Two of the three $3p$ Rydberg transitions have been located by two-photon REMPI [29]. In a subsequent polarization-selected two-photon REMPI experiment on a jet-cooled sample, one of these transitions was unambiguously assigned as the

A_g 3p ← X Rydberg transition [30]. The third 3p Rydberg state was tentatively located in the same energy region [30]. An alternative location for this transition around 7.2 eV, deduced from magnetic circular dichroism [31] and resonant Raman intensity measurements [32], is not supported by the more direct techniques.

The next higher energy region, 8.6–9.0 eV, should contain the 4s and five 3d Rydberg ← X transitions. The A_u 3d band is both one-photon and two-photons forbidden and is not expected to be observed. There are, therefore, five potentially observable, closely spaced, one-photon allowed transitions in this energy region of the spectrum of ethylene. Two origins are observed here in the optical spectrum—one at 8.6 and the other around 8.9 eV [23, 24, 26]. The magnetic circular dichroism spectrum of the higher energy of these origins has been interpreted as containing both the $3d\delta$ B_{2u} and $3d\delta$ B_{3u} electronic origins [33]. A 9.05-eV band, originally assigned as an electronic origin [26], is now assigned as unresolved vibrational subbands of the 8.9-eV pair [24, 33]. Thus, just three of the six 3d + 4s Rydberg bands have been observed. The symmetries of only the bands assigned as B_{2u} and B_{3u} $3d\delta$ have been experimentally determined. A summary of the experimental vertical transition energies from the ground to the excited states of ethylene is presented in Table 3. The symmetries of only four of the seven observed transitions—those to the 3s, 3p A_g and two 3d (the B_{2u} and one of the B_{3u}) Rydberg states—have been experimentally assigned.

Three recent high-level calculations of the vertical transition energies to most of the lower excited states of ethylene have been conducted [7, 34, 35]. In each calculation, the single-configuration configuration interaction (CI) transition energy (CIS in [7], CASSCF in [35]) is approximately 0.4 eV below the multiconfiguration CI energy (CIS/MP2 in [7], PT2F in [35].) Compared with each other, however, the calculations differ. For five of the nine states studied, in both calculations the single-configuration CI and multiconfiguration CI transition energies of [7] are approximately 0.3 eV lower than the corresponding single-configuration CI and multiconfiguration CI transition energies of [35]. For these five states, the single-configuration CI energies of [7] approximate the multiconfiguration CI energies of [35]. By contrast, for the other four states—the B_{1u} $\pi\pi^*$, A_g $p\pi$, A_u $d\pi$, and B_{1u} $d\pi$ states—the converged multiconfiguration CI energies are found to be similar in both calculations. The transition energies to a tenth state—the B_{2u} $d\delta$ state—is located above 10.47 eV in [7] (it was not calculated) and at 9.18 eV in [35]. The single-configuration CI energies of [7] and the multiconfiguration CI energies of [35] are presented in Table 3 for the $\pi\pi^*$ and lowest primary quantum number s, p, and d Rydberg states of ethylene. (The results of [34] are essentially the same as those of [7] and hence are not presented here.)

TABLE 3

Experimental and calculated transition energies and oscillator strengths of ethylene

State[a]	Symmetry[b]	Experiment		Calculations			
		Energy (eV)	Oscillator Strength	Energy[c] (eV)	Oscillator Strength[c]	Energy[d] (eV)	Oscillator Strength[d]
$\pi\pi^*$	B_{1u}	> 7.7	strong	7.74	.512	8.4	.16
$3s$	B_{3u}	7.11	strong	7.13	.091	7.17	.069
$3p\pi$	B_{1g}	7.8	0	7.71	g	7.85	g
$3p\sigma$	B_{2g}	(7.9)[e]	0	7.86	g	7.95	g
$3p\pi^f$	A_g	8.28		8.02	g	8.40	g
$3d\sigma$	B_{3u}	8.62	medium	8.63	.006	8.66	.001
$3d\pi$	A_u	not observed	—	8.77	g	8.94	g
$3d\pi$	B_{1u}	not observed	—	9.09	.054	9.31	.078
$3d\delta$	B_{3u}	8.9	strong	9.11	.100	9.03	< .001
$3d\delta$	B_{2u}	not observed	—	g	g	9.18	g

[a] Only the partially filled orbitals are indicated.
[b] The z-axis lies along the C=C bond. The x-axis is perpendicular to the molecular plane. This is the axis symmetry used in Table 1. The axes of [7] and [34] have been rotated to correspond with the symmetry convention given in Table 1.
[c] CIS results of [7]. The CIS results of [34] are essentially identical.
[d] PT2F results of [35].
[e] Less secure assignment.
[f] Out-of-plane p orbital.
[g] Was not calculated.

The two sets of calculated transition energies of ethylene presented in Table 3 are correlated by symmetry and s,p,d type. The experimental transitions are correlated with the calculated by symmetry and type to the extent possible. As mentioned earlier, the experimental value of the vertical transition energy to the $\pi\pi^*$ state is somewhat uncertain. The experimental identification of the $3s$ Rydberg band is indisputable. One of the $3p$ Rydberg bands, the 7.9-eV band, has been assigned as an electronic origin, but may be a vibrational subband of the 7.8-eV transition. Both of the other $3p$ Rydberg bands are observed, but only the 8.28-eV band can be rigorously assigned; it is the A_g band. Only two $3d$ Rydberg bands have been observed. There is, however, experimental support for assigning both the B_{2u} and B_{3u} bands close to the optically observed 8.9-eV band. One other band, assigned $3d$ on the basis of its term value, is observed at 8.62 eV in this energy region. No Rydberg transitions beyond those listed in Table 3 are experimentally identified as electronic origins in ethylene. While transitions to other states, such as a $\pi\sigma^*$ state [31], have been proposed to interpret experimental results, there is no uncontestable experimental support of the existence of such excited states. Table 3 thus presents the currently most accurate experimental–theoretical correlation of the lower excited states of ethylene. We shall discuss the accuracy of the theoretical results in Section VII, but it is clear from Table 3 that the numerical calculations reproduce the experimental transition energies well. The calculated intensities, by contrast, vary widely from the relative experimental values. Specifically, the transition to the B_{1u} $3d\delta$ Rydberg state, which is calculated to be very strong [35], is not observed. The transition to the B_{3u} $3d\sigma$ Rydberg state, which is correlated with an experimental band on the basis of energy, is calculated to be vanishingly small [35]. This, too, will be discussed in Section VII.

IV. BUTADIENE

Ground-state butadiene is a planar molecule, of C_{2h} symmetry. The symmetry species of its lower excited states are given in Table 2. As with ethylene, the spectrum of butadiene has been studied for almost 60 years [3]. Also as with ethylene, calculations of the energies of butadiene's excited states (or transition energies) have been carried out for almost as long [6]. A low-resolution survey spectrum of butadiene is presented in Figure 4.

The lowest energy optically allowed transition of butadiene, the extremely diffuse bands centered around 216, 210, and 204 nm, is assigned as the B_u one-electron promoted $\pi \rightarrow \pi^*$ transition. The 210- and 204-nm subbands were traditionally assigned as members of a progression in the a_g C=C stretching vibration. More recent results obtained by semi-empirical calculations [36] and the analysis of preresonance Raman profiles of $C_4 H_6$

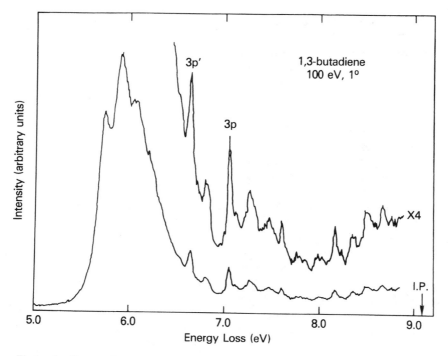

Figure 4. Survey electron energy loss spectrum of butadiene. (J. P. Doering and R. McDiarmid, unpublished.)

[37] suggest that the upper-state vibrational mode is a combination of the a_g C=C stretch and CH wagging vibrations. A reinvestigation of this portion of the spectrum of all four symmetrically deuterated isotopomers of butadiene [38] confirmed that in C_4H_6 and $C_4H_2D_4$ the CH wag mode is Franck–Condon active. This conclusion is supported by calculations of the vibrational substructures of the $\pi \rightarrow \pi^*$ transition of all four symmetric isotopomers, in which unresolved transitions to both modes were determined to constitute the "observed" bands of C_4H_6 and $C_4H_2D_4$ [39]. The experimental subbands of C_4D_6 and $C_4H_4D_2$, by contrast, appear to be predominantly progressions in the C=C stretching mode [38, 39]. Any further substructure of this transition is not resolved, even in jet-cooled samples [40]. The extreme breadth of the $\pi\pi^*$ transition remains to be fully interpreted. Because the upper state of this transition is believed to be more stable in the *s-cis* than in the *s-trans* configuration [41], *s-trans* → *s-cis* isomerization on the B_u (B_2 in the *s-cis* isomer) potential surface may shorten the lifetime of the excited state and hence broaden the spectral

bands. Alternatively, the origin of the transition to a two-electron promoted $\pi\pi^*$ state of A_g symmetry, which has been proposed to lie below the B_u state by analogy with assignments in longer polyenes [42], might provide a lifetime-shortening energy sink. This A_g transition has not been observed in a diene. The energy and properties of the two-electron promoted, A_g excited valence state will be discussed in Section VI.

The remainder of the bands present in Figure 4 arise from Rydberg $\leftarrow$ X transitions. The weak absorption bands superimposed on the descending tail of the diffuse V $\leftarrow$ N transition, first observed over 50 years ago, has been identified as a discrete, vibronically enabled transition and its vibrational subbands [43]. The transition to the electronic origin was shown to be allowed by two photon selection rules, and was assigned, from its quantum defect, as the $3s$ Rydberg $\leftarrow$ X transition [44]. That the transition has the correct B_g symmetry to be the $3s$ Rydberg $\leftarrow$ X transition was confirmed in a polarization-selection two-photon REMPI experiment [45]. The symmetry assignments of the transition's three optically enabling modes as a_u, in conjunction with its B_g electronic symmetry assignment, identifies the enabling electronic transition as B_u. Presumably, the enabling transition is the intense B_u $\pi\pi^*$ transition on which the $3s \leftarrow$ X transition is superimposed.

The higher energy transitions of butadiene have been studied by optical absorption [46, 47], by two- and three-photon REMPI [44, 48], and by electron impact [49] spectroscopies. Because one- and two-photon allowed transitions of butadiene are mutually exclusive, a comparison of its one- and two-photon spectra enables its $g \rightarrow g$, ns and nd Rydberg $\leftarrow$ X transitions to be distinguished from its $g \rightarrow u$, np and nf Rydberg $\leftarrow$ X transitions. Three Rydberg series have been identified in the optical spectrum. The first members of these series originate at 186, 175, and 151.5 nm (6.66, 7.07, and 8.18 eV) [44, 46], although there was initially some uncertainty about the origin of the third series [44, 46, 47]. The three origins are assigned as two $3p$ Rydberg and one $3f$ Rydberg $\leftarrow$ X transitions on the basis of their quantum defects (0.67, 0.42, and 0.06, respectively) and their u symmetries [44, 46]. The symmetries of the three $3p$ Rydberg $\leftarrow$ X transitions are A_u, A_u, and B_u. There are no experimental criteria to determine which of the three possible transitions are observed. The symmetries of the seven $4f$ Rydberg $\leftarrow$ X transitions are 5x A_u and 2x B_u. The transition to the $n = 9$ member of the optically strong nf Rydberg series—and thus, also its 8.18-eV first member—has been assigned A_u by rotational contour analysis [50]. No criteria exist for determining which A_u series is observed. A second $nf \leftarrow$ X Rydberg series has been observed by three-photon REMPI spectroscopy [44]. Its $4f$ member lies at 8.20 eV. No criteria exist for assigning it further.

The $nd \leftarrow$ X Rydberg series of butadiene are optically forbidden. Parameter-dependent electron impact spectroscopy indicates the presence of a forbidden transition around 7.4 eV, underlying the vibrational substructure of the 7.07-eV $3p$ Rydberg band [49]. Based on its quantum defect, 0.2, the 7.4-eV band is assigned as a $3d \leftarrow X$ Rydberg transition. Both the optical and three-photon REMPI spectra manifest discrete transitions around 7.33 eV. Presumably, these are vibronically enabled false origins of the $3d$ band. This assignment is supported by the observation of analogous, but much stronger, transitions in the three-photon REMPI spectra of the non-centersymmetric dienes piperylene and isoprene [48]. In butadiene, the displacement between the optical and the parameter-dependent electron impact bands is around $500 \, \mathrm{cm}^{-1}$. This vibrational interval approximates that of the a_u CH_2 twist that enables the B_g $3s$ Rydberg $\leftarrow X$ transition [43]. It is thus likely that the 8.4-eV $3d$ Rydberg band is also B_g.

One remaining Rydberg series has been observed in butadiene. After some confusion, it was definitively assigned as an A_g $nd \leftarrow X$ Rydberg series through the use of polarization-selected REMPI spectroscopy, a parity-selection rule, and symmetry [51]. The quantum defect of the series is an anomalous -0.1 if the first member is assumed to have a primary quantum number–of 3, or it is 0.9 if $n = 4$. The latter would require a $4s$ assignment, which is incompatable with the observed fragmentary ns Rydberg series. The former would indicate a perturbed d Rydberg series, with this being the $n = 3$ member. The occupied Rydberg orbital of the excited state is either $d\pi$ or $d\delta$ (Table 2). As discussed before, the $3d\pi$ Rydberg orbital and the highest occupied π orbital have the same nodal structure and have been proposed not to exist independently [7]. If this is so, the first member of the Rydberg series would have $n = 4$ as its principal quantum number and might have all members of the series displaced in energy from their unperturbed positions. The observed anomalous quantum defect of this series is consistent with such an interpretation. A summary of the experimental vertical transition energies from the ground to the excited states of butadiene is presented in Table 4. Just two of the generating transitions, the ns and $3d$ A_g Rydberg $\leftarrow X$ transitions, are assigned by symmetry. Only the parities of the other transitions have been experimentally determined.

Two recent calculations of the vertical transition energies to most of the lower excited states of butadiene and the oscillator strengths of the optically allowed transitions have been conducted [35, 52]. As with ethylene, the single-configuration CI calculation of the energy in [52] approximates the multiconfiguration CI calculation of [35]. These two sets of calculated energies are presented in Table 4. The results of the calculations are correlated by symmetry and orbital description, rotated where necessary to correspond with the axis convention of Table 2. The calculated transition energies agree closely, except for those to the B_u $3p\pi$ and A_g $3d\pi$ states. The

TABLE 4

Experimental and calculated transition energies and oscillator strengths of butadiene

State[a]	Symmetry[b]	Experiment		Calculation			
		Energy (eV)	Oscillator Strength	Energy[c] (eV)	Oscillator Strength[c]	Energy[d] (eV)	Oscillator Strength[d]
$\pi\pi^*$	B_u	5.91	very strong	6.21	.879	6.23	.686
$\pi\pi^*$	A_g	not observed	—	g	g	6.27	g
$3s$	B_g	6.27	0	6.11	0	6.29	0
$3p\sigma$	A_u	not observed	—	6.45	.005	6.56	.002
$3p\pi$	A_u	6.66	about equal	6.61	.052	6.69	.037
$3p\pi$[e]	B_u	7.07	intensity	6.99	.249	6.70	.080
$3d\pi$	A_g	7.61	—	7.19	0	7.47	0
$3d\sigma$	B_g	not observed	—	7.39	g	g	g
$3d\pi$	B_g	7.33	0	7.22	g	g	g
$3d\delta$	B_g	not observed	—	7.25	g	7.30	0
$3d\delta$	A_g	not observed	—	7.45	g	g	g

[a] Only the partially filled orbitals are indicated.
[b] The z-axis is out of plane. The axes of [52] have been rotated to correspond with the symmetry convention given in Table 1.
[c] CIS results of [52].
[d] PT2F results of [35].
[e] Out-of-plane p orbital.
[f] Possibly A_g $3d\delta$. See Section V.
[g] Was not calculated.

193

calculated oscillator strengths of all transitions except the B_u $3p\pi$ transition are similar: The latter strength differs significantly between the two calculations.

The experimental transition energies of butadiene presented in Table 4 are correlated with the calculated energies by excited state symmetry and type to the extent possible. As mentioned earlier, the identification of the $3s$ Rydberg band is indisputable. Two of the three $3p$ Rydberg bands are observed and have similar intensities. Because both calculations find the intensity of the transition to the A_u $3p\sigma$ Rydberg state less than 10% that of either of the other $3p\pi$ Rydberg ← X transitions, we assign the two observed transitions as to the A_u and B_u $3p\pi$ Rydberg states. Because both calculations locate the B_u $3p$ Rydberg state above both A_u states, the higher energy observed state is assumed to be the B_u state. Note, however, that the intensity of the transition to the B_u state is calculated to be either twice [35] or five times [52] that of the transition to the A_u state, whereas the two intensities are observed to be similar. We will return to this discrepancy in Section VII.

Transitions to only two $3d$ Rydberg states are observed. One of these, experimentally identified as A_g, has been observed to have an anomalous quantum defect. If this transition is to the $d\pi$ state, as discussed before, the angular and radial nodal patterns of the occupied d orbital are the same as those of the highest filled π orbital. In analogy to the situation with ethylene, the lowest $d\pi$ orbital of butadiene is predicted have a radial node. Unlike the case of ethylene, however, the prediction is not supported by the calculation. Nevertheless, this highly unusual Rydberg transition is tentatively assigned as the first member of a series of A_g $d\pi$ ← X Rydberg transitions. We will return to this transition in Section V. No criteria exist to assign the other observed $3d$ Rydberg transition or either of the two observed $4f$ Rydberg ← X transitions. The goodness of the calculations will be discussed in Section VII.

V. HEXATRIENE

Ground-state hexatriene is a planar molecule that exists in two stable geometries, *cis* and *trans* around the central double bond. As with the other polyenes, the spectra of the hexatrienes have been studied for a long time [53, 54]. The ionization potentials of the two isomers are very close (*trans* = 8.29 eV, *cis* = 8.32 eV [55]); hence, the transition energies to their respective Rydberg states are predicted to also be very close. Their $\pi\pi^*$ transition energies (*trans* = 4.95 eV, *cis* = 4.92 eV [56]) are very close as well. However, because *cis*- and *trans*-hexatriene belong to different symmetry groups, different selection rules apply to transitions of the two isomers that are identical in their orbital occupancies. (See Table 2) Consequently, the experimental spectra of the two isomers differ [15, 54, 56]. In addition, in *cis*-hexatriene, unlike the situation in ethylene, butadiene, or *trans*-hexatriene, all of the 3*p* Rydberg ← X transitions and three of the 3*d* Rydberg ←X transitions are experimentally assignable by the methods discussed in Section II. The spectra of the two isomers will be discussed separately.

A. *cis*-Hexatriene

A survey spectrum of *cis*-hexatriene is presented in Figure 5 [57]. Comparison optical, two-photon, and three-photon REMPI spectra of *cis*-hexatriene have been presented in Figure 1. The lowest energy optically allowed transition of *cis*-hexatriene is composed of several diffuse bands centered between 265 and 230 nm (4.7 and 5.4 eV). From its position,

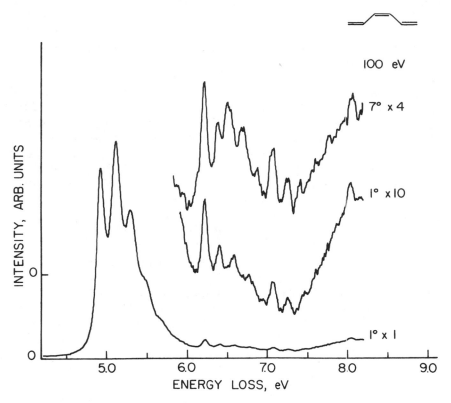

Figure 5. Survey electron energy loss spectra of *cis*-hexatriene. The zero level of the 7° spectrum has been shifted upwards. (Reproduced from [57] by permission of the American Chemical Society.)

breadth, and intensity, this transition has been assigned as a $V \leftarrow N$ transition, specifically to the lowest energy B_2 $\pi\pi^*$ valence state. Cooling the molecule sharpens these bands somewhat—the origin bandwidth (FWHM) is reduced from 290 to 155 cm^{-1} [40]—but their residual widths in the jet indicate that the lifetime of the upper state is very short. This deduction has been confirmed by time-resolved photoionization [58] and resonant Raman [59] spectroscopic investigations, in which the lifetime of the B_2 $\pi\pi^*$ state was observed to be around 20 fs. In *cis*-hexatriene, the origin of the A_1, two-electron promoted $\pi\pi^*$ transition has been directly observed [60, 61]. It lies below the B_2 valence origin. We will discuss this transition further in Section VI.

The remainder of the bands present in the *cis*-hexatriene spectra (see Figures 1 and 5) arise from Rydberg $\leftarrow$ X transitions. The discrete absorption bands superimposed on the descending tail of the $V \leftarrow N$ transition in the optical spectrum belong to a different transition. These bands are present in both the two- and three-photon REMPI spectra and hence are of A_1, B_1, or B_2 symmetry. On the basis of the effective quantum number of the origin band, the discrete bands are assigned as arising from the transition from the ground to the 3*s* Rydberg state and its vibrational subbands [17].

At higher energies, a very intense band is observed in the optical spectrum of *cis*-hexatriene. By polarization-selected two-photon REMPI spectroscopy (Figure 2), this band is determined to be A_1. Another two-photon allowed band, observed at a lower energy than the A_1 origin, is absent from the optical spectrum. Because this band is absent from the optical spectrum and present in the two-photon REMPI spectrum, it is deduced to be A_2. Based on their symmetries and effective quantum numbers, these bands are assigned as arising from two of the three possible 3*p* Rydberg $\leftarrow$ X transitions [17].

At still higher energies, a one-photon forbidden, two-photon allowed transition is observed at 6.51 eV. This transition is greatly enhanced in intensity by increasing the scattering angle in parameter-selected electron impact spectroscopy [57]. (See Figure 5.) It does not, however, appear to be enhanced by circularly polarized excitation in two-photon REMPI spectroscopy [15]. Its assignment will be discussed shortly. An adjacent strong one-photon allowed transition is observed to originate around 6.58 eV. Above this energy, at around 6.9 eV, a very weak, diffuse band is observed in the optical and electron impact spectra [15, 57]. A much stronger band approximately 560 cm^{-1} toward lower energy that this band is enhanced by parameter variation in electron impact spectroscopy. The transition giving rise to this band is assigned A_2. The 6.9-eV transition is assigned as a b_2 vibronically enabled false origin of the A_2 transition by analogy with vibrational assignments in other Rydberg states of *cis*-hexatriene. There thus appear to be two optically forbidden transitions in the 6.5–6.9 eV region of the spectrum of *cis*-hexatriene. However, this is the $3d$ Rydberg spectral region, and there is only one A_2 $3d$ Rydberg state. Other possible A_2 Rydberg states, such as the $4p$, $4d$, or $4f$ Rydberg states converging on the first ionization potential or the A_2 $3s$ Rydberg band converging on the second ionization potential, should lie above 7.2 eV unless they are severely perturbed. (The second ionization potential is approximately 2 eV greater than the first [55]; corresponding unperturbed Rydberg states should be similarly separated.) Hence, either the 6.51- or the 6.9-eV band is deduced to appear in the two-photon resonant spectrum, and not in the one-photon spectrum because it is one-photon weak, not because it is forbidden. Because the 6.51-eV band is not enhanced by circularly polarized excitation in two-photon resonant spectroscopy, it is assigned A_1. The 6.9-eV band is assigned A_2. The 6.51-, 6.58-, and 6.9-eV transitions have quantum defects characteristic of $3d$ Rydberg states and are, therefore, assigned as transitions from the ground to the A_1, A_2, and one of the three B-type $3d$ Rydberg states of *cis*-hexatriene. Note that the 6.9-eV band has an anomalous quantum defect (-0.1) similar to that observed for an A_g $3d$ band of butadiene. In that case, we argued that it might be to the $3d\pi$ state. However, it is the $d\delta$ state that belongs to both the A_1 symmetry species in *cis*-hexatriene and the A_g symmetry species in butadiene; the earlier assignment must, therefore, be reconsidered.

The next higher energy experimentally observed band, the 7.07-eV optical and electron impact band, has the same quantum defect as the $3s$ Rydberg transition. It is assigned as an unperturbed $4s \leftarrow X$ Rydberg transition converging on the first ionization potential. The observed transition energies of *cis*-hexatriene and their best possible experimental assignments are presented in Table 5.

Only one calculation of the transition energies of *cis*-hexatriene has been carried out [62]. The results of this calculation are presented in Table 5. They are correlated with the experimental transition energies by symmetry and type to the extent possible. The valence states will be discussed in Section VI. Concerning the experimental assignments, the identification of the $3s$ Rydberg band is indisputable. Two of the three $3p$ Rydberg bands are observed. Their symmetries have been experimentally determined, so the correlation with the calculation is unambiguous. Three of the $3d$ Rydberg bands have been observed. One has been experimentally demonstrated to be A_2 and another has tentatively been assigned A_1. The argument on which this assignment is based is consistent with the low intensity calculated for the transition. The other $3d$ Rydberg state is provisionally assigned B_1 for closest agreement with one of the calculated transition energies and intensities. All other bands that appear in the experimental spectra have

TABLE 5

Experimental and calculated transition energies and oscillator strengths of *cis*-hexatriene

State [a]	Symmetry [b]	Experiment		Calculation	
		Energy (eV)	Oscillator Strength	Energy [c] (eV)	Oscillator Strength [c]
$\pi\pi^*$	A_1	[d]	weak	5.04	0
$\pi\pi^*$	B_2	4.93	strong	5.00	.6170
$3s$	B_1	5.66	weak	5.69	.001
$3p\sigma$	A_2	6.08	—	6.11	0
$3p\pi$ [e]	A_1	6.22	strong	6.14	.011
$3p\pi$	B_1	not observed	—	5.91	.015
$3d\pi$	A_2	6.88	—	6.75	0
$3d\delta$	A_1	6.51	very weak	6.64	.0001
$3d\pi$	B_2	not observed	—	6.29	.070
$3d\delta$	B_1	6.58	strong	6.70	.015
$3d\sigma$	B_1	not observed	—	6.82	.0025
$4s$	B_1	7.07	medium	[f]	[f]

[a] Only the partially filled orbitals are indicated.

[b] The z-axis is in the molecular plane. The axes of [62] have been rotated to correspond with the symmetry convention given in Table 1.

[c] PT2F results of [62].

[d] See text.

[e] Out-of-plane p orbital.

[f] Was not calculated.

been satisfactorily assigned as the origins or vibrational subbands of the foregoing 3s, 3p, 3d ← X Rydberg transitions and do not constitute independent electronic transitions. The goodness of the fit will be discussed in Section VII.

B. *trans*-Hexatriene

A survey spectrum of *trans*-hexatriene is presented in Figure 6 [57]. (For comparison optical, two-photon, and three-photon REMPI spectra of *trans*-hexatriene, see [56].) The lowest energy optically allowed transition is composed of the several diffuse bands centered between 256 and 217 nm (4.8 and 5.7 eV). As in *cis*-hexatriene, from its position, breadth, and intensity, this transition has been assigned as a V ← N transition, specifically to the lowest energy B_u $\pi\pi^*$ valence state. Cooling the molecule sharpens the bands somewhat—the origin bandwidth (FWHM) is reduced from 290 to 155 cm^{-1} [40]—but their residual widths in the jet indicate that the lifetime of the upper state is very short. This deduction has been confirmed by time-resolved photoionization [63], photoelectron [58], and resonant Raman [64] spectroscopic investigations, in which the lifetime of the state was observed to be around 40 fs. Unlike the situation in *cis*-hexatriene, the origin of the

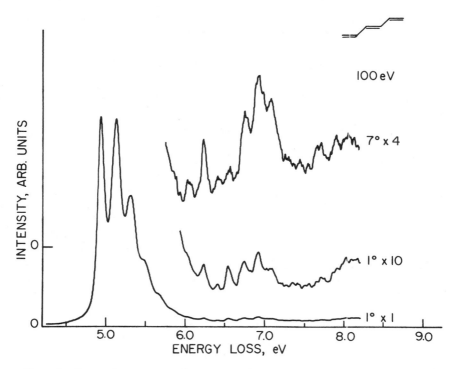

Figure 6. Survey electron energy loss spectra of *trans*-hexatriene. The zero level of the 7° spectrum has been shifted upwards. (Reproduced from [57] by permission of the American Chemical Society.)

A_g, two-electron promoted $\pi\pi^*$ transition has not been observed in *trans*-hexatriene. We will return to this topic in Section VI.

The remainder of the bands present in the *trans*-hexatriene spectrum arise from Rydberg $\leftarrow$ X transitions. The discrete absorption bands superimposed on the descending tail of the V $\leftarrow$ N transition in the optical spectrum belong to a different transition [56]. These bands are absent from the two-photon REMPI spectrum and present in the three-photon REMPI spectrum; hence, they arise from g $\rightarrow$ u transitions. Based on their u-g symmetry and the effective quantum number of the origin band, they are assigned as the transition from the ground to the $3s$ Rydberg state and its vibrational subbands.

At higher energies, a group of moderately weak bands is observed in the one- and three-photon resonant spectra of *trans*-hexatriene. A very intense group of bands is observed in the corresponding energy region of the two-photon resonant spectrum [65]. Polarization-selected two-photon REMPI spectroscopy indicates that the strong bands are A_g [66]. Another two-photon allowed band is observed toward lower energies than the A_g origin. This band was shown not to be A_g by polarization-selected REMPI spectroscopy. By elimination, the weak band is deduced to be B_g [56]. On the basis of their symmetries and effective quantum numbers, the diffuse bands are assigned as arising from two of the three $3p$ Rydberg ← X transitions. The weak bands observed in one- and three-photon resonant spectroscopies are an a_u enabled vibronic false origin of the A_g $3p$ Rydberg band and its vibrational subbands.

At still higher energies, there is a group of very strong, sharp bands in the optical spectrum that originates at 6.53 eV. These bands lie in the energy region of unperturbed $3d$ ← X Rydberg transitions and are so assigned. No further assignment is possible.

Further to the higher energy side of the spectrum there is a band at 6.9 eV, and still higher in energy two closely spaced bands that were originally assigned as vibrational subbands of the 6.9-eV transition [56]. Increasing the scattering angle in parameter-variation electron energy loss spectroscopy greatly increases the intensity of the signal around 6.9 eV, at or near these bands. The large parameter-dependent increase in intensity observed here resembles that observed around 6.5 eV in *cis*-hexatriene. The latter phenomenon was attributed to an enhancement of the optically weak A_1 $3d$ Rydberg band. An analogous effect may be occuring in *trans*-hexatriene, because this is the spectral region of the allowed $3d$ and $4s$ ← X Rydberg transitions. Were this effect, in fact, occurring, the enhanced band would have an anomalous quantum defect of -0.13, similar to that observed for the A_1 $3d\delta$ band of *cis*-hexatriene and the A_g $3d\delta$ band of butadiene. Clearly, more two-photon and polarization-selected two-photon REMPI investigations need to be conducted in this region of the spectra of the hexatrienes in order to permit a satisfactory experimental interpretation. The observed transition energies of *trans*-hexatriene and their best current assignments are presented in Table 6.

Several calculations have been carried out of the vertical transition energies to the lower excited states of *trans*-hexatriene and their one- and two-photon cross sections [35, 67, 68]. The results of these calculations are presented in Table 6, correlated with each other by symmetry and orbital description. The valence states will be discussed in Section VI. For the Rydberg transition, the two sets of transition energies agree closely for those states included in both calculations. Concerning the experimental assignments, the identification of the $3s$ Rydberg is indisputable. Two of the three $3p$ Rydberg bands are observed. One has been experimentally demonstrated to be A_g. Although both of the other $3p$ Rydberg states are B_g, the orbital assignment given in the table was determined by symmetry correlation with

TABLE 6
Experimental and calculated transition energies and oscillator strength of *trans*-hexatriene

State[a]	Symmetry[b]	Experiment		Calculation				
		Energy (eV)	Oscillator Strength	Energy[c] (eV)	Oscillator Strength[c]	Energy[d] (eV)	One-photon intensity[d]	Two-photon intensity[d]
$\pi\pi^*$	A_g	not observed	—	5.20	0	f	f	f
$\pi\pi^*$	B_u	4.95	very strong	5.01	.85	6.56	f	f
3s	A_u	5.67	weak	5.84	.0015	5.97	.056	f
$3p\pi$[e]	A_g	6.24	—	6.24	0	6.26	f	9×10^{-53}
$3p\sigma$	B_g	6.06	—	6.12	0	6.00	f	5×10^{-54}
$3p\pi$	B_g	not observed	—	f	f	6.20	f	7×10^{-54}
$3d\pi$	A_u	6.53	strong	f	f	6.68	.29	f
$3d\delta$	B_u	6.78	strong	f	f	6.81	.15	f
$3d\pi$	B_u	not observed	—	6.11	.071	6.27	2.35	f
$3d\delta$	A_u	not observed	—	f	f	7.07	.074	f
$3d\sigma$	A_u	not observed	—	f	f	6.72	.101	f

[a] Only the partially filled orbitals are indicated.
[b] The z-axis is out of plane. The axes used in [67] and [68] have been rotated to correspond with the symmetry convention given in Table 1.
[c] PT2F results of [35].
[d] See [67] and [68].
[e] Out-of-plane p orbital.
[f] Was not calculated.

the assignment of the same energy band of *cis*-hexatriene. Only two of the five 3*d* Rydberg bands are observed. One is assigned A_u by correlation with the assignment in *cis*-hexatriene. The other is assigned B_u 3*d*δ, as discussed earlier. The experimental transition energies are correlated with the calculated energies in the table on the basis of symmetry and type. Compared with experiment, both calculations overestimate the transition energy to the 3*s* Rydberg state and agree with the transition energy to the A_g 3*p* Rydberg state. The two-photon cross section of the A_g transition is calculated to be about 10 times as large as that of either of the B_g transitions [68]. One of the B_g transitions is calculated to be almost degenerate with the stronger A_g transition. It is, therefore, unlikely to be observed. The calculations are thus consistent with our observation of only two 3*p* Rydberg $\leftarrow$ X transitions and with their relative intensities. The calculated 3*d* Rydberg transition energies span the observed 3*d* Rydberg transitions. In the table, the experimentally observed origins are associated with the closest calculated transition energies of the correct symmetry. All other bands that appear in the experimental spectrum have been satisfactorily assigned as vibrational subbands of the 3*s*, two 3*p*, and two 3*d* transitions discussed above and do not constitute independent electronic transitions. The goodness of the fit will be discussed in Section VII.

C.   Comparison of *cis*- and *trans*-Hexatriene Results

The energy manifolds of the two geometric isomers of hexatriene have been shown to be very similar, despite the different appearances of their experimental spectra. Where corresponding transitions could be experimentally assigned, such as the A_g and A_1 3*p* Rydberg $\leftarrow$ X transitions, their energies agree closely. Where the corresponding transitions could not be independently experimentally assigned and the assignments were based on symmetry correlations of similar energy transitions, such as the A_2 and B_g 3*p* Rydberg transitions, the assignments for the *trans* isomer are consistent with the calculated energies and intensities. Anomalies, such as the observed quantum defect of -0.1 for the 3*d*δ Rydberg state, exist in both molecules. In both, the calculations are consistent with the experimental assignments; however, sufficient uncertainties exist in the higher energy spectra of the molecules to indicate a need for further experimentation.

VI.   VALENCE-EXCITED STATES

The excited states of the polyenes have several features in common. The polyenes all have relatively sharp Rydberg $\leftarrow$ X transitions that possess extensive vibrational substructure. Yet the molecules undergo little or no change in geometry on excitation, as is evidenced by the similarities between their adiabatic (0,0) and vertical transition energies. In addition, the polyenes all have rather diffuse low-energy $\pi\pi^*$ transitions. For these transitions, the molecules undergo large changes in geometry on excitation, as is evidenced by the large differences between their adiabatic and vertical transition energies. For *cis*- and *trans*-hexatriene, the breadths of the $\pi\pi^*$ transitions have been shown to be lifetime-limited. In *cis*-hexatriene, an excited state has been shown to lie below the optically allowed $\pi\pi^*$ transition [60, 61]. This state has been assigned as a two-electron promoted A_1 $\pi\pi^*$ state by correlation with assignments in *trans,trans*-octatetraene and dimethyldecapentaene. It is possible that decay from the higher, observed, to the lower, dark, state may be responsible for the observed short lifetime.

The spectra of *cis*-hexatriene [60, 61], *trans,trans*-octatetraenes [69–72], and dimethyldecapentaene [72], unlike those of the longer polyenes [42], have been studied in the isolated vapor phase, where the experimental and the computed transition energies are directly comparable. However, because the most accurate calculated transition energies are the vertical energies and the experimental transition energies are adiabatic, one or the other must be adjusted for the difference, the Franck–Condon (FC) energy. In the s-*trans* polyenes discussed here, the experimental Franck–Condon energies to the B_u state can be observed directly by absorption spectroscopy. This is not true in the s-*cis* polyenes, such as 1,3-cyclohexadiene, wherein the extreme weakness of the origin band makes its identification uncertain [73]. The $A_g \leftarrow A_g$ transitions, by contrast, have been detected only by REMPI or fluorescence excitation spectroscopies. Because REMPI and fluorescence excitation intensities reflect, in part, the lifetime of the upper state, and because the upper-state lifetimes of the polyenes decrease with increasing upper-state vibrational excitation, the vertical transition energies cannot be directly measured. The FC energies of the $2A_g \leftarrow 1A_g$ transitions are, therefore, experimentally unknown. The inverse situation exists for the theoretical investigations, in which the vertical excitation energies are accurately calculated, but because of state mixing in the lower symmetry of the relaxed excited states, the adiabatic transition energies are less well calculated. In this section, an attempt will be made to estimate the experimental FC energies to permit experimental and theoretical results to be compared. First, the available experimental data for the lower $\pi\pi^*$ states of butadiene, hexatriene, and octatetraene will be presented. Next, a reasonable extrapolation of the available experimental data will be used to estimate the experimentally unobtainable vertical transition energies. These results will then be compared with the corresponding calculated vertical transition energies. Experimental and computed equilibrium geometries of the $2A_g$ and B_u states of butadiene will also be discussed. On the basis of these investigations, the most probable energies and geometries of the valence-excited states of butadiene will be deduced.

The experimental adiabatic transition energies to the $2A_g$ and B_u excited states of gaseous polyenes are presented in Figure 7. For hexatriene, octatetraene, etc., the vertical and adiabation transition energies to the B_u states coincide [40, 69]. These data were extrapolated to butadiene. Note that the experimental butadiene B_u adiabatic transition energy, which was not used in the analysis, is above its extrapolated value. Butadiene also differs from the other polyenes in that the vertical transition energy to the B_u state is about 0.18 eV above its origin, not coincident with it.

The adiabatic and vertical transition energies to the $2A_g$ state differ for all the polyenes. To obtain these displacements, the following indirect route was taken: The A_g absorption and emission bands were assumed to be mirror images of each other [74]. The FC energy of the $2A_g \leftarrow A_g$ transition of octatetraene was determined from its $2A_g \rightarrow 1A_g$ fluorescence spectrum to be approximately 0.5 eV [71]. No simple experimental basis exists for

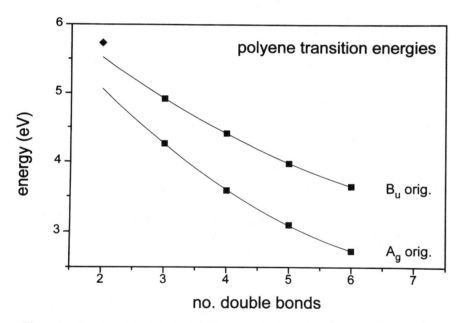

Figure 7. Experimental adiabatic transition energies from the ground to the B_u and A_g valence-excited states of the polyenes. The energies to the B_u states are directly observed. The energies to the A_g states are derived as discussed in the text. The regression curve for the B_u transition energies of $n = 3$–6 is below the adiabatic transition energy of butadiene, shown as ◆. Butadiene, B_u [40]; *cis*-hexatriene, B_u [40]; A_g [60,61]; octatetraene, B_u [69]; A_g [71]. Longer polyenes, B_u and A_g [72].

determining the existence or magnitude of a size dependence of the FC energy of isolated polyenes, because no emission spectra of unsubstituted polyenes of different lengths have been observed in the vapor phase. However, the emission spectra of octatetraene and decapentaene have been observed in 77° glasses [77]. There, the FC shift of the $2A_g \rightarrow A_g$ transition was observed to be approximately 0.4 eV in each molecule. In this phase, the FC shift of the $2A_g \leftarrow A_g$ transition appears, therefore, to be independent of molecular size for intermediate-sized molecules. As a first approximation, we assume that this is also true in the vapor phase. On the other hand, the FC energy of the $B_u \leftarrow X$ transition of butadiene is observed to be greater than those of the other polyenes [37]. This is a contradictory observation. To incorporate both observations, we propose that the FC energy of the $2A_g \leftarrow A_g$ transition of hexatriene is somewhat more than the 0.5 eV measured for octatetraene and that of the butadiene transition is somewhat larger, perhaps 0.7 eV. We further propose that the displacement of the experimental transition energy of the B_u state of butadiene from the polyene B_u regression curve also occurs for the A_g transition energy. The total FC energy of the $2A_g \leftarrow A_g$ transition of butadiene is thus estimated to be 0.9 eV [78]. The vertical transition energies to the A_g states of butadiene, hexatriene, and octatetraene, estimated as above, and the extrapolated adiabatic transition energy to the A_g state of butadiene are presented in Table 7. Notice that in each molecule the vertical transition energies to the $2A_g$ and B_u states are found to be very similar.

Our conclusion, that the extrapolated vertical transition energies to the $2A_g$ and B_u states of hexatriene are essentially degenerate, is consistent

TABLE 7

Experimental and calculated transition energies to the valence-excited states of polyenes (eV)

| Molecule | Experimental[a] | | | | Calculated[b] | |
| | Adiabatic | | Vertical | | Vertical | |
	A_g	B_u	A_g	B_u	A_g	B_u
Butadiene	5.07[c]	5.73[d]	6.0[c]	5.91	6.27	6.23
Hexatriene[e]	4.26	4.92	> 4.8[c]	4.93	5.04	5.00
Octatetraene	3.59	4.41	4.1	4.41	4.38	4.42

[a] Butadiene: See footnotes c and d. cis-Hexatriene: B_u [40] A_g [60, 61]. Octatetraene: B_u [69] A_g [71].

[b] Butadiene [35]. cis-Hexatriene [62]. Octatetraene [81].

[c] Estimated-see text.

[d] Experimental value. The value extrapolated from the longer polyenes is 5.53 eV.

[e] The experimental and theoretical transition energies of the cis isomer are presented because the experimental transition energy to the A_g state of the trans isomer is not known. In cis-hexatriene, the corresponding symmetries to B_u and A_g are B_2 and A_1, respectively.

with conclusions derived from solution resonance Raman and CARS experiments [79]. In these, the vertical transition to the $2A_g$ state of hexatriene was deduced to lie approximately 0.4 eV above that to the B_u state, a displacement that is about equal to the 0.4-eV solvent shift predicted for the B_u state on the basis of observations made on octatetraene [72]. It is also consistent with the observation of the maximum of a two-photon allowed transition in a sterically hindered, highly substituted triene approximately 0.6 eV above the observed one-photon maximum [80], a displacement that approximates the solvent and substitute shifts [72]. Our extrapolated value of the vertical transition energy to the $2A_g$ state of hexatriene is, thus, well supported. We assume that this is also true for the extrapolation procedure and, consequently, for our estimated vertical transition energy to the $2A_g$ state of butadiene.

Calculated multiconfiguration CI vertical transition energies to the $2A_g$ and $1B_u$ states of butadiene [35], cis-hexatriene [62], and trans,trans-octatetraene [81] are included in Table 7. The vertical transition energies to the $2A_g$ and B_u states are calculated to lie very close to each other in each molecule. This closeness concurs with the preceding estimates of the experimental vertical transition energies. Compared with the experimental values of these energies, the calculated transition energies to both states of butadiene are somewhat high. The calculated transition energies to both states of hexatriene are very close to their experimental values. For octa-tetraene, the calculated transition energy to the B_u state is very close to its experimental value, but the calculated transition energy to the A_g state is above its experimental value. These comparisons suggest that the quality of the calculations of the $2A_g$ and B_u states of butadiene and hexatriene are equivalent, but that may not be so for octatetraene.

Several theoretical investigations of the equilibrium geometries of the $2A_g$ and B_u valence-excited states of butadiene have been conducted [39,75,76,82]. Relevant results are presented in Table 8. Because these calculations are limited in either the size of the basis set, the number of included configurations, or both, they are not expected to be quantitatively accurate. Comparisons of the results with multiconfiguration CI vertical transition energies to either state confirms this expectation. However, the equilibrium geometries and relative excited-state energies, which are of interest here, are believed to be acceptably well determined in these types of calculations [83] and so will be discussed.

The vertical transition energy to the B_u state of butadiene is found to be about the same in all calculations, around 6.8 eV. The transition energy to the relaxed planar excited state is also found to be about the same, around 6.5 eV. The difference between these two energies, 0.3 eV, is in agreement with the Franck–Condon shift calculated from the changes in frequencies of

TABLE 8

Calculated transition energies to different configurations of the A_g and B_u states of butadiene (ev)

State	Calculated energy	Configuration		
		Vertical	Relaxed Planar	Optimized [a]
A_g	MCSCF [b]	6.76	5.54	5.34
	MRC [c]	6.82	5.55	5.26
	MCSCF [d]	6.67	5.55	5.42
	average	6.75	5.55	5.34
B_u	MRCI [c]	6.74	6.50	5.0 [e]
	MCSCF [d]	6.88	6.63	5.78
	average	6.81	6.56	5.4

[a] C_1, C_2, or S_2. All geometries have one or both CH_2 groups rotated relative to the molecular plane.

[b] See [75], 4–32 basis. There are three equal-energy minima for the A_g state.

[c] See [76], basis set A.

[d] See [82].

[e] MCSCF energy of [76] is 5.4 eV.

the a_1 modes of butadiene on excitation [39]. Twisting one of the terminal CH_2 groups out of plane is found to further lower the transition energy to the B_u state by at least 1 eV. This distortion yields a calculated energy difference between the vertical transition and the origin of over 1.3 eV [76,82]. This result is in disagreement with both conclusions drawn from a calculated fit to the Franck–Condon structure of the $B_u \leftarrow A_g$ absorption spectrum, in which nonplanar modes were found not to be significant [36,39], and with the experimental spectrum, in which the difference between the vertical transition energy and the origin is observed to be around 0.2 eV. We conclude that the calculated effect of twisting a CH_2 group on the energy of the B_u state is an artifact of the calculations and that the B_u excited state of butadiene is essentially planar.

The vertical transition energy to the $2A_g$ state of butadiene is also found to be about the same in all three calculations, around 6.7 eV. This energy is essentially degenerate with the vertical transition energy to the B_u state, in agreement with both our extrapolated transition energies and the result of the multiconfiguration CI calculations. The transition energy to the relaxed planar excited state is also found to be about the same in all three calculations, around 5.5 eV. The difference between the two, 1.2 eV, is significantly greater than that found for the B_u state. Distorting the molecule out of plane only slightly further lowers the energy of the $2A_g$ state, by about 0.2 eV, yielding a calculated energy difference between the vertical transition energy and the origin of around 1.4 eV. This value can be compared with the experimentally estimated value of 1.2 eV. The $2A_g$ state of butadiene is concluded to be nearly planar, but its relaxed geometry, is more displaced from that of the ground state than is the geometry of the (nearly planar) B_u state of butadiene.

In sum, the adiabatic transition energy to the $2A_g$ state of butadiene has been estimated from the observed adiabatic and vertical transition energies from the ground to the B_u states of butadiene, *cis*-hexatriene, and all *trans*-octatetraene combined with the adiabatic transition energies to the $2A_g$ valence-excited states of gas-phase hexatriene and octatetraene, and gas-phase transition energies of decapentaene and dodecahexaene extrapolated from solution-phase investigations. The vertical transition energies to the $2A_g$ states of butadiene and hexatriene were estimated from the observed Franck–Condon energies of the $2A_g \rightarrow A_g$ emission of gas-phase octate-traene and of solution-phase octatetraene and decapentaene combined with the assumption that the absorption and emission Franck–Condon energies are similar. For butadiene and hexatriene, the estimated experimental vertical transition energies to the $2A_g$ and B_u states were found to be similar. This result is in agreement with the best calculated vertical transition energies. Several calculations of the geometry dependence of the energies of the $2A_g$ and B_u states of butadiene were also evaluated. The reasonable conclusion that can be drawn from combining the results of these calculations and the experimental inferences is that butadiene is essentially planar in both its $2A_g$ and B_u excited states, but the equilibrium geometry of the $2A_g$ state is more distorted from that of the ground state than is that of the B_u state. The Franck–Condon displacement calculated for the planar B_u state approximates the observed displacement. The Franck–Condon displacement calculated for the equilibrium $2A_g$ state is comparable to that estimated from the experimental data. Notwithstanding these uncertainities, the origin of the $2A_g$ state in butadiene is concluded to lie at least 0.5 eV below the origin of the B_u state.

VII. EVALUATION OF NUMERICAL CALCULATIONS OF TRANSITION ENERGIES IN POLYENES

In the preceding sections, we have evaluated the existing experimental investigations of the spectra of ethylene, butadiene, and hexatriene and have compiled the most complete set of experimental assignments for the observed transitions of each molecule currently possible. For transitions to Rydberg states, the experimental transition energies have been correlated by symmetry and type with the best currently available numerically calculated corresponding transition energies. In this section, we wish to evaluate the strengths and weaknesses of the numerical calculations. The most accurate calculated transition energies are the vertical ones. For Rydberg states, the vertical energies are the adiabatic, so there are no difficulties in experimental–theoretical comparisons. Experimental–theoretical correlations, given in Tables 3–6, are possible for six Rydberg transitions in ethylene, four in butadiene, three in *trans*-hexatriene, and five in *cis*-hexatriene. The overall accuracies of the calculations are presented in Table 9, expressed as the standard deviation of the differences between the calculated and experimental transition energies to the Rydberg states for each molecule. Overall, the variance is 0.1 eV. This value is similar to what was previously estimated [35] using not-as-good experimental assignments—and hence, less stringent criteria—for experimental–theoretical correlation. An analogous comparison of the experimental and numerical transition energies was carried out for the valence-excited states using the results presented in Table 7. Except for butadiene, the calculated and experimental transition energies to the B_u states are less than 0.1 eV apart. The deviation of the extrapolated vertical experimental transition energies from the calculated ones to the A_g states is greater, almost 0.3 eV. Some of this discrepancy may arise from inaccuracies in the extrapolation procedure, but the discrepancy that is observed for octatetraene, in which the vertical transition energy to the A_g state is an observable, is essentially the same as for butadiene and hexatriene, which suggests that extrapolation-procedure inaccuracies are not, in fact, the source of the discrepancy. We conclude that the calculations for the B_u valence and Rydberg states are quite accurate, but those for the A_g valence-excited state are not as accurate.

There is, however, one inconsistency between the experimental and calculated Rydberg manifolds and that concerns the B_u Rydberg states. Transitions from the ground to the B_u Rydberg states are calculated to be very intense for all molecules studied here, but, except for butadiene, no such transitions are observed. The B_u Rydberg states are the states for which the calculations are expected to be more difficult because they are of the same symmetry as the valence state to which transitions are optically allowed. They are also the states whose independent existence has been

TABLE 9
Evaluation of polyene Rydberg calculations

Molecule	σ(eV)[a]	Number of transitions
Ethylene	.09	6
Butadiene	.07	4
trans-hexatriene	.10	3
cis-hexatriene	.13	6

[a] Standard deviation of the differences between the experimental and calculated transition energies given in Tables 3–6.

TABLE 10

Evaluation of polyene Rydberg calculations for transitions to states of $\pi\pi^*$ symmetry [a]

		Transition energy (eV)		Intensity	
Molecule	Type [b]	Typical	Rydberg state of $\pi\pi^*$ symmetry	Typical	Rydberg state of $\pi\pi^*$ symmetry
Ethylene	$3d$	8.9–9.2	9.31	.001	.08
Butadine	$3p$	6.7	6.7	.03	.08
trans-hexatriene	$3d$	?	6.11	?	.07
cis-hexatriene	$3d$	6.6–6.8	6.3	> .01	.07

[a] Data obtained from Tables 3–6.
[b] Occupied Rydberg orbital.

questioned [7]. The calculated transition energies to these states and the intensities of those transitions are compared with the energies and intensities to the other Rydberg states of the same s, p, d type in Table 10. Except for butadiene, the B_u states are calculated to be displaced in energy from other states of their corresponding type. Moreover, transitions to the B_u Rydberg states are calculated to be 3 to 10 times as intense as those to the other Rydberg states of their type. Were the calculation accurate, these transitions should be easily observable. Because they are not observed, we conclude that intensity calculations are not as accurate for these states as they are for the valence-excited or other Rydberg states.

VIII.   CONCLUDING REMARKS

This review has presented methods used for assignments of the electronic transitions of ethylene, butadiene, and hexatriene. Experimentally observed transitions have been assigned to the greatest extent possible, and the experimental transition energies have been compared with the best available numerical calculations of the corresponding transition energies. Fewer Rydberg states are observed than are calculated. The observed states manifest extensive vibrational substructures. Because many of the missing states are calculated to be weaker than, and close in energy to, other Rydberg states, it is likely that they are buried in the vibrational substructures of these other states and will never be observed. For the 18 observed Rydberg transitions, the agreement between the corresponding calculated and experimental transition energies was found to have a standard deviation of less than 0.1 eV. The energies of the B_u valence-excited states were also calculated to be very close to their observed energies. A similar comparison is more difficult for the A_g valence states, because the best calculations are of the vertical transition energies and, for some of the molecules, only the adiabatic transition energies are known. A method was developed here to estimate "experimental" vertical transition energies. These "experimental" energies were somewhat further from the calculated transition energies than are either the B_u valence-excited or the Rydberg states. While the extrapolation procedure may be the source of this discrepancy, the data for octatetraene indicates that is not true. In addition, the theoretical treatment does not yet appear to correctly handle Rydberg-valence mixing for Rydberg states of the same symmetry as the valence-excited state. Calculations of the equilibrium geometries of the $2A_g$ and B_u excited states of butadiene were also discussed. Comparisons of the results of these calculations with direct and indirect experimental results indicate that at equilibrium both states are planar. At the present time, the theoretical calculations are unable to accurately describe the geometry of the B_u $\pi\pi^*$ state of butadiene.

References

[1] R. S. Mulliken, *Phys. Rev.* **41**, 751 (1932).

[2] Ethylene: J. Stark and P. Lipp, *Z. Physik. Chem.* **86**, 36 (1913); G. Scheibe and H. Grienseisen, *Z. Physik. Chem.* **25B**, 52 (1934); C. P. Snow and C. B. Allsopp, *Trans. Faraday Soc.* **30**, 93 (1934).

[3] Butadiene: E. P. Carr, L. W. Pickett, and H. Stucklen, *Rev. Mod. Phys.* **14**, 260 (1942); W. C. Price and A. D. Walsh, *Proc. Roy. Soc. London* **174A**, 220 (1940).

[4] For a review of early investigations of ethylene, see A. J. Merer and R. S. Mulliken, *Chem. Rev.* **69**, 639 (1969); R. S. Mulliken, *J. Chem. Phys.* **66**, 2448 (1977).

[5] See also M. B. Robin, *Higher Excited States of Polyatomic Molecules*, Vol. II, Academic, New York, 1975, Sect. IV.A; Vol. III, Sect. X.A.

[6] R. S. Mulliken, *J. Chem. Phys.* **7**, 121 (1939).

[7] K. B. Wiberg, C. M. Hadad, J. B. Foresman, and W. A. Chupka, *J. Phys. Chem.* **96**, 10756 (1992) and references cited therein.

[8] R. Barrow, *Introduction to Molecular Spectroscopy*, McGraw-Hill, New York, 1962, p. 1.

[9] See G. Herzberg, *Molecular Spectra and Molecular Structure III: Electronic Spectra and Electronic Structure of Polyatomic Molecules*, Nostrand, Princeton, 1966, pp. 181–183.

[10] D. H. Levy, L. Wharton, and R. E. Smalley, in *Chemical and Biochemical Applications of Lasers*, Vol. II, C. B. Moore, Ed., Academic, New York, 1977, Chapter 1.

[11] S. H. Lin, Y. Fujimura, H. J. Neusser, and E. W. Schlag, *Multiphoton Spectroscopy of Molecules*, Academic, New York, 1984.

[12] J. I. Steinfield, *Molecules and Radiation*, Harper, New York, 1974, pp. 310–315.

[13] W. M. McClain and R. A. Harris, in *Excited States*, Vol. 3, E. C. Lim, Ed., Academic, New York, 1977, pp. 2–56.

[14] D. H. Parker, J. O. Berg, and M. A. El-Sayed, *Chem. Phys. Lett.* **56**, 197 (1978).

[15] A. Sabljic and R. McDiarmid, *J. Chem. Phys.* **84**, 2062 (1986).

[16] E. N. Lassettre, in *Chemical Spectroscopy and Photochemistry in the Vacuum Ultraviolet,* C. Sandorfy, P. J. Ausloss, and M. B. Robin, Eds., Reidel, Dordrecht, The Netherlands, 1974, pp. 43–74.

[17] A. D. Walsh, *J. Phys. Radium* **15**, 501 (1954); M. B. Robin , *Higher Excited States of Polyatomic Molecules,* Vol. I, Academic, New York, 1974, p. 51.

[18] R. G. Schmitt and R. K. Brehm, *Applied Optics* **5**, 1111 (1966).

[19] A. D. Walsh, *J. Chem. Soc.* **1953**, 2325.

[20] P. G. Wilkinson and R. S. Mulliken, *J. Chem. Phys.* **23**, 1895 (1955).

[21] R. McDiarmid and E. Charney, *J. Chem. Phys.* **47**, 1517 (1967).

[22] P. D. Foo and K. K. Innes, *J. Chem. Phys.* **60**, 4582 (1974).

[23] W. C. Price and W. T. Tutte, *Proc. Roy. Soc.* **A174**, 207 (1939).

[24] R. McDiarmid, *J. Phys. Chem.* **84**, 64 (1980).

[25] A. J. Merer and L. Schoonveld, *J. Chem. Phys.* **48**, 522 (1968); *Canad. J. Phys.* **47**, 1731 (1969).

[26] P. G. Wilkinson, *Canad. J. Phys.* **34**, 643 (1956).

[27] D. F. Dance and I. C. Walker, *J. Chem. Soc. Farad. II,* **71**, 1903 (1975).

[28] D. G. Wilden and J. Comer, *J. Phys. B: Atom. Molec. Phys.* **13**, 1009 (1980).

[29] A. Gedanken, N. A. Kuebler, and M. B. Robin, *J. Chem. Phys.* **76**, 46 (1982).

[30] B. A. Williams and T. A. Cool, *J. Chem. Phys.* **94**, 6358 (1991).

[31] M. Brith-Lindner and S. D. Allen, *Chem. Phys. Lett.* **47**, 32 (1977).

[32] R. J. Sension and B. S. Hudson, *J. Chem. Phys.* **90**, 1377 (1989).

[33] P. A. Snyder, P. N. Schatz, and E. M. Rowe, *Chem. Phys. Lett.* **110**, 508 (1984).

[34] J. B. Foresman, M. Head-Gordon, J. A. Pople, and M. J. Frisch, *J. Phys. Chem.* **96**, 135 (1992).

[35] L. Serrano-Andres, M. Merchan, I. Nebot-Gil, R. Lindh, and B. O. Roos, *J. Chem. Phys.* **98**, 3151 (1993).

[36] U. Dinur, R. J. Hemley, and M. Karplus, *J. Phys. Chem.* **87**, 924 (1983).

[37] R. J. Hemley, J. I. Dawson, and V. Vaida, *J. Chem. Phys.* **78**, 2915 (1983).

[38] R. McDiarmid and A.-H Sheybani, *J. Chem. Phys.* **89**, 1255 (1988).

[39] F. Zerbetto and M. Z. Zgierski, *Chem. Phys. Lett.* **157**, 515 (1989).

[40] D. G. Leopold, R. D. Pendley, J. L. Roebber, R. J. Hemley, and V.Vaida, *J. Chem. Phys.* **81**, 4218 (1984).

[41] R. S. Mulliken, *Reviews of Modern Physics* **14**, 265 (1942) and sequels.

[42] See B. S. Hudson, B. E. Kohler, and K. Schulten in *Excited States,* Vol. 6, E. C. Lim, Ed., Academic, New York, 1982, Chapter 1; and G. Orlandi, F. Zerbetto, and M. Z. Zgierski, *Chem. Rev.* **91**, 867 (1991).

[43] R. McDiarmid, *Chem. Phys. Lett.* **34**, 130 (1975).

[44] P. Johnson, *J. Chem. Phys.* **64**, 4638 (1976).

[45] J. O. Berg, D. H. Parker, and M. A. El-Sayed, *J. Chem. Phys.* **68**, 566 (1978).

[46] R. McDiarmid, *J. Chem. Phys.* **64**, 514 (1976).

[47] K. B. Wiberg, K. S. Peters, G. B. Ellison, and J. L. Dehmer, *J. Chem. Phys.* **66**, 2224 (1977).

[48] L. J. Rothberg, D. P. Gerrity, and V. Vaida, *J. Chem. Phys.* **73**, 5508 (1980).

[49] J. P. Doering, *J. Chem. Phys.* **70**, 3902 (1979); J. P. Doering and R. McDiarmid, *J. Chem. Phys.* **73**, 3617 (1980); *J. Chem. Phys.* **75**, 2477 (1981).

[50] B. D. Ranson, K. K. Innes, and R. McDiarmid, *J. Chem. Phys.* **68**, 3615 (1978).

[51] W. G. Mallard, J. H. Miller, and K.C. Smyth, *J. Chem. Phys.* **79**, 5900 (1983); P. H. Taylor, W. G. Mallard, and K.C. Smyth, *J. Chem. Phys.* **84**, 1053 (1986).

[52] K. B. Wiburg, C. M. Hadad, G. B. Ellison, and J. B. Foresman, *J. Phys. Chem.* **97**, 13586 (1993).

[53] W. C. Price and A. D. Walsh, *Trans. Farad. Soc.* **37**, 106 (1941).

[54] R. M. Gavin, Jr., S. Risenberg, and S. A. Rice, *J. Chem. Phys.* **58**, 3160 (1973); R. M. Gavin, Jr., and S. A. Rice, *J. Chem. Phys.* **60**, 3231 (1974).

[55] M. Beez, G. Bieri, H. Bock, and E. Heilbronner, *Helvetica Chim. Acta* **56**, 1028 (1973).

[56] A. Sabljic and R. McDiarmid, *J. Chem. Phys.* **82**, 2559 (1985).

[57] R. McDiarmid, A. Sabljic, and J. P. Doering, *J. Am. Chem. Soc.* **107**, 826 (1985).

[58] D. R. Cyr and C. C. Hayden, *J. Chem. Phys.* **104**, 771 (1996).

[59] X. Ci and A. B. Myers, *J. Chem. Phys.* **96**, 6433 (1992).

[60] W. J. Buma, B. E. Kohler, and K. Song, *J. Chem. Phys.* **92**, 4622 (1990); **94**, 6367 (1991).

[61] H. Petek, A. J. Bell, R. L. Christensen, and K. Yoshihara, *J. Chem. Phys.* **96**, 2412 (1992).

[62] L. Serrano-Andres, B. O. Roos, and M. Merchan, *Theoret. Chim. Acta* **87**, 387 (1994).

[63] C. C. Hayden and D. W. Chandler, *J. Phys. Chem.* **99**, 7897 (1995).

[64] A. B. Myers and K. S. Pranata, *J. Phys. Chem.* **93**, 5079 (1989).

[65] D. H. Parker, S. J. Sheng, and M. A. El-Sayed, *J. Chem. Phys.* **65**, 5534 (1976).

[66] D. H. Parker, J. O. Berg, and M. A. El-Sayed, *Chem. Phys. Lett.* **56**, 197 (1978).

[67] M. A. C. Nascimento and W. A. Goodard, III, *Chem. Phys. Lett.* **60**, 197 (1979); *Chem. Phys.* **53**, 265 (1980).

[68] M. A. C. Nascimento, *Can. J. Chem.* **63**, 1349 (1985).

[69] (a) R. M. Gavin, Jr., C. Weisman, J. K. McVey, and S. A. Rice, *J. Chem. Phys.* **68**, 522 (1978); (b) D. G. Leopold, V. Vaida, and M. F. Granville, *Ibid.* **81**, 4210 (1984).

[70] L. A. Heimbrook, J. E. Kenny, B. E. Kohler, and G.W. Scott, *J. Chem. Phys.* **75**, 4338 (1981); L. A. Heimbrook, B. E. Kohler, and I. J. Levy, *Ibid.* **81**, 1592 (1984).

[71] H. Petek, A. J. Bell, Y. S. Choi, K. Yoshihara, B. A. Tounge, and R. L. Christensen, *J. Chem. Phys.* **98**, 3777 (1993).

[72] W. G. Bouwman, A. C. Jones, D. Phillips, P. Thibodeau, C. Friel, and R. L. Christensen, *J. Phys. Chem.* **94**, 7429 (1990).

[73] M. Merchan, L. Serrano-Andres, L. S. Slater, B. O. Roos, R. McDiarmid, and X. Xing, *J. Phys. Chem.*, in press.

[74] This assumption is supported by a comparison of the B_u–A_g absorption and emission spectra of vapor-phase octatetraene [69] and one calculation of the Franck–Condon energies of the A_g–A_g absorption and emission bands of butadiene [76]. It is not supported by the results of another calculation of the Franck–Condon energies of the A_g–A_g absorption and emission bands of butadiene [75].

[75] F. Zerbetto and M. Z. Zgierski, *J. Chem. Phys.* **93**, 1235 (1990).

[76] P. G. Szalay, A. Karpfen, and H. Lischka, *Chem. Phys.* **130**, 219 (1989).

[77] K. L. D'Amico, C. Manos, and R. Christensen, *J. Am. Chem. Soc.* **102**, 1777 (1980).

[78] This is the 0.5-eV displacement observed in the A_g transition of octatetraene, plus the 0.2-eV Franck–Condon energy difference observed for absorption to the B_u state of butadiene.

[79] T. Fujii, A. Kamata, M. Shimizu, Y. Adachi, and S. Maeda, *Chem. Phys. Letters* **115**, 369 (1985).

[80] B. M. Pierce, J. A. Bennett, and R. R. Birge, *J. Chem. Phys.* **77**, 6343 (1982).

[81] L. Serrano-Andres, R. Lindh, B. O. Roos, and M. Merchan, *J. Phys Chem.* **97**, 9360 (1993).

[82] M. Aoyagi, Y. Osamura, and S. Iwata, *J. Chem. Phys.* **83**, 1140 (1985).

[83] K. B. Wiberg, P. R. Rablen, and M. Marquez, *J. Am. Chem. Soc.* **114**, 8654 (1992).

CHAPTER 3

UNDERSTANDING ELECTRON CORRELATION: RECENT PROGRESS IN MOLECULAR SYNCHROTRON PHOTOELECTRON SPECTROSCOPY

A. D. O. BAWAGAN

Ottawa-Carleton Chemistry Institute Carleton University, Ottawa, Ontario K1S 5B6, Canada

E. R. DAVIDSON

Department of Chemistry Indiana University, Bloomington, Indiana 47405, USA.

CONTENTS

Advances in Chemical Physics, Volume 110, Edited by I. Prigogine and Stuart A. Rice.
ISBN 0-471-33180-5. © 1999 John Wiley & Sons, Inc.

I. INTRODUCTION

The 1986 review of electron correlation effects in the valence ionization spectra of atoms and molecules reported by Cederbaum et al. (1986) is the springboard of the present work. It is commonly understood that the Hartree–Fock (HF) method for solving the many-body electronic Schrödinger equation is an approximation. The electron correlation energy is defined as $E_{corr} = E_{exact} - E_{HF}$ in the state for which the Hartree–Fock calculation was performed and has been widely studied by quantum chemists. In the frozen orbital, or Koopmans' (1933), approximation, the HF orbitals from the ground state of a molecule are used to form Slater determinants for describing other states of the system, including excited states of the ion. The extreme form of this approximation, known as Koopmans' theorem, assumes that the individual Slater determinants, or "configurations," are good approximations to the actual states of the ion. In this approximation, only single excitations from the Hartree–Fock ground state carry any intensity in photoelectron or electron momentum spectroscopy. Originally, there was little appreciation for the dramatic nonqualitative effects the error involved in this Koopmans' approximation would have on the predicted ionization spectra of atoms and molecules.

This appreciation changed drastically two decades ago, when photoelectron spectroscopy (PES) [Spears et al. (1974), Wuilleumier and Krause (1974), Svensson et al. (1988a)] and electron momentum spectroscopy (EMS) [Weigold et al. (1973), Leung and Brion (1983), Brion (1986)] clearly showed extra structure, or so-called satellites, in the experimental valence and core ionization spectra, which was not expected in the picture painted by Koopmans' theorem. In other words, experimental evidence pointed to a breakdown of the independent-particle (frozen orbital) picture of electronic structure. Several theoretical studies brought forth the proper interpretation of these satellite features as manifestations of electron correlation and orbital relaxation. Foremost among the theoretical methods presented were the Green's function method of von Niessen et al. (1984) and the configuration interaction (CI) method [Martin and Shirley (1976a,b), Bagus and Viinikka (1977), Martin and Davidson (1977)]. Other theoretical approaches [Hermann et al. (1978)] have been reported, but none of them have been applied to as wide a range of molecular systems as the Green's function and CI methods.

The simple theoretical interpretation that emerged is that the extra features in the ionization spectra appear (in addition to those expected from the one- orbital, one-peak picture) because of the strong interaction between single-hole configurations and two-hole, one-particle configurations. Many-body states generalize this concept to include the multitude of possible mixings and higher excitations that can be realized in a quantum system. The interaction between a single-hole configuration and a many-body configuration leads to a mixing of the configurations that redistributes the intensity among many states of the same symmetry, as shown in the following example for two electronic states Ψ_1 and Ψ_2:

$$\Psi_1 = c_{1a}\phi_{1-\text{hole}} + c_{1b}\phi_{2-\text{hole}-1-\text{particle}}, \tag{1}$$

$$\Psi_2 = c_{2a}\phi_{1-\text{hole}} + c_{2b}\phi_{2-\text{hole}-1-\text{particle}}. \tag{2}$$

Here, $c_{1a} \gg c_{1b}$, $c_{2b} \gg c_{2a}$, ϕ refers to a given electron configuration, and the c's correspond to the CI coefficients. This mixing of configurations leads to electronic states (e.g., Ψ_2) that will we refer to as correlation states, in addition to those expected from the simple Koopmans' theorem picture (e.g., Ψ_1), which are generally referred to as the "parent states." The term "correlation states" is used here because a proper account of electron correlation is necessary to predict the intensities of transitions to these states in photoelectron spectra, even though their energy may sometimes be estimated by simpler methods. Experimentally, this mixing redistributes the intensity between "parent peaks" associated with parent states and "satellite peaks" associated with correlation states.

A correlation state may be described by a wave function involving mixing of many configurations (configuration interaction). The mixing coefficients may be found either by variational (CI) or perturbational (Green's-function) methods. In the foregoing discussion, it is always assumed that the connection between experiment and theory is made through the square of the mixing coefficients (in general, the pole strengths in Green's-function methods [Cederbaum et al. (1986)] or S^2 in CI methods [Desjardins et al. (1995)]), which is directly proportional to the ionization probability. Much of the recent progress in this field is related to extensive studies of the range of validity of this connection.

Despite the tremendous progress made in the theoretical interpretation of the satellite features in PES and EMS, several serious questions remain, and these are the subject of this review. In particular, the following questions are pertinent:

- What are the various mechanisms by which correlation states manifest themselves in the PES spectra of molecules?
- What are the sources of error leading to our inability to quan-titatively predict the ionization spectra of molecules?
- Why are EMS intensities different from PES intensities?

The present review tries to bring up-to-date answers to the first two questions and offers clues to the third and, perhaps, most important question. The focus of the review is heavily weighted on our own experimental–theoretical experience and draws on substantial previous PES work by others since 1986. Furthermore, we limit our review to the valence region of the ionization spectra of molecular systems and exclude discussions of the core region as well as the ionization spectra of atomic systems, which have all recently been reviewed [Becker and Shirley (1990), Schmidt (1992)].

We highlight certain experimental and theoretical difficulties encountered in the accurate prediction of the energies and intensities of the ionization spectra of molecules. For example, the dependence of the theoretical calculations on the basis set are illustrated in the prediction of the main $(2a_g)^{-1}$ peak in ethylene at 23.7 eV. Previous theoretical calculations of the ionization spectrum [Murray and Davidson (1992)] did not show the "twinning" of the main poles, and it was only through exhaustive theoretical calculations that these were revealed by MRSDCI (multireference single and double excitation configuration interaction) [Desjardins et al. (1995)] and Green's-function calculations [Cederbaum et al. (1986), Deleuze and Cederbaum (1997)]. As seen in other molecular systems, basis-set effects are highly unpredictable with regard to the prediction of the energies, as well as the intensities, of correlation (satellite) states. Such is the case of pyridine [Moghaddam et al. (1996)], wherein improved basis sets (from double-zeta quality to triple-zeta quality) reveal shifts in energy and large changes in intensities. It is still not known whether the basis-set limit has been reached for pyridine. It is hoped that benchmark studies such as those conducted for ethylene will provide useful guidelines as to the necessary tools for the accurate prediction of the ionization spectra of even more complex molecular systems. Furthermore, truncating the CI spaces (or other computational schemes that make the calculations manageable) are also prone to problems, as is shown in the theoretical prediction of the ionization spectra of large alkenes [Fronzoni et al. (1994, 1995)].

From the experimental side, the problem has always been to reliably separate the different correlation peaks, especially those of different symmetries. This case is well illustrated by the ^2S satellite (Peak 7, according to the numbering scheme of Wuilleumier and Krause (1974)) of neon at 55.8 eV binding energy, which was usually considered a textbook case [Thomas (1984), Heimann et al. (1986), Wuilleumier and Krause (1974), Becker at al. (1986)]. However, it was shown by very high resolution synchrotron PES studies [Bawagan et al. (1991), Krause et al. (1992)] to be composed of at least two correlation states separated by less than 50 meV. The two correlation states are of different symmetries, $2s^2 2p^4(^1D)3p(^2P)$

and $2s^2 2p^4(^1S)3s(^2S)$, as can be clearly seen in the beta parameter measurements of Krause et al. (1992) and Pahler et al. (1993) and the latest synchrotron PES work of Kikas et al. (1996). In addition to these two states, other nearby states have previously been observed by optical spectroscopy [Persson (1971)], namely, $2s^2 2p^4(^1D)3p(^2D)$ at 55.95 eV and $2s^2 2p^4(^3P)3d(^4D)$ at 56.17 eV. Even if their pole strengths are zero by symmetry rules, these states can gain intensity via other mechanisms, such as conjugate shake-up and interchannel coupling. If it were not for the fact that the two states are satellites of different primary peaks (i.e., they have Dyson orbitals resembling different HF MOs), discussions regarding the photon energy dependence of Peak 7 (at 55.8 eV binding energy) of neon featured in the classic work of Wuilleumier and Krause (1974) would not be in error. What turns out to be the textbook illustration of "shake-up" photon energy dependence (i.e., increasing from a threshold and then reaching a plateau at high photon energies) is more the exception rather than the rule. This result illustrates the extreme caution that is necessary in interpreting the photon energy dependence of correlation peaks which are not clearly resolved in energy.

II. EXPERIMENTAL DETAILS

The interaction of radiation with neutral molecules leading up to the production of positive ions and the release of electrons has fascinated many people for a long time. The experimental analysis of the kinetic energy of the released electrons during the photon–matter interaction (using fixed-energy light sources) brought forth the field of PES. The advent of highly intense, energy-tunable photon sources, called synchrotrons, has ushered in the investigation of novel phenomena and very low cross-section processes. Consequently, PES has entered a phase of unprecedented growth. The basic principles of synchrotron radiation and monochromators [Winick and Doniach (1980)] and the various photoelectron energy analyzers [Sevier (1972), Roy and Tremblay (1990)] are well established. In this review, we only present, as an example of typical experimental procedures, the one we have used in our recent investigations. Particular attention is paid to calibration procedures that allow the accurate determination of satellite- to main-peak intensity ratios.

The synchrotron photoelectron spectra were obtained at the Aladdin Storage Ring of the University of Wisconsin at Madison's Synchrotron Radiation Center (SRC). The spectra were collected using the Canadian Synchrotron Radiation Facility's (CSRF's) McPherson photoelectron spectrometer equipped with a multichannel plate detector previously described by Bozek et al. (1990). Briefly, photoelectrons are collected at the pseudomagic angle that is calculated assuming 90% polarization of the synchrotron light. This ensures that the photoelectron intensities obtained are independent of well-known angular effects. Kinetic energy analysis is obtained using an electrostatic energy analyzer (a McPherson spectrometer with a 360-mm mean radius) with a measured energy resolution of better than 0.13% $\Delta E/E_0$ fwhm. Differential pumping of the gas cell is done to maintain a pressure differential of better than 10^4 between the cell and the energy analyzer.

The spectrometer is operated at zero retard mode. Therefore, the peak intensities in the PES spectra should be corrected for transmission effects to account for the different electron kinetic energies. In cases where the spectra are obtained in narrow energy ranges (≈ 5 eV), the transmission corrections are small. However, in the present application, where spectra are obtained over a wide energy range (30–40 eV), these corrections are important and necessary.

The total energy resolution (from the electron analyzer and the monochromator) for these experiments typically ranges from 140 meV fwhm at 50 eV photon energy to 600 meV fwhm at 200 eV photon energy. Better energy resolution can be attained, but is not really necessary because of the significant Franck–Condon widths involved in the inner valence ionization peaks of molecules, where most of the electron correlation effects are usually found. The large widths (typically 2–4 eV fwhm) are generally a consequence of transitions to molecular dissociative states.

The transmission function of the present spectrometer was calibrated routinely using neon as the calibrant gas [Bawagan et al. (1992)]. Neon PES spectra, covering the $2s$ and $2p$ ionization peaks, were obtained at several photon energies in the range 80–220 eV. The intensities or peak areas obtained in a typical photoelectron spectrum are given by

$$A_i = \text{constant} \times \sigma_i(h\nu)I_0(h\nu)T(\varepsilon_k)\Gamma(h\nu), \tag{3}$$

where $\sigma_i(h\nu)$ is the differential cross section for the particular process, $I_0(h\nu)$ is the photon intensity at photon energy $h\nu$, $T(\varepsilon_k)$ is the spectrometer transmission function at electron kinetic energy ε_k and $\Gamma(h\nu)$ is the monochromator shape function. The transmission function is found to be a linear function of ε_k; that is,

$$T(\varepsilon_k) = \text{constant} \times \varepsilon_k, \tag{4}$$

from previous calibration using Kr MNN Auger transitions [Werme et al. (1972)], to better than 5%. This result is in agreement with earlier theoretical and experimental studies of the transmission properties of dispersive electrostatic analyzers [Poole et al. (1973), Woodruff et al. (1977), Cross and Castle (1981)]. Note that even electron spectrometers operated at constant pass energy mode (i.e., with preretardation or preacceleration) will require similar transmission corrections because of lens effects. Both types of spectrometer operation are used in the field, and it simply becomes a matter of choice which to employ in a particular situation.

The incident photon flux was closely monitored using a calibrated gold mesh. The monochromator shape function accounts for the energy distribution of the photon source and allows for energy-dependent distortions of the optical properties of the monochromator as it is scanned. Distortions induced in the light pipe that separates the spectrometer gas cell from the beam line are also accounted for. In most applications in which relative photoionization cross sections are obtained at the same incident photon energy, the monochromator shape function need not be known exactly.

The tabulated Ne σ_{2p} and Ne σ_{2s} partial photoionization cross sections [Wuilleumier and Krause (1979), Bizau and Wuilleumier (1995)] are used. This allows two independent determinations of the shape function based on the assumed form of $T(\varepsilon_k)$:

$$\Gamma_{2p}(h\nu) = \frac{A_{2p}}{\sigma_{2p}(h\nu)I_0(h\nu)T(\varepsilon_k)},$$ (5)

$$\Gamma_{2s}(h\nu) = \frac{A_{2s}}{\sigma_{2s}(h\nu)I_0(h\nu)T(\varepsilon_k)}.$$ (6)

Both measurements of the monochromator shape functions were reasonably identical, suggesting that Eq. (4) is valid in the energy range $20\,\text{eV} \leq \varepsilon_k \leq 190\,\text{eV}$. Furthermore, the results indicated that optical distortions are not factors that could affect the interpretation of the recorded PES spectrum.

A further consistency check of the foregoing results is obtained by comparing Ne σ_{2s}/σ_{2p} ratios obtained in these calibration studies. These ratios are given by

$$\frac{\sigma_{2s}(h\nu)}{\sigma_{2p}(h\nu)} = \frac{A_{2s}T(h\nu - \text{IP}_{2s})}{A_{2p}T(h\nu - \text{IP}_{2p})},$$ (7)

where IP_{2s} and IP_{2p} are the 2s and 2p ionization potentials, respectively, of neon. The experimentally obtained σ_{2s}/σ_{2p} ratios agree very well with previously tabulated results [Wuilleumier and Krause (1979), Bizau and Wuilleumier (1995)]. This arduous calibration procedure is necessary for the critical comparison of the satellite–main peak intensity ratios with the theoretical ratios predicted by MRSDCI calculations.

All intensity ratios reported in this work have therefore been corrected for transmission effects. Based on the errors involved in measuring the transmission corrections, photon flux, and shape function, the error involved in obtaining transmission-corrected intensity ratios is approximately $\pm 5\%$. Systematic errors due to the assumed baseline and transmission function are estimated to be less than 10%. Special precautions are also undertaken to ensure that the spectral features observed are not contaminated by effects of higher order synchrotron radiation, which can be significant for some types of monochromators.

III.   THEORETICAL DETAILS

The prediction of the photoelectron spectrum of atoms and molecules generally proceeds via one of two theoretical approaches: Green's-function methods [Cederbaum et al. (1978, 1986)] or configuration interaction (CI) methods [Martin and Shirley (1976a&b), Martin and Davidson (1977)]. A central quantity involved in both methods is the pole strength, which is also referred to as the spectroscopic factor [McCarthy (1985), Bawagan (1993)]. While Green's functions are essentially a perturbation expansion in neutral ground-state Hartree–Fock orbitals, CI methods do not need to follow this approach. Many CI approaches begin with separate calculations of the HF orbitals for the neutral state and for several other states of the ion. This tack is feasible for small CI expansions. Other CI expansions expand all states in a common basis, because this makes the computation of pole strengths more convenient.

From Fermi's golden rule, the photoionization cross section is given by

$$\sigma \propto |\langle \Psi_f(N)|\hat{T}|\Psi_0(N)\rangle|^2, \tag{8}$$

where $|\Psi_f(N)\rangle$ and $|\Psi_0(N)\rangle$ are the final and initial states of the N-electron system and $\hat{T}$ is the general dipole transition operator. The calculation of the initial state, $\Psi_0(N)$, is straightforward; however, the accurate calculation of the final-state electron–ion-scattering wave function is extremely difficult [Winstead and McKoy (1996)] and requires several approximations. We approximate the final state as

$$|\Psi_f(N)\rangle \approx |\chi^f(k)\rangle|\Psi_f(N-1)\rangle, \tag{9}$$

where $\chi^f(k)$ is a continuum function of momentum k representing the ejected electron and $\Psi_f(N-1)$ is the final ionic-state wave function. The preceding approximation assumes that interchannel coupling is negligible and, further, that the continuum function $\chi(k)$ is strongly orthogonal to both the initial neutral state and the final ionic state; that is,

$$\hat{a}_\chi|\Psi_0(N)\rangle = \hat{a}_\chi|\Psi_f(N-1)\rangle = 0. \tag{10}$$

This allows us to write the photoionization amplitude as [Martin and Shirley (1976a&b)],

$$M_f = \sum_q \langle\chi^f|\mu|\phi_q\rangle\langle\Psi_f(N-1)|\hat{a}_q|\Psi_0(N)\rangle, \tag{11}$$

where we have used the complete MO basis ϕ_q. The operator $\hat{a}_q$ is the annihilation operator for orbital q, and μ is the one-electron dipole operator. The terms in Eq. (11) correspond to the "direct" transition.

The pole strength (probability) for the jth ionic state is defined by the square of the norm,

$$S_j^2 = \|\langle\Psi_j(N-1)|\Psi_0(N)\rangle_{N-1}\|^2 = \left\|\sum_{p,q} C_{pj}^* D_q s_{pq}\right\|^2, \tag{12}$$

where

$$s_{pq} = \langle\Phi^{(p)}(N-1)|\Phi^{(q)}(N)\rangle_{N-1} \tag{13}$$

and the subscript on the bracket indicates integration over only $N-1$ electrons. Here, the jth eigenstate of the cation can generally be written as

$$\Psi_j(N-1) = \sum_p C_{pj}\Phi^{(p)}(N-1), \tag{14}$$

where C_{pj} is a coefficient that describes the extent of configuration mixing and $\Phi^{(p)}(N-1)$ is the pth possible ion configuration ($0p$–$1h$, $1p$–$2h$, etc.). Similarly, the ground state of the neutral molecule can be represented as

$$\Psi_0(N) = \sum_q D_q\Phi^{(q)}(N), \tag{15}$$

where $\Phi^{(q)}(N)$ describes the qth possible configuration of the neutral molecule (i.e., HF, $1p$–$1h$, $2p$–$2h$, etc.) and D_q is the CI coefficient for the qth configuration.

Further, the Dyson orbital can be defined as

$$\varphi_{\text{Dyson}}^j = S_j^{-1}\langle\Psi_j(N-1)|\Psi_0(N)\rangle_{N-1} = S_j^{-1}\sum_{p,q} C_{pj}^* D_q s_{pq}. \tag{16}$$

Following the approximation in Eqs. (9–11), the transition moment

$$M_j = \langle \Psi_j(N-1)\chi^j(k)|\mu|\Psi_0(N)\rangle \tag{17}$$

can be written as

$$M_j = \langle \chi^j(k)|\mu|\varphi^j_{\text{Dyson}}\rangle S_j, \tag{18}$$

where $\chi^j(k)$ is the continuum function for the outgoing (ejected) electron associated with state j and μ is the dipole operator. These assumptions will never be exactly correct: Overlap of the outgoing continuum wave with the neutral wave function gives additional terms in M_j that are referred to as "conjugate effects" [Martin and Shirley (1976a&b)]. It is generally believed that the conjugate terms become important only near the threshold (within 10 eV). Since most of the synchrotron PES spectra reported in this review are taken well above the threshold, we shall neglect the conjugate term. Lack of factorization of the ion wave function [see Eq. (9)] will also lead to channel mixing and will give additional terms. The case of interchannel coupling is well-documented for atoms [Becker at al. (1986), Wills et al. (1990)] and becomes significant only within 10 eV of the threshold.

When Eq. (18) is valid, the intensity ratio of the jth satellite (correlation) peak to the pth primary peak is given by

$$\frac{I(j,k')}{I(p,k)} \approx \frac{M_j^2}{M_p^2} = \frac{|S_j\langle\chi^j(k')|\mu|\varphi^j_{\text{Dyson}}\rangle|^2}{|S_p\langle\chi^p(k)|\mu|\varphi^p_{\text{Dyson}}\rangle|^2}. \tag{19}$$

For most satellites (correlation states), it is possible to identify a primary hole state $(0p\text{–}1h)$ such that $\varphi^j_{\text{Dyson}} \approx \varphi^p_{\text{Dyson}}$. As we shall see later, this approximation is valid only in the target Hartree–Fock approximation, where correlation in the initial state can be ignored. The states j and p will necessarily be of the same symmetry in order for the primary hole configuration that dominates the pth state to also enter with a nonzero coefficient in the jth state. In this case,

$$\frac{I(j)}{I(p)} \approx \frac{S_j^2}{S_p^2} \tag{20}$$

if the comparison is made at the same outgoing photoelectron kinetic energy (i.e., $k' \approx k$) and if χ^j is sufficiently similar to χ^p in the region of space close to the molecule. If the satellite–main intensity ratio is computed at the same photon energy rather than the same photoelectron energy, then it must further be assumed that the dipole transition matrix elements do not change rapidly with k. For the current study, this condition holds favorably, since the kinetic energies are greater than 10 eV. Furthermore, the coupling of channels in the continuum is neglected. In the present approximation, the explicit form of the continuum function need not be specified; it is assumed only that the continuum functions are similar for the states being compared (i.e., for the relative intensity ratios). A key result of this derivation is that the predicted satellite-to-main peak intensity ratio (Eq. 20) would be approximately constant, independent of the outgoing photoelectron energy and thus independent of photon energy, if all the assumptions were approximately valid.

A further assumption critical to the comparison of experimental PES ratios ($I(j)/I(p)$) and spectroscopic factors (S_j^2/S_p^2) is that the experimental spectrum is well resolved, so that only one electronic state lies under each peak and the unresolved molecular-vibrational levels do not affect the result.

IV.   SURVEY OF SYNCHROTRON PHOTOELECTRON SPECTROSCOPY OF MOLECULES (POST-1986)

This survey illustrates the various mechanisms whereby electron correlation effects manifest themselves (in a nontrivial manner) in the PES spectra of molecules. The trivial case is that the absolute energies of the electronic states are not accurately predicted by Koopmans' theorem; however, their relative energies and intensities are predicted reasonably well. A nontrivial case would be the switching of the order of the primary hole ion states, compared with the order of the molecular orbital energies, as is observed, for example, in the outer valence MOs of N_2 [Kimura et al. (1981), Svensson et al. (1991)] or the appearance of additional peaks with intensities comparable to the primary hole peaks. We illustrate these mechanisms by presenting the PES spectra of representative molecules and compare them within families (i.e., increasing alkylation) and among families (i.e., different degrees of unsaturation). This approach follows from our traditional understanding of chemistry and results from a belief that functional groups are transferable and that molecules with similar functional groups have similar "chemistry" and therefore might also have similar spectroscopic properties. By that means, we illustrate the salient properties of electron correlations and unravel some of the "mechanisms" from which they arise.

A.   Alkanes and Alkenes

Hydrocarbons are categorized into two main classes: alkanes (saturated hydrocarbons) and alkenes (unsaturated hydrocarbons). Cederbaum et al. (1978) pointed out that these two classes of hydrocarbons are very distinct with regard to the severity of electron correlation effects on their ionization spectra. Their distinctness is illustrated for propene and propane in Figures 1 and 2 and also for 1-butene and n-butane in Figures 3 and 4, respectively. This type of comparison is further simplified by the fact that hydrocarbons are built up only by C and H atoms; thus, the inner valence MOs are dominated by C2s atomic contributions. The number of inner valence peaks is therefore expected to correlate with the number of carbon atoms in the hydrocarbon framework. The fact that there are additional distinct peaks and shoulders in the inner valence region of propene and 1-butene (see

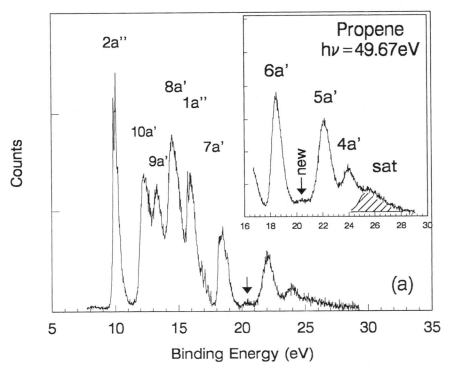

Figure 1. (a) Synchrotron PES spectrum of propene (C₃H₆) at 49.7 eV photon energy [Bawagan et al. (1997)]. As shown, the spectrum is not corrected for transmission effects. The estimated total energy resolution is 140 meV fwhm. The standard orbital notation assumes C_s symmetry. Calibrant gas (indicated by an asterisk) allows for absolute determination of the energy scale and experimental energy resolution. The high-statistics spectrum shown in the inset is taken without the calibrant gas. A new correlation peak (indicated by the arrow) is observed at 20.3 eV binding energy. (b) Theoretical PES of propene calculated using the MRSDCI energies and pole strengths convoluted with the experimental peak widths. The primary peak assignments are similar to that shown in (a). The 8a' and 1a" orbitals are noted especially, because previous theoretical calculations have misassigned these peaks. The correlation states (a, b, and c) that can also be seen in the experimental spectrum are highlighted.

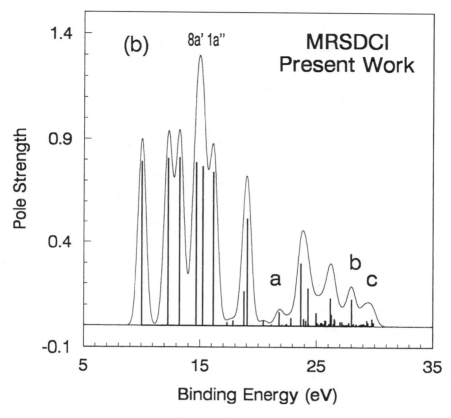

Figure 1. Continued.

Figures 1 and 3) clearly indicates a breakdown of the independent-particle picture of ionization for these simple alkenes. Quite the opposite is the case for propane and n-butane (see Figures 2 and 4), where the one-peak, one-orbital correspondence reasonably holds. There may be some very small features at higher binding energy corresponding to correlation states associated with the $^2A_1$ and $^2A'$ manifolds of propane and n-butane, respectively.

The ionization spectrum of propene predicted by using MRSDCI wave functions is shown in Figure 1(b). The quantitative agreement that can be obtained between the experimental PES spectrum and the theoretical MRSDCI spectrum is illustrated and represents perhaps the state of the art. What is quite interesting is that the correlation states arising in propene [Bawagan et al. (1997)] are very similar to those found in ethylene

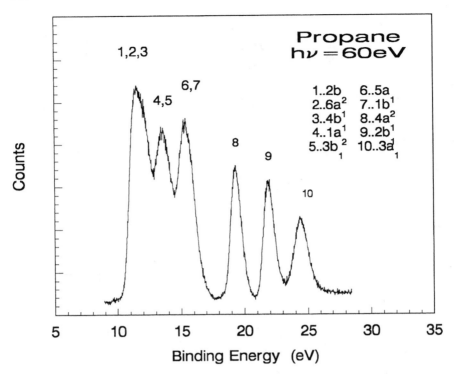

Figure 2. Synchrotron PES spectrum of propane (C_3H_8) at 60 eV photon energy [Bawagan, (1998)]. As shown, the spectrum is not corrected for transmission effects. The estimated energy resolution is 150 meV fwhm. The standard orbital notation assumes C_{2v} symmetry. Peaks 8, 9, and 10 correspond to the expected inner valence peaks from linear combinations of the C2s AOs.

[Desjardins et al. (1995)]. The dominant mechanism whereby these correlation states arise is often referred to as the "one-up, one-down" mechanism [Martin and Davidson (1977)] and is shown in Figure 5 for propene. The near degeneracy of the one-hole configuration, $(6a')^{-1}$, and the two-hole, one-particle configuration, $(8a')^{-1}(2a'')^{-1}(3a'')^1$, is quite apparent from the diagrams, since the $\pi \rightarrow \pi^*$ transition (one up) is balanced by the $8a' \rightarrow 6a'$ transition (one down). The one-up, one-down mechanism is quite general, especially for the unsaturated hydrocarbons, because of the presence of the characteristic π^* orbitals (LUMO). The π^* orbitals in these simple alkenes occur at low energies and are more localized than the Rydberg orbitals occupied in most of the ion excited states.

In ethylene (shown in Figure 6), the one-up, one-down mechanism takes a very prominent and evident role in which the satellite peak gains sufficient

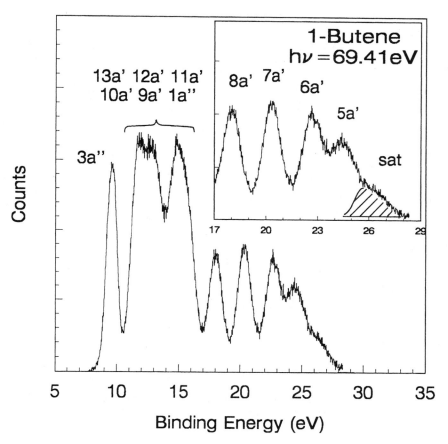

Figure 3. Synchrotron PES spectrum of 1-butene (C_4H_8) at 69.4 eV photon energy [Bawagan et al. (1997)]. As shown, the spectrum is not corrected for transmission effects. The estimated energy resolution is 160 meV fwhm. The standard orbital notation assumes C_s symmetry. The correlation peak is found at the high-binding-energy side of the $5a'$ primary peak.

intensity such that it is now more difficult to distinguish from the parent peaks. Martin and Davidson (1977) were the first to correctly discuss the origin of this intense PES satellite peak in ethylene in terms of the one-up, one-down mechanism. The correlation peak at 27.4 eV corresponds to a "cluster" of correlation states of ^2A_g symmetry in the energy range 25–30 eV. No other states in this energy region have any appreciable intensity, according to the MRSDCI calculations. The most prominent correlation state, at 27.8 eV, has the leading configuration, $(2b_{1u})^{-1}(1b_{3u})^{-1}(1b_{2g})^1$,

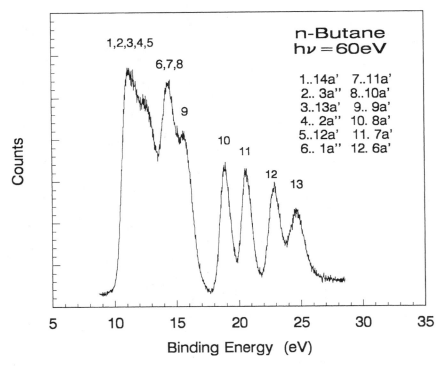

Figure 4. Synchrotron PES spectrum of n-butane (C_4H_{10}) at 60 eV photon energy [Bawagan (1998)]. As shown, the spectrum is not corrected for transmission effects. The estimated energy resolution is 150 meV fwhm. The standard orbital notation assumes C_s symmetry. Peaks 10, 11, 12, and 13 correspond to the expected inner valence peaks from linear combinations of the C2s AOs.

which, in common notation, is the $2h$–$1p$ configuration given by

$$0.37 \,^3(\pi, \pi^*)(2b_{1u})^{-1} - 0.52 \,^1(\pi, \pi^*)(2b_{1u})^{-1}. \tag{21}$$

The energies of the correlation peaks can often be simply estimated by writing down the configurations and summing the energies of the individual processes. For example, the excitation energies for $^3(\pi, \pi^*)$ and $^1(\pi, \pi^*)$ are 4.6 eV and 7.65 eV, respectively, while the ionization potential for the $2b_{1u}$ orbital is 19.2 eV. If we simply sum up the energies, we obtain 26.85 eV for $^1(\pi, \pi^*)(2b_{1u})^{-1}$ and 23.8 eV for $^3(\pi, \pi^*)(2b_{1u})^{-1}$. Neither of these estimated energies exactly matches the experimental energy (27.4 eV), but both are fairly close. The largest source of error in this estimate is the very large CI effect from mixing these configurations with the $2a_g^{-1}$ configuration. This

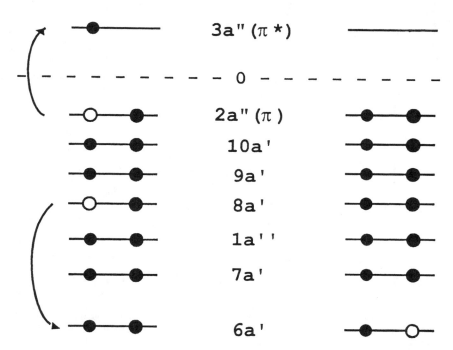

Figure 5. Schematic of the (one-up, one-down) mechanism [Bawagan et al. (1997)] for configuration interaction. The $(8a')^{-1}(2a'')^{-1}(3a'')^1$ $2h–1p$ configuration interacts strongly with the $(6a')^{-1}$ $1h$ configuration. The energy levels are not drawn to scale.

mixing stabilizes the $2a_g^{-1}$ hole state, but destabilizes the correlation state (at 27.4 eV) by several eV.

B. Intrinsic Correlations and Dynamic Correlations

Phenomenological schemes have been proposed to classify correlation states and their production mechanisms by studying the photon energy

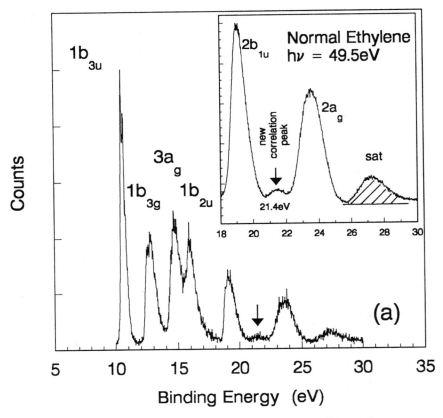

Figure 6. (a) Synchrotron PES spectrum of ethylene (C_2H_4) at 49.5 eV photon energy [Desjardins et al. (1995)]. As shown, the spectrum is not corrected for transmission effects. The estimated total energy resolution is 360 meV fwhm. The standard orbital notation assumes D_{2h} symmetry. The high-statistics spectrum shown in the inset highlights a new correlation peak (indicated by the arrow) at 21.4 eV binding energy. (b) Theoretical PES of ethylene calculated using the MRSDCI energies and pole strengths convoluted with the experimental peak widths. The primary peak assignments are similar to that shown in (a). The predicted twinning of the $2a_g^{-1}$ primary peak at 23.7 eV binding energy is evident.

dependence of the ratios of the satellite peak intensities to the main or "parent" peaks observed in the PES spectrum. One such scheme is that proposed by Becker and Shirley (1990) for atoms which we have successfully adopted to study the correlation states in molecules. Those correlation (satellite) peaks that exhibit a constant ratio with increasing photon energy are referred to as *intrinsic correlations* and are considered to be caused by initial- or final-state configuration interaction, whereas those correlation

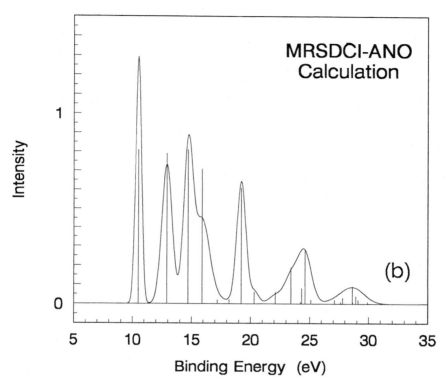

Figure 6. Continued.

(satellite) peaks that exhibit a strong photon energy dependence in the satellite-to-main peak intensity ratio are referred to as *dynamic correlations* and can be produced by shake-up processes, continuum state interactions, or interchannel coupling. The term "shake-up" is used here in a limited sense to refer only to correlation peaks that exhibit a particular photon energy dependence in their cross sections, as discussed by Becker and Shirley (1990). While the terminology used in this classification may be subject to debate, its main value is phenomenological. The classification is simply based on the empirical behavior of satellite cross sections as the photon energy is varied. If the ratio stays relatively constant, the correlation (satellite) peak is called "intrinsic"; if the ratio does not stay relatively constant, the correlation (satellite) peak is called "dynamic." This classification assumes that the satellite and main peak contain a single fully resolved transition and that the correct correspondence of satellite to main peak has been made. Although these two conditions are generally difficult to achieve, there are many cases of molecular systems in which correlation peaks belonging to different Dyson orbital manifolds are reasonably separated or only minimally overlap.

On the basis of the preceding simple classification scheme, the correlation peak in ethylene at 27.4 eV would be called an intrinsic correlation peak. Figure 7 illustrates the level of agreement between experiment and theory with regard to the prediction of the correlation (satellite) peak intensities of ethylene. The constant trend expected for "intrinsic" correlation states is shown by the dashed line and is in quantitative agreement with the synchrotron PES data over a wide photon energy range. With new higher energy monochromators slowly becoming available, PES data for $E_{h\nu} \geq$ 300 eV should further confirm the constant trend. Nevertheless, the present

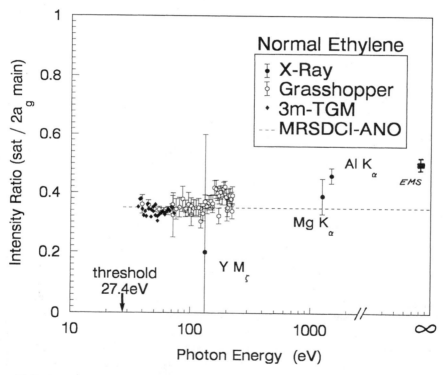

Figure 7. Photon energy dependence of the 27.4-eV major satellite–main $2a_g^{-1}$ peak intensity ratios derived from synchrotron PES measurements for ethylene [Desjardins et al. (1995)]. The experimentally derived intensity ratios are corrected for transmission effects. Previous discrete X-ray measurements are shown as solid circles. The dashed line is the predicted satellite–main intensity ratio from the MRSDCI calculation shown in Figure 6(b). The EMS data [Bawagan (1993)] are shown as solid squares at the infinite-energy scale.

level of agreement is unprecedented and serves as a useful benchmark for future comparisons of experiment and theory for the correlation states of molecules. The agreement also illustrates the importance of well-calibrated PES measurements where experimental uncertainties due to the transmission of energy analyzers are well known. In spite of the nice constant trend that superficially seems to verify that this is a case of a simple intrinsic correlation peak, there are actually many unresolved correlation states clustered under the main peak at 23.7 eV and under the correlation peak at 27.4 eV. Although most of these prominent states are of ^2A_g symmetry, one correlation state of $^2B_{1u}$ symmetry at 24.3 eV is within what we refer to as the $2a_g^{-1}$ primary peak.

On the other hand, the correlation peak at 21.4 eV would be called a dynamic correlation peak. The interesting photon energy dependence of this peak is shown in Figure 8. The overall feature of increasing cross section as the ionization threshold is approached is indicative of dynamic correlation if the new state at 21.4 eV is indeed a satellite of the $2b_{1u}^{-1}$ primary hole state, as has been assumed. The oscillations on top of this general trend have been noted [Desjardins et al. (1995)] and are believed to be a result of "resonance-type" interactions with the molecular framework. Further studies are still necessary to clarify the origins of these oscillations. The PES feature at 21.4 eV binding energy is believed to be a result of continuum-state configuration interaction otherwise known as conjugate shake-up [Ishihara et al. (1980), Ferrett et al. (1987), Schmidt (1992), Reich et al. (1994)].

Following the phenomenology of Becker and Shirley (1990), the term "shake-up" is limited to correlation peaks that exhibit an increasing cross section with increasing photon energy. These peaks eventually reach a plateau as the sudden limit is approached and are therefore classified as dynamic correlation. Historically, however, the term "shake-up" took a broader definition [Aberg (1967), Dyall and Larkins (1982a,b)] wherein it referred to a photoionization process in which an electron made a dipole transition from orbital i to continuum state f, accompanied by a "monopole" transition from orbital m to orbital n. The shake-up transition moment is therefore given by

$$M_{\text{shake-up}} = \langle f | \mu | i \rangle C_{(i^{-1}f)(i^{-1}m^{-1}nf)}, \tag{22}$$

where $C_{\alpha\beta}$ are the CI coefficients describing the contribution of configuration α to the ion in state β. This picture is akin to a common view that shake-up correlation states are due to "ionization plus excitation processes."

On the other hand, "conjugate shake-up" refers to a process whereby an electron makes a "monopole" transition from orbital i to continuum state f, accompanied by a dipole transition from orbital m to orbital n. The

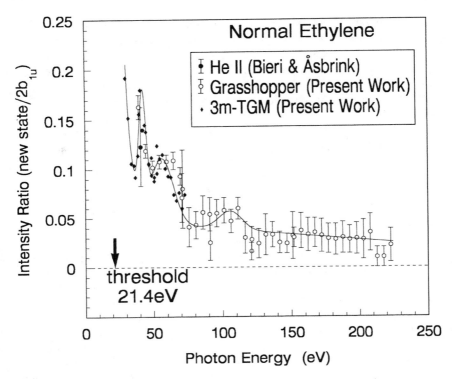

Figure 8. Photon energy dependence of the 21.4-eV new satellite–main $2b_{1u}^{-1}$ peak intensity ratios derived from synchrotron PES measurements for ethylene [Desjardins et al. (1995)]. The experimentally derived intensity ratios are corrected for transmission effects. A previous discrete HeII measurement is shown as a solid circle. The solid line serves to highlight the resonancelike features at photon energies of 42, 57, and 105 eV.

conjugate shake-up transition moment is given by

$$M_{\text{conjugate shake-up}} = \langle n|\mu|m\rangle C_{(m^{-1}n)(i^{-1}m^{-1}nf)}. \tag{23}$$

The relevant coefficients for calculating relative photoionization cross sections are, therefore,

$$C_{(i^{-1}f)(i^{-1}m^{-1}nf)} \approx \frac{\langle i^{-1}f|H|i^{-1}m^{-1}nf\rangle}{\varepsilon_n - \varepsilon_m} \tag{24}$$

for shake-up and

$$C_{(m^{-1}n)(i^{-1}m^{-1}nf)} \approx \frac{\langle m^{-1}n|H|i^{-1}m^{-1}nf\rangle}{\varepsilon_f - \varepsilon_i} \tag{25}$$

for conjugate shake-up. Notice in this case that the shake-up and conjugate shake-up configurations differ from the primary hole configuration by only single excitations, rather than the double excitation (i.e., one-up, one-down excitation) that characterizes strong electron correlation. Single-excitation configurations can be regarded as describing orbital relaxation rather than electron correlation effects.

In the lower energy primary hole, the mixing coefficient will be approximately equal and opposite, according to first-order perturbation theory. Hence,

$$C_{(i^{-1}m^{-1}nf)(i^{-1}f)} \approx -C_{(i^{-1}f)(i^{-1}m^{-1}nf)}, \tag{26}$$

and the two-configuration wave function

$$\phi_{(i^{-1}f)} + C_{(i^{-1}m^{-1}nf)(i^{-1}f)}\phi_{(i^{-1}m^{-1}nf)} \tag{27}$$

can be approximately factored into a single configuration by redefining the orbitals

$$g_{m'} = g_m + C_{(i^{-1}m^{-1}nf)(i^{-1}f)}g_n \tag{28a}$$

and

$$g_{n'} = g_n + C_{(i^{-1}f)(i^{-1}m^{-1}nf)}g_m. \tag{28b}$$

The primary hole state then becomes approximately $\phi_{(m^{-1}m'i^{-1}f)}$, which could be viewed as the result of the first cycle of an SCF optimization of the orbitals in this ion state in which the orbital g_m is replaced by an improved orbital $g_{m'}$. Similarly, the mixing with the conjugate shake-up configuration is equivalent to replacing the continuum orbital by a relaxed continuum orbital

$$\begin{aligned} g_{i'} &= g_i + C_{(i^{-1}m^{-1}nf)(m^{-1}n)}f, \\ f' &= f + C_{(m^{-1}n)(i^{-1}m^{-1}nf)}g_i, \end{aligned} \tag{29}$$

in the satellite state and replacing the bound orbital g_i by a relaxed bound orbital g_i' in the neutral bound excited state $\phi_{(m^{-1}n)}$.

Hence the three-configuration satellite state formed from mixing $(i^{-1}f)$ and $(m^{-1}n)$ with $(i^{-1}m^{-1}nf)$ should be given approximately by the single $(i^{-1}m^{-1}n'f')$ configuration. The transition moment is then

$$\begin{aligned} \langle \Psi | \mu | \Psi^{+*} \rangle &= \langle i | \mu | f' \rangle \langle m | n' \rangle - \langle i | \mu | n' \rangle \langle m | f' \rangle \\ &+ \langle i | f' \rangle \langle m | \mu | n' \rangle - \langle i | n' \rangle \langle m | \mu | f' \rangle. \end{aligned} \tag{30}$$

To first order in the C coefficients, this reduces to

$$\langle\Psi|\mu|\Psi^{+*}\rangle = \langle i|\mu|f\rangle C_{(i^{-1}f)(i^{-1}m^{-1}nf)} + C_{(m^{-1}n)(i^{-1}m^{-1}nf)}\langle m|\mu|n\rangle, \qquad (31)$$

which is like Eqs. (22) and (23). Note that all contributions to the transition moment are added, and phase information may cause coherence or cancellation in the intensity. This example illustrates clearly that relaxation effects may be treated equivalently either by CI or by separate SCF calculations if the outgoing wave is included in the ion wave function. Conjugate shake-up will be missed, however, if the ion is treated as only an (N-1) electron system, as indicated in Eqs. (9) through (18).

A similar approximate relation of the CI coefficients to monopole matrix elements holds whenever separate SCF optimized MOs are employed in the ion and the neutral, as, for example, in the relaxed Hartree–Fock approximation [Dyall and Larkins (1982a&b)]. It can be further shown that the peak intensity of the conjugate process decreases as k^{-2} at large k [Martin and Shirley (1976b)].

C. Acetylene, Allene, and Butadiene

Cederbaum et al. (1978) suggested that certain molecular properties "favor the breakdown phenomenon to occur" (p. 1600). They cited the presence of low-lying excited states, localized unoccupied orbitals, and large singlet–triplet splittings as indicators of the propensity of molecules to display a breakdown of the MO picture of ionization. While these may be qualitatively true, our results indicate that a quantitative understanding of these breakdown features require, in addition, a careful treatment of basis-set effects.

We illustrate these effects by considering the synchrotron PES spectrum of acetylene [Moghaddam et al. (1995)]. Figure 9 shows the PES spectrum of acetylene and its ^{13}C analog. As with ethylene [Desjardins et al. (1995)], isotope effects do not alter the intensities of the inner valence and satellite peaks, indicating quite clearly that this phenomenon is strictly of electronic origin. The basis-set effects are made evident by comparing the experimental PES spectrum in Figure 9 with various theoretical predictions of the ionization spectrum of acetylene, which is shown in Figure 10. The "double-hump" structure, which is very evident in the experiment, is predicted only by the MRSDCI calculations of Moghaddam et al. (1995). The Green's-function calculations of Weigold et al. (1991) and the SAC-CI calculations of Wasada and Hirao (1989) fail in this respect, not because of their methods, but because of inadequacies in their chosen basis sets. The $3d\delta_g$ Rydberg orbital of the ion is occupied in the dominant configuration of many of the correlation states of acetylene and needs to be included in the

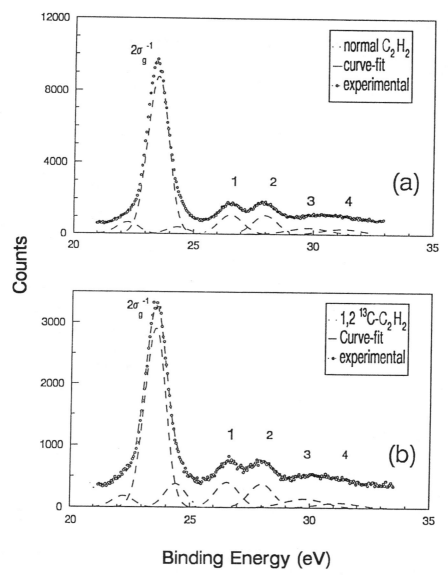

Figure 9. Synchrotron PES spectra of acetylene (C_2H_2) and isotopically labelled acetylene at (a) 68.8 eV photon energy and (b) 72.4 eV photon energy [Moghaddam et al. (1995)]. The total energy resolution is about 400 meV fwhm. As shown, the spectrum is not corrected for transmission effects. Gaussian curves are fitted to the correlation peaks (satellites 1–4) and the $2\sigma_g^{-1}$ main peak. The $1\pi_u^{-1}$ (11.43 eV binding energy), $3\sigma_g^{-1}$ (16.7 eV binding energy), and $2\sigma_u^{-1}$ (18.78 eV binding energy) main outer valence peaks are not shown.

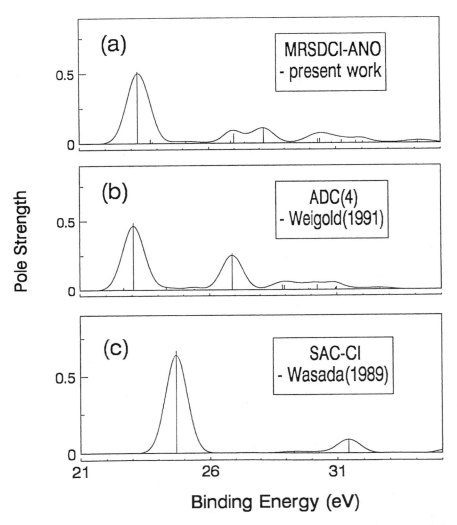

Figure 10. Comparison of the calculated PES spectra for acetylene: (a) MRSDCI calculation [Moghaddam et al. (1995)], (b) Green's-function calculation [Weigold et al. (1991)], and (c) SAC-CI calculation [Wasada and Hirao (1989)]. The calculated poles strengths (solid poles) are convoluted with the experimental peak widths to yield the solid curve, which can then be compared with experiment.

basis set. Often, basis sets are a minor consideration, especially in the calculation of the main peak ionization energies. However for correlation peaks, extreme care has to be exercised. Many correlation states of the $2h–1p$ type have a dominant configuration with an electron in a Rydberg orbital of the ion. These states will be missed unless diffuse basis functions of the correct symmetry are included. Unfortunately, ordinary basis sets for the neutral molecule do not include many of these basis functions.

It would be tempting to associate one satellite peak with every double bond if we were viewing only the PES spectra of ethylene and acetylene. This naive view is incorrect, as can be seen in the PES spectra of 1,3-butadiene (Figure 11) and allene, or 1,2-propadiene (Figure 12). The "double-hump" satellite peak, which is observed prominently in acetylene, is not seen in the PES spectra of allene or butadiene. It is interesting to note that the "double-hump" structure of acetylene was mistakenly thought to be the result of nonadiabatic (vibronic) coupling [Flores-Riveros et al. (1986)], as opposed to deficiencies in the basis set. Recent work has revealed that,

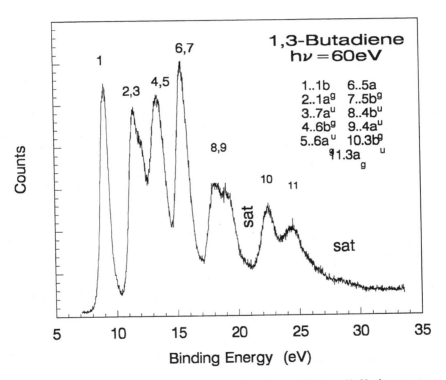

Figure 11. Synchrotron PES spectrum of 1,3-butadiene (C_4H_6) at 60 eV photon energy [Bawagan (1998)]. As shown, the spectrum is not corrected for transmission effects. The estimated energy resolution is 160 meV fwhm. The standard orbital notation assumes C_{2h} symmetry. Correlation peaks (26–30 eV binding energy) are found at the high-binding-energy side of the $3a_g$ primary peak. Another correlation peak is found at around 21 eV binding energy. These results are consistent with recent calculations by Deleuze and Cederbaum (1997).

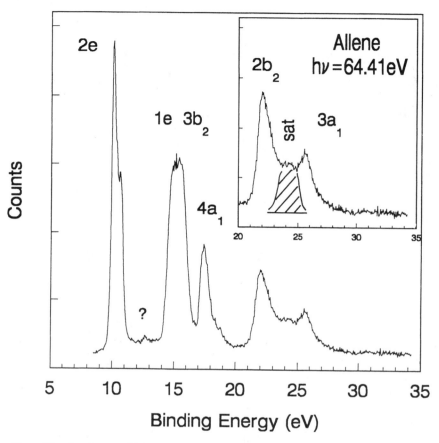

Figure 12. Synchrotron PES spectrum of allene (C_3H_4) at 64.4 eV photon energy [Bawagan et al. (1998)]. As shown, the spectrum is not corrected for transmission effects. The estimated energy resolution is 140 meV fwhm. The standard orbital notation assumes D_{2d} symmetry. The main $2e^{-1}$ peak is split by Jahn–Teller effects. Correlation peaks are indicated by the label "sat." The "mystery peak" at 12.7 eV binding energy is possibly a correlation peak that obtains its intensity from different primary holes as the torsional angle changes.

while nonadiabatic effects are not critical in the valence ionization of acetylene and ethylene [Koppel et al. (1978)], they may be more critical in the related molecular system of allene.

To illustrate this point, we recently investigated the origin of a small feature (12.7 eV binding energy) in the PES spectrum of allene [Bawagan et al. (1998)]. It appears that the 12.7 eV peak is a peculiar correlation (satellite) peak of ^2E symmetry. This correlation peak is peculiar in the sense that it gains intensity as a result of strong (Jahn–Teller-like) vibronic coupling between two different ^2E electronic states. The nonadiabatic mechanism whereby this phenomenon occurs is illustrated in Figure 13, where we show the calculated CASSCF energies of allene as a function of the twist (dihedral) angle. Photoionization is usually considered with the Franck–Condon principle in mind; that is, ionization proceeds via a vertical transition from the ground state of the neutral molecule to the ground or excited state of the ion. According to this picture, the correlation peak at

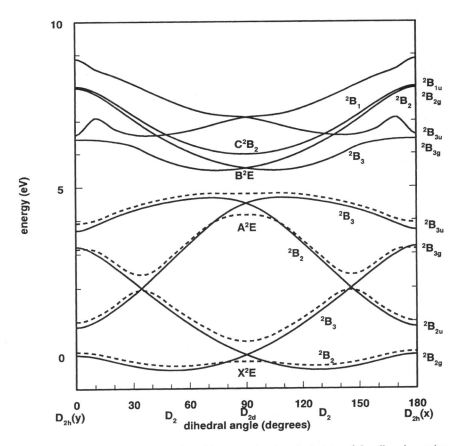

Figure 13. Total CASSCF energies of the ground and excited states of the allene ion, using 6-31G* basis as a function of dihedral angle [Bawagan et al. (1998)]. The total energy is plotted on a relative scale, with the total energy of the ground state of the ion (X^2E) at D_{2d} geometry taken as zero. For the D_2 symmetry scan (solid lines), the two C–C bonds are fixed at 1.38 Å, and all C–H bonds are fixed at 1.08 Å. The HCH angle is 121° and the CCC frame is located on the z-axis. For the C_2 symmetry scan (dashed lines), the two C–C bonds are 1.34 Å and 1.44 Å, respectively, and all other parameters are kept similar to the D_2 symmetry scan.

12.7 eV appears to be a result of a nonvertical transition to a strongly coupled electronic state illustrated by the crossing of the lower "Jahn–Teller" component of the A^2E state with the upper "Jahn–Teller" component of the X^2E state at precisely the same energy as that observed experimentally. This crossing of potential energy curves occurs at a dihedral twist angle of about 54 degrees from the neutral allene geometry.

Transitions to the lowest $^2B_2$ state of allene have also been extensively analyzed [Yang et al. (1990), Baltzer et al. (1995a)]. The high-resolution PES spectrum shows vibronic structure consistent with the Jahn–Teller splitting, which causes a twist of the lowest ionic state by about 49 degrees [Yang et al. (1990, 1993)]. The maximum intensity does not occur for the vertical transition. Similarly, the vibrational structure of the lowest π^{-1} state of ethylene has been resolved [Holland et al. (1997)]. Several vibrational progressions in the PES spectrum have been assigned. Notably, the maximum intensity occurs for the 0–0 transition and not for the "vertical" transition. These recent results indicate that nonadiabatic effects are quite important in photoionization experiments, contrary to what is generally believed. Experiments designed to probe the generality of these effects in other chemical systems are ongoing.

D. Other Molecular Systems

The high-resolution PES spectra of some molecular systems have revealed well-resolved vibrational structure. For example, the HeI PES spectrum of sulfur dioxide [Holland et al. (1994)] shows clear vibrational progressions in the satellite peak at 18 eV binding energy. The vibrational spacings are on the order of 90–120 meV, which is consistent with the ν_1 mode. Similarly, the HeII PES spectrum of carbon disulfide [Baltzer at al. (1996)] shows well-resolved vibrational progressions in four satellites at 14.09 eV, 17.0 eV, 18.81 eV, and 20.17 eV binding energy. These recent findings are in contrast to the satellite peaks of the alkenes, which do not show any clear vibrational structure. This is understood to be a result of the presence of strongly dissociative states among the excited states of the alkene ions, while SO_2 and CS_2 tend to form several very stable, excited ionic states.

Unlike the situation with the simple alkenes, it is not quickly obvious how simple models of electron correlation can be developed for an understanding of the correlation effects in the ionization of molecules such as CS_2, NO_2, and CS_2. Even for furan (C_4H_4O) and thiophene (C_4H_4S) [Bawagan et al. (1992)], which are chemically similar molecules, it is also difficult to develop such a scheme because the presence of inequivalent atoms brings about enormous complexity in the inner valence binding-energy region. Perhaps one exception is N_2 [Krummacher et al. (1980), Svensson et al. (1991)], which shows vibrational structure in its satellite peaks and also has equivalent atoms, thus making the analysis of electron correlation effects more transparent.

It is tempting to ask what the correlation states of polymers of conjugated hydrocarbons would look like. This is particularly important for the interpretation of the PES of organic polymers like polyacetylene or even polythiophene. Keane et al. (1991, 1992) recently reported the X-ray PES spectrum of poly (3-alkylthiophene) and their model compounds. Similar studies have also been reported by Fujimoto et al. (1990) on the PES spectra of a series of oligothiophenes. A recent X-ray PES study of polyacetylene and related compounds reported by Keane et al. (1992) showed that it is possible to make heuristic arguments and assign the correlation states of a semi-infinite solid. Further, Green's-function calculations supporting this heuristic extrapolation were made by Deleuze and Cederbaum (1997).

V. COMPARISON OF EMS AND PES INTENSITIES

An interesting historical account of the controversy between the ionization intensities (or spectroscopic factors) obtained in EMS and PES is provided by Brion et al. (1988). The controversy centers on the apparent differences between the Ar $3s^{-1}$ satellite intensities observed in EMS [Leung and Brion (1983), McCarthy and Weigold (1985)] and PES [Kossman et al. (1987), Svensson et al. (1987)]. Current understanding, however, seems to suggest that these differences are due to the different ionization mechanisms in photoionization and electron impact ionization. These differences were first discussed by Kheifets [(1993); Amusia and Kheifets (1985)] and McCarthy (1985) and are best illustrated by the very approximate relation derived by Amusia and Kheifets (1985) for atomic systems, viz.,

$$Z_{PES}^i = Z_{EMS}^i (1 + \frac{\varepsilon_i - E_0}{\Delta})^2, \tag{32}$$

where Z_{PES}^i and Z_{EMS}^i are the spectroscopic factors obtained in PES and EMS, respectively, ε_i is the exact energy of the correlated ion state i, E_0 is the Koopmans' energy of the corresponding ion primary hole state, and Δ is the average energy of the two-hole, one-particle states, estimated as the energy of the first strong satellite. The spectroscopic factors were defined in this work as the relative intensity of the peaks in the spectrum. The authors argued that, at low ion recoil momentum, the EMS spectroscopic factor (Z_{EMS}^i) would approach the probability of finding the single-hole configuration in the correlated final ion state i. This is different from the pole strengths, which we have defined in Eq. (12). According to Eq. (32), the PES spectroscopic factors (Z_{PES}^i) are not the true spectroscopic factors (Z_{EMS}^i)

and, in fact, can never approach the true spectroscopic factors, even if the photon energy is varied. The only possibility is by accident—that is, whenever $(\varepsilon_i - E_0)$ is negligible. Amusia and Kheifets suggested that spectroscopic factors depend significantly on the recoil momentum of the ion—which is large in PES and small in EMS. That Eq. (32) may be valid was corroborated by recent studies by Kheifets and Amusia (1992) on the spectroscopic factors of the Xe $5s^{-1}$ satellites. In the case of these satellites, $(\varepsilon_i - E_0)$ is small, and the satellite intensity ratios obtained in PES and EMS are fairly similar, in stark contrast to the situation of Ar $3s^{-1}$ satellites.

An important factor to consider in comparing spectroscopic factors in PES and EMS is the poorer energy resolution in EMS ($\geq 1000\,\mathrm{meV}$ fwhm) relative to PES using Al K_α radiation ($\geq 500\,\mathrm{meV}$ fwhm) or synchrotron radiation ($\approx 100\text{--}200\,\mathrm{meV}$ fwhm). The energy resolution is critical in the accurate determination of spectroscopic factors of closely spaced satellite or correlation states. In particular, a reanalysis of previously reported EMS Ar $3s^{-1}$ binding-energy spectra [Brion et al. (1988), Leung and Brion (1983)] indicated that the observed differences between spectroscopic factors obtained in PES and EMS can be reconciled if energy resolutions and detailed fine structure are taken into account. Note that this type of reanalysis does not contradict Eq. (32), but rather provides a more realistic assessment of the error limits of previous EMS studies. Clearly, a very high energy resolution EMS study of the noble gases, in the range 100–500 meV fwhm, would be most helpful in resolving whether Eq. (32) is, in fact, valid.

A partial clue to resolving this controversy is provided by the spectroscopic factors of ethylene [Bawagan (1993)]. Of particular interest is the correlation state at 27.4 eV binding energy, which has been previously observed by PES [Berndtsson et al. (1975)] and EMS [Dixon et al. (1978), Coplan et al. (1978), Hollebone et al. (1995)]. Figure 7 shows the transmission-corrected satellite/$2a_g^{-1}$ main peak intensity ratio obtained from a series of synchrotron PES studies [Desjardins et al. (1995)] with an energy resolution of better than 200 meV fwhm. The EMS satellite/$2a_g^{-1}$ main intensity ratios are at the "infinite" photon energy scale. Also shown are the calculated pole strengths from the cluster of correlation states (of the same symmetry) around 27.4 eV obtained from accurate MRSDCI calculations. Even more sophisticated MRSDCI calculations indicated that the calculated intensity ratio converges at a value of 0.35, similar to that shown in Figure 7. The important point to note is the quantitative agreement obtained between synchrotron PES intensity ratios and the pole strength ratios calculated by using very accurate MRSDCI wave functions.

For the ethylene satellite, the $\varepsilon_i - E_o$ energy difference is approximately 5 eV, which can be compared with the energy difference for the Xe $5s^{-1}$

satellite $(5s^2 5p^4({}^1D)5d({}^2S))$, which is about 2 eV and the energy difference for the Ar $3s^{-1}$ satellite $(3s^2 3p^4({}^1D)3d({}^2S))$, which is roughly 4 eV. The ethylene satellite at 27.4 eV is therefore an ideal test case for the Amusia and Kheifets (1985) theory of spectroscopic factors. The EMS studies all provide an intensity ratio of about 0.5, which is markedly different from the intensity ratios obtained from synchrotron PES and those predicted by MRSDCI calculations [Desjardins et al. (1995)]. Note that energy resolution considerations are not critical for this ethylene satellite, because most of the correlation states contributing to the intensity around 27.4 eV are of ${}^2A_g$ symmetry and have a Dyson orbital similar to $2a_g$.

VI. DEVELOPING PRACTICAL PICTURES OF ELECTRON CORRELATION EFFECTS

To gain an understanding of electron correlation effects in PES, let us consider a few simple examples. First, consider the effect of initial-state CI in the helium atom, which illustrates an example of shake-up due to orbital relaxation. The natural orbital expansion of helium [Davidson (1963), Lermer et al. (1996)] takes the form

$$\Psi = D_1[1s^2] + D_2[2p^2] + D_3[2s^2] + \cdots, \tag{33}$$

where $2s$ and $2p$ are of the same size as the $1s$ orbital, but with additional nodes. The ion ground-state wave function (labeled with the subscript 1) is

$$\Psi_1^+ = C_1[1s] + C_2[2s] + \cdots, \tag{34}$$

so the $\langle \Psi_1^+ | \Psi \rangle$ overlap gives

$$S_1 \phi_1 = D_1 C_1[1s] + D_3 C_2[2s] + \cdots. \tag{35}$$

The second term, with the coefficient $D_3 C_2$, is the result of radial electron correlation in the neutral and radial relaxation in the ion. The transition moment in this case is

$$M_1 = D_1 C_1 \langle \chi | \mu | 1s \rangle + D_3 C_2 \langle \chi | \mu | 2s \rangle, \tag{36}$$

while the pole strength is

$$S_1^2 = (D_1 C_1)^2 + (D_3 C_2)^2. \tag{37}$$

On the other hand, if we consider the excited state of He^+ (labeled with the subscript 2) in a truncated two-orbital expansion, this would be (by orthogonality)

$$\Psi_2^+ = -C_2[1s] + C_1[2s], \tag{38}$$

and the corresponding $\langle\Psi_2^+|\Psi\rangle$ overlap gives

$$S_2\phi_2 = -D_1C_2[1s] + D_3C_1[2s]. \tag{39}$$

The pole strength for the excited state of the helium ion is then given by

$$S_2^2 = (D_1C_2)^2 + (D_3C_1)^2, \tag{40}$$

and the corresponding transition moment is

$$M_2 = -D_1C_2\langle\chi'|\mu|1s\rangle + D_3C_1\langle\chi'|\mu|2s\rangle. \tag{41}$$

If we now take the satellite-to-primary hole intensity ratio at the same outgoing electron kinetic energy (considering the approximations in Section III), we obtain

$$\frac{M_2^2}{M_1^2} = \frac{|D_1C_2 - \lambda D_3C_1|^2}{|D_1C_1 + \lambda D_3C_2|^2}, \tag{42}$$

where $\lambda = \langle\chi|\mu|2s\rangle/\langle\chi|\mu|1s\rangle$. This is not exactly the same as S_2^2/S_1^2. Notice in particular that the ratio of transition moments [Eq. (42)] retains information about the signs of the coefficients that is lost in the simple ratio of the pole strengths [Eq. (20)]. Within the target Hartree–Fock approximation, D_3 would be zero, and the ratio of the transition moments would be equal to the ratios of the pole strengths.

As a second example, consider the $3s$ hole of argon, which leads to a strong satellite with a dominant $2h$–$1p$ configuration, $3s^23p^43d$. For simplicity, consider only the configuration interaction wave function mixing two configurations of ^1S symmetry for the neutral, i.e.,

$$\Psi = D_1[3s^23p^6] + D_2[3s^23p^43d4s], \tag{43}$$

and the two orthogonal ion CI wave functions with ^2S symmetry, viz.,

$$\Psi_1^+ = C_1[3s3p^6] + C_2[3s^23p^43d] \tag{44}$$

and

$$\Psi_2^+ = -C_2[3s3p^6] + C_1[3s^23p^43d].$$

This is the dominant term considered by Amusia and Kheifets (1985) in deriving Eq. (32). The order of magnitude of the coefficients, C_2^2/C_1^2, has been computed to be about 0.3. As in Eq. (35),

$$S_1\phi_1 = D_1 C_1 [3s] + D_2 C_2 [4s], \tag{45}$$

and we also have the following equations:

$$
\begin{aligned}
S_2\phi_2 &= -D_1 C_2 [3s] + D_2 C_1 [4s], \\
S_1^2 &= (D_1 C_1)^2 + (D_2 C_2)^2, \\
S_2^2 &= (D_1 C_2)^2 + (D_2 C_1)^2, \\
M_1 &= D_1 C_1 \langle \chi | \mu | 3s \rangle + D_2 C_2 \langle \chi | \mu | 4s \rangle, \\
M_2 &= -D_1 C_2 \langle \chi' | \mu | 3s \rangle + D_2 C_1 \langle \chi' | \mu | 4s \rangle.
\end{aligned}
$$

Again, $M_2^2/M_1^2 \neq S_2^2/S_1^2$, because of the D_2 term representing initial-state correlation. The ratio of M_2^2/M_1^2 is the same quantity as $Z_{\text{PES}}^{(2)}/Z_{\text{PES}}^{(1)}$ for which Eq. (32) is supposed to provide an approximate relation. In their derivation, Amusia and Kheifets (1985) introduced approximate closure relations at this point in order to obtain Eq. (32).

For EMS, the formula for the cross section is proportional to the spherically averaged square of the Dyson orbital in momentum space. In the plane-wave impulse approximation, the ion recoil momentum balances the momentum of the incoming and outgoing free electrons. For S states, the maximum cross section is at zero recoil momentum. The Dyson orbital in momentum space, evaluated at zero momentum ($\tilde{\phi}_D(0)$), provides the EMS cross section at zero recoil momentum, i.e.,

$$\sigma = K |\tilde{\phi}_D(0)|^2, \tag{46}$$

where K represents experimental geometric and energy factors. In comparing the EMS satellite-to-primary peak intensity ratio, this gives

$$\frac{\sigma_2}{\sigma_1} = \frac{|\tilde{\phi}_{D_2}(0)|^2}{|\tilde{\phi}_{D_1}(0)|^2}. \tag{47}$$

For the argon example, the ratio is

$$\frac{\sigma_2}{\sigma_1} = \frac{|-D_1 C_2 [3\tilde{s}(0)] + D_2 C_1 [4\tilde{s}(0)]|^2}{|D_1 C_1 [3\tilde{s}(0)] + D_2 C_1 [4\tilde{s}(0)]|^2}. \tag{48}$$

In the target Hartree–Fock approximation, this would be the same as M_2^2/M_1^2, but with the initial-state correlation included,

$$\left(\frac{\sigma_2}{\sigma_1}\right)_{EMS} \neq \left(\frac{M_2^2}{M_1^2}\right)_{PES} \neq \left(\frac{S_2^2}{S_1^2}\right)_{theory}. \tag{49}$$

While these relations do not entirely agree with Eq. (32), they do agree with Amusia and Kheifets (1985) in that the inequality is caused by a combination of initial- and final-state correlation effects.

A third example is provided by the excited states of the ethylene ion. The ground-state electronic configuration of neutral ethylene is

$$2a_g^2(\sigma)2b_{1u}^2(\sigma^*)1b_{2u}^2 3a_g^2 1b_{3g}^2 1b_{3u}^2(\pi)1b_{2g}^0(\pi^*). \tag{50}$$

The orbital energies of the $2a_g$ and $3a_g$ orbitals give Koopmans' ionization potentials of 28.1 eV and 15.9 eV, respectively. These energies are higher than the experimental binding energies of the primary peaks attributed to the $2a_g$ (23.7-eV) and $3a_g$ (14.7-eV) ionization processes in the PES spectrum. [See Figure 6(a)]. A useful catalog of the different states in ethylene and their corresponding PES labels and spectral labels is given by Davidson and Wang (1996).

Double ionization can be estimated from Koopmans' theorem for π ionization from the neutral (10.2 eV) plus π ionization from the cation (20.4 eV), giving an energy of 30.6 eV above the neutral for the doubly charged ion. If the second electron is now placed back into the doubly charged cation in a Rydberg orbital with $n = 3$ and an effective charge of $+2e$, the binding energy will be about 6 eV. Hence, the excited states of the ion (with configuration $\pi^{-2}nl, n = 3$) will have energies around 24 eV. The actual CI calculation finds the following energies:

$$\begin{aligned}
E_{\pi^{-2}3s} &\approx 22 \text{ eV}, \\
E_{\pi^{-2}3p} &\approx 23 \text{ eV}, \\
E_{\pi^{-2}3d} &\approx 24 \text{ eV}, \\
E_{\pi^{-2}\pi^*} &\approx 17 \text{ eV}.
\end{aligned} \tag{51}$$

Notice that most of these states are actually of lower energy than the (unresolved) experimental $(2a_g)^{-1}$ primary peak (23.7 eV). The $\pi^{-2}3da_g$ configuration, however, is the one that mixes most strongly with $2a_g^{-1}$ to give the twin correlation states at 23.3 eV and 24.5 eV [Desjardins et al. (1995)]. These two states plus the $3a_g^{-1}1b_{3u}^{-1}1b_{2g}$ correlation state of $^2B_{1u}$ symmetry contribute most of the intensity to the unresolved peak centered at 23.7 eV that is referred to as the $2a_g^{-1}$ primary peak.

Simultaneous ionization from the π and the b_{3g} orbital is expected to be about 2 eV higher in energy than the second IP from the π orbital. Similarly, ionization from the $3a_g$ orbital should be about 4 eV higher, while ionization from the b_{2u} orbital should be 5 eV higher. Adding a Rydberg electron to these doubly charged cores will produce many additional configurations in the energy range of interest (20–30 eV binding energy). For example, the configuration $\pi^{-1}3a_g^{-1}\pi^*$ with three unpaired electrons gives $^2B_{1u}$ states near 20.3 eV and 24.3 eV. These correlation states appear to be hidden by the stronger primary hole peaks.

The $2a_g$ satellite experimental binding energy (27.4 eV) is similar to the predicted Koopmans' energy for the $2a_g$ primary hole. Before configuration interaction, there are also several ^2A_g configurations in this energy region, such as $\pi^{-1}b_{1u}^{-1}\pi^*$ and $\pi^{-1}3a_g^{-1}3pb_{3u}$. Thus, this region of the PES spectrum is expected to have many Rydberg excited states of the ion with every possible symmetry. In the target Hartree–Fock approximation, the $2a_g^{-1}$ configuration would carry all the intensity. However, this configuration will mix with all the ^2A_g Rydberg states that differ from the $2a_g^{-1}$ hole configuration by double excitations. That is, starting from $a_{2ag}|\psi_{SCF}\rangle$, where a_x is an anni-hilation operator for orbital x, one can recover the other configurations by double excitations

$$a_R^+ a_{2ag}^+ a_\phi a_\pi (a_{2ag}|\psi_{SCF}\rangle), \qquad (52)$$

where ϕ is any occupied orbital and R is any Rydberg orbital. In second-order perturbation theory, double excitations are used to describe electron correlation, so these configurations provide a description of electron correlation in the $2a_g^{-1}$ configuration. The mixing matrix element connecting the $1h$ configuration to the $2h$–$1p$ configuration will be small because of the very diffuse Rydberg orbital. However, the mixing coefficient is large since the energy denominator is also small. Conversely, the $2a_g^{-1}$ configuration is doubly excited relative to the $a_R^+ a_\phi a_\pi$ ($2h$–$1p$) configuration,

$$a_{2ag}|\psi_{SCF}\rangle = a_\phi^+ a_\pi^+ a_R a_{2ag}(a_R^+ a_\phi a_\pi|\psi_{SCF}\rangle), \qquad (53)$$

so the $2a_g^{-1}$ configuration mixes with these Rydberg $2h$–$1p$ configurations to form states with the Rydberg configuration as the dominant term. In short, configuration mixing works both ways.

Configurations having a π^* orbital instead of a Rydberg orbital give a large mixing matrix element, but a smaller mixing coefficient, since they are further away in energy. Hence, in the diagram shown for ethylene (Figure 14), it is the $\pi^{-2}3d$ Rydberg configuration that mixes very strongly with $2a_g^{-1}$ because of the near degeneracy. The $\pi^{-1}b_{1u}^{-1}\pi^*$ configuration mixes

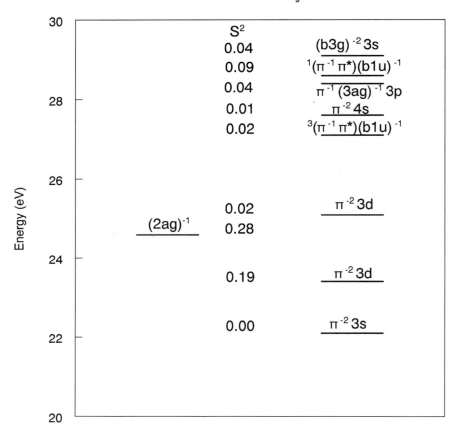

Figure 14. Calculated MRSDCI energies and pole strengths (S^2) for the ^2A_g correlation states of ethylene [Davidson and Wang (1996)].

somewhat less with the $2a_g^{-1}$ configuration in spite of a larger mixing (connecting) matrix element.

We can obtain a qualitative understanding of the electron correlation effects in ethylene by considering some simple pictorial models. The one-up, one-down mechanism was illustrated earlier for propene and depicted also for ethylene in Eq. (52). The way in which electron correlation manifests itself in the excited states of the ethylene ion can be seen by considering the shapes of the molecular orbitals involved, as shown in Figure 15.

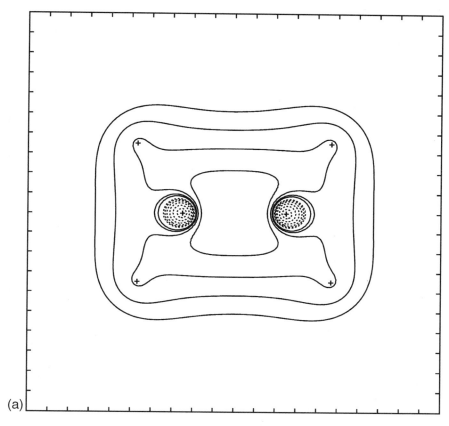

(a)

Figure 15. Contour plots of the valence MOs of ethylene ion that are involved in the J_1 and J_2 integrals defined in the text. The tick marks are spaced 0.25 Å apart. The molecule lies in the xz-plane with z along the C–C axis. (a) $2a_g(\sigma)$ MO in yz-plane; (b) $2b_{1u}(\sigma)^*$ MO in yz-plane; (c) $1b_{3u}(\pi)$ MO in xz-plane; (d) $1b_{2g}(\pi)^*$ MO in xz-plane.

Ignoring the accidental degeneracy that leads to splitting of the peaks (at 23.7 eV) by mixing with nearby Rydberg states, the electron correlation effect can be explained in terms of the valence states that generate large mixing matrix elements. As with argon, this is a one-up, one-down mechanism that mixes the primary hole configuration $D_1 = (2a_g)^{-1}$ with the three Slater determinants corresponding to $\pi^{-1}(b_{1u})^{-1}\pi^*$:

$$D_2 = |2a_g\alpha, 2a_g\beta, 1b_{1u}\alpha, \pi\alpha, \pi^*\beta|,$$
$$D_3 = |2a_g\alpha, 2a_g\beta, 1b_{1u}\alpha, \pi\beta, \pi^*\alpha|, \qquad (54)$$
$$D_4 = |2a_g\alpha, 2a_g\beta, 1b_{1u}\beta, \pi\alpha, \pi^*\alpha|.$$

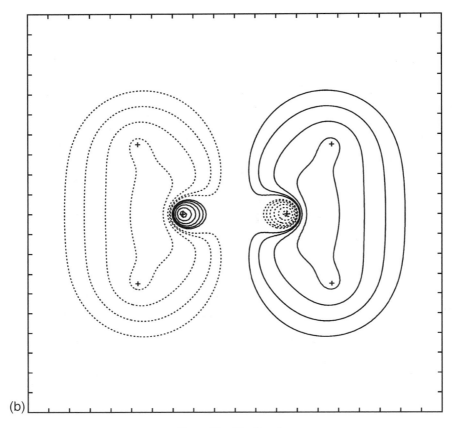

(b)

Figure 15. Continued.

The electron repulsion integrals mixing D_1 with these Slater determinants involve

$$J_1 = \langle 1b_{1u}(1)\pi(2)|1/r_{12}|\pi^*(1)2a_g(2)\rangle$$

and

$$J_2 = \langle 1b_{1u}(1)\pi(2)|1/r_{12}|2a_g(1)\pi^*(2)\rangle. \tag{55}$$

The D_1 configuration differs from these other configurations by a double excitation, $2b_{1u}, \pi \rightarrow 2a_g, \pi^*$ with an energy difference ΔE. To the lowest order in perturbation theory, the state corresponding to the $2a_g^{-1}$ primary

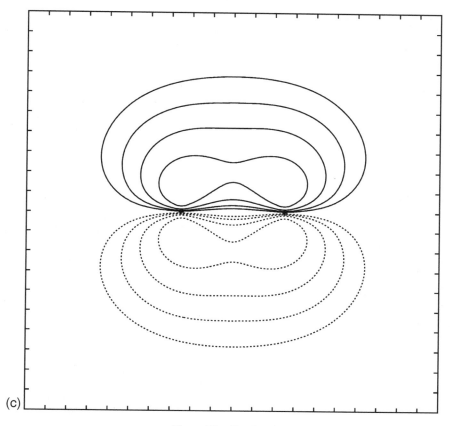

(c)

Figure 15.   Continued.

hole has a wave function

$$D_1 - \frac{J_1}{\Delta E}(D_2 - D_4) - \frac{J_2}{\Delta E}(D_3 - D_2). \tag{56}$$

The $D_2 - D_4$ term is just the $^1(2b_{1u}^{-1}, \pi^*)\pi^{-1}$ configuration and describes "top–bottom" correlation between the $2b_{1u}(\sigma^*)$ and π electrons. Top–bottom correlation increases the probability that the $2b_{1u}$ and π electrons are on opposite sides of the plane of the molecule, as illustrated in Figure 16(a). The $D_3 - D_2$ term, which is the $^1(\pi^{-1}, \pi^*)2b_{1u}^{-1}$ configuration, describes "left–right" correlation between the $2b_{1u}(\sigma^*)$ and π electrons. Left–right correlation increases the probability that the $2b_{1u}$ and π electrons are on opposite sides of a mirror plane bisecting the C–C bond, as illustrated in

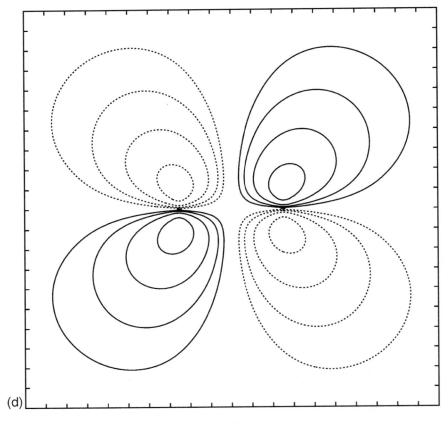

(d)

Figure 15. Continued.

Figure 16(b). As noted by Martin and Davidson (1977) and Wendin (1981), when expressed in valence bond language, left-right correlation can be regarded as a dynamic polarization [Borden and Davidson (1996)] of the π electrons around the localized $2s$ hole. It is tempting to postulate that the dynamic polarization of the $2s$ hole leads to an actual inner valence hole that oscillates between the two carbon nuclei, as suggested by Wendin (1981), who interpreted figures like Figure 16 as actual time-dependent fluctuations. However, one should remember that the CI calculations and the Green's-function calculations all pertain to time-independent wave functions, and the interpretation is more properly made in terms of modification to the static electron–electron pair distribution function.

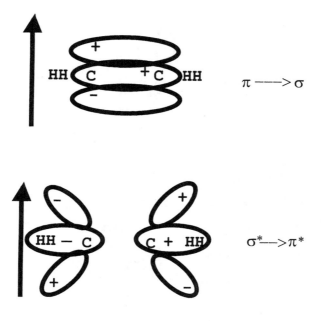

$\pi \longrightarrow \sigma$

$\sigma^* \longrightarrow \pi^*$

(a) **Top-Bottom Correlations**

Figure 16. (a) Schematic of "top–bottom" electron correlation. (b) Schematic of "left–right" electron correlation. The arrows indicate the directions along which the π and σ^* electrons in the σ^{-1} state partially avoid each other because of mixing with the indicated one-up, one-down double excitation.

Diagonalization of the CI matrix in this set of four configurations yields higher energy states (such as that at 27.4 eV) that carry most of the intensity into the satellite region. The energies of these higher states are raised by configuration interaction with the $2a_g^{-1}$ hole configuration. In a sense, the high energy states are "anti-correlated"; that is, the electron–electron repulsion energy is increased by the mixing. For example, relative to the dominant configuration of the upper state at 27.4 eV, the $2a_g^{-1}$ hole configuration represents a $(2a_g, \pi^*) \rightarrow (2b_{1u}, \pi)$ double excitation. The J_1 integral enters in terms that increase the probability that the $2a_g$ and π^* electrons are on the same side of the molecular plane. The J_2 integral enters in terms that increase the probability that the $2a_g$ and π^* electrons are on the same end of the molecule.

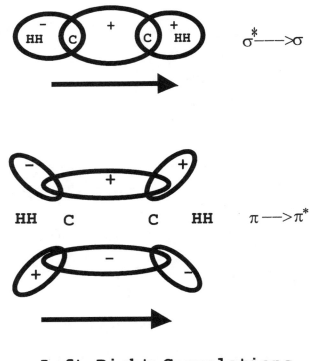

Left–Right Correlations

Figure 16. Continued.

VII. CONCLUSIONS

Many experimentalists and some theoreticians consider electron correlation an enigma. It is our hope that we have elucidated some aspects of this mechanism for intensity borrowing as manifested in the synchrotron photoelectron spectra (PES) of molecules. We have presented sample PES spectra of the alkanes and alkenes that illustrate the one-up, one-down mechanism responsible for major electron correlation effects in the ionization spectra. This mechanism is quite general. We have also presented methods that help visualize the two-hole, one-particle correlation states. This is in keeping with the tradition pioneered by Fano (1976) for visualizing electron–electron correlations in helium.

Many questions remain for the future, and much work must still be done. The following are only a sampling of possible questions:

- How are electron correlation effects manifested very close to the ionization threshold? Can simple pictures of these processes be developed?
- How do different molecular conformers affect correlation states?
- In which molecular systems will nonadiabatic effects be prominent? What will be the best means for observing these effects?
- What do these experiments and theoretical studies tell us with regard to the electronic structure of highly extended molecular systems such as polymers?

While this list is not exhaustive, it points to many significant areas for future research. The feverish development of even more intense synchrotron sources and free-electron lasers (so-called third- and fourth-generation sources) will allow orders-of-magnitude improvements in energy resolution, as well as the conduct of novel multicoincidence experiments. In turn, these may lead to answers to the preceding questions and to a deeper understanding of electron correlation effects.

Acknowledgments

We would like to acknowledge the assistance of the staff of the Canadian Synchrotron Radiation Facility and the Synchrotron Radiation Center at the University of Wisconsin. We thank J. Logan for assistance with some of the diagrams. We wish to acknowledge discussions with Y. Wang. This work was funded by the Natural Sciences and Engineering Research Council of Canada. The work at Indiana University was supported by the National Science Foundation.

References

Aberg, T. (1967). *Phys. Rev.* **156**, 35.

Adam, M. Y., Cauletti, C., Piancastelli, M. N., and Svensson, A. (1990). *J. Elect. Spect. Relat. Phenom.* **50**, 219.

Adam, M. Y., de Brito, A. N., Keane, M. P., Svensson, S., Karlsson, L., Kallne, E., and Correia, N. (1991). *J. Elect. Spect. Relat. Phenom.* **56**, 241.

Amusia, M. Y. and Kheifets, A. S. (1985). *J. Phys. B* **18**, L679.

Bagus, P. S. and Viinikka, E. K. (1977). *Phys. Rev. A* **15**, 1486.

Baltzer, P., Wannberg, B., Lundquist, M., Karlsson, L., Holland, D. M. P., MacDonald, M. A., and von Niessen, W. (1995a). *Chem. Phys.* **196**, 551.

Baltzer, P., Karlsson, L., Lundquist, M., Wannberg, B., Holland, D. M. P., and MacDonald, M. A. (1995b). *Chem. Phys.* **195**, 403.

Baltzer, P., Wannberg, B., Lundquist, M., Karlsson, L., Holland, D. M. P., MacDonald, M. A., Hayes, M. A., Tomasello, P., and von Niessen, W. (1996). *Chem. Phys.* **202**, 185.

Baltzer, P., Karlsson, L., Wannberg, B., Ohrwall, G., Holland, D. M. P., Hayes, M. A., and von Niessen, W. (1997). *Chem. Phys.* **224**, 95.

Bawagan, A. D. O. (1993). *J. Physique Colloque C6* **3**, 175.

Bawagan, A. D. O. (1998). Unpublished work.

Bawagan, A. D. O., Olsson, B. J., Tan, K. H., Chen, J. M., and Bancroft, G. M. (1991). *Chem. Phys. Lett.* **179**, 344.

Bawagan, A. D. O., Olsson, B. J., Tan, K. H., Chen, J. M., and Yang, B. X. (1992). *Chem. Phys.* **164**, 283.

Bawagan, A. D. O., Desjardins, S. J., Dailey, R. S., and Davidson, E. R. (1997). *J. Chem. Phys.* **107**, 4295.

Bawagan, A. D. O., Ghanty, T. K., Davidson, E. R., and Tan, K. H. (1998). *Chem. Phys. Lett.* **287**, 61.

Becker, U. and Shirley, D. A. (1990). *Phys. Scrip.* **T31**, 56.

Becker, U., Holzel, R., Kerkhoff, H. G., Langer, B., Szostak, D., and Wehlitz, R. (1986). *Phys. Rev. Lett.* **56**, 1120.

Becker, U., Hemmers, O., Langer, B., Menzel, A., Wehlitz, R., and Peatman, W. B. (1992). *Phys. Rev. A* **45**, R1295.

Berndtsson, B., Basilier, E., Gelius, U., Hedman, J., Klassen, M., Nilsson, R., Nordling, C., and Svensson, S. (1975). *Phys. Scrip.* **12**, 235.

Bizau, J. M. and Wuilleumier, F. J. (1995). *J. Elect. Spect. Relat. Phenom.* **71**, 205.

Bozek, J. D., Cutler, J. N., Bancroft, G. M., Coatsworth, L. L., Tan, K. H., Yang, D. S., and Cavell, R. G. (1990). *Chem. Phys. Lett.* **165**, 1.

Borden, W. T. and Davidson, E. R. (1996). *Acc. Chem. Res.* **29**, 67.

Brion, C. E. (1986). *Int. J. Quantum Chem.* **29**, 1397.

Brion, C. E., Bawagan, A. D. O., and Tan, K. H. (1988). *Can. J. Chem.* **66**, 1877.

Carlsson-Gothe, M., Wannberg, B., Karlsson, L., Svensson, S., Baltzer, P., Chau, F. T., and Adam, M. Y. (1991). *J. Chem. Phys.* **94**, 536.

Cederbaum, L. S., Domcke, W., Schirmer, J., von Niessen, W., Diercksen, G. H. F., and Kraemer, W. P. (1978). *J. Chem. Phys.* **69**, 1591.

Cederbaum, L. S., Domcke, W., Schirmer, J., and von Niessen, W. (1986). *Adv. Chem. Phys.* **65**, 115.

Coplan, M. A., Migdall, A. L., Moore, J. H., and Tossell, J. A. (1978). *J. Am. Chem. Soc.* **100**, 5008.

Cross, Y. M. and Castle, J. E. (1981). *J. Elect. Spect. Relat. Phenom.* **22**, 53.

Davidson, E. R. (1963). *J. Chem. Phys.* **39**, 875.

Davidson, E. R. and Wang, Y. (1996). *Aust. J. Phys.* **49**, 247.

Deleuze, M. S. and Cederbaum, L. S. (1996). *J. Chem. Phys.* **105**, 7583.

Deleuze, M. S. and Cederbaum, L. S. (1997). *Int. J. Quantum Chem.* **63**, 465.

Desjardins, S. J., Bawagan, A. D. O., and Tan, K. H. (1992). *Chem. Phys. Lett.* **196**, 261.

Desjardins, S. J., Bawagan, A. D. O., Tan, K. H., Wang, Y., and Davidson, E. R. (1994). *Chem. Phys. Lett.* **227**, 519.

Desjardins, S. J., Bawagan, A. D. O., Liu, Z. F., Tan, K. H., Wang, Y., and Davidson, E. R. (1995). *J. Chem. Phys.* **102**, 6385.

Dixon, A. J., Hood, S. T., Weigold, E., and Williams, G. R. J. (1978). *J. Elect. Spect. Relat. Phenom.* **14**, 267.

Dyall, K. G. and Larkins, F. P. (1982a). *J. Phys. B* **15**, 203.

Dyall, K. G. and Larkins, F. P. (1982b). *J. Phys. B* **15**, 219.

Fano, U. (1976). In Wuilleumier, F. J. (Ed.), *Photoionization and Other Probes of Many-Electron Interactions*. Plenum, New York, p. 11.

Ferrett, T. A., Lindle, D. W., Heimann, P. A., Brewer, W. D., Becker, U., Kerkhoff, H. G., and Shirley, D. A. (1987). *Phys. Rev. A* **36**, 3172.

Flores-Riveros, A., Agren, H., Brammer, R., and Jensen, H. J. A. (1986). *J. Chem. Phys.* **85**, 6270.

Fronzoni, G., Decleva, P., Lisini, A., and de Alti, G. (1994). *J. Elect. Spect. Relat. Phenom.* **69**, 207.

Fronzoni, G., De Alti, G., Decleva, P., and Lisini, A. (1995). *Chem. Phys.* **195**, 171.

Fujimoto, N., Nagashima, U., Inokuchi, H., Seki, K., Cao, Y., Nakahara, N., Nakayama, J., Noshino, M., and Fukuda, K. (1990). *J. Chem. Phys.* **92**, 4077.

Heimann, P. A., Becker, U., Kerkhoff, H. G., Langer, B., Szostak, D., Wehlitz, R., Lindle, D. W., Ferrett, T. A., and Shirley, D. A. (1986). *Phys. Rev. A* **34**, 3782.

Hermann, M. F., Yeager, D. L., and Freed, K. F. (1978). *Chem. Phys.* **29**, 77.

Holland, D. M. P., MacDonald, M. A., and Hayes, M. A. (1990). *Chem. Phys.* **142**, 291.

Holland, D. M. P., MacDonald, M. A., Hayes, M. A., Baltzer, P., Karlsson, L., Lundquist, M., Wannberg, B., and von Niessen, W. (1994). *Chem. Phys.* **188**, 317.

Holland, D. M. P., MacDonald, M. A., Baltzer, P., Karlsson, L., Lundquist, M., Wannberg, B., and von Niessen, W. (1995). *Chem. Phys.* **192**, 333.

Holland, D. M. P., MacDonald, M. A., Hayes, M. A., Baltzer, P., Wannberg, B., Lundquist, M., Karlsson, L., and von Niessen, W. (1996). *J. Phys. B* **29**, 3091.

Holland, D. M. P., Shaw, D. A., Hayes, M. A., Shpinkova, L. G., Rennie, E. E., Karlsson, L., Baltzer, P., and Wannberg, B. (1997). *Chem. Phys.* **219**, 91.

Hollebone, B. P., Neville, J. J., Zheng, Y., Brion, C. E., Wang, Y., and Davidson, E. R. (1995). *Chem. Phys.* **196**, 13.

Hu, Y. F., Bancroft, G. M., Liu, Z., and Tan, K. H. (1995). *Inorg. Chem.* **34**, 3716.

Ishihara, T., Mizuno, J., and Watanabe, T. (1980). *Phys. Rev. A* **22**, 1552.

Keane, M. P., Lunell, S., de Brito, A. N., Carlsson-Gothe, M., Svensson, S., Wannberg, B., and Karlsson, L. (1991). *J. Elect. Spect. Relat. Phenom.* **56**, 313.

Keane, M. P., de Brito, A. N., Correia, N., Svensson, S., Karlsson, L., Wannberg, B., Gelius, U., Lunell, S., Salaneck, W. R., Logdlund, M., Swanson, D. B., and MacDiarmid, A. G. (1992). *Phys. Rev. B* **45**, 6390.

Kempgens, B., Koppel, H., Kivimaki, A., Neeb, M., Cederbaum, L. S., and Bradshaw, A. M. (1997). *Phys. Rev. Lett.* **79**, 3617.

Kheifets, A. S. (1993). *J. Phys. B* **26**, L641.

Kheifets, A. S. and Amusia, M. Y. (1992). *Phys. Rev. A* **47**, 1261.

Kikas, A., Osborne, S. J., Ausmees, A., Svensson, S., Sairanen, O. P., and Aksela, S. (1996). *J. Elect. Spect. Relat. Phenom.* **77**, 241.

Kimura, K., Katsuwata, S., Achiba, Y., Yamazaki, T., and Iwata, S. (1981). *Handbook of HeI Photoelectron Spectra of Fundamental Organic Molecules* (New York, Halsted).

Koopmans, T. A. (1933). *Physica* **1**, 104.

Koppel, H., Domcke, W., Cederbaum, L. S., and von Niessen, W. (1978). *J. Chem. Phys.* **69**, 4252.

Kossman, H., Krassig, B., Schmidt, V., and Hansen, J. E. (1987). *Phys. Rev. Lett.* **58**, 1620.

Krause, M. O., Whitfield, S. B., Caldwell, C. D., Wu, J. Z., van der Meulen, P., de Lange, C. A., and Hansen, R. W. C. (1992). *J. Elect. Spect. Relat. Phenom.* **58**, 79.

Krummacher, S., Schmidt, V., and Wuilleumier, F. (1980). *J. Phys. B* **13**, 3993.

Lermer, N., Todd, B. R., Cann, N. M., Brion, C. E., Zheng, Y., Chakravorty, S., and Davidson, E. R. (1996). *Can. J. Phys.* **74**, 748.

Leung, K. T. and Brion, C. E. (1983). *Chem. Phys.* **82**, 87.

Li, X., Bancroft, G. M., and Puddephatt, R. J. (1997). *Acc. Chem. Res.* **30**, 213.

Liebsch, T., Plotzke, O., Hentges, R., Hempelmann, A., Hergenhahn, U., Heiser, F., Viefhaus, J., Becker, U., and Yu, Y. (1996). *J. Elect. Spect. Relat. Phenom.* **79**, 419.

Liegener, C., Svensson, S. and Agren, H. (1994). *Chem. Phys.* **179**, 313.

Martin, R. L. and Davidson, E. R. (1977). *Chem. Phys. Lett.* **51**, 237.

Martin, R. L. and Shirley, D. A. (1976a). *Phys. Rev. A* **13**, 1475.

Martin, R. L. and Shirley, D. A. (1976b). *J. Chem. Phys.* **64**, 3685.

Mathers, C. P., Ying, J. F., Gover, B. N., and Leung, K. T. (1994a). *Chem. Phys.* **184**, 295.

Mathers, C. P., Gover, N. N., Ying, J. F., Zhu, H., and Leung, K. T. (1994b). *J. Am. Chem. Soc.* **116**, 7250.

McCarthy, I. E. (1985). *J. Elec. Spec. Relat. Phenom.* **36**, 37.

McCarthy, I. E. and Weigold, E. (1985). *Phys. Rev. A* **31**, 160.

Moghaddam, M. S., Desjardins, S. J., Bawagan, A. D. O., Tan, K. H., Wang, Y., and Davidson, E. R. (1995). *J. Chem. Phys.* **103**, 10537.

Moghaddam, M. S., Bawagan, A. D. O., Tan, K. H., and von Niessen, W. (1996). *Chem. Phys.* **207**, 19.

Murray, C. W. and Davidson, E. R. (1992). *Chem. Phys. Lett.* **190**, 231.

Nordfors, D., Nilsson, A., Martensson, N., Svensson, S., Gelius, U. and Agren, H. (1991). *J. Elect. Spect. Relat. Phenom.* **56**, 117.

Ohno, M., Zakrzewski, V. G., Ortiz, J. V., and von Niessen, W. (1997). *J. Chem. Phys.* **106**, 3258.

Olney, T. N., Cooper, G., Chan, W. F., Burton, G. R., and Brion, C. E. (1994). *Chem. Phys.* **189**, 733.

Olney, T. N., Cooper, G., Chan, W. F., Burton, G. R., Brion, C. E., and Tan, K. H. (1996). *Chem. Phys.* **205**, 421.

Olney, T. N., Cooper, G., Chan, W. F., Burton, G. R., Brion, C. E., and Tan, K. H. (1997). *Chem. Phys.* **218**, 127.

Ortiz, J. V. and Zakrzewski, V. G. (1994). *J. Chem. Phys.* **100**, 6614.

Pahler, M., Caldwell, C. D., Schaphorst, S. J., and Krause, M. O. (1993). *J. Phys. B* **26**, 1617.

Persson, W. (1971). *Phys. Scrip.* **3**, 133.

Poole, R. T., Leckey, R., Liesegang, J., and Jenkin, J. (1973). *J. Phys. E* **6**, 226.

Reich, T., Heimann, P. A., Petersen, B. L., Hudson, E., Hussain, Z., and Shirley, D. A. (1994). *Phys. Rev. A* **49**, 4570.

Roy, D. and Tremblay, D. (1990). *Rep. Prog. Phys.* **53**, 1621.

Roy, P., Nenner, I., Millie, P., and Morin, P. (1986). *J. Chem. Phys.* **84**, 2050.

Roy, P., Nenner, I., Millie, P., and Morin, P. (1987). *J. Chem. Phys.* **87**, 2536.

Sevier, K. D. (1972). *Low Energy Electron Spectrometry* (New York, Wiley).

Schmidt, V. (1992). *Rep. Prog. Phys.* **55**, 1483.

Spears, D. P., Fishbeck, H. J., and Carlson, T. A. (1974). *Phys. Rev. A* **9**, 1603.

Stranges, S., Adam, M. Y., Cauletti, C., de Simone, M., Furlani, C., Piancastelli, M. N., Decleva, P., and Lisini, A. (1992). *J. Chem. Phys.* **97**, 4764.

Svensson, S., Helenlund, K., and Gelius, U. (1987). *Phys. Rev. Lett.* **58**, 1624.

Svensson, S., Eriksson, B., Martensson, N., Wendin, G., and Gelius, U. (1988a). *J. Elect. Spect. Relat. Phenom.* **47**, 327.

Svensson, S., Karlsson, L., Baltzer, P., Wannberg, B., Gelius, U., and Adam, M.Y. (1988b). *J. Chem. Phys.* **89**, 7193.

Svensson, S., Zdansky, E., Gelius, U., and Agren, H. (1988c). *Phys. Rev. A* **37**, 4730.

Svensson, S., Carlsson-Gothe, M., Karlsson, L., Nilsson, A., Martensson, N., and Gelius, U. (1991). *Phys. Scrip.* **44**, 141.

Thomas, T. D. (1984). *Phys. Rev. Lett.* **52**, 417.

Von Niessen, W., Schirmer, J., and Cederbaum, L. S. (1984). *Comp. Phys. Rep.* **1**, 57.

Wasada, H. and Hirao, K. (1989). *Chem. Phys.* **138**, 227.

Weigold, E., Hood, S. T., and Teubner, P. J. O. (1973). *Phys. Rev. Lett.* **30**, 457.

Weigold, E., Zhao, K., and von Niessen, W. (1991). *J. Chem. Phys.* **94**, 3468.

Wendin, G. (1981). *Structure and Bonding* **45**, 1.

Werme, L. O., Bergmark, T., and Siegbahn, K. (1972). *Phys. Scrip.* **6**, 141.

Wills, A. A., Cafolla, A. A., Svensson, A., and Comer, J. (1990). *J. Phys. B* **23**, 2013.

Winick, H. and Doniach, S. (Eds.), (1980). *Synchrotron Radiation Research* (New York, Plenum).

Winstead, C. and McKoy, V. (1996). *Adv. Chem. Phys.* **46**, 103.

Woodruff, P. R., Torop, L., and West, J. B. (1977). *J. Elect. Spect. Relat. Phenom.* **12**, 133.

Wuilleumier, F. and Krause, M. O. (1974). *Phys. Rev. A* **10**, 242.

Wuilleumier, F. and Krause, M. O. (1979). *J. Elect. Spect. Relat. Phenom.* **15**, 15.

Yang, Z. Z., Wang, L. S., Lee, Y. T., Shirley, D. A., Huang, S. Y., and Lester, W. A. (1990). *Chem. Phys. Lett.* **171**, 9.

Yang, Z. Z., Niu, S. Y., and Wang, L. S. (1993). *Chinese Science Bull.* **38**, 1722.

APPENDIX A

TABLE 1

Molecular synchrotron PES work since 1986, depicting correlation states

Molecule	Formula	Experiment (E) or Theory (T)	$E_{h\nu}$ (eV)	Reference
acetaldehyde	C_2H_4O	E	1,487	Keane et al. (1991)
		E	1,487	Nordfors et al. (1991)
acetone	C_3H_6O	E	1,487	Nordfors et al. (1991)
		E	1,487	Keane et al. (1991)
acetylene	C_2H_2	E/T	32–72	Moghaddam et al. (1995)
		E/T	1,487	Svensson et al. (1988c)
allene	C_3H_4	E/T	64	Bawagan et al. (1998)
		E/T	15–120	Baltzer et al. (1995a)
benzaldehyde	C_7H_6O	E	1,487	Nordfors et al. (1991)
benzene	C_6H_6	E/T	12–120	Baltzer et al. (1997)
benzyl alcohol	C_6H_6O	E	1,487	Nordfors et al. (1991)
1,3-butadiene	C_4H_6	E	1,487	Keane et al. (1992)
		T		Fronzoni et al. (1994)
		E	60	Bawagan (1998)
		E/T	12–120	Holland et al. (1996)
n-butane	C_4H_{10}	T		Deleuze and Cederbaum (1996)
		E	60	Bawagan (1998)
1-butene	C_4H_8	E	1,487	Liegener et al. (1994)
		E	69	Bawagan et al. (1997)
		T		Fronzoni et al. (1995)
cis-butene	C_4H_8	T		Fronzoni et al. (1995)
		E	120	Mathers et al. (1994a)
iso-butene	C_4H_8	T		Fronzoni et al. (1995)
		E	120	Mathers et al. (1994b)
trans-butene	C_4H_8	T		Fronzoni et al. (1995)
		E	120	Mathers et al. (1994b)
t-butanol	$C_4H_{10}O$	E	1,487	Nordfors et al. (1991)
carbon clusters	C_{2n+1}	T		Ohno et al. (1997)
	$n = 1$ to 6			Ortiz and Zakrzewski (1994)
carbon dioxide	CO_2	E/T	30–55	Roy et al. (1986)
carbon disulfide	CS_2	E/T	20–90	Baltzer et al. (1996)
		E/T	35–75	Roy et al. (1987)
carbon monoxide	CO	E	1,487	Svensson et al. (1991)
		E	50–120	Becker et al. (1992)
cyclobutane	C_4H_8	T		Deleuze and Cederbaum (1996)
cyclohexane	C_6H_{12}	T		Deleuze and Cederbaum (1996)

TABLE 1
Continued

Molecule	Formula	Experiment (E) or Theory (T)	$E_{h\nu}$ (eV)	Reference
cyclopentane	C_5H_{10}	T		Deleuze and Cederbaum (1996)
ethanol	C_2H_6O	E	1,487	Nordfors et al. (1991)
ethylene	C_2H_4	E/T	298–338	Kempgens et al. (1997)
		E	1,487	Keane et al. (1992)
		E/T	30–220	Desjardins et al. (1995)
		E/T	40.8	Holland et al. (1997)
		E/T	70–220	Desjardins et al. (1992)
		E/T	30–220	Desjardins et al. (1994)
formaldehyde	CH_2O	E	1,487	Keane et al. (1991)
		E	1,487	Nordfors et al. (1991)
fullerene	C_{60}	E	21–108	Liebsch et al. (1996)
furan	C_4H_4O	E/T	90–200	Bawagan et al. (1992)
n-hexane	C_6H_{14}	T		Deleuze and Cederbaum (1996)
hexatriene	C_6H_8	E	1,487	Keane et al. (1992)
		T		Fronzoni et al. (1994)
hydrogen chloride	HCl	E	120	Adam et al. (1990)
		E	20–65, 1,487	Svensson et al. (1988b)
hydrogen sulfide	H_2S	E	120	Adam et al. (1990)
		E	1,487	Adam et al. (1991)
		E	14–120	Baltzer et al. (1995b)
metal carbonyls	$W(CO)_6$	E	40–80	Hu et al. (1995)
	$Re(CO)_5Cl$			
	$Re(CO)_5Br$			
	$Re_2(CO)_{10}$			
	$CpRe(CO)_3$			
	$CpMn(CO)_3$			
methane	CH_4	E	40, 65, 1,487	Carlsson-Gothe et al. (1991)
methanol	CH_4O	E	1,487	Nordfors et al. (1991)
methyl bromide	CH_3Br	E	21–68	Olney et al. (1997)
methyl chloride	CH_3Cl	E	21–72	Olney et al. (1996)
methyl fluoride	CH_3F	E	21–72	Olney et al. (1994)
nitrogen	N_2	E	1,487	Svensson et al. (1991)
nitrous oxide	N_2O	E	40–120	Holland et al. (1990)
n-octanol	$C_8H_{17}O$	E	1,487	Nordfors et al. (1991)
octatetraene	C_8H_{10}	T		Fronzoni et al. (1994)
oligothiophenes	$-[C_4H_4S]-_n$ $n = 4$ to 8	E/T	2–150	Fujimoto et al. (1990)
organometallics	$M(\eta^3 - C_3H_5)_2$ $M = $ Ni, Pd, Pt	E	20–60	Li et al. (1997)

TABLE 1
Continued

Molecule	Formula	Experiment (E) or Theory (T)	$E_{h\nu}$ (eV)	Reference
pentadecaene	$C_{10}H_{12}$	T		Fronzoni et al. (1994)
n-pentane	C_5H_{12}	T		Deleuze and Cederbaum (1996)
1-pentene	C_5H_{10}	E	1,487	Liegener et al. (1994)
1-pentyne	C_5H_8	E	1,487	Liegener et al. (1994)
phosphine	PH_3	E	120	Adam et al. (1990)
polyacetylene	$-[C_2H_2]-_n$	E	1,487	Keane et al. (1992)
n-propane	C_3H_8	E	60	Bawagan (1998)
n-propanol	C_3H_8O	E	1,487	Nordfors et al. (1991)
1-propene	C_3H_6	E/T	40–70	Bawagan et al. (1997)
		E	1,487	Liegener et al. (1994)
pyridine	C_5H_5N	E/T	50–100	Moghaddam et al. (1996)
sulfur dioxide	SO_2	E/T	20–120	Holland et al. (1994)
sulfur hexafluoride	SF_6	E/T	20–120	Holland et al. (1995)
thiophene	C_4H_4S	E/T	90–200	Bawagan et al. (1992)
tin chloride	$SnCl_2$	E/T	21–52	Stranges et al. (1992)
water	H_2O	E	1,487	Nordfors et al. (1991)

CHAPTER 4

DEVELOPMENTS IN PARALLEL ELECTRONIC STRUCTURE THEORY

G. D. FLETCHER, M. W. SCHMIDT, and M. S. GORDON

Iowa State University, Ames, IA

CONTENTS

Abstract

A primary goal of parallel electronic structure theory is the exploitation of large-scale distributed memory platforms. This paper describes scalable distributed data algorithms that target the increasingly large aggregate memories available with such architectures. The design of distributed data algorithms is enhanced considerably by the availability of one-sided forms of communication that facilitate remote data access in a shared-memory style of environment. One example, a distributed-data Møller–Plesset second-order (MP2) gradient [1] algorithm, which achieves good scalability from 1 to 256 nodes of a Cray T3E, is described. The calculation of a numeric MP2 Hessian and a large-basis-set geometry optimization demonstrate the effectiveness of the approach in real applications.

Advances in Chemical Physics, Volume 110, Edited by I. Prigogine and Stuart A. Rice.
ISBN 0-471-33180-5. © 1999 John Wiley & Sons, Inc.

I. INTRODUCTION

How is progress to be made in accurately predicting the properties and behavior of large systems? Barring a fundamental breakthrough in mathematics, it is hard to imagine how the analysis of many-body systems will become generally more efficient. Until relatively recently, the time consumed by such analyses has depended largely on the speed with which operations can be executed in sequence. Ultimately, the rate of sequential execution is limited by such factors as the speed of light and the size of the atom. On the other hand, there is, in principle, no physical limit to the number of operations that can take place simultaneously. In this sense, parallelism is at the forefront of scientific computing. For instance, in the field of computational chemistry, there is an impressive array of sophisticated approximations to the solution of the Schrödinger equation through which these ever-increasing computational resources can be exploited.

The present work chiefly concerns the parallelization of four-index transformations with distributed data, a critical step for the application of correlated methods to larger systems. A broader issue is the best long-term scenario for computational chemistry and what investment, in terms of software development, is required to combine scalable data and time requirements with scalable, cost-effective hardware.

Parallel computing originally arose in response to the desire for cost-effective alternatives to "supercomputing", then dominated by vector processing. Today's parallel computers fall into two broad categories: those in which the processors share a common pool of memory (shared memory) and those in which the processors are physically separate (distributed memory) and communicate by sending and receiving messages. Shared-memory architectures have the advantage that they are straightforward to program, but as the number of processors is increased, the communication to main memory becomes a bottleneck. Distributed-memory architectures are more scalable, as evidenced by the fact that their nodes already number in the hundreds to thousands. Massively parallel processors (MPPs) now routinely deliver performance in excess of a hundred times that of serial technology [2]. Furthermore, the recent exploration into low-cost alternatives constructed from off-the-shelf components—for instance, clusters of PCs linked by an ethernet network capable of gigabit-per-second transmission—promises to significantly reduce the price–performance ratio associated with parallel computing [3]. Of particular interest here is the benefit of this performance, applied to the costly higher levels of theory in computational chemistry that account for the effects of electron correlation. However, exploiting the full potential of distributed-memory machines for correlation energy calculations requires a significant investment in programming in order to pass messages efficiently.

In previous reports [4–6], the parallelization of the energy and analytical gradient for most types of ab initio reference wave function (RHF, UHF, ROHF, GVB, and MCSCF) was described, together with the analytical RHF, ROHF, and GVB Hessians and the RHF MP2 energy. The algorithms are incorporated into the General Atomic and Molecular Electronic Structure System (GAMESS) program [7]. This first step and those of others [8–13] were vital for confirming the importance of parallel computing to chemistry. The majority of these early efforts viewed the issue of parallelization as being largely a porting exercise.

The widespread adoption of the single-program, multiple-data (SPMD) model, in which each processor executes the same code, provides the starting point for porting. The desire to minimize the effort of porting favors "replicated-data" schemes, in which each processor stores all of the data and therefore the per-processor memory requirements are the same as in the serial calculation. Simple forms of task distribution, dependent on the knowledge of a processor identifier number, cause each processor to compute a subset of the data elements or the contributions to them. Interprocessor communications, limited to global operations such as synchronizations, broadcasts, and reductions, are inserted as required to equalize the data on each processor for the parts of the algorithm that are not parallel. Another common ingredient of this approach is to account for the bottleneck associated with input/output (I/O) devices. The divergence in the performance of computation versus I/O is severe enough with conventional hardware to have prompted the development of so-called direct methods [14] in which the electron repulsion integrals (ERIs), which present the biggest storage problem, are recomputed as required and not written to disk. For parallel computers, the I/O-to-computation performance ratio generally worsens, particularly if the processors lack local disk space, making the use of direct methods even more relevant [15].

Statically load-balanced direct methods, with global communications inserted as appropriate, are satisfactory for many problems in computational chemistry. Significant enhancements in performance have been achieved in, for instance, the self-consistent-field (SCF) method [16]. Vendor implementations of the message-passing interface (MPI) standard [17] now deliver portability across a wide range of platforms for this approach. However, replicated-data strategies are limited by the memories of individual nodes, and affording the capability to handle larger memory requirements with a greater number of nodes is a fundamental goal of parallelism.

One reason for the increased difficulty of programming distributed-memory architectures is the presence of an additional level in the data storage hierarchy—that of remote memory. In order of increasing data-transfer time, this hierarchy begins with the CPU registers, proceeds through various levels of cache to main memory and remote memory, and, in principle, includes the more permanent storage devices (those with moving parts), such as disk and tape. (See Figure 1.) The MEMORY hierarchy is mostly hidden from the programmer, and decisions such as what data should be stored in cache are normally shared between compile time and run time. It is a matter of debate whether programmers should have more control over the upper regions of the hierarchy; however, the difference in speed between local and remote memory can have a profound effect on scalability.

A. Porting versus Parallelization

Limitations of the porting style of parallelization, so far described, become evident in calculations of the correlation energy and the evaluation of analytical derivatives. The latter are far easier to evaluate in the molecular orbital (MO) basis than in the atomic orbital (AO) basis, but doing so necessitates a four-index transformation of the ERIs that are originally evaluated in the AO basis; that is, one needs

$$(pq \mid rs) = \sum_{\mu\nu\lambda\sigma} C_{\mu p} C_{\nu q} C_{\lambda r} C_{\sigma s} (\mu\nu \mid \lambda\sigma), \tag{1}$$

where μ, ν, λ, and σ index AOs, p, q, r, and s index MOs, $C_{\mu p}$ are the MO coefficients, and the integrals in the AO basis,

$$(\mu\nu \mid \lambda\sigma) = \int \int \eta_\mu(1) \eta_\nu(1) \frac{1}{r_{12}} \eta_\lambda(2) \eta_\sigma(2) \partial \tau_1 \partial \tau_2, \tag{2}$$

are evaluated over the n basis functions η_μ. In conventional calculations, Eq. (1) presents problems arising from the n^4 storage requirement of the ERIs and the formally n^5 computational effort.

For parallelization, Eq. (1) presents two additional problems. First, there is high data dependency—every MO integral needs every AO integral—possibly leading to a high communication overhead. In this respect, the high cost of computing the ERIs is favorable; communication time is likely to remain small by comparison. Such is not always the case, however; for example, matrix diagonalization combines high data dependency with a fairly low computational cost, making good scalability very difficult to achieve. Second, the computation of ERIs is most efficient when all those arising from complete shells of basis functions sharing the same radial function are done together. Thus, load-balancing problems may ensue, since

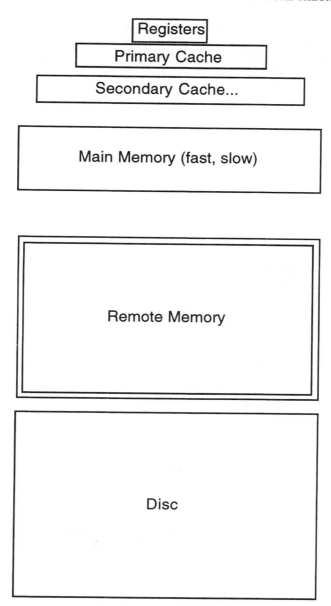

Figure 1.   Parallel data storage hierarchy.

the number of such integrals is highly dependent on the angular momenta of the shells. Moreover, it is evident that a multiple-instruction, multiple-data (MIMD) form of parallelism is required to handle all the integral types concurrently. Single-instruction, multiple-data (SIMD) parallelism, of the kind associated with vector processing, is of limited value unless the integral code can be radically restructured.

In a parallel implementation of Eq. (1), based on the direct approach [18], the ERIs are computed and transformed for all values of three MO indices, but only a subset of the fourth [5]. The process takes memory of order n^3 and is repeated for each subrange of the fourth index; the number of MO labels that can be treated in one batch is further determined by memory constraints. For instance, to compute the MP2 energy

$$E^{(2)} = \sum_{i,a,j,b} T_{ij}^{ab} (ia|jb), \qquad (3)$$

where i, j, k, and l index the n_o occupied, or active, MOs, a, b, c, and d index the n_ν virtual MOs, and

$$T_{ij}^{ab} = \frac{2(ia|jb) - (ib|ja)}{e_i + e_j - e_a - e_b} \qquad (4)$$

are the MP2 double-substitution amplitudes, the number of integral computation/transformation batches tends to the number of active orbitals. The parallel scheme involves distributing the batches to the nodes and combining the contributions to the MP2 energy at the end in a global sum. The advantages of this scheme are that the load balancing is simple and no interprocessor communication is required for the transformation. It is also assumed that a complete transformation is not required, and a direct method is used. The disadvantages are that scalability is limited—the number of nodes, P, must evenly divide the number of batches if an extreme load imbalance is to be avoided—and, more importantly, the integral computation itself is not parallel.

Efforts to address the issues surrounding the four-index transformation on distributed-memory MIMD machines span more than a decade. Whiteside et al. [19] proposed a distributed-data scheme, implemented for the Intel Hypercube, that assumed n^4 global memory in which to store all the ERIs. This, coupled with the small node memory of the hardware, meant that the largest application involved 32 basis functions. However, good speedups were achieved with a fixed load balance due, in part, to the use of nonblocking forms of message passing that could effectively overlap communication with computation. The term "nonblocking" is used here in place of "asynchronous", the word originally used by Whiteside and coworkers in reference to send/receive subprograms for the Hypercube that could return immediately, without blocking the calling process. While the former term follows the terminology adopted in MPI, the latter has more recently become associated with interprocessor communication in the absence of matching point-to-point operations.

Other studies proposed improved methods for handling the synchronous communication phases [20] and a form of dynamic load balancing [21] to distribute the computation of ERIs. However, a truly scalable solution due to Wong, Harrison, and Rendell [22] combines several key features. First is the recognition that much of post-SCF computational chemistry requires only a subset of the MO integrals. In particular, several post-SCF methods, including MP2, never require the MO integral class with four virtual indices. Given the size of basis set typically employed in such calculations, this eliminates one of the largest storage problems. In addition, the computational effort of the transformation is reduced from n^5 to $n_o n^4$. Bearing in mind that MPPs with global memories in the gigabyte range are now common, it is quite feasible that the smaller MO integral classes can be stored in memory even for very large problems. Thus, global memory on the order of $n_o^2 n^2$ is assumed. Since n^4-order memory is not assumed, the integrals over AOs are treated in a direct fashion and are not themselves stored.

Optimal load balancing is achieved by means of a shared counter to dynamically allocate the computation of ERIs to the nodes as they become idle. (See Figure 2.) In practice, dynamic load balancing (DLB) is implemented as a condition ("Is this my task?") in which a local counter, which is incremented for all values of the task index, is compared with the global counter, whose value is incremented and obtained through an exchange of messages. The overhead associated with DLB (exchange of short messages) must, nonetheless, be weighed against the degree of granularity achieved. A large number of small tasks, achieving a fine-grained parallelism, exact a substantial communication overhead. On the other hand, distributing a small number of much larger tasks may be too coarse grained to balance the load evenly.

Most importantly, the movement of data in global memory makes use of one-sided forms of communication that do not require matching send and receive calls. In this way, point-to-point synchronizations that could adversely affect the load balance are avoided. Such one-sided, or non-cooperative, message passing permits data copy operations directly to and from remote memory—often referred to as PUT and GET, respectively. The capability to access remote memory asynchronously to remote processes circumvents the problem of high data dependency inherent in Eq. (1) by allowing communication to overlap with computation and not interfere with

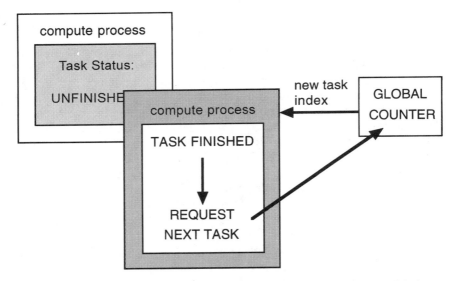

Figure 2. Illustration of dynamic load balancing involving update of global task index.

the load-balancing process. Moreover, the DLB itself can be implemented in this way if the shared counter is made a global data object.

PUT and GET can form the basis of a *virtual shared-memory interface*, such as the "Global Array" (GA) tool kit [23], which brings some of the "feel" of a shared-memory environment, accessed through library calls, to the distributed-memory architecture. In effect, a compromise has been struck between the desire for shared-memory programming and the greater scalability of distributed-memory hardware. However, this compromise will be successful only with an awareness of the penalty of remote memory access, so it is important for the programmer to have control over the distribution of data and be able to get information about it anytime. In addition, the integrity of data in the shared-memory region must be seen as the responsibility of the programmer. To this end, appropriate measures should be taken to ensure that nodes do not attempt to access global data ahead of time or out of step with the other nodes. A further consideration in the design of distributed-data algorithms is the minimization of local, or replicated, data requirements because they are not scalable. All of these considerations are vital parts of the GA approach.

It should now be clear that the achievement of a scalable parallel algorithm is not always a simple porting exercise. To combine the dual goals of load balancing and a distributed-data scheme, it is necessary for the parallel four-index transformation for all post-Hartree–Fock methods and for analytical derivatives to be radically different from the serial one, although the existence of software interfaces for manipulating distributed data objects greatly assists the code development process. Moreover, it is hard to imagine such a scheme arising automatically without significant insights on the part of the developers. As will be demonstrated in this work, the foregoing approach has led to highly successful applications [24]. In the next section, the techniques described here are applied to the evaluation of the gradient of the MP2 energy on distributed-memory platforms.

II. PARALLEL MP2 GRADIENT ALGORITHM

A distributed-data closed-shell MP2 gradient algorithm is described in this section; further details can be found elsewhere [2]. Using notation established earlier, we see that the MP2 energy is the sum of the SCF and second-order correction energies (Eq. (3)):

$$E = E^{\text{SCF}} + E^{(2)}. \tag{5}$$

Equation (5) can be differentiated analytically with respect to a perturbation x to yield the gradient expression in terms of sums over the derivative integrals times the relevant density matrices, to be defined shortly, plus the nuclear repulsion potential derivative V^x:

$$E^x = \sum_{\mu\nu} H^x_{\mu\nu} P^{\text{MP2}}_{\mu\nu} + \sum_{\mu\nu} S^x_{\mu\nu} W^{\text{MP2}}_{\mu\nu} + \sum_{\mu\nu\lambda\sigma} (\mu\nu|\lambda\sigma)^x \, \Gamma^{\text{MP2}}_{\mu\nu\lambda\sigma} + V^x \tag{6}$$

Here, $H^x_{\mu\nu}$ are the core-Hamiltonian derivative integrals, $S^x_{\mu\nu}$ are the overlap derivative integrals, and $(\mu\nu|\lambda\sigma)^x$ are the derivatives of the ERIs. If $\{x\}$ are nuclear Cartesian coordinates, Eq. (6) gives the residual force due to bonding between atoms, which can be used to efficiently locate stationary points on the nuclear potential-energy surface, such as equilibrium geometries and transition states, at the MP2 level.

The MP2 density matrix P^{MP2} and "energy-weighted" density matrix W^{MP2} in Eq. (6) are sums of their SCF analogues and a second-order correction term:

$$P^{\text{MP2}}_{\mu\nu} = P^{\text{SCF}}_{\mu\nu} + P^{(2)}_{\mu\nu}; \tag{7}$$

$$W^{\text{MP2}}_{\mu\nu} = W^{\text{SCF}}_{\mu\nu} + W^{(2)}_{\mu\nu}. \tag{8}$$

Although the final evaluation of Eq. (6) takes place in the AO basis, it is computationally more convenient to derive $P^{(2)}$ and $W^{(2)}$ in the MO basis, in which expressions for their occupied–occupied, occupied–virtual, and virtual–virtual blocks (assuming that there are no frozen occupied orbitals), are

$$P_{ij}^{(2)} = -2 \sum_{kab} T_{ik}^{ab} \frac{(ja|kb)}{e_j + e_k - e_a - e_b}, \tag{9}$$

$$P_{ab}^{(2)} = 2 \sum_{ijc} T_{ij}^{ac} \frac{(ib|jc)}{e_i + e_j - e_b - e_c}, \tag{10}$$

$$W_{ij}^{(2)} = -2 \sum_{kab} T_{ik}^{ab}(ja|kb) - P_{ij}^{(2)} e_i - \sum_{pq} P_{pq}^{(2)}[2(ij)|pq) - (ip|jq)], \tag{11}$$

$$W_{ab}^{(2)} = -2 \sum_{ijc} T_{ij}^{ac}(ib|jc) - P_{ab}^{(2)} e_a, \tag{12}$$

and

$$W_{ia}^{(2)} = -4 \sum_{jkb} T_{jk}^{ab}(ik|jb) - P_{ia}^{(2)} e_i \tag{13}$$

where e_p are the canonical orbital energies. It is customary to symmetrize the one-particle density matrices prior to their contraction with the derivative integrals. The occupied–virtual block of $P^{(2)}$ is obtained from the iterative solution of the Z-vector equation [25]

$$\sum_{jb} [A_{iajb} + \delta_{ab}\delta_{ij}(e_b - e_i)]P_{bj}^{(2)} = L_{ai}, \tag{14}$$

where

$$A_{pqrs} = 4(pq|rs) - (pr|qs) - (ps|qr) \tag{15}$$

and

$$L_{ia} = -4 \sum_{jbc} T_{ij}^{bc}(jc|ab) + 4 \sum_{jkb} T_{jk}^{ab}(ij|kb) - \sum_{jk} P_{jk}^{(2)} A_{aijk} - \sum_{bc} P_{bc}^{(2)} A_{iabc} \tag{16}$$

is the so-called MP2 Lagrangian.

Like P^{MP2} and W^{MP2}, the two-particle density is the sum of SCF (Γ^{SCF}) and second-order-corrections ($\Gamma^{(2)}$). The latter is the sum of so-called separable (superscript S) and non-separable (superscript NS) terms, i.e.,

$$\Gamma^{(2)} = \Gamma^S + \Gamma^{NS}, \tag{17}$$

where

$$\Gamma^S_{\mu\nu\lambda\sigma} = P^{(2)}_{\mu\nu} P^{SCF}_{\lambda\sigma} - \frac{1}{2} P^{(2)}_{\mu\lambda} P^{SCF}_{\nu\sigma} \tag{18}$$

and

$$\Gamma^{NS}_{\mu\nu\lambda\sigma} = \sum_{iajb} C_{\mu i} C_{\nu a} C_{\lambda j} C_{\sigma b} T^{ab}_{ij} \tag{19}$$

is a back-transformation of the amplitudes.

First, we note that ERIs with at least one occupied MO index are required to evaluate Eqs. (9)–(19). That is, using "v" to denote a virtual MO and "o" to denote an occupied MO, and placing the expressions in order of increasing size, we must compute the following classes of MO ERI:

1. $(oo|oo)$
2. $(vo|oo)$
3. $(vv|oo)$
4. $(vo|vo)$
5. $(vv|vo)$

The first four classes fall within the assumption of $n_o^2 n^2$ global memory, so these integrals may be stored in a distributed fashion following a transformation of the AO integrals. Data of order n^2 and lower are replicated across the nodes and include the one-particle density matrices, MO coefficients, and orbital energies.

The fifth class, with three virtual MO indices, is closer in size to $n_o n^3$, but is required for only the third and fourth terms of Eq. (16). Therefore, one should not store this class, but rather compute the terms involving the $(vv|vo)$ integrals separately and in a direct fashion. To summarize, the principal concerns here are how best to handle the steps involving transformations and the subsequent storage of integrals with a view toward maximizing data locality and minimizing communication overhead. That is, the goals are to generate integral classes 1 through 4, to compute terms involving class 5, and to achieve the back-transformation in Eq. (19).

A.   Integral Classes 1–4

The first step is to establish four distributed data structures to store the integrals. The distribution is over a triangular pairing of two occupied indices (See Figure 3), with the full range of the remaining indices present on a given node. (these can also be triangular for classes 1 and 3.) One consequence of this distribution scheme is that the long summations over virtual indices are made local. The "supermatrix" symmetry of class 1 is not exploited for the convenience of having a regular rectangular structure at the cost of a twofold redundancy in storage, although the cost is not great considering that this is the smallest integral class.

The parallel transformation of Wong et al. [22] is based on a fourfold arrangement of loops over shells with DLB inserted inside the second loop. (See Figure 4.) A DLB task consists of all iterations of the inner two loops, computing ERIs and transforming them to occupied and virtual MO indices in one position, but only occupied MO indices are required for the other position. This process generates (aa|oo) and (aa|vo) integrals, where, in the nomenclature for integral classes used here, "a" denotes an AO index from the shells indexed by the outer two loops. Thus, the first half-transformation is local, in the sense that only locally held data are required, and hence, there is no communication. The local, or replicated, data requirement here is $l^2 n n_o (n = n_o + n_\nu)$, where l is the length of a shell (1 for s, 3 for p, 6 for Cartesian d, and so on).

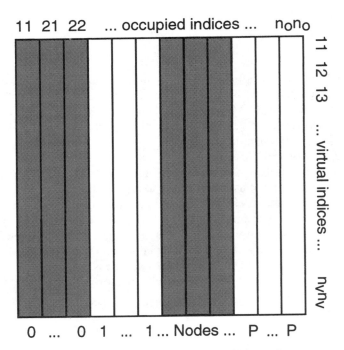

Figure 3.   Distribution of the (VO|VO) integrals.

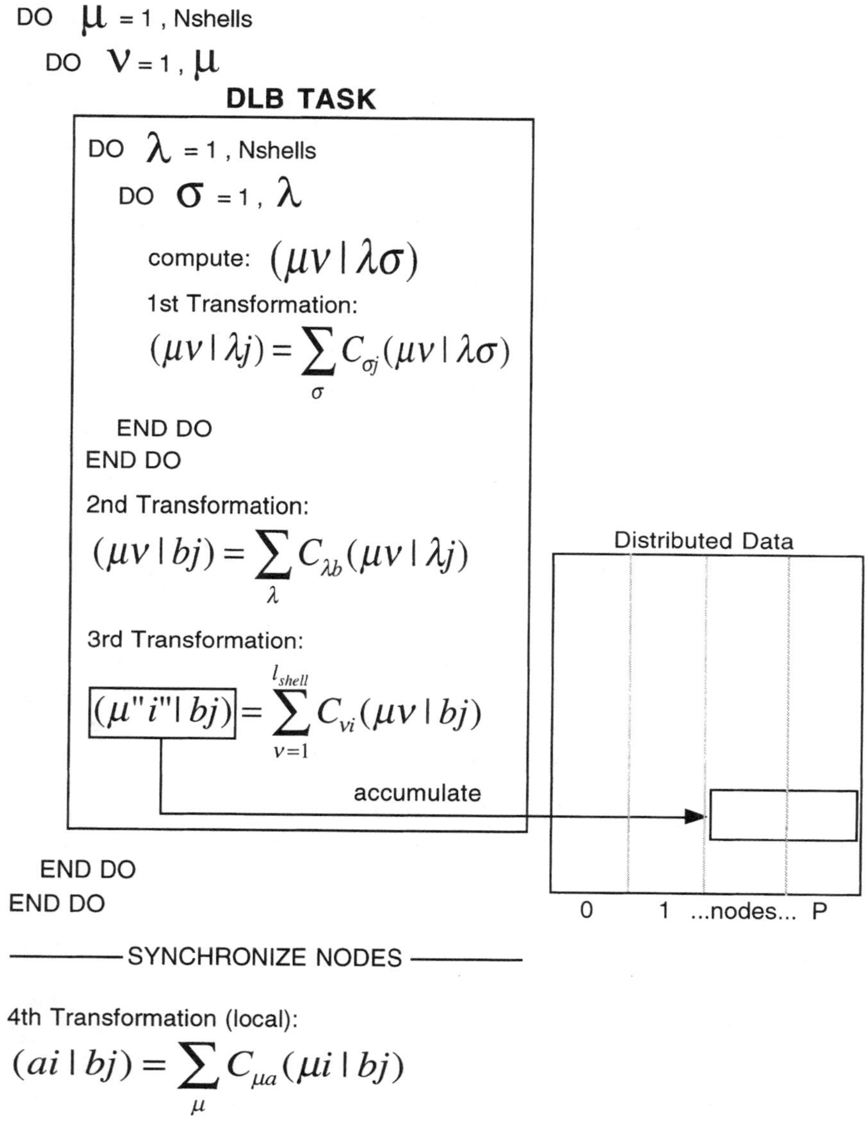

DO μ = 1 , Nshells

DO ν = 1 , μ

DLB TASK

DO λ = 1 , Nshells

DO σ = 1 , λ

compute: $(\mu\nu \mid \lambda\sigma)$

1st Transformation:

$$(\mu\nu \mid \lambda j) = \sum_{\sigma} C_{\sigma j}(\mu\nu \mid \lambda\sigma)$$

END DO
END DO

2nd Transformation:

$$(\mu\nu \mid bj) = \sum_{\lambda} C_{\lambda b}(\mu\nu \mid \lambda j)$$

3rd Transformation:

$$\boxed{(\mu''i'' \mid bj)} = \sum_{v=1}^{l_{shell}} C_{vi}(\mu\nu \mid bj)$$

accumulate

Distributed Data

0   1 ...nodes... P

END DO
END DO

——————— SYNCHRONIZE NODES ———————

4th Transformation (local):

$$(ai \mid bj) = \sum_{\mu} C_{\mu a}(\mu i \mid bj)$$

Figure 4. Distributed-data four-index transformation.

The (aa|oo) integrals are copied to their locations in the distributed-data region by means of a PUT operation. So that the first two quarter-transformations are complete, the exploitation of supermatrix symmetry between coulomb pairs in the AO ERIs, (aa|aa), is lifted, representing a twofold redundancy relative to the minimal list. The (aa|vo) integrals are further transformed to occupied MO indices, (a"o"|vo). However, the third transformation is incomplete within the DLB task (as indicated by the quotation marks), since the range of AO indices spans only those of the shell defined in the second loop index. A DLB task finishes with summing the (a"o"|vo) integrals into the appropriate global data region and requesting the next task index.

When all tasks—corresponding to all iterations of the four loops—have been completed, the finished (aa|oo) and (ao|vo) integrals for all values of each index are held in two distributed data structures. Thus, the net effect of summing the (a"o"|vo) contributions over all nodes is to complete the third quarter-transformation. Since the third transformation is a global event, a synchronization point must be placed immediately after the fourfold loop over shells to prevent any processes from continuing before the transformation is complete. Finally, the (aa|oo) integrals can be half-transformed to give the (oo|oo), (vo|oo), and (vv|oo) classes, and the (ao|vo) integrals can have the remaining AO index transformed to give the (vo|vo) class. These last operations are local, since the distribution leaves the full range of the (aa| and (ao| indices present on each node for a subset of the occupied index pairs.

Storing classes 1–4 permits most terms of the one-particle density matrices to be completed without further evaluation of integrals. In particular, all of Eqs. (9) and (10), followed by Eq. (12), and the first two terms of Eq. (11) can be done using the (vo|vo) class. Here, explicit storage of the amplitudes is avoided by recomputing them when needed from the stored (vo|vo) integrals. The choice of distribution allows the virtual blocks of $P^{(2)}$ and $W^{(2)}$ of Eqs. (10) and (12) to be formed simply from locally held (vo|vo) integrals and to be globally summed, while the occupied blocks of Eqs. (9) and (11) require remote GET operations. The total data traffic in this step is formally fifth order, scaling as $n_o^3 n_v^2$, and is sufficient to warrant the use of DLB.

The first term of Eq. (13) and first two terms of Eq. (16) require (vo|oo) and (vo|vo) integrals. The third term of Eq. (11) uses the (oo|oo) integrals when p and q index occupied MOs (this is the only term involving these integrals) and uses the (vv|oo) and (vo|vo) classes when p and q index virtual MOs. Since the occupied block of $P^{(2)}$ is replicated, this term requires no communication.

B. Three-Virtual Terms

The three-virtual terms of Eq. (16) can be rearranged as

$$L'_{ai} = \sum_{jbc} (jb|ac) T_{ij}^{bc}$$

$$= \sum_{jbc} \sum_{\nu\lambda\sigma} C_{\nu b} C_{\lambda a} C_{\sigma c} (j\nu|\lambda\sigma) T_{ij}^{bc}$$

$$= \sum_{j} \sum_{\nu\lambda\sigma} C_{\lambda a} (j\nu|\lambda\sigma) \left(\sum_{bc} C_{\nu b} C_{\sigma c} T_{ij}^{bc} \right),$$

$$\sum_{\lambda} C_{\lambda a} L'_{\lambda i} = \sum_{\lambda} C_{\lambda a} \sum_{j\nu\sigma} (j\nu|\lambda\sigma) T_{ij}^{\nu\sigma},$$

$$L'_{\lambda i} = \sum_{j\nu\sigma} (j\nu|\lambda\sigma) T_{ij}^{\nu\sigma}, \tag{20}$$

and

$$L'_{ai} = \sum_{bc} (ia|bc) P_{bc}^{(2)}$$

$$= \sum_{bc} \sum_{\nu\lambda\sigma} C_{\nu a} C_{\lambda b} C_{\sigma c} (i\nu|\lambda\sigma) P_{bc}^{(2)}$$

$$= \sum_{\nu\lambda\sigma} C_{\nu a} (i\nu|\lambda\sigma) \left(\sum_{bc} C_{\lambda b} C_{\sigma c} P_{bc}^{(2)} \right),$$

$$\sum_{\nu} C_{\nu a} L'_{\nu i} = \sum_{\nu} C_{\nu a} \sum_{\lambda\sigma} (i\nu|\lambda\sigma) P_{\lambda\sigma}^{(2)},$$

$$L'_{\nu i} = \sum_{\lambda\sigma} (i\nu|\lambda\sigma) P_{\lambda\sigma}^{(2)}, \tag{21}$$

$$L_{ai} \leftarrow \sum_{\mu} C_{\mu a} L'_{\mu i}. \tag{22}$$

The use of Eqs. (20) and (21) has the advantage that the quarter-transformed ERIs with one occupied MO index and three AO indices can be computed and utilized immediately without storing them. The intermediate half-transformed amplitudes can be formed from locally held $(vo|vo)$ integrals whose distribution is such that a back-transformation of virtual MO indices requires no communication. In addition, $P^{(2)}$ must be back-transformed, and the intermediate quantity L' is transformed and added to the Lagrangian in Eq. (22), but these are trivial two-index transformations and are not parallel.

A second fourfold loop over shells, again assuming $l^2 n n_o$ local memory, is required to compute and transform the ERIs. As in the first parallel transformation, a DLB task consists of all iterations of the inner two loops. (See Figure 5.) Likewise, the supermatrix symmetry is not exploited, so that the quarter-transformation to occupied MO indices, yielding ERIs of the

1) HALF-TRANSFORMATION OF AMPLITUDES

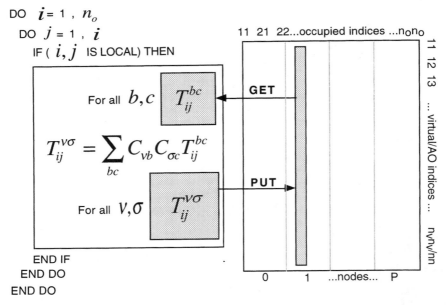

Figure 5. Three-virtual terms of the MP2 Lagrangian: (1) Half-transformation of amplitudes; (2) Lagrangian $(VV|VO)$ terms.

form $(aa|ao)$, has a full range of AO indices to complete. The back-transformed amplitudes can be thought of as integrals of the form $(ao|ao)$, with two occupied MO indices and two AO indices. To combine the integrals and amplitudes in a load-balanced way, the second and third loops over shells are, effectively, interchanged so that the outer loops run over AO indices from different charge clouds to match the AO indices of the amplitudes.

To perform the summation in Eq. (20), a block of amplitudes spanning all i, j indices for the AO indices associated with the outer two shell loops must be obtained. Unlike the virtual MO indices, whose distribution permitted a communication-free back-transformation, the occupied MO indices are distributed across all the nodes. Consequently, a substantial overhead cost may be incurred, as a single GET operation must communicate to all nodes. However, the number of GET operations can be minimized by exploiting a triangular symmetry of the outer-shell loops. If this is done, a further inner-loop symmetry must be lifted to allow the computation of $(ao|aa)$-type ERIs, which can be combined with the same

2) LAGRANGIAN (VV|VO) TERMS

DO V = 1 , Nshell

  DO σ = 1 , V

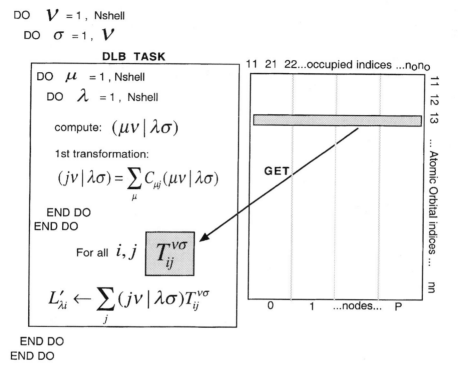

Figure 5.   Continued.

block of amplitudes to complete Eq. (20). Although this represents a fourfold redundancy in ERI computation relative to the minimal list, communication is minimized at the expense of further computation. The present algorithm is characterized by such decisions, which are intended to bias the algorithm ultimately in favour of scalability. Finally, the evaluation of Eq. (21) is more straightforward, since the back-transformed density matrix is small enough to be replicated.

C.   Solution of Z-Vector Equations

Like the (vo|vo) integrals, the back-transformed amplitudes require storage of the order $n_o^2 n^2$. In practice, further storage at this level is avoided by copying the amplitudes over the (vo|vo) integrals in memory. However, the (vo|vo) integrals are still needed for the orbital Hessian in Eq. (14) to solve the Z-vector equations. This problem is circumvented by simply inverting the formation of the back-transformed amplitudes—that is, by transforming the amplitudes with the inverse MO coefficient matrix ($C^{-1} = C^T S$, where S is the overlap matrix over AOs) to restore the virtual MO indices and then combining the amplitudes with the inverse of the orbital-energy factor in Eq. (4) to yield the original (vo|vo) integrals. The formation of back-transformed amplitudes and their inverses requires no communication and hence does not compromise the scalability of the algorithm, while at the same time consuming only a minor portion of the time.

To solve the Z-vector equations, the coincidence of the (vo|vo) and (vv|oo) distributions allows blocks of the orbital Hessian in Eq. (14), of size n_v^2, to be computed on the fly. For each iteration, the blocks multiply appropriate elements of the trial vector, and the contributions from all nodes are globally summed. (See Figure 6.) In this way, explicit storage of the orbital Hessian is avoided. To further avoid a possible storage bottleneck, the trial vectors, which consume $n_o n_v$ memory per iteration, are distributed over the nodes. The remaining arithmetic of the iterative process is much less significant and is, therefore, left sequential.

Having obtained the occupied-virtual block of $P^{(2)}$, we can evaluate the second term of Eq. (13) and delete the (vv|oo) integrals. The contribution to the third term of Eq. (12), when p and q index an occupied and a virtual MO, completes the use of the (vo|oo) integrals. As described earlier, $P^{(2)}$ and $W^{(2)}$ can be back-transformed to the AO basis, added to their SCF counterparts, and contracted with the one-electron derivative integrals to complete the one-electron gradient.

D.   Back-Transformation

The four-index back-transformation of the amplitudes of Eq. (19) to complete the two-particle gradient has much in common with the three-virtual terms of Eq. (20). Indeed, the process begins with a repetition of the half-back-transformation described earlier. A fourfold loop algorithm similar to that described in Section B is required, though for slightly different reasons, to compute derivative ERIs. Again, DLB occurs within the second loop, and a block of back-transformed amplitudes spanning all occupied indices and the AO indices of the outer-loop shells is obtained prior to the inner-loop iterations and the second step of the back-transformation. (See Figure 7.) However, Γ^{NS} lacks the permutational symmetry associated with the interchange of indices in the bra or ket:

$$\Gamma^{NS}_{\mu\nu\lambda\sigma} \neq \Gamma^{NS}_{\nu\mu\lambda\sigma} \neq \Gamma^{NS}_{\mu\nu\sigma\lambda} \neq \Gamma^{NS}_{\nu\mu\sigma\lambda}. \tag{23}$$

Normally, this fourfold asymmetry is removed by symmetrizing the density matrix prior to combining it with a minimal list of derivative integrals. In

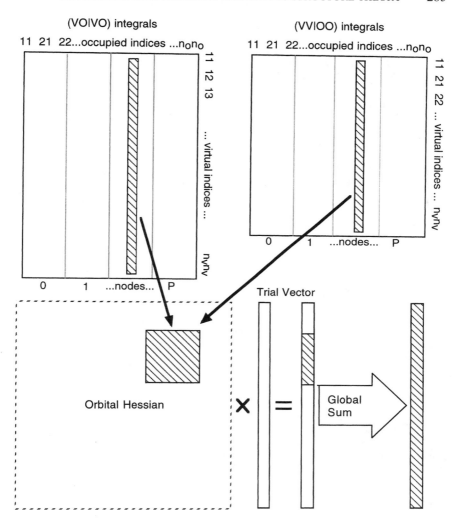

Figure 6. Product of the orbital Hessian and the trial vector.

the absence of order-n^4 storage in which to carry out the symmetrization, a fourfold redundancy in the computation of derivative ERIs allows the second half-back-transformation of the amplitudes to be done locally, in l^3n memory, for each DLB task. That is, inside the third loop over shells, the intermediate amplitudes are further back-transformed to all AO indices in one range, but only those of the third-loop shell in the other. In this way, n^4 storage and increased n^5 effort in the transformation are avoided at the

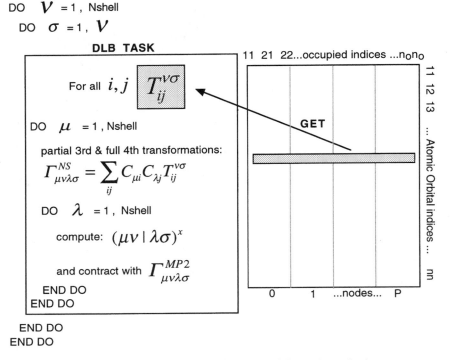

Figure 7. Second half-back-transformation of the MP2 amplitudes.

expense of increased n^4 computation. Finally, the Γ^{NS} elements are combined with Γ^S terms, computed from data held in local memory, before contraction with the derivative ERIs to yield the two-electron gradient.

E. Summary

The parallel algorithm just described computes the regular ERIs six times and the derivative ERIs four times. An alternative distributed-data MP2 gradient due to Nielsen [26] closely parallels the direct approach of Frisch, Head-Gordon, and Pople [18] by assuming global memory of order $n_o n^2$. The advantage of the latter scheme is that MP2 calculations with more than a thousand basis functions can be considered on large MPPs. However, the number of *derivative* ERI computations tends to n_o in the limit of $n_o n^2$ memory and more than twice this number for regular ERIs. For efficiency, the present algorithm can exploit molecular symmetry. Global memory demands are reduced by freezing the core orbitals in a calculation of the MP2 energy, but remain the same for the evaluation of the gradient.

To assist in the management of distributed data objects, the approach used here closely follows that of GA [23]. For ease of maintainance and customization, a minimal functionality is implemented, consisting of CREATE, DESTROY, and DISTRIBUTION, for managing and getting information about distributed data structures; GET, PUT, and ACC, to implement the global data copy and summation operations (See Figure 8); and DLBNEXT and DLBRESET, facilitating DLB. (We call this functionality 'distributed data interface'.) Together, the subprograms consume approximately 600 lines of FORTRAN and translate to MPI, sockets or vendor extensions, for handling one-sided message passing, on most platforms.

III.   TIMINGS AND APPLICATIONS

In this section, we examine the scalability of the MP2 gradient algorithm described in Section 2. Timing analysis is presented for typical benchmark

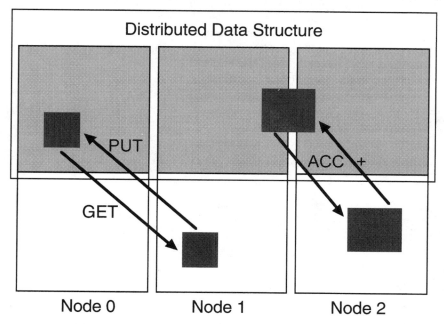

Figure 8. Illustration of the principal virtual shared-memory functionality. The boxes labeled **Node 0**, **Node 1**, and **Node 2** represent the memory local to three processors. Each processor allocates memory (light shading) to **CREATE** the contiguous "Distributed Data Structure". Data can be copied from any region of this structure, regardless of the physical distribution (dark shading) of the data, into a local workspace using **GET**. (**PUT** performs the inverse of **GET**.) Data can also be atomically accumulated using **ACC**.

calculations repeated on different numbers of nodes. In addition, two applications of chemical interest in the spirit of proof of principle are presented: the numeric Hessian of tri-amino-trinitro-benzene (TATB), which has a straightforward comparison to the serial performance, and a large-basis-set geometrical optimization, which would be impractical without MPP technology. All calculations were performed on the Cray T3E located at CEWES MSRC, Vicksburg, MS.

Table 1 contains elapsed times, in seconds, for the evaluation of the MP2 gradient on node partitions P, which yield sufficient global memory; the SCF and setup times are not included. Two molecular systems are considered: silicocene [27], a silicon analogue of ferrocene, with formula $(C_5H_5)_2Si$ (See Figure 9), and an octa-azide candidate high-energy-density material (HEDM), $C_2N_{12}O_4$ (See Figure 10). Basis sets chosen to study the scalability of the algorithm for a range of small, medium, and large problems include the split-valence basis sets (6-31G and 6-31G*) of Pople [28] and the correlation-consistent basis sets (pVDZ, aug-pVDZ, and pVTZ) of Dunning [29,30]. For instance, the silicocene calculation that uses a 6-31G basis (123 functions) provides the broadest study of scalability, because the global memory requirements are met by a single node. Moreover, since small calculations are less dominated by higher order computation, to which most of the effort of parallelization is devoted, they, in fact, test scalability more exhaustively than large cases.

The last row of Table 1 contains an analysis of the scaling for each case in terms of the "percentage parallelism". This can be thought of as the

TABLE 1
MP2 gradient step on Cray T3E; wall times are in seconds

	Silicocene					HEDM		
n (basis)	123 (6-31G)	189 (6-31G*)	219 (pVDZ)	369 (aug-pVDZ)	539 (pVTZ)	270 (pVDZ)	450 (aug-pVDZ)	630 (pVTZ)
P								
1	1,335							
2	718							
4	369	1,258	3,866					
8	189	638	1,935					
16	96	325	984	6,998		4,438		
32	52	171	573	3,492	7,496	2,374	21,122	
64	30	97	271	1,814	4,119	1,176	10,926	
128	23	67	152	983	2,179	618	5,791	13,926
256	18	54	109	555	1,283	339	2,888	7,429
$100Tp/T(1)$	98.87	99.52	99.69	99.90	99.80	99.87	99.92	99.94

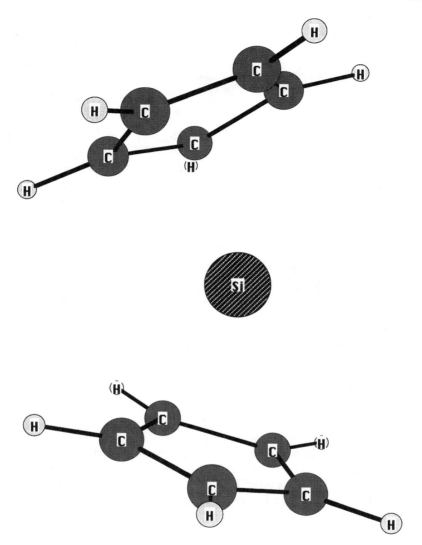

Figure 9.   A C_2 conformation of silicocene.

percentage of the elapsed time in a sequential run that is scalable; the remaining, "serial," time is a constant for any number of nodes. An estimate of the parallel time is obtained by assuming "Amdahl" scaling [31] and fitting an equation of the form, $T(P) = Ts + Tp/P$ to the total timings $T(P)$, where Ts and Tp are the serial and parallel times, respectively, and P is the

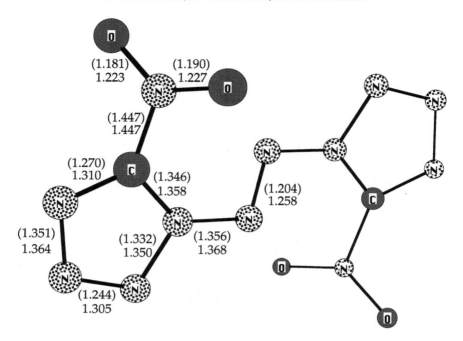

Figure 10. Structure of the High Energy Density Material showing MP2 bond lengths in Angstroms (RHF bond lengths are given in parentheses).

number of processors. The percentage parallelism is then given by $100Tp/T(1)$. In this way, the scalability over a range of node partitions is assessed, unlike simple parallel efficiencies, which compare only two partitions. The disadvantage is that the simple model does not account for effects due to load balancing or communication patterns. For instance, if the load balances more evenly on a larger partition or if the communication pattern is more efficient, then the percentage parallelism can become exaggerated. However, such effects are usually small, especially for timings that are bounded by synchronization points, and can be more or less eliminated by studying a sufficiently broad range of cases. These results show a scalability in the region of 98.8 to greater than 99.9 percent on the T3E, with a loose correlation to the size of the job, in accordance with the tendency of higher order steps to dominate the computing time.

In Table 2, elapsed times for the three main steps that involve integral transformations are given. The timings are taken from a breakdown of the total timings entered in the first column of Table 1. The three steps, labeled

TABLE 2
Main steps of the silicocene 6-31G calculation; wall times are in seconds

P	1	2	4	8	16	32	64	128	256
1	374.9	200.7	102.7	51.8	26.5	13.9	7.4	5.1	3.6
ACC	29.55	20.94	11.63	6.30	3.26	1.84	1.00	0.57	0.32
2	548.4	296.0	154.1	75.5	38.6	20.0	11.6	9.2	5.9
GET	2.78	1.18	0.55	0.25	0.12	0.07	0.04	0.02	0.03
3	313.2	170.8	85.6	44.8	22.1	11.7	6.4	5.0	3.4
GET	2.45	1.01	0.49	0.23	0.11	0.06	0.04	0.03	0.02

1, **2**, and **3**, are described in Sections A, B, and D, respectively. With each step, the time spent in communication activities is given; note that these timings are averaged over all nodes. For steps **2** and **3**, the communication time arises from the access of data blocks distributed across all nodes and is labeled "GET" in each case. For step **1**, the time consumed by summing contributions to data in global memory—labeled "ACC"—involves both communication and the computation associated with the addition operation. Because these times are averaged over all nodes, the communication time decreases with increasing numbers of nodes, although the total communication overhead (estimated by multiplying the average time by the number of nodes) certainly increases. Note that the GET operation has a nonzero "latency," even on one node, where, in fact, no messages are sent. Of the three main steps, step **2** turns out to be the most time consuming, even though the fourfold computation of derivative ERIs in the 2-PG step takes longer than its counterpart for regular ERIs in step **2**. The overriding factor is the greater cost of integral transformations in step **2**, which effectively generate the largest integral class. Overall, the preceding calculation scales well up to 64 nodes, with some speedup remaining on 128 and even 256 nodes.

A.   Numeric MP2 Hessian of TATB

With the use of a 6-31G* basis set [28] with 282 functions, calculation of the numeric MP2 Hessian of TATB by single-differencing of 73 gradients took 6.7 CPU hours on 200 nodes of the Cray T3E. For comparison, a serial single-point gradient was performed on a DEC workstation with an EV5 500-MHz CPU, 512 Mb memory, and 18 Gb of disk storage. By conventional and direct methods, both involving a segmented integral transformation, the gradient took 16 and 17 hours, respectively, implying almost 49 days to complete the numeric Hessian using only serial technology.

B. HEDM Molecular Structure

The strained octa-azide shown in Figure 10 was suggested by Schmitt as a potential high-energy species [32]. Preliminary calculations at the SCF level indicate a very high heat of formation, leading to a specific impulse which exceeds that of existing fuels. The next step is to determine the equilibrium geometry with greater accuracy, by including electron correlation effects through MP2 theory. The calculation performed here involves a large correlation-consistent basis set with polarization functions up to f. With the use of a cc-pVTZ basis set due to Dunning [29], with 630 basis functions, the MP2 geometry was converged to a maximum gradient of 0.0004 after 37 steps, taking 92 hours on 256 nodes of the Cray T3E. Notable here is the marked change in certain bond lengths on going from the geometry determined at the SCF level (with a 6-31G* basis [28]) to the MP2 geometry. At the same time, rather little change is observed in the bond angles, which are omitted for clarity. A single-point gradient calculation for this system would require more than 1 Gb of core memory by a sequential direct method, not to mention a considerable investment in time.

IV. CONCLUSION

This work demonstrates how the use of one-sided message passing allows the problems of high data dependency and load balancing to be solved for a distributed-data algorithm on a distributed-memory platform. Future plans are to reuse the present methodology for other levels of theory and applications that require scalable integral transformations, including SCF Hessian, ROMP2, UMP2, and MRMP2 energies and gradients, CIS gradients, and MP2 Hessians, which require many quantities similar to the MP2 gradient in terms of integrals and density matrices.

It is likely that computing will, in general, become more scalable in the coming years, with a growing need for scalable software. Whether this need will be met depends on how issues such as data distribution and load balancing affect the software development process. A useful tool in this process is the one-sided, or asynchronous, message-passing model. However, despite the availability of MPI, full portability for such a model awaits vendor implementations of MPI-2 [33] for the support of one-sided messages. The time taken to establish a synchronous communication standard will, no doubt, also apply to the asynchronous one. Advances in interprocessor communications have far-reaching benefits, not the least of which will be the prospects for the type of distributed-data parallelism described here. If performance can be delivered through scalability, there is less need for expensive highly optimized serial technology.

Acknowledgments

Software development was supported by a CHSSI grant from the Department of Defense, administered by the Air Force Office of Scientific Research (AFOSR). The allocation of time on the Cray-T3E came from the DoD Grand Challenge project. The authors would like to thank Howard Pritchard of Cray Research for help on the T3E and James Shoemaker for the MP2 Hessian calculation.

References

[1] G. D. Fletcher, A. P. Rendell, and P. Sherwood, *Mol. Phys.* **91**, 431 (1997).

[2] R. A. Kendall, R. J. Harrison, R. J. Littlefield, and M. F. Guest, *Rev. Comput. Chem.* **6**, 209 (1995).

[3] "In Search of Clusters' G. F. Pfister, Prentice-Hall 1998.

[4] T. L. Windus, M. W. Schmidt, and M. S. Gordon, *Chem. Phys. Lett.* **216**, 375 (1993).

[5] T. L. Windus, M. W. Schmidt, and M. S. Gordon, *Theor. Chim. Acta.* **89**, 77 (1994).

[6] T. L. Windus, M. W. Schmidt, and M. S. Gordon, "*Parallel Computing in Computational Chemistry*," ACS Symposium Series 592, T. G. Mattson, Ed., (1995), p. 16.

[7] M. W. Schmidt, K. K. Baldridge, J. A. Boatz, S. T. Elbert, M. S. Gordon, J. H. Jensen, S. Koseki, N. Matsunaga, K. A. Nguyen, S. Su, T. L. Windus, M. Dupuis, and J. A. Montgomery, Jr., *J. Comp. Chem.* **14**, 1347 (1993).

[8] H. P. Lüthi, J. E. Mertz, M. W. Feyereisen, and J. Almlöf, *J. Comput. Chem.* **13**, 160 (1992).

[9] A. Burkhardt, U. Wedig, and H. G. v. Schnering, *Theor. Chim. Acta.*, **84**, 497 (1993).

[10] H. P. Lüthi and J. Almlöf, *Theor. Chim. Acta.*, **84**, 443 (1993).

[11] S. Brode, H. Horn, M. Ehrig, D. Moldrup, J. E. Rice, and R. Ahlrichs, *J. Comput. Chem.* **14**, 1142 (1993).

[12] M. W. Feyereisen and R. A. Kendall, *Theor. Chim. Acta*, **84**, 289 (1993).

[13] M. F. Guest, P. Sherwood, and J. H. Van Lenthe, *Theor. Chim. Acta*, **84**, 423 (1993).

[14] J. Almlöf, K. Faegri, and K. Korsell, *J. Comput. Chem.* **3**, 385 (1982).

[15] M. W. Feyereisen, R. A. Kendall, J. Nichols, D. Dame, and J. T. Golab, *J. Comput. Chem.* **14**, 818 (1993).

[16] M. E. Colvin, C. L. Janssen, R. A. Whiteside, and C. H. Tong, *Theor. Chim. Acta* **84**, 301 (1993).

[17] Draft document for a standard message-passing interface. This document is available via electronic mail in PostScript at netlib@ornl.gov. To find out more about the MPI standardization effort, send electronic mail to Prof. J. Dongarra at the University of Tennessee, Knoxville (dongarra@cs.utk.edu).

[18] M. J. Frisch, M. Head-Gordon, and J. A. Pople, *Chem. Phys. Lett.* **166**, 275 (1990).

[19] R. A. Whiteside, J. S. Binkley, M. E. Colvin, and H. F. Schaefer, III, *J. Chem. Phys.* **86**, 2185 (1987).

[20] L. A. Covick and K. M. Sando, *J. Comp. Chem.* **11**, 1151 (1990).

[21] A. C. Limaye and S. R. Gadre, *J. Chem. Phys.* **100**, 1303 (1994).

[22] A. T. Wong, R. J. Harrison, and A. P. Rendell, *Theor. Chim. Acta* **93**, 317 (1996).

[23] J. Nieplocha, R. J. Harrison, and R. J. Littlefield, *Supercomputing '94* (New York, IEEE), p. 330.

[24] R. Kobayashi and A. P. Rendell, *Chem. Phys. Lett.* **265**, 1 (1997).

[25] N. C. Handy and H. F. Schaefer, *J. Chem. Phys.* **81**, 5031 (1984).

[26] I. M. B. Nielsen, *Chem. Phys. Lett.* **255**, 210 (1996).

[27] P. Jutzi, F. Kohl, P. Hofmann, C. Krüger, and Y.-H. Tsay, *Chem. Ber.* **113**, 757 (1980).

[28] See P. C. Hariharan and J. A. Pople, *Theor. Chim. Acta* **28**, 213 (1971), and references therein.

[29] T. H. Dunning, *J. Chem. Phys.* **90**, 1007 (1989).

[30] D. E. Woon and T. H. Dunning, *J. Chem. Phys.* **98**, 1358 (1993).

[31] G. Amdahl, *AFIPS Comput. Conf.* **30**, 483 (1967).

[32] R. Schmitt, S. R. I., private communication.

[33] Draft document titled "MPI-2: Extensions to the Message-Passing Interface"; available via electronic mail in PostScript from netlib@ornl.gov.

CHAPTER 5

EXPERIMENTAL AND THEORETICAL BUBBLE DYNAMICS

W. LAUTERBORN, T. KURZ, R. METTIN, and C. D. OHL

Universität Göttingen, Göttingen, Germany

CONTENTS

I. INTRODUCTION

Bubble dynamics lies at the heart of many a phenomenon in physics, chemistry, biology, and medicine [1–9]. Moreover, the knowledge of bubble

Advances in Chemical Physics, Volume 110, Edited by I. Prigogine and Stuart A. Rice.
ISBN 0-471-33180-5. © 1999 John Wiley & Sons, Inc.

behavior may foster technical applications in many areas, from pumps and pipelines, to propellers and hydrofoils, to cooling systems and chemical reactions. The importance of bubble dynamics was first noticed with respect to ship propellers and given a theoretical description, pertaining to single, spherical bubbles, by Rayleigh [10]. Since then, interest in bubbles and their dynamics has constantly grown, albeit not on too fast a pace, because of difficulties in making, handling, and controlling them and in following their at-times exceedingly fast dynamics. These days, bubbles have become a fashionable subject of investigation because many of the former limitations have ceased. In particular, techniques of working with isolated single bubbles have been developed [11, 12] that make possible detailed studies of bubble life down to the picosecond time scale [13].

Bubbles in liquids can be formed in a variety of ways. They may come off a wall in liquids oversaturated with some gas, as in beer or champagne. These are "soft" bubbles, usually not doing any harm unless formed in blood by depressurization, as it sometimes happens with divers. But they can also be forced to appear by tearing the liquid with brute force. This type of bubble formation is called cavitation, and the bubbles are termed *cavitation bubbles*, because, when formed by the rupture of a liquid, they are essentially empty, i.e., just cavities. Later in their life, they may fill with vapor and gas to become soft bubbles. However, when driven into large excursions, they may acquire a state of relative emptiness, preparing them for rapid collapse with all the strange effects of shock waves and light radiation. These "strong" bubbles that clean and disinfect water, destroy the hardest materials, and initiate or enhance chemical reactions are the scope of the present paper.

II. CLASSIFICATION OF CAVITATION BUBBLE GENERATION

The generation of cavitation bubbles may be classified according to the scheme given in Figure 1. There are essentially two physical mechanisms whereby cavitation bubbles appear: by setting up tension in the liquid and by depositing energy. The main disciplines in which tension in a liquid plays a role are hydrodynamics and acoustics. In hydrodynamics, fluid flow may produce tension — i.e., a negative pressure — by Bernoulli forces, the same forces that let airplanes fly. In liquids, these may become so large as to rupture the liquid, giving rise to what is called *hydrodynamic cavitation*, of main importance for ship propellers, turbines, hydrofoils, and pumps. In acoustics, sound waves consist of pressure and tension phases, and thus, strong sound waves are able to rupture a liquid in their tension phase, giving rise to the phenomenon of *acoustic cavitation*. Sonochemistry makes use of this phenomenon, for instance.

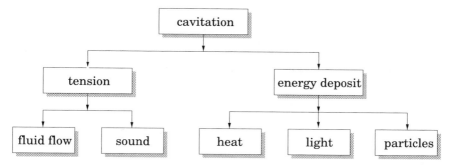

Figure 1.   Classification scheme for the different types of cavitation.

Cavitation bubbles can also be formed by depositing energy into the liquid. This can be done with light waves, heat, or particles. Light of sufficiently high intensity can rapidly heat up absorbing impurities or dielectrically break down the liquid to yield bubbles. This *optic cavitation* opens up a special way to study details of bubble dynamics not obtainable in any other way, as exemplified subsequently. Heating up a liquid by heat conduction or bringing the liquid in contact with a hot body leads to boiling, or *heat cavitation*. Nature makes use of this effect in the geyser and in volcano eruptions. Elementary particles leave energy when passing liquids, giving rise to bubble formation, as seen in the bubble chamber used for particle detection.

III.   CAVITATION BUBBLE PRODUCTION DEVICES

A variety of devices exists to produce cavitation bubbles for investigating them and for working with them. The "Mason horn" (Figure 2) is used for erosion tests, sonoluminescence investigations, and sonochemistry. It consists of a transducer with an attached amplitude transformer (the horn, usually made of titanium) that is dipped into the liquid to be cavitated. The transducer driving the horn consists of a sandwich of nickel plates making use of magnetostriction or plates of piezoelectric material—for instance, barium titanate ($BaTiO_4$) or, nowadays, usually PZT ceramics (lead zirconate/lead titanate). Masan horns are mainly used around 20 kHz and achieve high pressure amplitudes in the liquid. Water up to a hydrostatic pressure of 12 bar could be cavitated to investigate erosion and sonoluminescence [14].

In the horn device, the ultrasound is focused in a solid and released into the liquid at the tip that gains a high amplitude, up to several micrometers. The ultrasound can also be focused in the liquid by using a container that is

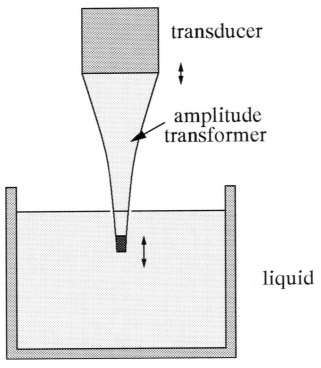

Figure 2. Mason horn to produce cavitation bubbles for erosion, luminescence, and sonochemical studies.

driven by one or more transducers from the outside, as seen in Figure 3. A glass container, usually a sphere or a cylinder, but also a rectangle or ovoid, is driven by several transducers to generate a standing-wave pattern in the liquid at the fundamental or some upper resonance of the whole system. The fundamental has the advantage that only one maximum of the sound pressure (and of the tension at the same place at a shifted time) occurs in the liquid. Bubbles placed there can be levitated for extended studies [12, 15]. The device then acts as a bubble trap. Higher order standing waves open up the possibility of investigating bubble interaction by trapping different bubbles in different pressure antinodes, as shown later.

A variant of this type of cavitation device is shown in Figure 4. It consists of a hollow cylinder of piezoelectric material simply submerged in the liquid to be cavitated. When the system is driven sinusoidally to excite the fundamental resonance, maximum sound pressure and tension occur at the center of the cylinder. The device is used for studying the dynamics of

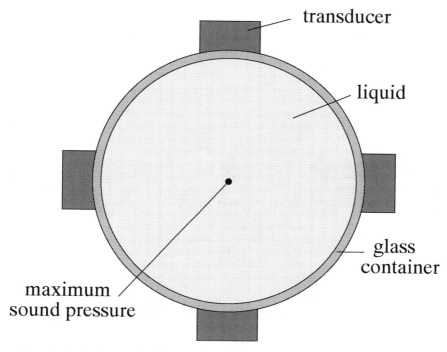

Figure 3. Spherical or cylindrical container filled with liquid for cavitation threshold, bubble oscillation, and sonoluminescence studies.

cavitation bubble clouds, cavitation noise emission (acoustic chaos [16]), and multibubble sonoluminescence.

Besides sound, light provides a convenient means of generating single and multiple bubbles in a transparent liquid, through optic cavitation [7, 11, 17–19]. For that purpose, a short laser pulse is focused into the liquid. Figure 5 depicts an experimental arrangement for observing the bubble or bubbles and the emitted shock waves by high-speed photography. The hydrophone is placed there to monitor sound and shock waves. Almost any laser emitting short pulses can be used with pulse durations on the order of a few nanoseconds or less and energies per pulse in the range of a few millijoules. So far we have used ruby, Nd:YAG and Ti:sapphire lasers.

IV. CAVITATION THRESHOLD

Sound waves in liquids need a certain intensity for cavitation bubbles to appear. The inception of cavitation is thus a threshold process. It is

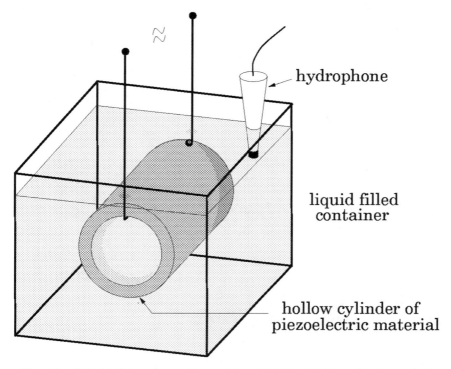

Figure 4. Cylindrical transducer submerged in a liquid for bubble oscillation, cavitation noise, and sonoluminescence studies.

commonly agreed that nuclei must mediate the inception in the case of acoustic cavitation. The nuclei are tiny bubbles being stabilized against dissolution by some mechanism. There is the crevice model, in which gas is trapped in conical pits of solid impurities [20], and the skin model, with organic or surface-active molecules occupying the bubble wall [21]. When the nuclei encounter a sound field, they are set into oscillation and may grow by a process called rectified diffusion [22]. A threshold process, rectified diffusion is the net effect of mass diffusion across the bubble wall due to varying bubble wall area and concentration gradients during oscillation. When a nucleus grows, it eventually reaches a size such that the oscillations turn into large excursions of the bubble radius, with strong and fast collapse shattering the bubble into tiny fractions. This then marks the cavitation threshold, and acoustic cavitation sets in. However, strong bubble oscillations may take place with shock waves and the emission of light, yet without the bubbles shattering. This phenomenon has been observed earlier

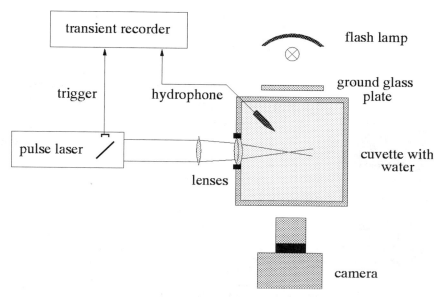

Figure 5. Typical experimental arrangement for optic cavitation with high-speed photography of bubble dynamics and a hydrophone for monitoring the sound and shock wave output.

[23] and finally led to the bubble trap, wherein single, small bubbles are kept in a strong sound field for extended periods of time emitting a shock wave and a light flash every cycle of the sound field [12, 15].

The mechanism for producing bubbles with light is different and may start without preformed bubble nuclei. Tiny particles floating in the liquid can absorb light, as may the liquid molecules themselves. Also, a dielectric breakdown with avalanche ionization may occur due to the high electric field strength. Any of these mechanisms leads to a plasma and an explosive growth, because of the locally high pressure compared with the ambient pressure. A cavity opens up, expanding to some maximum volume until its energy is used up, and then shrinks under the action of the ambient pressure. That way, a cavitation bubble is born.

V. SINGLE-BUBBLE DYNAMICS IN A SOUND FIELD

A. Experimental Observations

A single bubble can be trapped in the pressure antinode of a standing sound field. Figure 6 shows an arrangement for stable bubble levitation that has

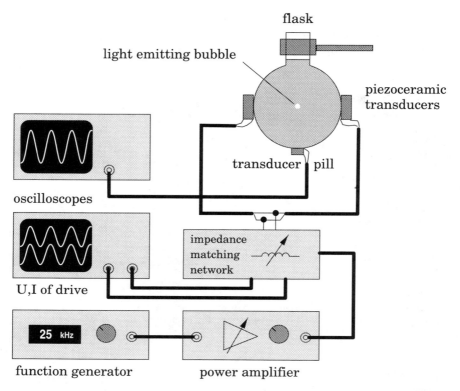

Figure 6. Experimental arrangement for the investigation of single-bubble sonoluminescence.

become widely used in the investigation of single-bubble sonoluminescence (SBSL). Instead of the spherical flask (Figure 7), a cylindrical or rectangular container (Figure 8) can be used; we also succeeded in producing SBSL in ovoids. The flask is driven by piezoceramic transducers glued to its sides. A small transducer pill is used as a microphone to detect the acoustic emission of the bubble. By inspection of the acoustic signal alone, it is possible to tell whether a bubble is stably trapped. Its presence reveals itself by small ripples or spikes in the microphone signal due to the acoustic emission of the bubble, which is driven far below its linear resonance frequency (see Sect. V.B.9). For stable trapping, it is essential to control the dissolved-gas content of the liquid. Therefore, a closed flask is preferable to an open one, with control over the ambient pressure. Furthermore, the driving frequency and driving amplitude have to be adjusted carefully. For nonoptimal

Figure 7. Photograph of a spherical flask with light-emitting bubble. (Courtesy of R. Geisler.)

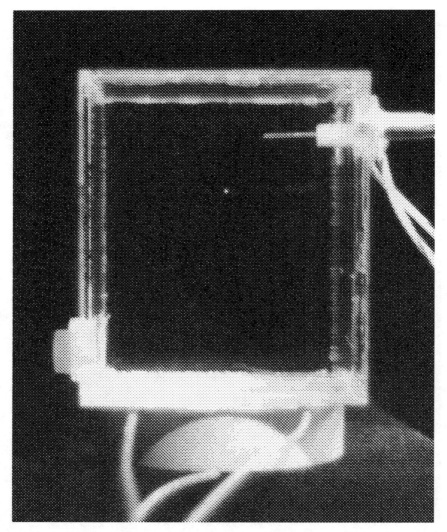

Figure 8. Photograph of a rectangular flask with light-emitting bubble. There is only one transducer glued to the bottom plate of the container. (Courtesy of R. Geisler.)

parameters, especially too high a gas content, the bubble's position is unstable, and a dancing motion will be observed.

The discovery of bright single-bubble luminescence has spurred much research, both theoretical and experimental, aimed at uncovering the hydrodynamical aspects of the phenomenon and the properties of the emitted light. In experiments, SBSL bubbles were attacked with hydrophones, with photomultipliers, with Mie scattering of cw or pulsed laser light, or with streak cameras. SBSL pulses have been found to have remarkable properties, straining current experimental techniques to their limit. First of all, the SBSL flashes are of very short duration in the range of 60 to 250 ps [13]. No clear evidence for spectrally differentiated pulse evolution has been found. Each pulse consists of about 10^5 to 10^6 photons, the intensity being strongly dependent on the temperature of the liquid, the type of dissolved gas, and its concentration [24]. The pulse spectrum extends well into the UV region and is compatible with a blackbody or bremsstrahlung spectrum, indicating temperatures in excess of 10,000 K [25]. The radial dynamics of sonoluminescing bubbles was investigated by laser light scattering (see, e.g., [26]) and photography [27, 28]; bubble stability was assessed by means of phase diagrams in parameter planes spanned by acoustic pressure and the equilibrium radius [29].

In brief, a large amount of experimental facts about single-bubble sonoluminescence has been gathered during the past years [30]. On the theoretical side, an equal amount of modeling and speculation took place. There are quite a number of competing theories that, e.g., invoke different collapse modes, posit different causes of the light emission, etc. One of the most widely accepted models assumes the generation of a converging shock wave upon bubble collapse [31], giving rise to a hot spot and a partially ionized plasma in the bubble center [32, 33]. This would imply a certain degree of sphericity of the collapse as a prerequisite for SBSL emission.

One approach to attack the question of sphericity experimentally that up to now was not exploited sufficiently is the direct optical observation of the surface dynamics and, in particular, the collapse of a trapped sonoluminescing bubble at high spatial and temporal resolution. As an example, Figure 9 shows the radial oscillation of such a bubble obtained by photography with a gated CCD camera and flash illumination. The gating time was set to 5 ns, the smallest possible time with the device used. Of course, this sequence was not acquired within one oscillation cycle, but a stably oscillating bubble was illuminated stroboscopically, and the time delay between the driving signal and the camera gating signal was incremented before each exposure.

The same experimental arrangement, with proper parallel back-illumination of the bubble, was used to obtain a photographic series of the shock wave emission of the SBSL bubble. In Figure 10, 12 frames are shown, starting at about the moment of shock wave formation at bubble collapse and extending in steps of approximately 30 ns to about 330 ns later. The shock wave can be seen expanding as a perfect ring, with the tiny spot in the

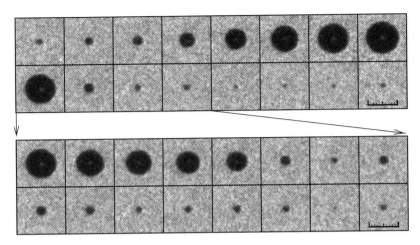

Figure 9. Photographic series of a trapped sonoluminescing bubble driven at 21.4 kHz. The top row presents the bubble dynamics at an interframe time of approximately 2.5 μs. The bottom row shows the bubble collapse with fivefold temporal resolution (500 ns interframe time). The scale of the image is indicated by the ruler (100 μm). (Courtesy of R. Geisler.)

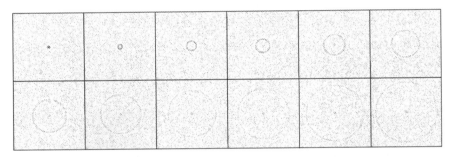

Figure 10. Spherical shock wave emitted by a sonoluminescing bubble into the surrounding liquid. The temporal separation of adjacent frames is approximately 30 ns; the frame size corresponds to an area of 1.6 mm × 1.6 mm in the object plane. (Courtesy of R. Geisler.)

center being the bubble that expands on a much slower time scale. The measurements give no indication of any deviation from spherical symmetry.

B.   Theoretical Description

A spherical bubble in a liquid can be viewed as an oscillator that can be set into radial oscillations by a sound field. For very small sound pressure amplitudes, these oscillations can be considered to be linear about the equilibrium radius of the bubble. The response then is that of a harmonic oscillator. At larger oscillation amplitudes, however, this oscillation must become nonlinear, because the bubble can be elongated from its equilibrium radius to arbitrarily large radii, but can be compressed only down to near zero radius. Bubbles of different rest radii R_n respond differently to the same sound field. Here we give some responses of bubbles in the range 1 µm $\leq R_n \leq 500$ µm for sound field frequencies in the near ultrasonic range and sound pressure amplitudes up to 150 kPa.

1. Bubble Models

In the course of time, several mathematical models of different sophistication for the oscillation of a spherical bubble in a liquid have been developed. The simplest one is the Rayleigh model [10],

$$\rho R\ddot{R} + \frac{3}{2}\rho\dot{R}^2 = p_i - p_e, \tag{1}$$

where R is the bubble radius, a dot over a variable means differentiation with respect to time, ρ is the density of the liquid, and p_i and p_e are the internal and external pressure, respectively—i.e., the pressure in the bubble and the pressure in the liquid, including the surface of the bubble. The difference in pressure drives the bubble's motion. The form of the inertial terms on the left-hand side is due to the spherical three-dimensional geometry that is transformed to one radial dimension in the differential equation. Both p_i and p_e become functions of radius R and time t when gas and vapor fill the bubble and the surface tension σ, (kinematic) liquid viscosity μ, and a sound field are taken into account. Figure 11 presents the quantities involved in a more realistic bubble model.

With these features included, the Rayleigh model takes the form [34–37]

$$\rho R\ddot{R} + \frac{3}{2}\rho\dot{R}^2 = p_{gn}\left(\frac{R_n}{R}\right)^{3\kappa} + p_v - p_{stat} - \frac{2\sigma}{R} - \frac{4\mu}{R}\dot{R} - p(t), \tag{2}$$

with

$$p_{gn} = \frac{2\sigma}{R_n} + p_{stat} - p_v, \tag{3}$$

Here, p_{gn} denotes the equilibrium gas pressure in the bubble, and R_n is the equilibrium radius, p_{stat} is the static pressure, p_v is the (constant) vapor pressure, and

$$p(t) = -\hat{p}_a \sin \omega t \tag{4}$$

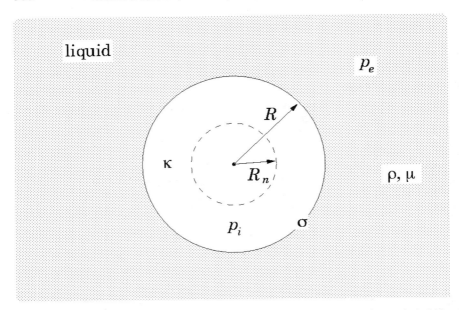

Figure 11. Quantities of importance for a spherical bubble model. R and R_n are the bubble radius and the bubble rest radius, respectively, p_e and p_i are the pressure in the liquid and in the bubble, respectively, ρ, μ, and σ are the density, viscosity, and surface tension of the liquid, respectively, and κ is the polytropic exponent of the gas in the bubble.

is the acoustic pressure of angular frequency ω and pressure amplitude $\hat{p}_a$. This model and some variants are called Rayleigh–Plesset models.

Here we use the Gilmore model [38], which incorporates sound radiation into the liquid, augmented by a van der Waals hard-core law [39] to account for a noncompressible volume of the inert gas inside the bubble. The bubble model reads

$$\left(1 - \frac{\dot{R}}{C}\right)R\ddot{R} + \frac{3}{2}\left(1 - \frac{\dot{R}}{3C}\right)\dot{R}^2 = \left(1 + \frac{\dot{R}}{C}\right)H + \frac{\dot{R}}{C}\left(1 - \frac{\dot{R}}{C}\right)R\frac{dH}{dR}, \quad (5)$$

$$H = \int_{p|_{r\to\infty}}^{p|_{r=R}} \frac{dp(\rho)}{\rho}, \quad (6)$$

$$p(\rho) = A\left(\frac{\rho}{\rho_0}\right)^n - B, \quad (7)$$

$$p|_{r=R} = \left(p_{\text{stat}} + \frac{2\sigma}{R_n}\right)\left(\frac{R_n^3 - bR_n^3}{R^3 - bR_n^3}\right)^\kappa - \frac{2\sigma}{R} - \frac{4\mu}{R}\dot{R}, \quad (8)$$

$$p|_{r \to \infty} = p_{\text{stat}} + p(t), \tag{9}$$

$$C = \sqrt{c_0{}^2 + (n-1)H}, \tag{10}$$

where the variables and constants appearing in the equation have the following meaning (numbers in parantheses are for water): R is the bubble radius, R_n is the bubble rest radius, ρ is the density of the liquid, ρ_0 is the density of the liquid at normal conditions ($998\,\text{kg/m}^3$), p is the pressure in the liquid, σ is the surface tension ($0.0725\,\text{N/m}$), μ is the viscosity of the liquid ($0.001\,\text{Ns/m}^2$), c_0 is the sound velocity in the liquid at normal conditions ($1{,}500\,\text{m/s}$), C is the velocity of sound in the liquid at the wall of the bubble, p_{stat} is the static ambient pressure ($100\,\text{kPa}$), κ is the polytropic exponent (chosen as $5/3$ for a monoatomic gas), b is the van der Waals constant (taken to be 0.0016 to model some artificial gas), A of the Tait equation (7) is $321.4\,\text{MPa}$, B is $321.3\,\text{MPa}$, and n is 7. This set of equations is solved for a sinusoidal sound field $p(t) = -\hat{p}_a \sin 2\pi\nu t$ with $\nu = 20\,\text{kHz}$ for several sound pressure amplitudes $\hat{p}_a$ up to $150\,\text{kPa}$. Other bubble models have been developed to better incorporate a time-varying pressure, as given by an external field, for example the model formulated by Keller and Miksis [40], that is described later in the context of Bjerknes forces.

2. Response Curves at Low Driving

When bubbles of different sizes are subject to only a very small acoustic pressure amplitude, they will respond with small oscillations about their equilibrium radius. Going up in the driving amplitude will bring out the effects of nonlinearity, which manifest themselves in a nonsinusoidal bubble oscillation and in the occurrence of several resonances. Figure 12 shows a radius–time curve of a steady-state solution for a bubble with an equilibrium radius of $R_n = 120\,\mu\text{m}$ and a pressure amplitude of $\hat{p}_a = 70\,\text{kPa}$. The typical long expansion phase and rapid collapse phase of a bubble oscillating in a sound field is noticeable.

   Figure 13 shows resonance curves—i. e., the maximum relative response $(R_{\max} - R_n)/R_n$ for a driving frequency of $20\,\text{kHz}$ and for sound pressure amplitudes reaching up to $70\,\text{kPa}$. The emerging peaks are resonances that can be labeled with two numbers [37]. Two sets of resonances are seen: the large peaks that fall off towards lower radii, labeled with a "1" in the denominator, and the smaller ones in between, labeled with a "2" in the denominator. The first set constitutes the main resonance (resonance of order $1/1$) and the harmonic resonances of order $n/1$ with $n = 2, 3, \ldots$. The second set belongs to the set of subharmonic resonances, starting with the main subharmonic resonance, $1/2$, to the right of the main resonance, and going on with the $3/2$, $5/2$, $7/2$ resonances (also called ultrasubharmonic

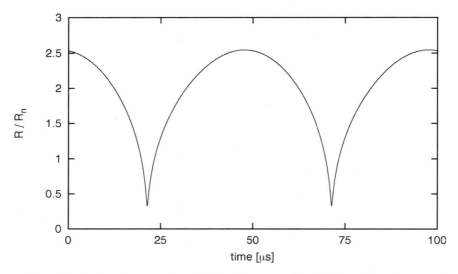

Figure 12. Radius–time curve for a bubble with a rest radius of 120 μm driven by a sound field of frequency 20 kHz and amplitude 70 kPa. Steady-state solution after the transients have died out. The radius is normalized to its rest radius R_n.

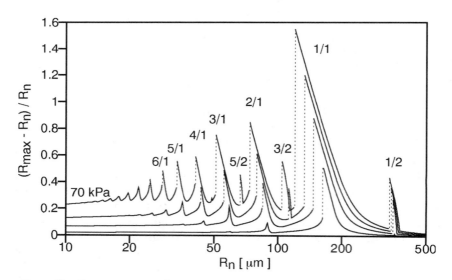

Figure 13. Resonance curves for bubbles of different sizes subject to a fixed frequency of 20 kHz. The driving amplitudes are 10 (bottom curve), 30, 50, and 70 kPa (upper curve). The initial condition is the respective bubble at rest.

resonances), where the last two resonances in the figure occur only with a driving pressure amplitude of 70 kPa. Straight, vertical, dotted lines have been plotted in the diagram. At these radii, the steady-state oscillation behaves nonmonotonic with the bubble radius at rest. The reason is the overturning of the resonances, leading to a small- or a large-amplitude oscillation, depending on the initial condition chosen. In the language of nonlinear dynamical systems theory [41], there are two coexisting attractors spanning a region of hysteresis — a region in which two different motions are stable. Each of the attractors then has a set of initial conditions from where it is approached, called its basin of attraction. The diagram was calculated with the respective bubble starting from its rest position ($R = R_n, \dot{R} = 0$). Then just one attractor will be reached, and the jump occurs when the boundary between the two basins of attraction sweeps over the rest position.

In Figure 14, for $\hat{p}_a = 80$ kPa, two attractors are obtained in hysteresis regions by the following computing technique: The curve is started at a small bubble rest radius R_n (here, 10 μm), and the radius is increased in small steps (here, 600 steps) to a final R_n (here, 200 μm), whereby the initial condition at the next R_n is taken from the final condition at the previous R_n. That way, an attractor can be followed until its basin shrinks to zero or is

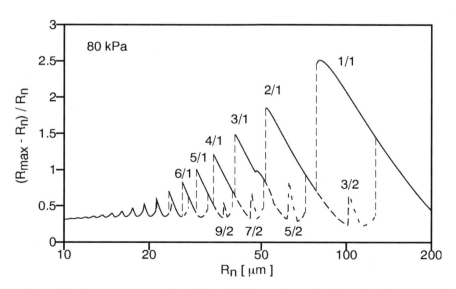

Figure 14. Resonance curves for bubbles of different sizes subject to a fixed frequency of 20 kHz. The driving amplitude is 80 kPa. The two curves belong to calculations with a stepwise decreasing bubble radius (solid line) and stepwise increasing bubble radius (bold dashed line) using 600 steps.

otherwise destroyed. Subsequently, the same procedure is repeated, but going backwards, i.e., decreasing R_n, with the last final condition again being the new initial condition. We see that the high amplitude oscillation of the main resonance leans over to lower bubble radii so strongly that it does not stop until beyond the 3/2 ultrasubharmonic resonance. Similarly the 2/1 harmonic resonance shadows the 5/2 resonance and so forth, in the case of the 80 kPa curve up to the 9/2 resonance.

3.   Response Curves at High Driving

If we continue to increase the driving level to $\hat{p}_a = 120\,\text{kPa}$, a dramatic increase in the response of small bubbles is observed (Figure 15). The response gets higher than even that of the main resonance and quite suddenly drops down to low values towards small R_n. The harmonic resonances of high order are still present and still decay towards smaller bubble radii. But they ride on a giant response. Physically, this giant response comes about through the instability of the bubble when the static pressure p_{stat} is overcome by the sound pressure during part of the sound cycle (actually an overall tension then). It is also related to the surface tension via the pressure $2\sigma/R$ that comes with a bubble of radius R in a liquid with surface tension σ. Because the surface tension pressure $2\sigma/R_n$ increases with lowering R_n for fixed σ, there exists a radius R_n such that a

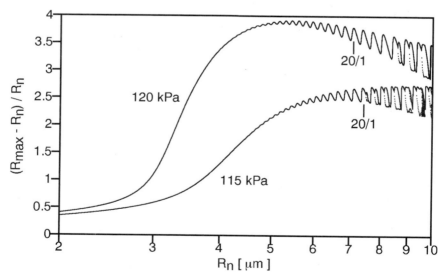

Figure 15. Response curves for bubbles of different sizes subject to a fixed frequency of 20 kHz. The driving amplitudes are 115 and 120 kPa.

given driving pressure is no longer strong enough to overcome the surface tension pressure, and the bubble oscillation drops to small values. This response gets more pronounced at driving pressures beyond 120 kPa. In Figure 16, the response curves are given for $\hat{p}_a = 130$, 140, and 150 kPa. The curves are remarkably smooth. The harmonic resonances are damped out below about 5 μm. Figure 17 gives the typical radius–time curve of a bubble in the giant response. The large excursion is followed by a steep and fast collapse, with a number of smaller afterbounces damping out before the cycle is repeated. This type of motion is encountered in a single, sonoluminescing bubble in a pressure antinode, as shown in Figure 9. A direct comparison with measurements will be given shortly.

Surface tension has a strong influence in this region, through forcing the sudden, but smooth, decay that occurs at lower bubble radii. Without surface tension, this decay would be missing. To put the giant response at small bubble radii into perspective, Figure 18 gives the response for $\hat{p}_a = 130$ kPa and 70 kPa at 20 kHz for bubbles from 1 μm to 100 μm on a linear radius scale. The far-greater relative amplitude of the giant response with respect to the main resonance now can easily be noticed. The relation gets even more pronounced at higher driving amplitudes.

In Figure 16 and even more so in Figure 18, not only the jumps from hysteresis can be seen, but also, so can regions of scattered dots often

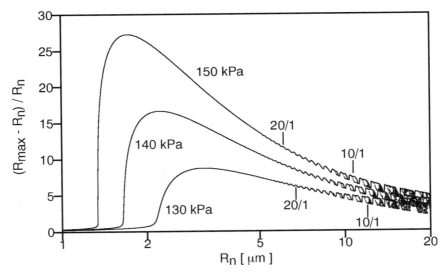

Figure 16. Response curves for bubbles of different sizes subject to a fixed frequency of 20 kHz. The driving amplitudes are 130, 140, and 150 kPa.

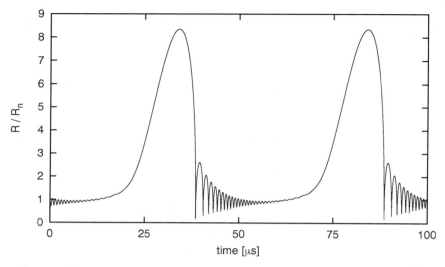

Figure 17.   Steady-state solution in the giant-response region. The bubble rest radius is 5 μm, the driving frequency is 20 kHz, and the driving amplitude is 130 kPa.

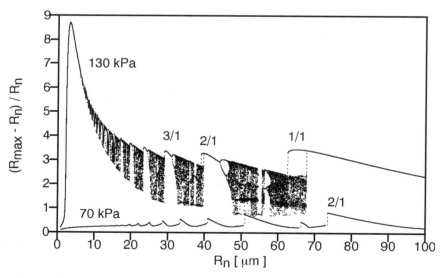

Figure 18.   Complete view of the response for bubbles of different sizes subject to a fixed frequency of 20 kHz. The driving amplitude is 130 kPa. For comparison, the response for 70 kPa is given.

beginning or ending with branched lines. These are chaotic bubble oscillations not pursued further here. They show up in the diagrams through the following plotting procedure: Bubble oscillations are calculated until a steady state is reached, and then the maximum radius value per period of the driving is plotted for several cycles. If the oscillation does not repeat before two periods of the driving, two points appear in the diagram, indicating a subharmonic oscillation of period 2. When there is no steady state after a few hundred oscillations, then the radius values alter from one cycle to the next, and scattered points pile up above the respective rest radii when they are plotted. We see from Figure 18 that chaotic dynamics covers a substantial part of the diagram at this elevated driving pressure.

4. Frequency Dependence

The giant response strongly depends on frequency (Figure 19). The peak reduces with increasing frequency, as expected, because the duration of the inertial instability gets smaller in proportion to the smaller period of the driving. Simultaneously, the figure demonstrates that, with lowering the driving frequency at a fixed driving amplitude, the chaotic bubble oscillations are expelled by the giant response from the small-bubble region towards larger bubbles. High frequencies thus suppress the giant response and attract chaotic oscillations. This may be of importance in sonoluminescence

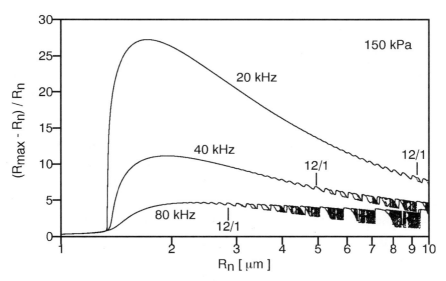

Figure 19.   Response for bubbles of different sizes subject to different frequencies of 20, 40, and 80 kHz. The driving amplitude is 150 kPa.

and sonochemical experiments for finding the optimal operating conditions. Low frequencies more easily lead to strong collapse, high frequencies to more collapses in a fixed amount of time.

As a result of the theoretical investigation, we can state that the response of bubbles to periodic acoustic driving shows a very complex behavior, even in the case of just spherical oscillations. Small bubbles develop a giant response much larger than the main resonance at fixed frequency through inertial instability during part of the acoustic cycle. Surface tension stops the giant response towards smaller bubble sizes. Chaotic bubble oscillations cease in the giant-response region, the region of single-bubble sonoluminescence.

5. Comparison of Experiment with Theory

In Figure 20, the dependence of the bubble radius on time, derived from experiment, is compared with a numerical calculation based on the foregoing extended Gilmore equation (5) with parameters R_n (the equilibrium radius), p_a (the acoustic drive amplitude), and κ (the polytropic exponent of

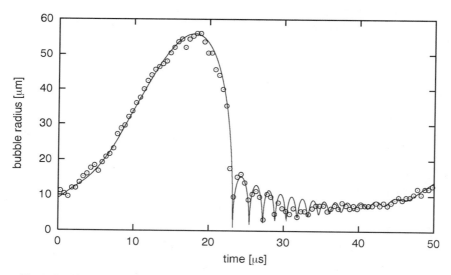

Figure 20. Radius–time curve of a trapped bubble in a water–glycerine mixture derived from photographic observations. A numerically calculated curve (Gilmore model) is superimposed on the experimental data points (open circles). The calculation is based on the following parameters: driving frequency $f_0 = 21.4\,\text{kHz}$, ambient pressure $p_0 = 100\,\text{kPa}$, driving pressure $p_0 = 132\,\text{kPa}$, vapor pressure $p_v = 0$, equilibrium radius $R_n = 8.1\,\mu\text{m}$, density of the liquid $\rho = 1000\,\text{kg/m}^3$, viscosity $\mu = 0.0018\,\text{Ns/m}^2$ (measured), and surface tension $\sigma = 0.0725$ N/m. The gas within the bubble is assumed to obey the adiabatic equation of state for an ideal gas with $\kappa = 1.2$. (Measured points courtesy of R. Geisler.)

the gas inside the bubble) chosen for best fit. Within the error bounds of the measurement, the data set coincides quite well with the numerical curve. The figure illustrates the characteristic dependence of the radius on time for a trapped SBSL bubble with its many fast afterbounces after the main collapse, when the bubble would "prefer" to oscillate at its own resonance frequency.

6. Shock Wave Radiation

Collapsing bubbles radiate shock waves, as shown in Figure 10 and in further examples to follow. The strength of these shock waves can be measured from schlieren pictures [42] and with hydrophones [42–44]. But a shock wave is predicted to also radiate into the interior of the bubble [31–33]. This, however, cannot be measured yet.

To support the view that shock waves may be radiated into the interior of the bubble, new results from molecular dynamics calculations are presented here. The art of molecular dynamics simulation [45] and the computing power have improved considerably since the first works on the dynamics of hard spheres in a collapsing spherical container [46], so that the formation of shock waves now can be obtained in reasonable computing time for a reasonable resolution. Figure 21 gives an example. A spherical bubble has

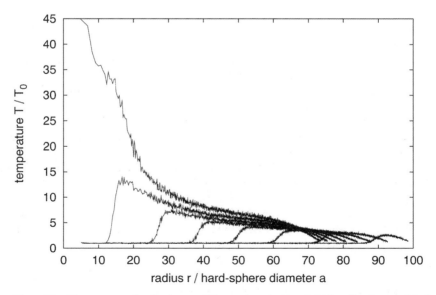

Figure 21. Temperature distributions at different times inside a bubble collapsing at a Mach number of about 1.5. About 1 million hard spheres are used in a molecular dynamics calculation. (Courtesy of B. Metten.)

been filled with $N = 1,008,281$ nonreacting hard, spherical atoms in a way that the normalized number density is $\rho_{N0} = 0.197189$. The density is normalized by dividing the number N by V/a^3, where V is the bubble volume and a^3 is the volume of one atom. The calculation is started by moving the bubble–liquid interface at a Mach number of about 1.5 relative to that of an ideal gas. The temperature is calculated in the interior in dependence on the normalized radius of the bubble (normalization with the diameter, a, of one hard sphere atom) for number density steps of 0.05 (increasing from ρ_{N0}). The series of eight curves starts at r/a about 100 with $\rho_{N0} + 0.05$. The eight radial temperature distribution curves thus belong to unequal time intervals. It is seen that a shock wave in temperature is formed that steepens and peaks at the bubble center where it is reflected. The reflected wave is not plotted here for not to obscure the picture. The reflection, however, can already be seen as the shoulder on the peak. The temperature numbers are relative to the ideal gas compression temperature T_0. The number 45 thus means a 45 fold increase over normal compression for an ideal gas by shock wave focusing. To give specific numbers, these normalized results predict a temperature of 45,000 K, if the shock wave build-up starts when the bubble has reached a compression temperature of 1,000 K during collapse. These are first results. A more complete study is under way.

7. Spherical Stability

Static gas or vapor bubbles in a liquid and also rain drops or soap bubbles in air, for example, attain a spherical shape because of their surface tension. Indeed, the very notion of bubble is often associated with the spherical form. The bubble models described above all rely on the assumption that the spherical shape is maintained when a bubble oscillates so that the fluid dynamical relations can be reduced to simple ordinary differential equations describing the radial bubble dynamics.

In a strongly collapsing bubble, however, the curved interface between the surrounding dense liquid and the less dense gas phase within the bubble, up to the final moments of collapse, is strongly accelerated in the inward direction. This situation is similar to the scenario of classical Rayleigh–Taylor instability, wherein an initially plain interface between two fluids develops perturbations—i.e., kinks and folds of growing size—upon acceleration of the fluids towards the less dense medium. A corresponding mechanism is at work in collapsing bubbles, and thus, shape instabilities will occur under certain conditions [47–49].

Figure 22 shows an example in which a gas bubble has been put into our cylindrical transducer and subjected to a sound field of high amplitude. Surface oscillations start and grow until the bubble disintegrates into several pieces.

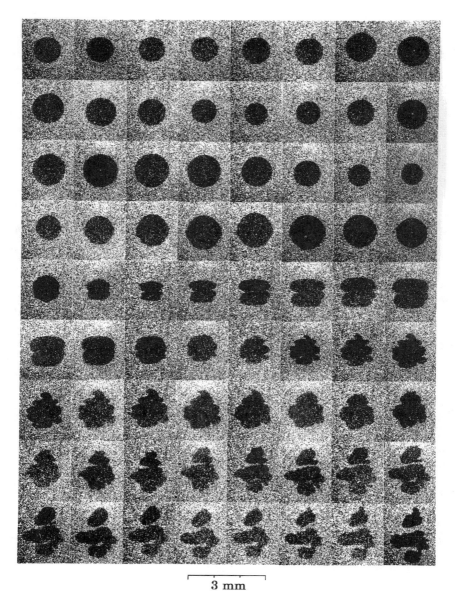

3 mm

Figure 22. Gas bubble oscillating in a sound field of growing amplitude.

The shape instability of bubbles is counteracted by the smoothing effect of surface tension and of energy dissipation by viscosity. Furthermore, shape distortions often occur in characteristic forms (surface modes) that have definite oscillation frequencies assigned to them. Therefore, the onset of shape instabilities takes place, in general, beyond a definite threshold of bubble oscillation amplitude that depends on the fluid parameters, the resonance frequency of the bubble, and the frequency of external driving. Shape instabilities are encountered in large or strongly oscillating bubbles, being enhanced at low surface tension or small viscosity. It is particularly easy to excite shape instabilities when radial motion and surface oscillations are in near resonance.

The description of nonspherical bubbles and the assessment of their stability presents a quite difficult mathematical problem. For an analytical treatment, one usually assumes a nearly spherical bubble, with the surface perturbed only slightly to allow for a linearization of the fluid-dynamical equations around the state of spherical symmetry. The radial dynamics is then described to a good approximation by the well-known, previously discussed models—for example, the Rayleigh–Plesset equation (2). Let $r_S(\Theta, \varphi; t)$ denote the location of the surface in a spherical coordinate system with its origin at the bubble center, and let $R(t)$ be the time-dependent radius of the associated sphere (Figure 23). The surface perturbation is expanded into spherical harmonics,

$$r_S(\Theta, \varphi; t) = R(t) + \sum_{N=1}^{\infty} \sum_{m=-N}^{N} a_{N,m}(t) Y_{N,m}(\Theta, \varphi), \tag{11}$$

with time-dependent coefficients $a_{N,m}(t)$. The spherical harmonics are given by

$$Y_{N,m} = C_{N,m} P_N^{(m)}(\cos \Theta) \exp(im\varphi), \tag{12}$$

where $C_{N,m}$ denotes a normalization factor and $P_N^{(m)}$ is the associated Legendre polynomial of the first kind. Each spherical harmonic describes a surface mode with an oscillation pattern having $2m$ node lines in the azimuthal (φ-)direction and N nodes in the polar (Θ-)direction. The coefficients $a_{N,m}$ are the oscillation amplitudes of the corresponding modes.

For small amplitudes $a_{N,m}$, the linearized fluid-dynamical equations yield a system of mutually uncoupled, linear ordinary differential equations that are parametrically coupled to the radial dynamics $R(t)$:

$$\ddot{a}_{N,m} + 3\frac{\dot{R}}{R}\dot{a}_{N,m} + (N-1)\left(\frac{\sigma(N+1)(N+2)}{\rho R^3} - \frac{\ddot{R}}{R}\right)a_{N,m} = 0. \tag{13}$$

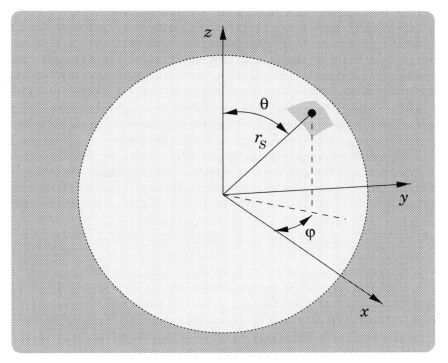

Figure 23. Spherical polar coordinates used in the description of surface perturbations.

This model assumes incompressible, irrotational, and inviscid flow and thus does not take into account damping due to viscosity. The inclusion of dissipation and vorticity gives a far more complicated description involving integro-differential equations [50, 51] that, for the sake of simplicity, will not be considered here any further.

For a static bubble, i.e., for $R(t) = R_n = $ constant, Eq. (13) reduces to an uncoupled collection of linear, undamped oscillators corresponding to the surface modes, with their eigenfrequencies

$$\Omega_{N,m} \equiv \Omega_N = \sqrt{\frac{(N^2 - 1)(N + 2)\sigma}{\varrho R_n^3}} \qquad (14)$$

being independent of m and thus $(2m + 1)$-fold degenerate.

Since the differential equations describing radial bubble dynamics are strongly nonlinear, no analytic evaluation of shape stability at large excursions of the bubble radius is possible, and one has to resort to numerical integration of the system of Eq. (13). As an example, Figure 24 shows the radial oscillation and the temporal behavior of the $N = 2$ surface oscillation of a freely oscillating bubble, as calculated with the Rayleigh–Plesset model, Eq. (2), together with Eq. (13). The example demonstrates that shape stability depends sensitively on the oscillation amplitude of the bubble, since the bubble's radial dynamics acts as a parametric drive of the shape oscillation. It is also clear from the figure that the growth of the surface perturbation covers several oscillation periods of the bubble in this case, which is a typical feature of a parametric instability.

Since the equations governing the evolution of $a_{N,m}$ are linear, it is the growth of $a(t)$ within one oscillation cycle relative to the initial value $a(0)$ that determines the stability of the bubble. This growth is not uniform over the cycle and can attain large values immediately after bubble collapse. Figure 25 gives an example of this behavior. The figure shows the free radial oscillation of a bubble with an equilibrium radius of $R_n = 10\,\mu m$, starting from an initial radius of $R(0) = 20\,\mu m$, and the amplitude of the $N = 4$

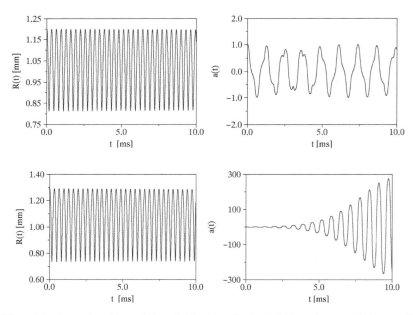

Figure 24. Loss of surface stability of a freely oscillating bubble with an equilibrium radius of $R_n = 1\,mm$. *Top row, left:* radial oscillation $R(t)$ of the bubble, starting from initial conditions $R(0) = 1.2$ mm and $\dot{R} = 0$. The associated amplitude of the $N = 2$ surface mode (right graph) remains bounded. Its initial conditions are $a(0) = 1$, $\dot{a}(0) = 0$. *Bottom row:* Corresponding graphs for the same bubble at a slightly larger initial radius of $R(0) = 1.291$ mm. The $N = 2$ surface mode is linearly unstable. The further parameters that enter the calculation were given in Section V.B.1.

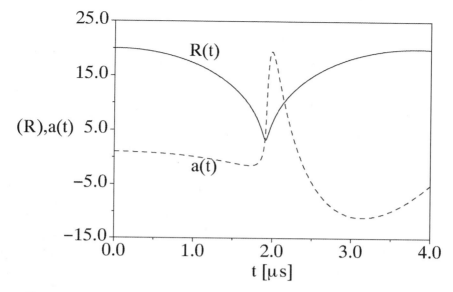

Figure 25. Free radial oscillation (solid line) and evolution of the $N = 4$ surface mode (dashed line) of a bubble with equilibrium radius $R_n = 10\,\mu m$, starting from the initial conditions $R(0) = 20\,\mu m$, $\dot{R}(0) = 0$, $a(0) = 1$, and $\dot{a}(0) = 0$.

surface mode as a function of time. Here, the radial oscillation is large enough to yield strong shape instability. The mode amplitude grows rapidly immediately after the first collapse. This is the instant at which bubble breakup is also observed experimentally. Thus, it is perfectly possible for a bubble to lose its integrity after one oscillation cycle only. This type of behavior is not well characterized by the term "parametric instability," and thus, in the literature a distinction has been made between Rayleigh–Taylor instability and parametric instability [51]. The nomenclature, however, is not very fortunate because it suggests two different physical mechanisms at work, while it is just the strength of instability that distinguishes the two scenarios.

By detailed numerical analysis based on the mode equations (13), bubble stability can be assessed as it relates to the physical parameters, yielding Matthieu-type stability charts, or "phase diagrams". In Figures 26 and 27, we give just one example, for the simple case of a freely oscillating bubble without damping. For a given fluid (water at 20 °C), the bubble's stability depends only on its equilibrium radius and the oscillation amplitude, which, for zero viscosity, is determined by the initial radius $R(0)$ when the bubble is started from rest. In accordance with the procedure described in [49], the

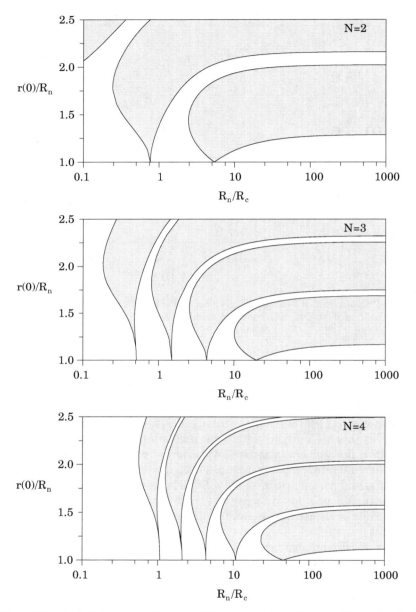

Figure 26. Part of the phase diagram of the parametric surface instability of a freely oscillating bubble. In the calculation of the diagram, viscous effects have been ignored. The shaded areas depict parameter regions where the bubble is unstable against perturbations in the $N = 2$ mode (upper diagram), the $N = 3$ mode (middle diagram) and the $N = 4$ mode (lower diagram).

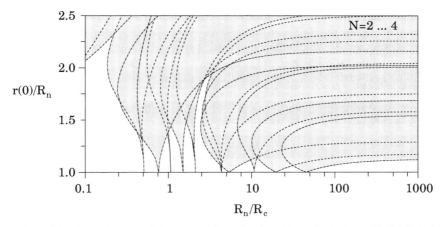

Figure 27. Phase diagram of the parametric surface instability of a freely oscillating bubble without viscous damping. The shaded areas depict parameter regions where the bubble is unstable against perturbations in the $N = 2$, the $N = 3$, and the $N = 4$ mode. The figure is an overlay of the areas in the previous figure in order to depict, roughly, the area of stability (white) of a spherical bubble. Viscous effects will increase this area.

boundaries of parametric instability have been determined numerically, yielding a collection of overlapping instability regions with characteristic cusp-shaped resonance tongues. In the figure, the abscissa measures the bubble equilibrium radius, R_n/R_c, in units of the critical radius

$$R_c = \frac{2\sigma}{p_{\text{stat}}}, \tag{15}$$

which, for water and normal conditions, is $R_c = 1.45\,\mu\text{m}$. The ordinate gives the initial relative elongation $R(0)/R_n$ of the bubble. The phase diagram presented is incomplete, however. For the sake of clearness, the individual instability regions were plotted only up to mode index $N = 4$. A multitude of further regions exists for larger N, all reaching down in cusps to the horizontal axis, leaving as their complement a fairly complicated parameter region of stable bubbles.

Admittedly, the phase diagram of Figure 27 is not very realistic, in that viscous effects were not taken into account and only freely oscillating bubbles were considered. With the inclusion of viscosity, the instability tongues will detach from the abscissa and render a contiguous region of bubble stability for sufficiently low oscillation amplitudes. When, further-more, an external sound field is present, resonances between the surface mode oscillations and the driven bubble oscillation come into play. Also, the response of the bubble need not be periodic anymore.

Quite naturally, the problem of the shape stability of driven bubbles has come up again in the context of bubble trapping and single-bubble sonoluminescence, where the parameter range of stable light emission appears to be bounded by the onset of shape oscillations [51]. In this case the bubbles are very small, possessing an equilibrium radius of only a few micrometers, and the influence of surface tension is accordingly large. Then shape instability does not yet occur for acoustic driving strong enough to cause light emission.

8. Rectified Diffusion

The content of noncondensable gas inside a bubble is normally not constant in time: Diffusion processes from and to the liquid lead to a change in the amount of gas in the bubble. The saturation concentration c_0 of a gas species in the liquid is usually given by Henry's law, which relates the concentration linearly to the partial pressure p_0 of the gas above the solution, $p_0 = kc_0$, with Henry's constant k. The surface tension σ of the liquid will always cause a pressure $2\sigma/R$ that adds to the partial gas pressure inside the bubble in pressure equilibrium. Consequently, spherical non-oscillating bubbles submerged in a saturated liquid under pressure p_0 will always tend to *dissolve*; the gas concentration in the liquid would need to be higher than c_0 to allow for a diffusive equilibrium.

For oscillating bubbles, however, the situation differs, as a phenomenon called *rectified diffusion* sets in [52–55]. Then, the following factors contribute to the possibility of a net *growth* of an oscillating bubble: During the oscillation, the bubble volume, bubble surface, and pressure inside the bubble change periodically and cause, via Henry's law, a variable gas concentration in the liquid layer at the bubble wall. These variations contain a certain asymmetry: The surface area during expansion (and low pressure inside the bubble) is larger than during contraction (and high pressure). Because the diffusion rate is proportional to the interface area, the inward mass flow is greater than the outward diffusion. Additionally, the nonlinearity of the radial oscillation leads to a longer fraction of the cycle during which the bubble is large (and gas diffuses inward) and only short intervals during which the bubble is small (and gas is pushed out). Furthermore, the radial fluid motion near the bubble alternately steepens and shallows the concentration gradient in the liquid at the bubble wall. Since the mass transport through the bubble wall is also proportional to this concentration gradient, such a "shell effect" also contributes to diffusion rates and may intensify the bubble's growth.

In a recent analytical work, Fyrillas and Szeri [55] treat the problem of diffusion for a spherically oscillating bubble in a liquid of arbitrary initial saturation; that is, the gas content in the liquid far from the bubble, c_∞, may differ from the saturation concentration c_0 under normal pressure p_0. Fyrillas and Szeri take into account all the mechanisms already mentioned and consider the case where the (slow) time scale of diffusion can be separated from the (fast) time scale of the bubble oscillation (i.e., for a large Peclet number $Pe = R_n^2 \omega/D \gg 1$, where D is the diffusion constant). One important result is that the mass flow is governed by the quantity

$$\langle p[R(t)]\rangle_\tau = \frac{1}{\int_0^T R^4(t)dt} \int_0^T R^4(t)p[R(t)]dt, \tag{16}$$

where $p[R(t)]$ denotes the pressure inside the bubble at time t, depending on the instantaneous bubble radius $R(t)$, $\langle \ldots \rangle_\tau$ abbreviates the indicated time averaging weighted by $R^4(t)$, and T is the period of oscillation. In particular, a diffusional equilibrium of the oscillating bubble in the liquid occurs for

$$\frac{\langle p[R(t)]\rangle_\tau}{p_0} = \frac{c_\infty}{c_0}. \tag{17}$$

Whenever $\langle p[R(t)]\rangle_\tau > p_0 c_\infty/c_0$, the bubble shrinks, and for $\langle p[R(t)]\rangle_\tau < p_0 c_\infty/c_0$, it grows. Thus, one recognizes again that a static bubble with $p[R(t)] = p[R_n] = (p_0 + 2\sigma/R_n)$ can be in equilibrium only with oversaturated liquid, $c_\infty > c_0$. With respect to the stability of a diffusional equilibrium, it has to be ensured that a *growing* bubble leads to a *larger* $\langle p[R(t)]\rangle_\tau$, and vice versa. This can be fulfilled only on the right-hand side of a resonance curve. In particular, the right-hand side of the giant-response curve for small bubbles (see Figure 16) provides a possibility for stable equilibria with respect to gas diffusion.

In Figure 28, curves of equilibrium radii are shown for different ratios c_∞/c_0 in the R_n–$\hat{p}_a$ plane ($p_0 = 100$ kPa). In this depiction, a positive slope of a curve indicates stable equilibria, while a negative slope indicates unstable equilibria. Liquid saturated or oversaturated with gas ($c_\infty/c_0 \geq 1$) provides only unstable equilibrium points, since the curves are monotonically decreasing. However, reducing the gas content with respect to normal saturation can lead to arcs with positive slope, emerging near nonlinear resonances in the undulating portions of the curves [56] and in a pronounced manner for very low gas content, high pressure amplitude, and small bubbles in the region of the giant response [57]. Two examples are included in Figure 28 to show how to read the graph. For saturated liquid, $c_\infty/c_0 = 1$, and an excitation pressure amplitude of 60 kPa, we find one unstable equilibrium radius at $R_n \approx 8\,\mu m$, denoted by a circle. The arrows indicate the direction of bubble growth or shrinkage. For reduced gas content, $c_\infty/c_0 = 10^{-4}$, and a pressure amplitude of 150 kPa, there are two

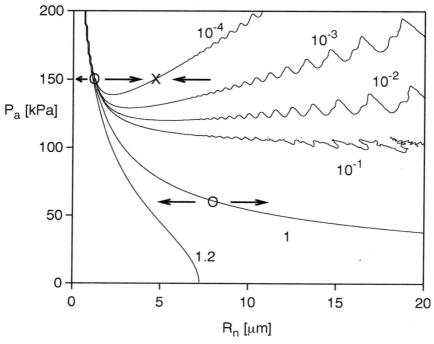

Figure 28. Curves of diffusional equilibrium in a plane of rest radius R_n and driving pressure amplitude P_a ($= \hat{p}_a$). The labels indicate c_∞/c_0, the gas concentration in the liquid far from the bubble, normalized with respect to the saturation concentration under normal air pressure. Only equilibrium points on curve segments with positive slope are stable. As examples, the arrows give the direction of change of the rest radius for a constant pressure amplitude $P_a = 60\,\text{kPa}$ (one unstable equilibrium, denoted by an open circle) and $P_a = 150\,\text{kPa}$ (one unstable and one stable equilibrium, the latter denoted by a cross). The undulation of the curves in the upper right corner is caused by nonlinear resonances and chaotic oscillations of the bubble. (The undulations in the upper left corner are due to a plotting artefact of coarse graining.)

equilibrium points, at $R_n \approx 1\,\mu\text{m}$ and $R_n \approx 5\,\mu\text{m}$. Only the latter rest radius is diffusionally stable, denoted by a cross. It is these conditions that allow for stable single-bubble sonoluminescence in degassed water.

9.   Bjerknes Forces

A body in an inhomogeneous pressure field experiences a force in the direction of lower pressure. In a gravitational field, for instance, this leads to the buoyancy force, which is not considered here any further. For bubbles in a sound field, forces appear due to the inevitable pressure gradient and, moreover, due to the oscillations of the bubble volume, which also induce pressure gradients. If the bubble is small compared with the typical spatial scale of pressure variations (the wavelength $\lambda = 2\pi c/\omega$), the Bjerknes force can be written as

$$\mathbf{F}_B = -\langle V(t)\nabla p(t)\rangle_t, \tag{18}$$

where $V = (4\pi/3)R^3$ is the bubble volume, $\nabla p(t)$ denotes the gradient of the pressure at the bubble's position, and $\langle\ldots\rangle_t$ indicates a time average.

In general, Bjerknes forces are separated into primary and secondary forces, depending on the origin of the pressure gradient [58]. Primary Bjerknes forces relate to the primary sound field originally causing the bubble to oscillate. Secondary Bjerknes forces are due to the sound emitted from other bubbles via their oscillations, which is a secondary effect when the oscillation is caused by the primary sound field. The primary Bjerknes forces thus act on a bubble through the externally imposed acoustic field, while the secondary Bjerknes forces act between oscillating bubbles.

The sign and magnitude of the forces depend on details of the bubble oscillation. The discussion that follows examines Bjerknes forces occurring for strongly nonlinear oscillations and compares the results with approximations obtained for bubbles oscillating harmonically. The nonlinear bubble dynamics is simulated using the Keller–Miksis model [40],

$$\left(1 - \frac{\dot{R}}{c}\right)R\ddot{R} + \frac{3}{2}\dot{R}^2\left(1 - \frac{\dot{R}}{3c}\right) = \left(1 + \frac{\dot{R}}{c}\right)\frac{p_l}{\rho} + \frac{R}{\rho c}\frac{dp_l}{dt}, \tag{19}$$

$$p_l = \left(p_0 + \frac{2\sigma}{R_n}\right)\left(\frac{R_0}{R}\right)^{3\kappa} - p_0 - \frac{2\sigma}{R} - \frac{4\mu}{R}\dot{R} - p_a(t), \tag{20}$$

$$p_a(t) = \hat{p}_a\cos(\omega t),$$

for air bubbles in water at 20°C with the polytropic exponent $\kappa = 1.4$. The other quantities were given the same values as before in the response curve calculations.

For bubbles whose equilibrium radius R_n fulfills the condition $4\mu/\rho c \ll R_n \ll c/\omega = \lambda/2\pi$, and for small amplitudes $\hat{p}_a \ll p_0$ of the external sound field, linearization of the Keller–Miksis model with $R(t) = R_n + R'(t)$ yields

$$\ddot{R}' + \alpha\dot{R}' + \omega_0^2 R' = -\frac{\hat{p}_a}{\rho R_n}\cos(\omega t), \tag{21}$$

where

$$\omega_0^2 = \frac{1}{\rho R_n^2}\left[3\kappa p_0 + \frac{2\sigma}{R_n}(3\kappa - 1)\right], \quad \alpha = \frac{4\mu}{\rho R_n^2} + \frac{\omega_0^2 R_n}{c}. \tag{22}$$

As can be seen from Eq. (1.22), the resonance frequency and the equilibrium radius are directly related: For a fixed bubble size R_n, we can speak of a linear resonance frequency $\nu_r = \omega_0/2\pi$, and for fixed frequency ν, we can find a linear resonance bubble radius $R_r = R_n$ via Eq. (1.22). The relationship is approximated (for normal air pressure and for water, neglecting surface tension) by the easily memorizable form $\nu_r R_r \approx 3\text{ms}^{-1}$. The radius and the associated frequency can thus often be exchanged with one another—for instance, with respect to resonance curves. Resonance curves of oscillatory systems are usually given as a function of the driving frequency. To make them more intuitive, they were given in the preceding discussion as a function of the bubble equilibrium radius.

The inhomogeneous solution of the linear ordinary differential equation (21) is given by $R'(t) = R'_A \cos(\omega t + \varphi)$, which is characterized by an amplitude $R'_A = R'_A(\omega, R_n)$ and a phase shift $\varphi = \varphi(\omega, R_n)$. The linearized model tells us that bubbles larger than the linear resonance radius R_r oscillate in such a way that they have a large volume at high-pressure phases ($\varphi \in [0, \pi/2]$), and smaller bubbles have a large volume during low-pressure times ($\varphi \in [\pi/2, \pi]$).

The *primary Bjerknes force* acting on the harmonically oscillating bubble in a standing sound field $p(\mathbf{x}, t) = p_0 + \hat{p}_a(\mathbf{x}) \cos(\omega t)$ is given by

$$\mathbf{F}_{B1} = -\tfrac{1}{2} V_A \nabla \hat{p}_a \cos(\varphi), \qquad (23)$$

where $V_A = 4\pi R_n^2 R'_A$ denotes the amplitude of the linear volume oscillation, which is phase shifted with respect to the pressure and is itself given by $V(t) = V_0 + V_A \cos(\omega t + \varphi)$. With the preceding phase considerations in mind, the following result is obtained: For bubbles smaller than the linear resonance radius, the force $\mathbf{F}_{B1}$ acts in a direction towards the pressure antinode ($\cos(\varphi) < 0$), while larger bubbles ($\cos(\varphi) > 0$) are attracted by the pressure node.

This well-known result, however, is no longer valid as soon as nonlinear oscillations occur [59, 60]. Let us consider the vicinity of a pressure antinode. Figure 29 shows that smaller and smaller bubbles are repelled for a stepwise increment of the driving amplitude. The reason is that the relative phase between exciting pressure and bubble response is affected by the amplitude of oscillation. The occurrence of nonlinear resonances leads to a certain zigzag course of the border between attractive and repulsive primary Bjerknes forces. Near the center of the diagram, chaotic bubble oscillations and coexisting attractors cause a complicated pattern that is not fully resolved. Note that above a pressure amplitude of about 180 kPa, only very small bubbles in the micrometer range are still attracted by the antinode, in striking contrast to the linear theory [60].

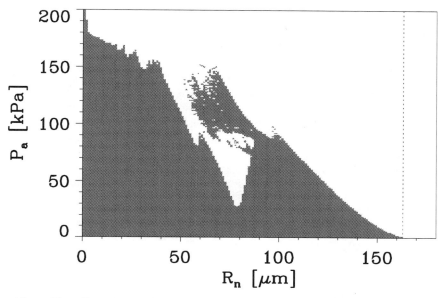

Figure 29. Effect of nonlinear bubble oscillations on the primary Bjerknes force near a pressure antinode ($\omega = 2\pi \cdot 20\,\mathrm{kHz}$). The dark regions in the plane of pressure amplitude P_a ($= \hat{p}_a$) and the bubble equilibrium radius R_n indicate attraction, the bright regions repulsion. The linear resonance radius (approx. $163\,\mu\mathrm{m}$) is indicated by the vertical dashed line. In the linear theory, all parameters to the left of the dashed line yield attraction of the bubble towards the pressure antinode.

Similar strong effects of nonlinear oscillations have been found for the *secondary Bjerknes forces* [61–63]. The force of an oscillating bubble "1" on a neighboring bubble "2" is given, to some approximation, by

$$\mathbf{F}_{B2} = -\frac{\rho}{4\pi}\langle \dot{V}_1 \dot{V}_2 \rangle \frac{\mathbf{x}_2 - \mathbf{x}_1}{\|\mathbf{x}_2 - \mathbf{x}_1\|^3}, \tag{24}$$

where $\mathbf{x}_1$ and $\mathbf{x}_2$ denote the locations of the interacting bubbles. For harmonic bubble oscillations, we obtain

$$\mathbf{F}_{B2} = -\frac{\rho\omega^2}{8\pi} V_{1A}V_{2A}\cos(\varphi_1 - \varphi_2)\frac{\mathbf{x}_2 - \mathbf{x}_1}{\|\mathbf{x}_2 - \mathbf{x}_1\|^3}, \tag{25}$$

where V_{1A}, V_{2A} and φ_1, φ_2 are the amplitudes and the phases of the volume oscillations $V_i(t) = V_{i0} + V_{iA}\cos(\omega t + \varphi_i)$ ($i = 1, 2$), respectively. According to this result, a bubble smaller than the linear resonance radius and a bubble larger than the linear resonance radius repel each other, while pairs of smaller or larger bubbles are subject to an attracting secondary Bjerknes force.

Including the nonlinearity, a coupling of the oscillations, or a coupling of shape distortions, for instance, can lead to a considerable change in this situation. We just sketch the influence of strong nonlinear oscillation in Eq. (24) on small spherical bubbles [62]. It is found that the magnitude of the secondary Bjerknes force increases by orders of magnitude in comparison to that predicted by linear theory and that unforeseen mutual repulsion of bubbles may occur. A depiction in the $R_{1n}-R_{2n}$ plane is shown in Figure 30. For a fixed driving amplitude and frequency, the attraction and repulsion between bubbles is coded by dark and bright areas,

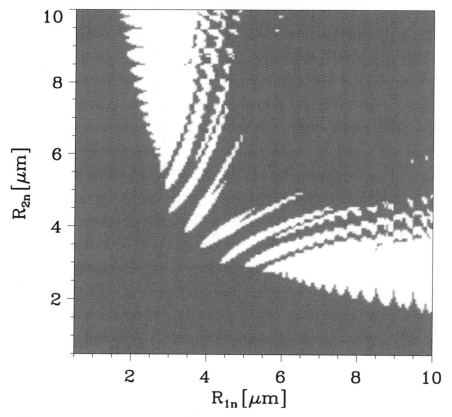

Figure 30. Secondary Bjerknes forces between two neighboring bubbles in the sound field ($\hat{p}_a = 112\,\text{kPa}$, $\omega = 2\pi \cdot 20\,\text{kHz}$). The axes denote the equilibrium radii; dark areas indicate mutual attraction, bright areas mutual repulsion.

respectively. The bubble radii below 10 µm are chosen because of their supposed relevance in cavitation structures.

Further modifications of the forces between oscillating bubbles emerge because of additional neighbors, the motion of the bubbles relative to the liquid, or a delay of the mutual action because of the finite speed of sound. All these effects render the subject of bubble–bubble interaction still open to further investigation.

VI. SINGLE LASER-INDUCED BUBBLE DYNAMICS

A convenient method of producing a single bubble in a liquid is to focus a short pulse of laser light into the liquid, as shown in Figure 5. Depending on the focal spot size, the transverse mode structure of the laser, the duration of the pulse, and the intensity of the light, a small volume or several small volumes of liquid are heated up rapidly in nanoseconds, picoseconds, or femtoseconds, according to the laser and its pulse width used. A bright light-emitting plasma is formed of obviously high pressure, because a shock wave is radiated out into the liquid. The high-pressure plasma expands, forming a bubble from where the shock wave detaches. This sequence of events is demonstrated in Figure 31. To stop the rapid expansion of the cavity and the propagation of the shock, the photographs were taken with single laser pulses of about 100 fs duration.

An example of multiple breakdown from one laser pulse is given in Figure 32. Here, a Q-switch pulse from a ruby laser has been used, and holography has been employed to get rid of the bright breakdown light obscuring the bubble outline and shock wave detachment.

Not only the breakdown process, but also the subsequent bubble dynamics, is exceedingly fast. In particular, the collapse of a bubble can be

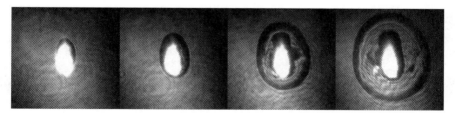

Figure 31. Bubble generation and shock wave emission upon laser-induced breakdown caused by a focused Nd:YAG laser pulse ($\lambda = 1064$ nm) in water. The photographs were taken with 100-fs pulses for back-illumination, obtained via a pulse picker from a Ti:Sa laser. The bright spot in the middle of each frame in this case is due to the breakdown. The frame size is 660 µm × 660 µm. From left to right, the following time delays between the generating pulse and the illumination pulse apply (measured from peak to peak): 9.5 ns, 13 ns, 69 ns, and 107 ns.

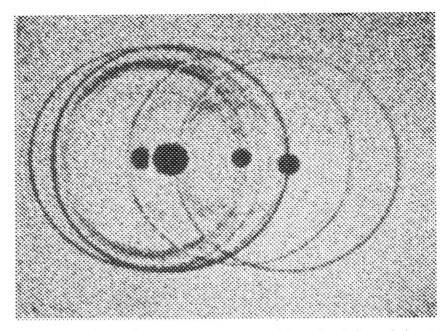

Figure 32. Bubbles and shock waves upon multiple breakdown of a ruby laser pulse in water ($\lambda = 694.3$ nm). This reconstructed image from a hologram is taken 1.75 μs after breakdown with a second ruby laser of long coherence length. The size of the picture is 1.1 mm × 1.6 mm.

captured only with ultrahigh-speed framing cameras, as it depends sensitively on details of the breakdown process—for instance, its degree of sphericity. Thereby, the instant of collapse is shifted so that it is difficult to reach high sampling rates, as with the trapped bubbles in a sound field. A minimum framing rate of 20 million frames per second [64] proved necessary to partly resolve the events encountered upon the collapse of a cavitation bubble (e.g., the emission of multiple shock waves).

A. Spherical Bubble Dynamics

Figure 33 gives an example of a spherical bubble produced in water and its subsequent dynamics, taken at 75,000 frames per second with a rotating mirror camera. The series starts with the bright spot of the laser light from the breakdown site (and some reflexes). The bubble is seen as a dark disk with a bright spot in its center because the illuminating diffuse backlight is deflected off the bubble wall, except for the central part, where it goes undeflected. During expansion, work is done against the ambient pressure. Therefore, bubble expansion stops at some maximum radius. From there

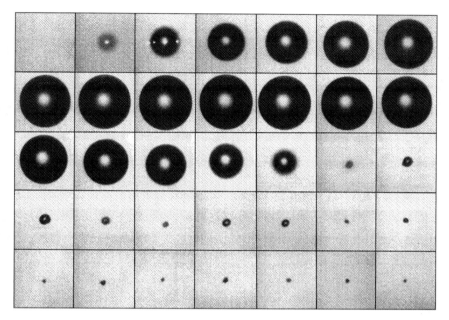

Figure 33. Dynamics of a laser-produced spherical bubble in water, observed at 75,000 frames per second. The maximum bubble radius is about 1.3 mm.

on, the bubble starts to shrink, gaining speed at an ever-increasing rate driven by the ambient pressure and leading to a collapse with compression of the bubble's contents (gas and vapor) in the final phase. This way, the gas acts as a spring and drives the bubble back into expansion, called rebound. Due to the ever-present damping—in this case, mainly sound radiation or the emission of shock waves—subsequent oscillations are highly damped. Also, some signs of asymmetric collapse, which is expected to lead to extra damping, are present.

It is a peculiar effect that the bubble oscillations undergo less damping in higher viscosity liquids, at least in a certain range of viscosity higher than that of water. This phenomenon is exemplified in Figure 34, where a spherical bubble has been produced in silicone oil with a viscosity of 4.85 poise. Again the framing rate is 75,000 frames per second. It is evident that the decay of the oscillations is slower than in the case of water. There are several explanations for this behavior. One invokes the larger gas content in the bubble, preventing the bubble from collapsing strongly, losing less energy in a shock wave. This may overcompensate for the losses due to the higher viscosity. Another explanation invokes nonsphericity. The bubble shape is unstable during collapse, and the different disconnected volumes of

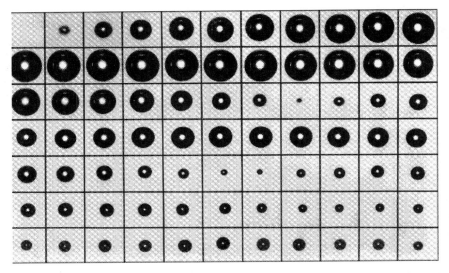

Figure 34. Dynamics of a laser-produced spherical bubble in silicone oil of viscosity 4.85 poise, observed at 75,000 frames per second. The maximum bubble radius is about 2 mm.

gas and vapor can dissipate energy more easily into the surrounding liquid. A bubble collapsing more strongly will suffer from increased instability and thus will undergo increased damping.

The shock waves radiated upon collapse propagate at about the velocity of sound—i.e., at about 1,500 m/s in water. To capture them, including their propagation, requires a short exposure and very high speed framing. A suitable framing rate for this purpose is 20 million frames per second. Then a shock wave can be seen on several frames, and multiple shock waves, as indeed are encountered, can be timed to bubble dynamics events. A (nearly) spherical bubble collapse with the emission of one shock wave is seen in Figure 35. This photographic series has been taken at 20.8 million frames per second with an image converter camera. The maximum number of frames per shot is limited to eight. Therefore, four different shots have been combined into and close one series. This is made possible by the excellent reproducibility of the bubble size. The single shock wave shown here is from the first collapse and rebound. A similar shock wave is radiated during breakdown, and weaker ones are emitted during the subsequent collapses and rebounds of the bubble. This series of shock waves can be captured with a hydrophone [43]. Figure 36 shows the acoustic emission of a spherical laser-induced bubble into water. Four peaks can be seen. The first one belongs to the shock wave connected with the breakdown process. The

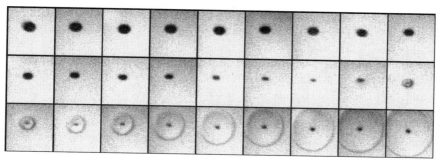

Figure 35. Collapse of a laser-produced spherical bubble in water, observed at 20.8 million frames per second (48-ns interframe time) with radiated shock wave. The size of the picture is 1.5×1.8 mm.

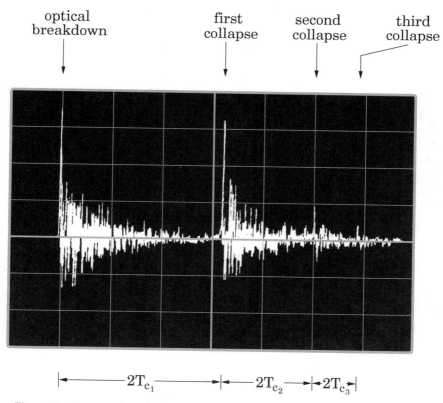

Figure 36. Pressure–time signal of an oscillating laser-induced bubble in water. T_c is the collapse time—i.e., the time from maximum to the next minimum of the bubble radius.

further three peaks belong to the first, second, and third collapse of the bubble. Note that the first and second peaks are of about equal amplitude; that is, the breakdown and collapse pulses are of similar strength. At times, the second peak deriving from the first bubble collapse has been found to even be higher than the breakdown pulse. In this case, a hydrophone with a bandwidth of 1.5 MHz has been used, and the signal has been digitally recorded at a sampling rate of 10 MHz.

The time from optical breakdown to first collapse is denoted by $2T_{c_1}$, the time from first to second collapse by $2T_{c_2}$, and the time from second to third collapse by $2T_{c_3}$. Here, T_{c_1} to T_{c_3} are the Rayleigh collapse times [10], which are proportional to the maximum bubble radius R_{max} before the respective collapse. The Rayleigh collapse time is given by

$$T_c = 0.915 R_{max} \sqrt{\frac{\rho}{p_{stat} - p_v}} \qquad (26)$$

and has been derived by Rayleigh for the case of an empty bubble without surface tension and viscosity collapsing under a constant pressure (here, $p_{stat} - p_v$). It has been found that, for spherical laser-produced bubbles expanding and contracting under the action of the static ambient pressure in water under normal conditions, the expansion phase and the contraction phase are symmetrical to a high degree, so that the time from generation to first collapse and from one collapse to the next is given by twice the Rayleigh collapse time, or $2T_c$. Because of the relationship between T_c and R_{max}, the maximum bubble radius can be determined from the time between breakdown and first collapse via the formula for the Rayleigh collapse time. This has been done in cases where only the bubble collapse has been photographed, to determine the maximum bubble size from the hydrophone signal—e. g., in the luminescence studies.

B. Spherical Laser Bubble Luminescence

To photograph the luminescence event, the experimental arrangement was augmented by a long-distance microscope and an intensified CCD (ICCD) camera, as shown in Figure 37 [65]. This camera had a high contrast ratio between the shuttered and the opened state, which is a favorable property for suppressing the intense continuum light emission from the dielectric breakdown process. An optical resolution of the luminescence image better than 3 μm was achieved. Figure 38 shows an image of the luminescence that occurs during the spherical bubble collapse taken with the ICCD camera. The gating time was adjusted to 5 μs, and an attenuated flash illuminated the bubble. The bright SCBL spot in the left part of the figure is in the center of

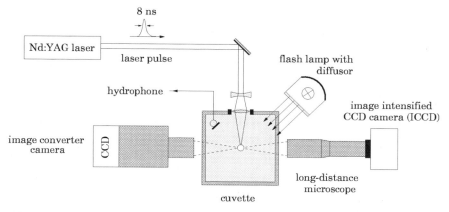

Figure 37. Extended experimental arrangement for cavitation bubble luminescence studies.

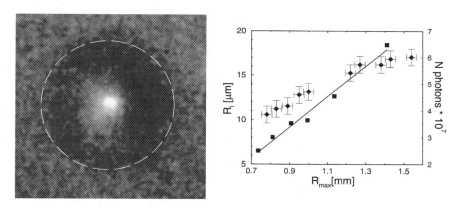

Figure 38. Single cavitation bubble luminescence. *Left:* Light emission by a laser-generated bubble upon first collapse. Superimposed on the image is an outline of the maximum bubble size before collapse at the beginning of the 5 µs open time. *Right:* Radius of emission spot (diamonds) and number of photons emitted (squares) as a function of maximum radius of the bubble.

the bubble outline. Light emission is observed when the gating time of the ICCD covers the bubble collapse, monitored with a hydrophone.

By measuring the size of the spot, integrating the collected light over the area of the spot, and, in the same run, determining the maximum radius, we obtained the graph at the right of the figure. The graph shows the dependence of the radius of the light-emitting region within the bubble and the number of photons emitted on the maximum bubble radius R_{max} that is attained. The figure demonstrates clearly that the size of the bubble at maximum expansion is strongly correlated with the light output, the dependence being linear to a good approximation. Obviously, the larger the bubble radius R_{max} the more violent is the collapse, and the more light is emitted, illuminating a larger volume.

C.   Aspherical Bubble Dynamics

When a bubble collapses in an environment that is not spherically symmetric, the collapse changes in a remarkable way. Of utmost interest is the case of a bubble near a solid boundary [66, 67], because bubbles are the source of cavitation erosion [68–70]. The use of the normalized distance $\gamma = s/R_{max}$, where s is the distance of the laser focus from the boundary and R_{max} is the maximum bubble radius attained (Figure 39), has proven advantageous in classifying the bubble dynamics near a plane rigid boundary. Bubbles with different R_{max}, but the same γ-value, exhibit similar dynamics, thus affording the chance of specifying the degree of asymmetry of bubble collapse: cavitation bubbles with a small value of γ are more influenced by the boundary, collapsing with a more pronounced shape variation, than those with a large value, which collapse in a more spherelike manner. This statement, however, does not apply to bubbles too close to the

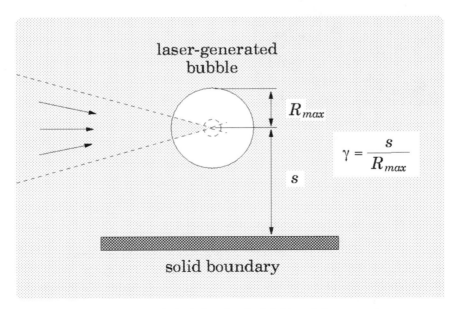

Figure 39.   Geometrical relations for the definition of the parameter γ.

boundary, where $\gamma \approx 0$ and the bubble adopts a hemispherical shape—i.e., approaches spherical symmetry again.

1. Jet Formation

A flat, solid surface nearby causes the bubble to involute from the top (the surface below the bubble) and to develop a high-speed liquid jet towards the solid surface. When the jet hits the opposite bubble wall from the inside, it pushes the wall ahead, causing a funnel-shaped protrusion with the jet inside. Figure 40 is a high-speed photographic series of a bubble collapsing in water near a flat, solid wall, taken at 75,000 frames per second with a rotating mirror camera. The jet is best visible in the first rebound phase as the dark line inside the bright central spot of the bubble, where the back light can pass undisturbed through the smooth surface of the bubble. The funnel-shaped protrusion downwards is the elongated bubble wall containing the jet that drives the elongation until its energy is used up. Then the long tube of gas and vapor gets unstable and decays into many tiny bubbles. The main bubble surface snaps back to its former locally spherical shape.

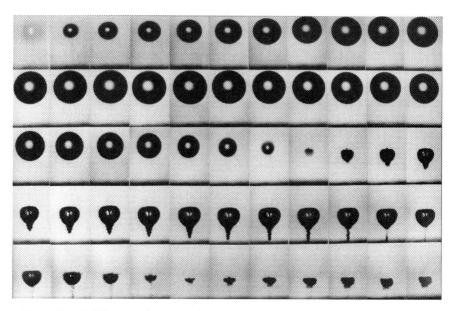

Figure 40. Bubble dynamics near a flat, solid boundary, shown in a photographic series taken at 75,000 frames per second. The frame size is 7.2 mm × 4.6 mm, the maximum bubble radius is 2.0 mm, and the distance of the center of the bubble from the boundary when the bubble is maximum size is $s = 4.9$ mm, i.e., $\gamma = 2.45$.

Figure 41 is an enlargement of a bubble with a jet and its protrusion pointing to the solid boundary.

2.   Shock Wave Radiation

Shock wave radiation is much more involved when jet formation occurs. Usually, at least three shock waves are radiated, two from the jet and the third or others when the bubble attains (or nears) its minimum shape. This sequence is documented in Figure 42, which shows a sequence of a collapsing bubble with jet formation, taken at 20.8 million frames per second with an image converter camera. The maximum bubble radius R_{max} is 1.29 mm, attained about 90 μs before the first picture starts. The normalized distance $\gamma = s/R_{max}$ to the boundary is 2.4. The first shock wave is radiated when the jet hits the (inside moving) opposite wall of the bubble. The jet is so broad at its "tip," that it contacts the lower bubble wall at a ring above the lowest point, giving rise to a toruslike shock wave. This "jet torus shock wave" later evolves into a single outgoing shock wave, as the shock torus must close upon expansion. The jet torus shock thereby surrounds the bubble, becomes very weak, and soon ceases to be seen in the frames. The toruslike shock wave from the jet implies that, in addition to the bubble becoming a torus by the jet impact, a separate tiny bubble ("tip bubble") must be created between the jet "tip" and the curved lower bubble surface. This bubble will be compressed further by the jet and the ongoing bubble collapse, giving rise to a second shock wave, seen in frame 10 of Figure 42 and in the subsequent frames. The "tip bubble shock wave" definitely emanates from the lower bubble wall, as is seen by the asymmetric propagation in relation to the bubble shape. The collapse of the tip bubble is much faster than the collapse of the main bubble. In frame 13, the bubble is at or near its minimum volume and emits a third shock wave seen detaching from the bubble in the subsequent frames. The collapse of the main bubble thus is the latest in this series of shock waves. The main bubble collapses in the form of a torus, whose stability upon collapse may be questioned. Thus, several shock waves may emanate from the bubble torus. The broad shock "front" seen in the last row of the figure is an indication that this, in fact, may have happened.

The hydrophone, of course, is also capable of capturing the shock waves in the liquid in the case of aspherical bubbles. Figure 43 shows two tracks, for a spherical bubble and a bubble of the same size of 8.8 mm in diameter at maximum expansion, but collapsing near a solid wall with $\gamma = s/R_{max} = 1.4$. The collapse times are substantially prolonged for a bubble near a solid wall. In this case, the first collapse is weakened and the second collapse is enhanced. A complete study of the relationship has not yet been performed.

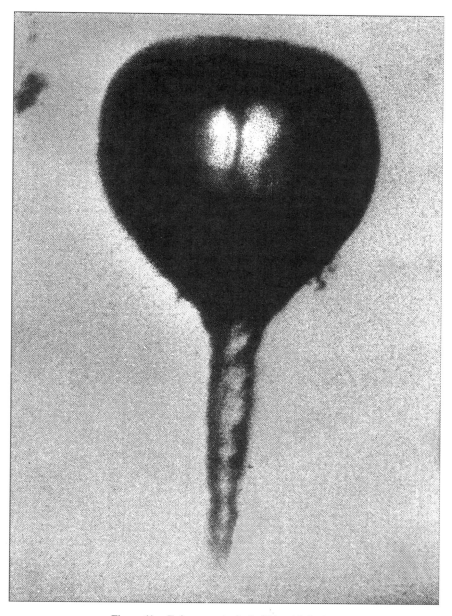

Figure 41. Enlargement of a bubble with its jet.

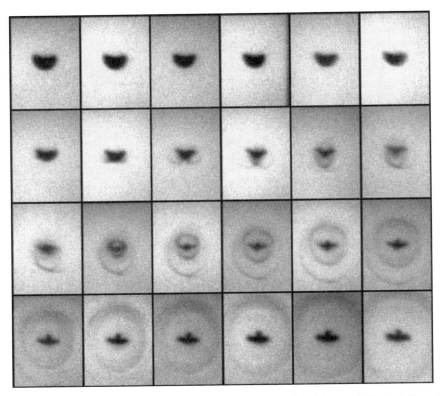

Figure 42. Collapse of a bubble near a solid boundary (outside, below each frame), taken at 20.8 million frames per second. The maximum bubble size is 1.29 mm. The relative distance to the boundary is $\gamma = 2.4$. Radiation of three shock waves: jet torus shock wave (frames 8, 9, 10), tip bubble shock wave (frames 10 and subsequent ones), and main bubble shock wave (frames 13 and subsequent ones). The picture size is 2.0 mm × 1.4 mm.

3. Counterjet Formation

The protrusion sticking upwards out of the bubble (see the last several frames of Figure 42), formerly called counterjet by us [17, 18], presumably is the result of microcavitation inside the jet. The jet shock waves propagate not only into the liquid below the bubble, but also backwards through the jet. Thereby, they are reflected off the jet wall as tension waves. These waves are conjectured to produce cavitation inside the jet. Therefore, almost no outgoing shock is seen above the bubble that actually has the shape of a torus after the jet has hit the inside bubble wall. Counterjets are observed from γ about 1 to about 2.5 [18]. Figure 44 gives a prominent example of a

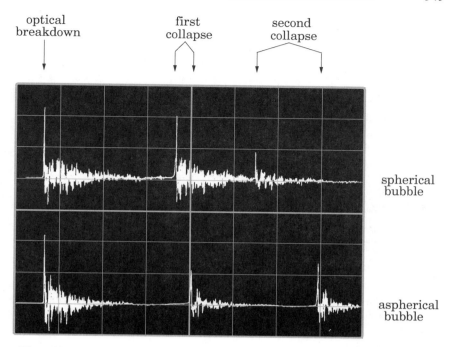

Figure 43. Pressure–time signal of an oscillating laser-induced bubble in water. The top row is from a spherical bubble, the bottom row from a bubble near a solid boundary with $\gamma = 1.4$. In both cases, the maximum bubble radius is 4.4 mm.

counterjet for $\gamma = 1.8$ and $R_{max} = 1.45$ mm. The sequence starts just before the first collapse and covers the first rebound and second collapse phase up to the third collapse. The counterjet appears immediately after the first collapse, outside the bubble and earlier than the jet makes its way to elongate the bubble surface downwards. The jet reaches the boundary, and the second collapse already takes place there. The bubbles from the counterjet reappear after the second collapse. Presumably, they are activated again by the shock wave(s) from the second collapse.

4. Erosion

Cavitation bubbles are able to damage solid surfaces; in fact, no material has been found to resist their attack. This seems peculiar, as, intuitively, bubbles are registered as soft objects. Yet they pierce holes even into steel. Mainly two characteristic effects are believed to be responsible for the destructive action of cavitation bubbles: the emission of shock waves upon the collapse of the bubble and the generation of a high-speed liquid jet

Figure 44. Counterjet formation upon collapse of a laser-produced cavitation bubble in the vicinity of a solid boundary. The maximum bubble radius R_{max} = 1.45 mm, γ = 1.8, and the framing rate is 56,500 frames/s.

directed towards the solid boundary. The main results obtained so far by investigation of laser-produced bubbles are the following: Damage is observed when the bubble is generated at a distance less than twice its maximum radius from a solid boundary ($\gamma \leq 2$). The impact of the jet contributes to the damage only at small initial distances ($\gamma \leq 0.7$). The largest erosive force is caused by the collapse of the bubble in direct contact with the boundary, where pressures of up to several GPa act on the material of the surface. Therefore, it is essential for the damaging effect that bubbles be accelerated towards the boundary during the collapse phases due to Bjerknes forces. The bubble touches the boundary at the moment of second collapse, when $\gamma < 2$, and already at the moment of first collapse, when $\gamma < 1$. In the range $\gamma = 1.7$ to 2, where the bubble collapses mainly down to a single point, one pit below the bubble center is observed. At $\gamma \leq 1.7$, the bubble has become toroidal, a shape induced by the jet flow through the bubble center. Corresponding to the decay of this bubble torus into multiple tiny bubbles, each collapsing separately along the circumference of the torus, the observed damage is circular as well. Bubbles in the ranges $\gamma \leq 0.3$ and $\gamma = 1.2$ to 1.4 cause the greatest damage. To give an example of how bubble dynamics and the pattern of erosion are mutually related, Figure 45 shows the dynamics of one of the 100 bubbles that led to the damage shown in Figure 46. It is seen that the torus collapses with the emission of several shock waves within a microsecond and as quickly reconstitutes after collapse. Comparing the size of the torus on collapse directly on the surface with the pattern of damage, we see that both sizes coincide. This coincidence has been found on any occasion, giving rise to the statement that only

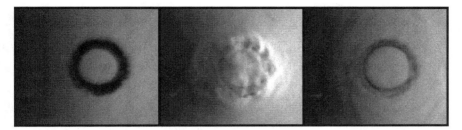

Figure 45. Bubble dynamics at first collapse, in a photograph taken at a million frames per second for $\gamma = 0.5$. The size of the frames is 4.2×4.2 mm. The photographs were taken in top view (perpendicular to the boundary), where the torus formed after jet formation and the torus shock waves are better seen.

Figure 46. Picture showing damage on aluminum after application of 100 bubbles with $\gamma = 0.5$.

bubbles collapsing in direct contact with the solid surface lead to damage and that the form of the collapse directly on the surface of the solid determines the pattern of damage. More information on this topic can be found in [70].

The complex series of events just described, with jet and counterjet formation, shock wave radiation, and erosion patterns, is typical for the asymmetrical collapse of a bubble in the vicinity of a flat, solid boundary.

D.   Aspherical Laser Bubble Luminescence

We now turn to the luminescence of asymmetrically collapsing cavities [65]. A corresponding picture series is shown in Figure 47. The series is derived from a combination of two photographic devices, namely, an image converter camera and a separate intensified CCD, as shown in Figure 37. Frames 1 to 7 show the bubble dynamics for $\gamma = 4.7$, photographed with the image converter camera at 220,000 frames per second. Prior to collapse (frames 1 to 4), the bubble has a spherical shape. In frame 5a, the image taken with the high-speed camera was blended over a photograph of the luminescence spot, obtained with the ICCD (with the same experimental parameters, but from a different run). After the collapse, the liquid jet again forms a protrusion towards the solid boundary, which is out of view below the frames.

During bubble collapse, the bubble center translates towards the boundary. Thus, the light emission is located 135 µm below the center of the bubble prior to collapse, and the position of the light emission is at the geometrical center of the main body of the reexpanding bubble. The sequence does not resolve the shape of the bubble at the instant of light emission, as even a framing rate of 20 million frames per second is too slow to capture the details of multiple shock wave emission for γ-values larger

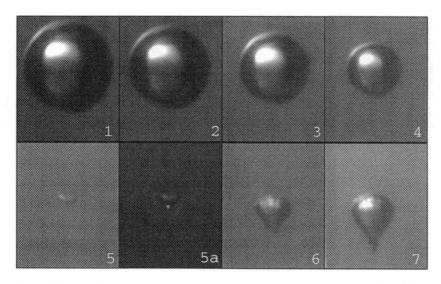

Figure 47.   Luminescence of an asymmetrically collapsing laser-generated bubble. Photographic sequence indicating maximum bubble size and location of light-emitting spot with $\gamma = 4.7$.

than about 3. Indirect evidence for an almost spherical collapse is the missing counterjet (see the previous section) above the reexpanding bubble. In our current research project, we set out to resolve the final collapse stages by photography with ultrashort laser pulses. An example of the method is given in Figure 31.

How does the asphericity of collapse influence the bubble's luminescence? By proper placing of the laser focus, we are able to control γ and thus the degree of asphericity. Figure 48 reports the result of the corresponding experiment. The graph reveals the dependence of the radiated energy on the parameter γ. The points give the normalized (with respect to the spherical case) energy for a fixed R_{max} as a function of the normalized distance γ. For $\gamma \leq 3.5$, the light output is not distinguishable from the dark signal at the sensitivity of our equipment. The integrated luminescence decreases rapidly with smaller γ-value and is not detectable for too strong an asphericity of bubble collapse.

We found that, for an aspherical bubble collapse, shock wave emission is not necessarily connected with a luminescence event. There is only a certain

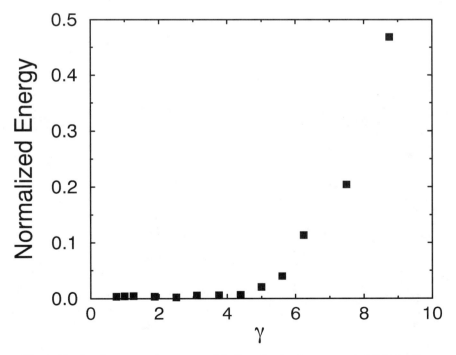

Figure 48. Luminescence of an asymmetrically collapsing laser-generated bubble. Dependence of light intensity (a.u.) on the distance parameter γ.

regime of asphericity where luminescence occurs. Here, an opportunity to gain more insight into the processes inside the bubble arises by the study of aspherical collapse with state-of-the-art 3D numerical algorithms. Comparing these results with the bubble outline obtained from experiments could determine the conditions for light emission and, finally, provide evidence of the mechanisms that are responsible for cavitation bubble luminescence.

VII.   TWO-BUBBLE SYSTEMS

Bubbles usually do not form alone. When there are more bubbles in a liquid, they interact and alter their dynamics. The next step in bubble dynamics, therefore, is the analysis of two bubbles and then more, up to whole clouds. For this purpose, both methods discussed in the case of single-bubble dynamics can be extended or even combined to study few-bubble systems, beginning with two bubbles separated by a certain distance. For example, it is possible to confine two or more bubbles in different pressure antinodes of a large-aspect-ratio cuvette. (See Figure 49.) This arrangement is most suitable for the investigation of the effect of a shock wave emitted by one bubble on the dynamics of a second one.

If we would like to study the interactions of closely spaced bubbles, optic cavitation is the method of choice. Figure 50 shows two bubbles that have been generated by one laser pulse. Each develops a jet directed towards the

Figure 49.   Cylindrical bubble trap with two levitated sonoluminescing bubbles. The bubbles blink alternately, each with the frequency of the sound field of 25.8 kHz. (Courtesy of R. Geisler.)

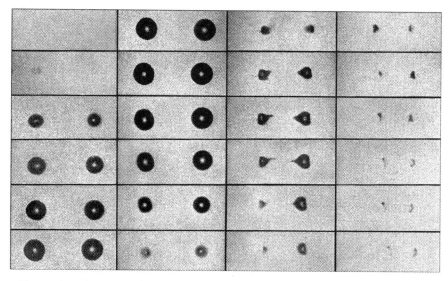

Figure 50. High-speed photographic series of two laser-produced bubbles in water taken at 58, 000 frames per second. The series runs from top to bottom and from left to right. The frame size is 3 mm × 7 mm. The bubble distance at production of the bubbles is 3.6 mm.

other bubble. This is as expected, because symmetry arguments immediately reveal that the case of a bubble in front of a solid boundary is equivalent to two identical bubbles having double the distance from each other than a single one to the boundary.

Also, two bubbles of different sizes can be generated this way from one laser pulse. Strongly different phenomena are observed, depending on the relative size of the cavities and their distance. A cavity of smaller, but comparable, size at a distance such that both bubbles would touch during growth develops a jet towards the bigger one. An example is shown in Figure 51. The sequence has been taken at 75,000 frames per second. The smaller cavity becomes flattened at the side of the bigger one in the expansion phase and then collapses on the surface of the bigger one with strong jet formation. The jet penetrates the bigger bubble. The smaller bubble itself is destroyed by this action, leaving only a few tiny bubbles attached to the flattened side of the bigger bubble.

When the two bubbles are of strongly different sizes and at a distance such that the small bubble does not touch the bigger one at maximum expansion, the smaller one undergoes an interesting fate, as shown in Figure 52. The small bubble first flattens on the side of the big bubble and then divides upon collapse, giving two bubbles with two jets, one in the direction of the big bubble and one away from it.

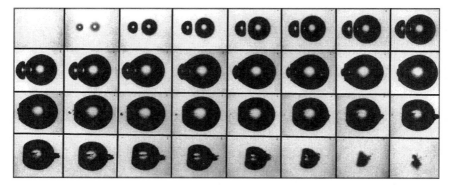

Figure 51. High-speed photographic series of two laser-produced bubbles in water taken at 75,000 frames per second. The frame size is 5 mm × 6 mm.

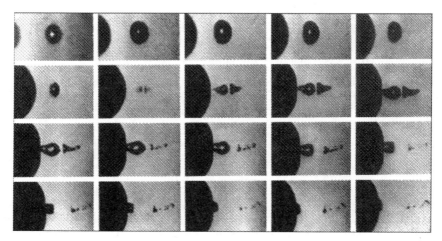

Figure 52. High-speed photographic series of two laser-produced bubbles in water taken at 75,000 frames per second. The frame size is 2.25 mm × 3.5 mm.

When two bubbles are collapsing aside each other parallel to the boundary, it is found that the jets are inclined toward the boundary and in the direction of the other bubble. Thus, a second bubble nearby can deflect the jet away from the boundary and also deflect the migration of the bubble towards the boundary away from it. This has implications for the erosion process: Only bubbles collapsing sufficiently undisturbed or under very favorable conditions (concerted collapse) will cause damage.

With two lasers, two bubbles can be generated, with the additional freedom that they can be produced at different times to study even more sophisticated configurations of two-bubble interaction, as has been done in [71, 72].

VIII. FEW-BUBBLE SYSTEMS

With suitable focusing, e.g., by holographic-optical elements, of short laser pulses, almost any initial configuration of bubbles is producible. A quick start for realizing few-bubble systems may be gotten by a focusing lens in combination with a grating [7]. An example is given in Figure 53. The five-bubble system has been photographed when the oscillations and the interaction of the bubbles have been stopped by damping processes. Because of the decomposition products of the liquid (silicone oil) produced by the laser light, gas bubbles are left that also stick to their places for some time due to the higher viscosity of the oil with respect to water.

IX. MANY-BUBBLE SYSTEMS

Many-bubble systems are naturally encountered in acoustic cavitation experiments and in many practical situations, e.g., sonochemistry. The experiments and also the theory get even more demanding when many-bubble systems are considered. To give an idea of the bubble fields that are encountered, Figure 54 shows a bubble cloud inside our cylindrical transducer, submerged in water as viewed along the axis under backlight

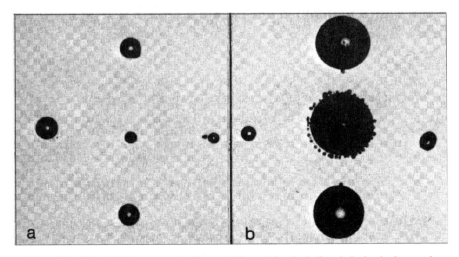

Figure 53. Five-bubble systems in silicone oil formed by single Q-switched ruby laser pulses multifocused by a lens–grating combination. The picture on the right shows that significant bubble interaction has taken place. Note the appearance of satellite microbubbles.

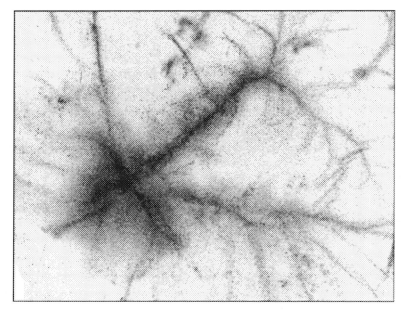

Figure 54.   Filamentary structure of bubbles in sonically induced cavitation. (Courtesy of A. Billo.)

conditions. The bubbles form a branched structure ("streamers") also called acoustic Lichtenberg figures by us, in reminiscence of the electric Lichtenberg figures. A homogeneous cloud of bubbles is never observed. The bubbles always organize themselves into dendritic branches of filaments. A further example of a peculiar, but obviously ordered, structure of bubbles is given in Figure 55. This configuration looks like an example of the repulsion of bubbles oscillating in a sound field.

A.   High-Speed Cinematography

As with single bubbles, the dynamics of many-bubble systems can be captured only with high-speed photography and cinematography. However, the situation is more involved, as now there are more and different centers at which high-speed processes take place. It is therefore more difficult to trigger on some collapse of one of the different bubbles. To study bubble clouds in sound fields, one cycle of the sound field should be about sufficient to cover the essential events possible. It has been found that the filamentary structure normally oscillates with the driving sound field—i.e., the bubbles collapse every cycle. This can be seen in Figure 56, which has been taken at 200,000 frames per second with a rotating-mirror camera. The frequency of the

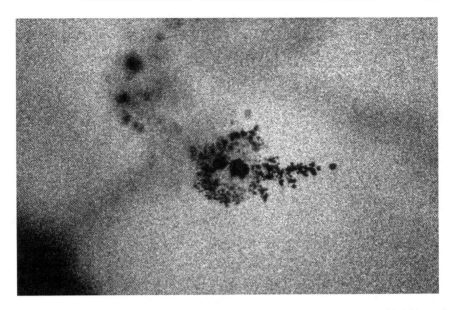

Figure 55.   Bubble cluster in a sound field. Two bubbles surrounded by a ring of bubbles and a bubble tail.

driving sound field is 13 kHz. Slightly more than one cycle of the driving is covered by the 20 frames. The disappearance and reappearance of the complex filamentary structure is remarkable and points to a tightly coupled bubble system with mechanisms for synchronizing the bubbles' oscillations.

Synchronous oscillations of bubbles with the period of the driving sound field are not the only possible form of collective dynamics in acoustic cavitation. In fact, subharmonic response and the period-doubling route to chaos have also been observed experimentally [74, 75], as demonstrated subsequently.

B. Cavitation Noise

When acoustic cavitation sets in in a liquid, a hissing noise is heard. Obviously, this noise is generated by the superposition of a multitude of acoustic emissions from bubbles within the container and hence is expected to be a stochastic signal. It therefore came as a great surprise that the acoustic signal shows clear-cut dynamical features—for example, a strong line spectrum with subharmonics of the sound frequency [76]. This finding gives strong support to the hypothesis that the individual bubbles oscillate in the sound field in a largely synchronous fashion. Figure 57 shows a sequence

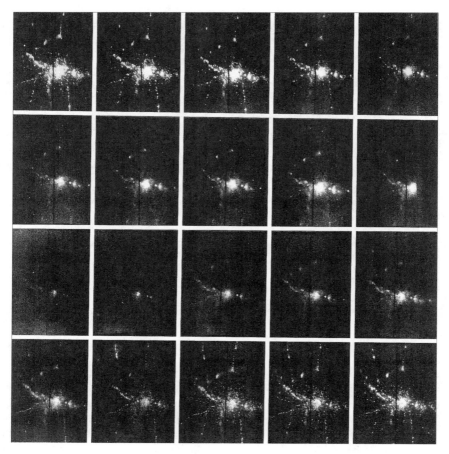

Figure 56. Forced oscillations of a filamentary structure of cavitation bubbles in water inside a cylindrical piezoelectric transducer driven at 13 kHz. The framing rate is 200,000 frames per second.

of spectra from a number of sections of a measured hydrophone signal. The signal was recorded with a cavitation cylinder being driven by a carefully adjusted, slowly amplitude-swept signal. The figure demonstrates the occurrence of periodic and then subharmonic (period-2) oscillation, followed by period-4 oscillation. This sequence is indicative of a period-doubling sequence commonly encountered in nonlinear dynamical systems when one of the system parameters is varied. It also abounds in the parameter space of single-bubble radial motion [77].

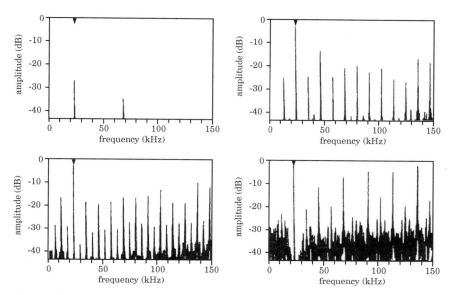

Figure 57. Power spectra of cavitation noise, taken at a driving frequency of $f_0 = 23\,\text{kHz}$, marked in the figures by a small triangle. The driving pressure was increased from one graph to the next, giving first a periodic, anharmonic signal (*top left*), then subharmonic motion with lines at multiples of $f_0/2$ (period-2 motion, *top right*) then period-4 motion with lines at multiples of $f_0/4$ (*bottom left*), and, finally, a broadband spectrum indicative of chaotic behavior (*bottom right*).

It is well-known that the observed bifurcation sequence constitutes one road to chaotic behavior. In fact, the power spectrum of the acoustic signal beyond the period-doubling regime has a broadband character that may well originate in deterministic-chaotic dynamics. (See the last graph of Figure 57.) Fourier spectra are not very helpful here for distinguishing true noise from chaos. Therefore, methods of nonlinear time-series analysis—for instance, attractor reconstruction by embedding, calculation of fractal dimensions, and Lyapunov spectra—have been employed to assess the deterministic origin of the signal [16]. For example, its phase-space reconstruction yields an irregular-looking trajectory with all features of a chaotic attractor.

These results demonstrate that much can be learned about acoustic cavitation from the measurement of even a single, averaged quantity. This is a very fortunate property of bubble ensembles that may be attributed to their tendency to oscillate coherently and thus to exhibit low-dimensional dynamics, at least on a coarse-grained scale.

C. Light Transmission

If the noise emission really is related to the dynamics of the bubbles, then another integral quantity—that is, the extinction of light passing through the bubble cloud—should reveal the same properties as the noise signal does. This has been checked in a light transmission experiment in conjunction with noise measurements. Figure 58 shows the experimental arrangement. The beam of a He–Ne laser is expanded and directed along the axis of a piezoceramic cylinder, where it penetrates the cavitation zone. The laser light is scattered, the amount of scattering being dependent on the bubble sizes. The transmitted light, which is collected by a lens and focused onto a photodiode, decreases in proportion to the total scattering cross section.

In Figure 59, the acoustic spectrum derived from the hydrophone signal is compared with the spectrum of the transmitted light signal for two measurements at a driving frequency of 23.1 kHz. Note that the qualitative features of both signals—e.g., the appearance of harmonics or subharmonics in the spectrum—are the same for the two types of signals. The sound emission thus is clearly correlated with the dynamics of the bubbles in the sound field.

D. High-Speed Holographic Cinematography

The observation and the visualization of the dynamics of three-dimensional motion is a difficult problem. The visualization problem is presently approached by virtual reality in many labs, with quickly improving results. The observation of real physical systems by measurement, however, has attracted much less attention and is intrinsically difficult to perform in real time. Should the optical measurements be done with a similar time

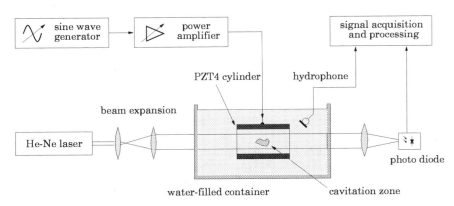

Figure 58. Arrangement for light transmission experiment with cavitation noise processing.

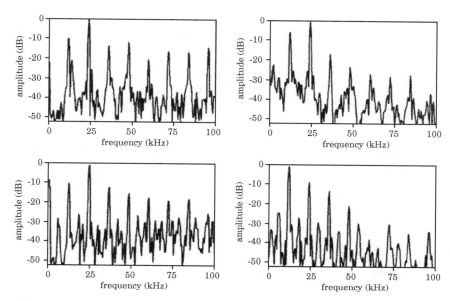

Figure 59. Comparison of the acoustic spectra of cavitation noise (*left*) and the power spectra of the light transmission signal obtained with the arrangement of Figure 58 for two different driving pressures at frequency $f_0 = 23.1$ kHz. *Top row:* Subharmonic response giving multiples of $f_0/2$. *Bottom row:* Response giving multiples of $f_0/4$.

resolution as the acoustic ones—for instance, at 1-MHz sampling rate—at least a few Tbyte-per-second data rate would be needed, as a single three-dimensional image contains about 1 Gbyte of voxels (volume picture elements). While this cannot yet be done in real time, high-speed holographic cinematography comes close to it off-line [73]. By taking holograms at a rate of up to 69,300 holograms per second, it could be shown that the whole filamentary structure is undergoing a doubling of its period simultaneously with the doubling of the period of the sound output [75]. The acoustic source has thus been traced back to bubble oscillations.

The potential of high-speed holographic methods is demonstrated in Figures 60 and 61, which contain reconstructed images from several high-speed holographic series [75]. The images were taken of a cavitation field in synchrony with the applied driving frequency, at different stages of a period-doubling sequence. The holographic series reveals that the bubble field oscillates as a whole in the periodic mode indicated by the acoustic spectrum. The spectra corresponding to the image rows of Figures 60 and 61 are reproduced in Figure 62. As far as the limited optical resolution admits, one can state that all bubbles seem to oscillate in synchrony, repeating their

Period 1

Period 2

Period 4

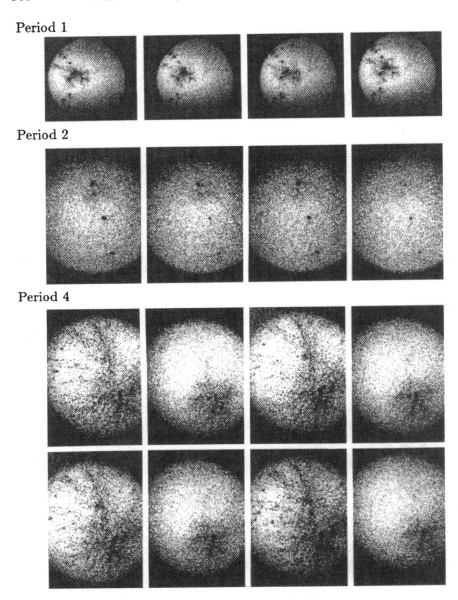

Figure 60. High-speed holographic images of a bubble field driven at 23.1 kHz. One hologram was taken per period of the driving. The bubble cloud oscillates as a whole periodically, with the indicated period.

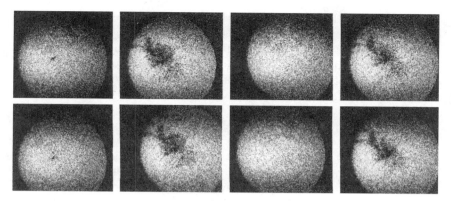

Figure 61. Continuation of the previous figure, showing period-8 motion of the bubble cloud.

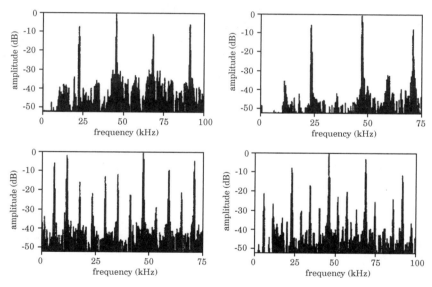

Figure 62. Acoustic spectra recorded simultaneously with the holographic series of Figures 60 and 61. The spectra belong to different stages in a period-doubling cascade. *Top left:* period 1, *top right:* period 2, *bottom left:* period 4, *bottom right:* period 8.

pattern after one driving period (the first row of Figure 60), after two driving periods (the second row), after four (the third and fourth rows), and after eight (Figure 61) periods, in accordance with the associated acoustic spectrum. High-speed holography has thus given us direct evidence of the coherence of bubble oscillations, which we have postulated previously from the results of cavitation noise analysis.

For these investigations, a cavity-dumped argon ion laser has been used. Due to the limited power per light pulse for taking a hologram, however, only holograms of a few square millimeters could be exposed sufficiently, giving unsatisfactory resolution in the reconstructed three-dimensional images.

Therefore, an improved arrangement has been developed that utilizes a dye laser pumped by a copper vapor laser for producing a series of high-power light pulses [78]. The device is capable of taking holograms at rates up to 20,000 holograms per second, with an area per hologram of up to 1 square centimeter and a total capacity of about 40 holograms or more, depending on the actual size chosen for a hologram. The arrangement for taking in-line holograms with a rotating holographic plate for spatial separation of the individual holograms is depicted in Figure 63. The copper vapor laser is capable of emitting a quasi-infinite series of short light pulses with a duration of about 15 ns and an energy of about 1 mJ each. The light consists of two spectral lines, green and yellow. The dichroic beam splitter selects the green line to pump the dye laser and dumps the yellow line into a beam stop. The electric shutter opens for the time of one revolution of the

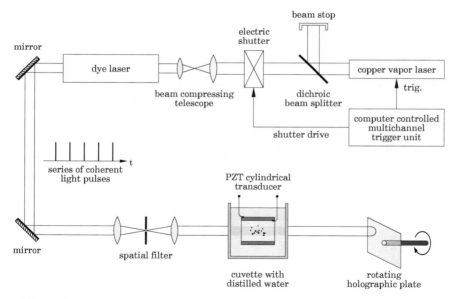

Figure 63. Arrangement for high-speed holographic cinematography with a copper vapor laser and a pumped-dye laser in an in-line geometry.

rotating holographic plate on which the series of holograms is recorded. The dye laser serves as a coherence transformer, boosting the coherence length from the few millimeters of the copper vapor laser light pulses up to about 7 cm. Simultaneously, the wavelength is changed from $\lambda = 510.6$ nm to $\lambda = 571$ nm (using Rhodamine 6G dye). The spatial filter improves the quality of the beam by removing interference fringes formed by diffraction on dust particles on the mirrors. An in-line holographic arrangement is used. In this geometry, the light passing undisturbed through the liquid serves as the reference beam for the light scattered at the bubbles. A series of 16 holographic images is shown in Figure 64, taken photographically from a series of in-line holograms recorded at 5,700 holograms per second. The driving sound field had a frequency of about 7 kHz in this case. The filaments to be seen are quite stable and are mainly made up of bubbles with a typical size of 50 μm. In-line holography has the disadvantage that large and small objects cannot be recorded simultaneously with sufficient accuracy in depth location, because differently sized objects demand a different optimum distance of the holographic plate from the objects. Therefore, off-axis holography has been employed, with diffuse illumination to spread the interference pattern all over the holographic plate. Figure 65 shows the arrangement for off-axis holographic cinematography. This arrangement is more involved than that of the in-line case, as a separate reference beam is needed and special precautions have to be taken to adjust the lengths of the reference and object beams to near equality. The two prisms serve as an adjustable delay line for that purpose. The two lenses in the 4f arrangement image the bubble cloud in front of the rotating holographic plate for improved recording via an adjustable distance of the cloud to the holographic plate. The images of the holograms have been subjected to three-dimensional image processing, described next.

E.   Digital Image Processing

A series of holograms contains an enormous amount of information that cannot be properly analyzed manually. Therefore, a digital image–processing system for three-dimensional images from holograms has been developed to handle the large amount of data [79]. The basic arrangement is depicted in Figure 66. The hologram is illuminated with the phase-conjugate reference beam for projecting the real image into space. An image dissector camera with $4,096 \times 4,096$ addressable pixels scans the three-dimensional image plane by plane, typically for 50 to 100 planes, depending on the depth of the investigated volume, and feeds the data into a computer system. A large software package has been written for detecting and locating bubbles in the three-dimensional real image. Figure 67 gives a three-dimensional view of the output of the holographic image-processing system for one

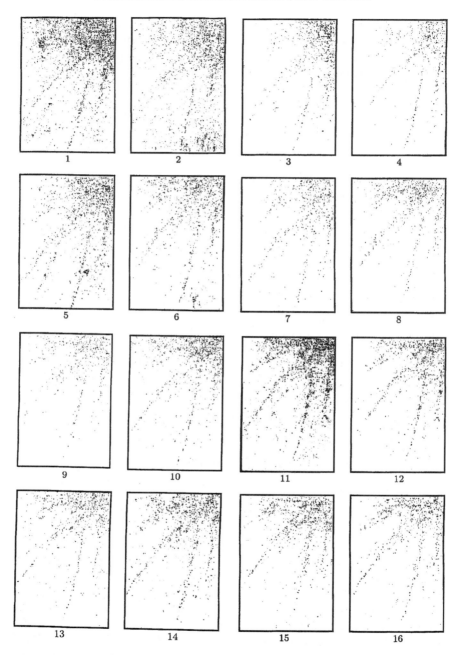

Figure 64. Reconstructed holographic images from a series of holograms taken at a rate of 5,700 holograms per second of a filament bubble pattern. (Courtesy of A. Judt.)

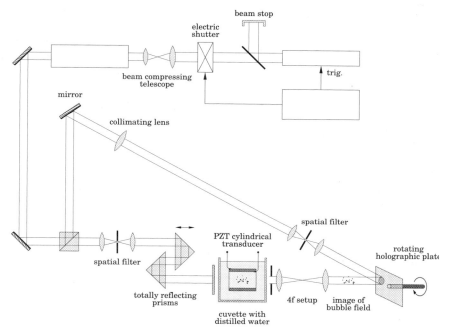

Figure 65. Arangement for high-speed holographic cinematography with a copper vapor laser and pumped-dye laser in an off-axis geometry.

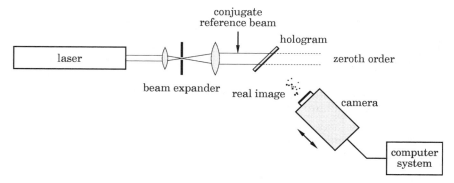

Figure 66. Arrangement for digital image processing from holograms, using the phase conjugate image, high-resolution image dissector camera, and parallel computing.

hologram of a series. The filaments are best seen in the z-projection. It is notoriously difficult to visualize three-dimensional point distributions. The solution, which, however, cannot be printed, is virtual reality, wherein the observer can look around in the distribution. As a first step toward

3D Holographic Image Reconstruction

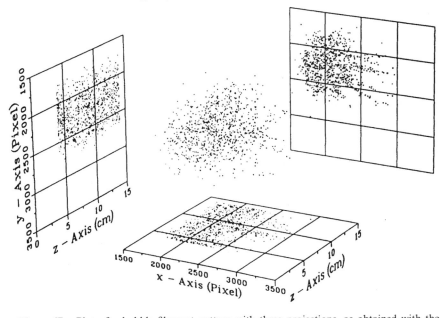

Figure 67. Plot of a bubble filament pattern with three projections, as obtained with the digital holographic image-processing system. (Courtesy of E. Schmitz.)

implementing a virtual-reality system for this purpose, we have stored a database of three-dimensional images of bubble distributions obtained from a holographic series. One of the images is given in Figure 68, with the bubbles plotted as spheres according to their size. The series can be viewed on a computer monitor with stereo capabilities and shutter eyewear. The holographic image-processing system is continuously improved for better performance, with the ultimate goal to be able to study space-time dynamical systems.

F.   Multibubble Sonoluminescence

We have seen before that, if they collapse sufficiently undisturbed, single bubbles emit light. Light emission, later termed sonoluminescence, was first discovered in multibubble systems in 1933. It was observed that photographic plates immersed in a water container are blackened by ultrasonic irradiation [80]. It soon became clear that the effect is due to light emission that accompanies acoustic cavitation [81].

Figure 68. Three-dimensional plot of a bubble distribution. (Courtesy of H. Chodura.)

In a standing field of plane acoustic waves, small bubbles accumulate at the location of the pressure antinodes through the action of primary Bjerknes forces. In a sufficiently intense field of sound, they can be set into such large vibrations that they emit light, thereby indicating the location of the antinodes. Figure 69 gives a historic example [82].

Figure 70 shows a photograph of multibubble sonoluminescence (MBSL) taken with modern equipment, a highly sensitive image-intensified CCD camera. The faint glow of a bubble cloud in a cavitation cylinder was observed, with no further illumination at a long exposure time to collect

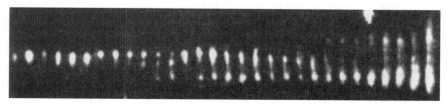

Figure 69. Sonoluminescence in a standing-wave field of 260 kHz in water with krypton added for brighter light emission. The exposure time on the photographic film was 15 minutes.

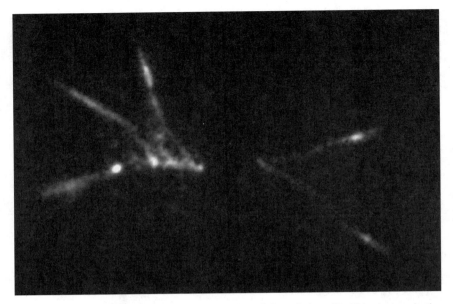

Figure 70. Long-time exposure revealing the faint light emission of bubbles in a cavitation cylinder. The image-intensified CCD picture shows that the spatial distribution of sonoluminescence is correlated with the bubble structure (streamers) in the sound field.

sufficient light. The SL emission reflects quite well the distribution of bubbles, obviously so stable, on the average, as to be visible in the picture.

After the discovery of sonoluminescence, subsequent studies in the following decades, at a moderate pace in this not-too-overpopulated branch of science then, revealed many properties of the phenomenon—in particular, the fact that the light emission is closely related to violent bubble collapse [83]. The result of an experiment demonstrating the phase relationship between light emission and driving pressure (thus, bubble collapse), conducted with a gated ICCD camera, is presented in Figure 71. It is seen that in a multibubble field, the single luminescence events are not distributed evenly over the acoustic cycle, but all take place within a narrow phase window of the driving period. For multibubble sonoluminescence, the influence of parameters such as the composition and temperature of the liquid, its gas content, acoustic pressure, dissolved substances, etc., was thoroughly studied, as were the spectral features of the light. For a detailed account of the development of the subject and of experimental facts, the reader is referred to the literature, in particular to the review by Walton and Reynolds [84]. In comparing the experimental findings on single- and multi-

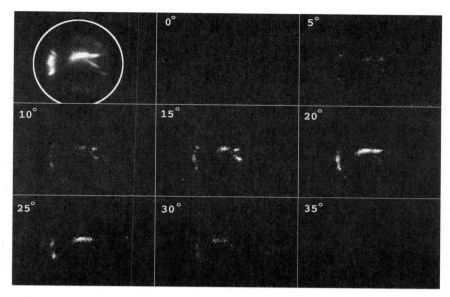

Figure 71. The light emission from a bubble ensemble takes places in a narrowly defined phase window with respect to the driving signal. The images are ICCD images gated a large number of times at the indicated phase (the zero phase being chosen arbitarily).The leftmost top picture gives the integrated view of the luminescing bubble field with an outline of the cavitation cylinder.

bubble sonoluminescence, certain differences between the two phenomena become evident. For example, spectroscopic measurements have revealed that MBSL can be assigned a temperature of about 5,000 to 6,000 K [85, 86], obviously smaller than the value found for SBSL. Furthermore, unlike SBSL spectra, the MBSL light features spectral lines and band emission, e.g., by OH^- radicals or from dissolved salt [87]. On the other hand, both phenomena share a number of properties—for example, the strong influence of noble gases on light intensity.

Thus, the question arises, how are single-bubble and multibubble sonoluminescence related? This important subject is still under debate. It is closely connected with the question of the role of spherical symmetry in bubble luminescence. Crum [88] has put forward an interesting conjecture about the difference between MBSL and SBSL. He suggests that SBSL is intrinsically different from MBSL in that SBSL is fueled by an interior shock wave of a symmetrically collapsing bubble, while MBSL is due to incandescence of the host liquid by the liquid jet generated during asymmetric collapse. On the other hand, there are theories like that of Prosperetti [89] which invoke asymmetric bubble collapse also in the case of

SBSL. The results on single-bubble dynamics collected in this paper provide strong evidence for a certain degree of spherical symmetry of the collapsing bubble as a prerequisite for light generation.

While laser-generated bubbles are different from stably levitated bubbles in some respects, we are confident that a further exploration of cavitation bubble luminescence in optical cavitation can shed light also on the sonoluminescence phenomenon. After all, in both cases it is a collapsing bubble that produces the light, and the findings that are presented are consistent with the presence of converging shock waves within the bubble. With single laser-generated bubbles near a boundary or with few-bubble systems, it will be possible to simulate the environment of MBSL locally and to study the light emission there with high temporal and spatial resolution. Furthermore, with laser-generated bubbles, we are not restricted to the parameter space accessible to small bubbles in diffusive equilibrium and thus may achieve considerably more light by a more violent collapse. This facilitates spectroscopic studies, for example.

Thus, a wealth of experiments lies ahead in optical cavitation and bubble luminescence. With high-speed and ultrafast optical instruments, we hopefully will gather new insights into the fascinating, but complex, world of bubbles.

X. THEORY OF MANY-BUBBLE SYSTEMS

Bubbles in a liquid can be viewed as individual interacting objects. A description of their dynamics then proceeds as an extension of single-bubble dynamics with the inclusion of interaction terms. This idea is a natural extension if one is interested in the motion of a few bubbles only. But it may even be extended to a system of many bubbles, as is encountered in acoustic cavitation.

To arrive at an appropriate theory of this particular, new type of many-body problem, the different forces that bubbles are subject to have to be found and formulated. To begin with, the following forces acting on each bubble are considered: an added-mass force $\mathbf{F}_M$, a primary Bjerknes force $\mathbf{F}_{B1}$, a secondary Bjerknes force $\mathbf{F}_{B2}$, and a drag force $\mathbf{F}_D$. The added-mass force may need some explanation. A moving object in a liquid possesses inertia not only according to its mass, but also because of the liquid set into motion. This inertia appears as an *added mass* attached to the object. For a rigid, spherical object, it has the amount of half the mass of the displaced liquid [1]. For a bubble, this is by far the main inertia with respect to the inertia of the gas content, so that the gas part usually is neglected. Further, we assume a stationary, nonstreaming liquid in a rectangular resonator containing a standing wave $p_a(\mathbf{x}; t) = P_a(\mathbf{x}) \cos(\omega t)$. The model is restricted to spherical bubbles of a common equilibrium size R_n. However, we allow for strongly nonlinear radial bubble oscillations, which is an essential point. The time-varying radii $R(t)$ are computed by the Keller–Miksis model given in Eqs. (19) and (20), for the local driving pressure at the bubble's position. We suppose slowly moving bubbles; that is, the bubbles do not encounter different sound field amplitudes during one radial oscillation period (which is assumed to be equal to the sound field oscillation period $T = 2\pi/\omega$ for all bubbles). Then, the forces are determined as follows, involving time averaging over T:

$$\mathbf{F}^i_M = \frac{\rho}{2} \langle V_i(t) \rangle_T \dot{\mathbf{v}}_i, \tag{27}$$

$$\mathbf{F}^i_{B1} = - \langle \nabla p_a(\mathbf{x}_i; t) V_i(t) \rangle_T, \tag{28}$$

$$\mathbf{F}^i_{B2} = \sum_{j \neq i} \frac{\rho}{4\pi} \langle \dot{V}_i(t) \dot{V}_j(t) \rangle_T \frac{\mathbf{d}_{ij}}{|\mathbf{d}_{ij}|^3} \approx f^i_{B2} \sum_{j \neq i} \frac{\mathbf{d}_{ij}}{|\mathbf{d}_{ij}|^3}, \quad f^i_{B2} = \frac{\rho}{4\pi} \langle \dot{V}_i^2(t) \rangle_T, \tag{29}$$

$$\mathbf{F}^i_D = -(\beta_1 \langle R(t) \rangle_T + \beta_2 \langle R(t) \rangle_T^2 |\mathbf{v}_i|) \mathbf{v}_i . \tag{30}$$

Here, i indexes the bubbles with positions $\mathbf{x}_i$, velocities $\mathbf{v}_i$, and volumes V_i, and $\mathbf{d}_{ij} = \mathbf{x}_j - \mathbf{x}_i$ is the vector from bubble i to bubble j. The drag force $\mathbf{F}_D$ is fitted to an experimentally based formula from Crum [58], with $\beta_1 = 0.015 \, \text{Ns/m}^2$ and $\beta_2 = 4,000 \, \text{Ns}^2/\text{m}^3$. The equations of motion $\mathbf{F}^i_M = \mathbf{F}^i_{B1} + \mathbf{F}^i_{B2} + \mathbf{F}^i_D$ are solved by a semi-implicit Euler method for N bubbles.

According to the standing pressure wave in the container, the driving amplitude varies in space. Due to this sound field variation, $R(t)$ and the resulting forces can change dramatically when a bubble moves to a different position. To keep the computations simple and fast, we introduced the approximation of equal bubble volumes for the summation of $\mathbf{F}_{B2}$ in Eq. (29). Additionally, the time-averaged values in Eqs. (27)–(30) are tabulated on a grid in space, and linear interpolation is used for points lying in between. Figure 72 illustrates the strong quantitative and even qualitative variation of the primary and secondary Bjerknes forces with increasing pressure amplitude. The calculations have been done for a cubic resonator (with edge length $a = 6 \, \text{cm}$ and $\omega = 2\pi \cdot 21.66 \, \text{kHz}$, according to the (111) mode) and the fixed bubble size of $R_n = 5 \, \mu\text{m}$. The first component of the primary Bjerknes force, $F_{B1,1}$, shows increasing attraction (negative values) towards the pressure antinode (the origin) for an increasing driving pressure, up to 160 kPa. (The negative values associated with 100 kPa are very close to zero in this scaling.) The sign of the force near the origin changes, however,

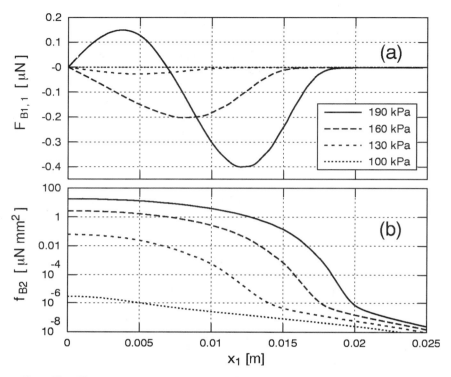

Figure 72. First component $F_{B1,1}$ of the primary Bjerknes force (a) and secondary Bjerknes force coefficient f_{B2} (b) vs. the first coordinate x_1 in a cubic resonator. (See text; $x_2 = x_3 = 0$ in this picture.) The maximum pressure amplitudes P_a range from 100 to 190 kPa. Positive values of $F_{B1,1}$ indicate repulsion from the antinode. Note the logarithmic scaling in (b).

when the amplitude is further increased up to 190 kPa: The antinode becomes repulsive for the bubble size considered. Since the force is still attractive in the outer regions of the standing wave, a stable equilibrium surface forms around the antinode. This evolution is accompanied by an increase of the secondary Bjerknes forces by orders of magnitude.

In the model, the creation of bubbles takes place near some randomly chosen sites outside the center. This is similar to the experimental observation of where the bubbles occur. Coalescence is modeled by a certain chance of annihilation after each time step if another bubble is located closer than $2\langle R(t)\rangle_T$. If a bubble vanishes, a new one appears at a creation site. Thus, the total number N of bubbles is kept constant here.

In Figure 73, we compare typical results from the model with structures obtained experimentally. The left column of pictures corresponds to a central pressure amplitude of $P_a = 130$ kPa, the right column to

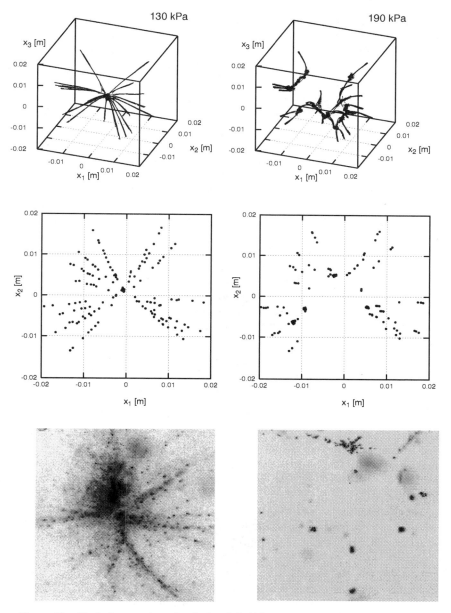

Figure 73. Typical examples of calculated bubble traces and snapshots opposed to experimental structures. Left column: medium pressure amplitude (130 kPa); right column: high pressure amplitude (190 kPa). Top row: simulated bubble tracks (150 bubbles, originating from 20 fixed creation points). Middle row: snapshots from the preceding simulations. Bottom row: experimental photographs at approximately the same conditions and dimensions as in the simulations. (Photos courtesy of S. Luther.)

$P_a = 190 \, \text{kPa}$. The upper pictures show simulated bubble tracks in three dimensions, and the middle row depicts snapshots from the model projected onto two dimensions. The lower pictures present snapshots from the experiment, which have been inverted with respect to black and white for better visibility. The parameters of the simulation correspond to Figure 72 and strongly resemble the experimental cubic resonator arrangement. Only the pressure amplitudes in the experiment might have differed slightly from the indicated values, and the cuvette has been chosen a little larger in the simulation in order to come close to the experimentally observed resonance frequency (in the range of 21 kHz). The simulated bubbles ($N = 150$, $R_0 = 5 \, \mu\text{m}$) originate at 20 fixed creation sites at a 2-cm distance from the center (the antinode). The bubble traces in the upper row of the figure cover a total simulation time of 0.4 s (130 kPa, left) and 0.1 s (190 kPa, right), respectively. These periods of time seem short enough to justify nonmoving creation sites.

For the lower pressure, all simulated bubbles move more or less straight to the center at velocities not exceeding 0.1 m/s. The geometry corresponds to the impression from the experimental snapshot, although the nebulous center cannot yet be captured by simulation.

For the increased pressure, the situation has changed. At 190 kPa, bubbles move faster (up to 0.5 m/s) and cluster off the center—the antinode is void, and the creation sites appear interconnected by shortcuts. The experiment indeed shows a corresponding transition for increasing driving from one central to many fast-drifting, non-central smaller clusters. This scenario is difficult to image experimentally, but several small bubble clusters can be recognized as darker spots on the experimental snapshot (Figure 73, bottom right). Two effects contribute to this transition phenomenon: The antinode becomes repulsive with increasing pressure (Figure 72(a)), and the secondary Bjerknes force increases by several orders of magnitude (Figure 72(b)). Therefore, the inward-moving bubbles attract each other earlier and cluster near the stable equilibrium surface around the antinode.

It is interesting to note that, if many bubbles were placed in the center, where f_{B2} is largest, their mutual attraction could outrange the primary Bjerknes repulsion and lead to a bistability of a central and off-central clustering structure. Indeed, an intermittency on a time scale of seconds between a populated and a void center can be observed in our resonator experiments for certain parameters (without changing the pressure from the outside). As mechanisms for transitions between multistable patterns, we suspect the influence of the bubble distribution on the sound field, leading to effective pressure amplitude changes by impedance mismatch, or streaming of the liquid in the container.

The particle model with a few fixed bubble sources apparently can mimic the emergence of different types of structures in an acoustic resonator standing-wave experiment, as the comparison of simulated and experimental patterns in Figure 73 suggests. Furthermore, we have found that the simulated pattern of bubble tracks does not vary much within a certain range of the bubble quantity N if a sparse number of creation sites is used that stays fixed in time. (A pattern transition might be modeled by variable creation sites, which is beyond the scope of this article.) Therefore, one gets a good impression of the formation of structure just by looking at the bubble traces, although the full process is spatiotemporal and thus demands a representation by a three dimensional (holographic) movie.

We tried to use observed or estimated real-world mechanisms, dimensions, and quantities in the particle model wherever possible, but this approach is still rather a cartoon than a one-to-one reproduction of the formation of structure in acoustic cavitation. Variations in the bubble equilibrium size, shape oscillations and the shedding of microbubbles, liquid streaming, and the exact treatment of very close bubbles are still not included (e.g., we have modified the secondary Bjerknes force law for very near distances in a heuristic manner to avoid unnatural divergencies in the simulation). In its simple form, however, the model can already reproduce gross features of different types of pattern. From our particle model studies, we therefore draw the main conclusion that a strong spatial variation of Bjerknes forces, according to nonlinear spherical bubble oscillations in a standing pressure wave, is consistent with experimental observations. This holds both qualitatively (the types of emerging patterns) and quantitatively (the involved time scales and pressure amplitude values) and thus gives an indication of a correct treatment of the Bjerknes forces of strongly oscillating bubbles.

XI. SUMMARY

Bubbles in liquids develop peculiar properties when set into motion. Starting from single, spherical bubbles, the dynamics of aspherical bubbles, two and few bubbles, and, finally, bubble clouds is reviewed. Empty or almost empty spherical bubbles collapse at speeds not yet finally measured. On collapse, the bubbles' contents is compressed so strongly that shock waves and short flashes of light are emitted. Single bubbles can be trapped in standing sound fields emitting shock waves and light every period of the sound field. Photographic series of a sonoluminescing bubble and its emitted shock wave are given. Very flexible experiments with bubbles are possible with focused laser light. High-speed photography reveals many details of the dynamics of bubble interaction, be it with a solid surface, or another bubble, or both.

Fundamentally, a high-speed liquid jet is formed in all cases of spherical asymmetry, and the spherical case itself often is unstable. Bubble clouds have been visualised via high-speed cinematography, holography, and also holographic cinematography. In high-pressure sound fields, bubbles form filamentary structures that oscillate periodically with the sound field or also chaotically via a period doubling route to chaos. A theoretical description of the dynamics of a single, spherical bubble in a sound field is given, including response curves and stability diagrams for surface oscillations, rectified diffusion, and Bjerknes forces. A theory of many-bubble systems is formulated as a new type of many-body problem. Numerical results reproduce the formation of filaments in bubbles in a sound field.

Acknowledgments

We thank the Deutsche Forschungsgemeinschaft, Bonn, for long and continued support, most recently via the Graduiertenkolleg "Strömungsinstabilitäten und Turbulenz." We also have the pleasure of thanking U. Parlitz and our numerous coworkers who helped to gather the information presented here—in particular, R. Geisler, O. Lindau, S. Luther, and B. Metten. Thanks also go to the Fraunhofer Gesellschaft for the loan of equipment.

References

[1] C. E. Brennen, *Cavitation and Bubble Dynamics*, Oxford University Press, Oxford, U.K., 1995.

[2] T. G. Leighton, *The Acoustic Bubble*, Academic Press, London, 1994.

[3] J. R. Blake, J. M. Boulton-Stone, and N. H. Thomas (Eds.), *Bubble Dynamics and Interface Phenomena*, Kluwer, Dordrecht, The Netherlands, 1994.

[4] F. R. Young, *Cavitation*, McGraw-Hill, London, 1989.

[5] K. S. Suslick (Ed.), *Ultrasound: Its Chemical, Physical and Biological Effects*, VCH Publishers, New York, 1988.

[6] L. van Wijngaarden (Ed.), *Mechanics and Physics of Bubbles in Liquids*, Martinus Nijhoff Publishers, The Hague, 1982.

[7] W. Lauterborn (Ed.), *Cavitation and Inhomogeneities in Underwater Acoustics*, Springer, Berlin, 1980.

[8] E. A. Neppiras, "Acoustic Cavitation," *Phys. Rep.* **61**, 159–251 (1980).

[9] H. G. Flynn, "Physics of Acoustic Cavitation," in *Physical Acoustics*, W. P. Mason, Ed., Academic Press, New York, 1964, pp. 57–172.

[10] Lord Rayleigh, "On the Pressure Developed in a Liquid during the Collapse of a Spherical Cavity," *Phil. Mag., Ser. 6* **34**, 94–98 (1917).

[11] W. Lauterborn, "Kavitation durch Laserlicht," *Acustica* **31**, 51–78 (1974).

[12] D. F. Gaitan, L. A. Crum, C. C. Church, and R. A. Roy, "An Experimental Investigation of Acoustic Cavitation and Sonoluminescence from a Single Bubble," *J. Acoust. Soc. Am.* **91**, 3166–3183 (1992).

[13] B. Gompf, R. Günther, G. Nick, R. Pecha, and W. Eisenmenger, "Resolving Sonoluminescence Pulse Width with Time-Correlated Single Photon Counting," *Phys. Rev. Lett.* **79**, 1405–1408 (1997).

[14] N. Dezhkunov, G. Iernetti, A. Francescutto, M. Reali, and P. Ciuti, "Cavitation Erosion and Sonoluminescence at High Hydrostatic Pressures," *Acustica* **83**, 19–24 (1997).

[15] L. Crum, "Sonoluminescence," *Physics Today* **47**, No. 9, 22–29 (1994).

[16] W. Lauterborn and J. Holzfuss, "Acoustic Chaos," *Int. J. Bifurcation and Chaos* **1**, 13–26 (1991).

[17] W. Lauterborn and H. Bolle, "Experimental Investigations of Cavitation-Bubble Collapse in the Neighbourhood of a Solid Boundary," *J. Fluid Mech.* **72**, 391–399 (1975).

[18] A. Vogel, W. Lauterborn, and R. Timm, "Optical and Acoustic Investigations of the Dynamics of Laser-Produced Cavitation Bubbles near a Solid Boundary," *J. Fluid Mech.* **206**, 299–338 (1989).

[19] Y. Tomita and A. Shima, "High-Speed Photographic Observation Of Laser-Induced Cavitation Bubbles in Water," *Acustica* **71**, 161–171 (1990).

[20] L. A. Crum, "Nucleation and Stabilization of Microbubbles in Liquids," *Appl. Sci. Res.* **38**, 101–115 (1982).

[21] D. E. Yount, E. W. Gillary, and D. C. Hoffmann, "A Microscopic Investigation of Bubble Formation Nuclei," *J. Acoust. Soc. Am.* **76**, 1511–1521 (1984).

[22] L. A. Crum, "Rectified Diffusion," *Ultrasonics* **22**, 215–223 (1984).

[23] L. A. Crum and G. T. Reynolds, "Sonoluminescence Produced by "Stable" Cavitation," *J. Acoust. Soc. Am.* **78**, 137–139 (1985).

[24] B. B. Barber, C. C. Wu, R. Löfstedt, P. H. Roberts, and S. J. Putterman, "Sensitivity of Sonoluminescence to Experimental Parameters," *Phys. Rev. Lett.* **72**, 1380–1383 (1994).

[25] R. Hiller, S. J. Putterman, and B. P. Barber, "Spectrum of Synchronous Picosecond Sonoluminescence," *Phys. Rev. Lett.* **69**, 1182–1184 (1992).

[26] K. R. Weninger, B. P. Barber, and S. J. Putterman, "Pulsed Mie Scattering Measurements of the Collapse of a Sonoluminescing Bubble," *Phys. Rev. Lett.* **78**, 1799–1802 (1997).

[27] Y. Tian, J. A. Ketterling, and R. E. Apfel, "Direct Observation of Microbubble Oscillations," *J. Acoust. Soc. Am.* **100**, 3976–3978 (1996).

[28] T. R. Stottlemyer and R. E. Apfel, "The Effects of Surfactant Additivesion the Acoustic and Light Emissions from a Single Stable Sonoluminescing Bubble," *J. Acoust. Soc. Am.* **102**, 1418–1423 (1997).

[29] R. G. Holt and D. F. Gaitan, "Observation of Stability Boundaries in the Parameter Space of Single Bubble Sonoluminescence," *Phys. Rev. Lett.* **77**, 3791–3794 (1996).

[30] B. P. Barber, R. A. Hiller, R. Löfstedt, S. J. Putterman, and K. R. Weninger, "Defining the Unknowns of Sonoluminescence," *Phys. Rep.* **281**, 65–143 (1997).

[31] E. Heim, "Über das Zustandekommen der Sonolumineszenz," in *Proc. Third Int. Congr. on Acoustics, Stuttgart, 1959* (ed. L. Cremer), pp. 343–346, Elsevier, Amsterdam, 1961.

[32] C. C. Wu and P. H. Roberts, "A Model of Sonoluminescence," *Proc. Roy. Soc. Lond.* **A445**, 323–349 (1994).

[33] W. C. Moss, D. B. Clarke, and D. A. Young, "Calculated Pulse Widths and Spectra of a Single Sonoluminescing Bubble," *Science* **276**, 1398–1401 (1997).

[34] B. E. Noltingk and E. A. Neppiras, "Cavitation Produced by Ultrasonics," *Proc. Phys. Soc. Lond.* **B63**, 674–685 (1950).

[35] H. Poritsky, "The Collapse or Growth of a Spherical Bubble or Cavity in a Viscous Fluid," in *Proc. First U. S. Nat. Congr. Appl. Mech., New York, 1952* (ed. E. Sternberg), pp. 813–821.

[36] M. S. Plesset and A. Prosperetti, "Bubble Dynamics and Cavitation," *Ann. Rev. Fluid Mech.* **9**, 145–185 (1977).

[37] W. Lauterborn, "Investigation of Nonlinear Oscillations of Gas Bubbles in Liquids," *J. Acoust. Soc. Am.* **59**, 283–293 (1976).

[38] F. R. Gilmore, "The Growth or Collapse of a Spherical Bubble in a Viscous Compressible Liquid," Report No. 26-4, Hydrodynamics Laboratory, California Institute of Technology, Pasadena, California, USA (1952).

[39] R. Löfstedt, B. P. Barber, and S. J. Putterman, "Toward a Hydrodynamic Theory of Sonoluminescence," *Phys. Fluids* A**5**, 2911–2928 (1993).

[40] J. B. Keller and M. Miksis, "Bubble Oscillations of Large Amplitude," *J. Acoust. Soc. Am.* **68**, 628–633 (1980).

[41] W. Lauterborn and U. Parlitz, "Methods of Chaos Physics and Their Applications to Acoustics," *J. Acoust. Soc. Am.* **84**, 1975–1993 (1988).

[42] H. Kuttruff and U. Radek, "Messungen des Druckverlaufs in kavitationserzeugten Druckimpulsen," *Acustica* **21**, 253–259 (1969).

[43] A. Vogel and W. Lauterborn, "Acoustic Transient Generation by Laser-Produced Cavitation Bubbles," *J. Acoust. Soc. Am.* **84**, 719–731 (1988).

[44] T. J. Matula, I. M. Hallaj, R. O. Cleveland, L. A. Crum, W. C. Moss, and R. A. Roy, "The Acoustic Emissions from Single-Bubble Sonoluminescence," *J. Acoust. Soc. Am.* **103**, 1377–1382 (1998).

[45] D. C. Rapaport, *The Art of Molecular Dynamics Simulation*, Cambridge University Press, Cambridge, U.K., 1995.

[46] T. Vladimiroff, Y. P. Carignan, A. K. Macpherson, and P. A. Macpherson, "The Dynamics of Hard Spheres in a Collapsing Spherical Container," *Molecular Phys.* **71**, 441–451 (1990).

[47] M. Kornfeld and L. Suvorov, "On the Destructive Action of Cavitation," *J. Appl. Phys.* **15**, 495–506 (1944).

[48] M. S. Plesset, "On the Stability of Fluid Flows with Spherical Symmetry," *J. Appl. Phys.* **25**, 96–98 (1954).

[49] H. W. Strube, "Numerische Untersuchungen zur Stabilität nichtsphärisch schwingender Blasen," *Acustica* **25**, 289–303 (1971).

[50] A. Prosperetti, "Viscous Effects on Perturbed Spherical Flow," *Quart. Appl. Math.* **34**, 339–352 (1977).

[51] S. Hilgenfeldt, D. Lohse, and M. P. Brenner, "Phase Diagrams for Sonoluminescing Bubbles," *Phys. Fluids* **8**, 2808–2826 (1996).

[52] D.-Y. Hsieh and M. S. Plesset, "Theory of Rectified Diffusion of Mass into Gas Bubbles," *J. Acoust. Soc. Am.* **33**, 206–215 (1961).

[53] A. Eller and H. G. Flynn, "Rectified Diffusion during Nonlinear Pulsations of Cavitation Bubbles," *J. Acoust. Soc. Am.* **37**, 493–503 (1965).

[54] L. A. Crum, "Measurements of the Growth of Air Bubbles by Rectified Diffusion," *J. Acoust. Soc Am.* **68**, 203–211 (1980).

[55] M. M. Fyrillas and A. J. Szeri, "Dissolution or Growth of Oscillating Bubbles," *J. Fluid Mech.* **277**, 381–407 (1994).

[56] M. P. Brenner, D. Lohse, D. Oxtoby, and T. F. Dupont, "Mechanisms for Stable Single Bubble Sonoluminescence," *Phys. Rev. Lett.* **76**, 1158–1161 (1996).

[57] I. Akhatov, N. Gumerov, C. D. Ohl, U. Parlitz, and W. Lauterborn, "The Role of Surface Tension in Stable Single-Bubble Sonoluminescence," *Phys. Rev. Lett.* **78**, 227–230 (1997).

[58] L. A. Crum, "Bjerknes Forces on Bubbles in a Stationary Sound Field," *J. Acoust. Soc. Am.* **57**, 1363–1370 (1975).

[59] T. J. Matula, A. M. Cordry, R. A. Roy, and L. A. Crum, "Bjerknes Force and Bubble Levitation under Single-Bubble Sonoluminescence Conditions," *J. Acoust. Soc. Am.* **102**, 1522–1527 (1997).

[60] I. Akhatov, R. Mettin, C. D. Ohl, U. Parlitz, and W. Lauterborn, "Bjerknes Force Threshold for Stable Single Bubble Sonoluminescence," *Phys. Rev. E***55**, 3747–3750 (1997).

[61] H. N. Oguz and A. Prosperetti, "A Generalization of the Impulse and Virial Theorems with an Application to Bubble Oscillations," *J. Fluid Mech.* **218**, 143–162 (1990).

[62] R. Mettin, I. Akhatov, U. Parlitz, C. D. Ohl, and W. Lauterborn, "Bjerknes Forces between Small Cavitation Bubbles in a Strong Acoustic Field," *Phys. Rev. E***56**, 2924–2931 (1997).

[63] N. A. Pelekasis and J. A. Tsamopoulos, "Bjerknes Forces between Two Bubbles: Part 2. Response to an Oscillatory Pressure Field," *J. Fluid Mech.* **254**, 501–527 (1993).

[64] C. D. Ohl, A. Philipp, and W. Lauterborn, "Cavitation Bubble Collapse Studied at 20 Million Frames per Second," *Ann. Phys.* **4**, 26–34 (1995).

[65] C. D. Ohl, O. Lindau, and W. Lauterborn, "Luminescence from Spherically and Aspherically Collapsing Laser Induced Bubbles," *Phys. Rev. Lett.* **80**, 393–396 (1998).

[66] T. B. Benjamin and A. T. Ellis, "The Collapse of Cavitation Bubbles and the Pressure Thereby Produced against Solid Boundaries," *Phil. Trans. Roy. Soc. Lond. A***260**, 221–240 (1966).

[67] J. R. Blake and D. C. Gibson, "Cavitation Bubbles near Boundaries," *Ann. Rev. Fluid Mech.* **19**, 99–123 (1987).

[68] Y. Tomita and A. Shima, "Mechanism of Impulsive Pressure Generation and Damage Pit Formation by Bubble Collapse," *J. Fluid Mech.* **169**, 535–564 (1986).

[69] N. K. Bourne and J. E. Field, "A High-Speed Photographic Study of Cavitation Damage," *J. Appl. Phys.* **78**, 4423–4427 (1995).

[70] A. Philipp and W. Lauterborn, "Cavitation Erosion by Single Laser-Produced Bubbles," *J. Fluid Mech.* **361**, 75–116 (1998).

[71] J. R. Blake, P. B. Robinson, A. Shima, and Y. Tomita, "Interaction of Two Cavitation Bubbles with a Rigid Boundary," *J. Fluid Mech.* **255**, 707–721 (1993).

[72] K. Jungnickel and A. Vogel, "Interaction of Two Laser-Induced Cavitation Bubbles," in J. R. Blake, J. M. Boulton-Stone, and N. H. Thomas (Eds.), *BubbleDynamics and Interface Phenomena*, Kluwer, Dordrecht, The Netherlands, 1994, pp. 47–53.

[73] W. Hentschel and W. Lauterborn, "High Speed Holographic Movie Camera," *Opt. Eng.* **24**, 687–691 (1985).

[74] W. Lauterborn and E. Cramer, "Subharmonic Route to Chaos Observed in Acoustics," *Phys. Rev. Lett.* **47**, 1445–1448 (1981).

[75] W. Lauterborn and A. Koch, "Holographic Observation of Period-Doubled and Chaotic Bubble Oscillations in Acoustic Cavitation," *Phys. Rev. A***35**, 1974–1976 (1987).

[76] R. Esche, "Untersuchung der Schwingungskavitation in Flüssigkeiten," *Acustica* **2**, AB208–AB218 (1952).

[77] U. Parlitz, V. Englisch, C. Scheffczyk, and W. Lauterborn, "Bifurcation Structure of Bubble Oscillators," *J. Acoust. Soc. Am.* **88**, 1061–1077 (1990).

[78] W. Lauterborn, A. Judt, and E. Schmitz, "High-Speed Off-Axis Holographic Cinematography with a Copper-Vapor Pumped Dye Laser," *Opt. Lett.* **18**, 4–6 (1993).

[79] G. Haussmann and W. Lauterborn, "Determination of Size and Position of Fast Moving Gas Bubbles in Liquids by Digital 3-D Image Processing of Hologram Reconstructions," *Appl. Opt* **19**, 3529–3535 (1980).

[80] M. Marinesco and J. J. Trillat, "Action Des Ultrasons Sur Les Plaques Photographiques," *Compt. Rend.* **196**, 858–860 (1933).

[81] H. Frenzel and H. Schultes, "Luminescenz im ultraschallbeschickten Wasser," *Z. Phys. Chem.* **B27**, 421–424 (1934).

[82] W.-U. Wagner, "Phasenkorrelation von Schalldruck und Sonolumineszenz," *Z. Angew. Phys.* **10**, 445–452 (1958).

[83] E. Meyer and H. Kuttruff, "Zur Phasenbeziehung zwischen Sonolumineszenz und Kavitationsvorgang bei periodischer Anregung," *Z. Angew. Phys.* **11**, 325–333 (1959).

[84] A. J. Walton and G. T. Reynolds, "Sonoluminescence," *Adv. Phys.* **33**, 595–660 (1984).

[85] P. Günther, E. Heim, and H. U. Borgstedt, "Über die kontinuierlichen Sonolumineszenzspektren wäßriger Lösungen," *Z. Elektrochemie* **63**, 43–47 (1959).

[86] E. B. Flint and K. S. Suslick, "The Temperature of Cavitation," *Science*, **253**, 1397–1399 (1991).

[87] T. J. Matula, R. A. Roy, P. D. Mourad, W. B. McNamara III, K. S. Suslick, "Comparison of Multibubble and Single-Bubble Sonoluminescence Spectra," *Phys. Rev. Lett.* **75**, 2602–2605 (1995).

[88] L. A. Crum, "Sonoluminescence, Sonochemistry, and Sonophysics," *J. Acoust. Soc. Am.* **95**, 559–562 (1994).

[89] A. Prosperetti, "A New Mechanism for Sonoluminescence," *J. Acoust. Soc. Am.* **101**, 2003–2007 (1997).

CHAPTER 6

ACID–BASE PROTON TRANSFER AND ION PAIR FORMATION IN SOLUTION

KOJI ANDO

Institute of Materials Science, University of Tsukuba, Tsukuba, Ibaraki 305-8573, Japan

JAMES T. HYNES

Department of Chemistry and Biochemistry, University of Colorado, Boulder, Colorado 80309-0215, USA

CONTENTS

I. INTRODUCTION

The theoretical and experimental investigation of chemical reactions in solution [1, 2] has been the focus of an enormous effort by many workers in the past several decades via a range of modern ideas, methodologies, and techniques, and much knowledge about, and insight into, the rates and mechanisms of various reactions have been gained as a result. Yet this is not

Advances in Chemical Physics, Volume 110, Edited by I. Prigogine and Stuart A. Rice.
ISBN 0-471-33180-5. © 1999 John Wiley & Sons, Inc.

at all to say that everything is now understood or is predictable, either in an
a priori fashion or even in an empirical one. Among the central elementary
reaction classes of interest and importance to chemists and biochemists such
that one can fairly say that much remains to be clarified, proton-transfer
(PT) reactions [3–7] can be singled out. And among proton-transfer
reactions, one can further single out the acid ionization

$$HA(aq) \rightleftharpoons A^-(aq) + H_3O^+(aq) \tag{1}$$

in aqueous solution of the acid AH, where water acts both as the proton-
accepting base and the solvent medium. These ionizations are of obvious
and traditional importance per se, as well as in connection with acid–base
catalysis [5–7]. Indeed, the generalization of Eq. (1) to a general base and
solvent, given by the reaction

$$HA(solvated) + B(solvated) \rightarrow A^-(solvated) + HB^+(solvated), \tag{2}$$

is of similar interest and importance [3–7], particularly if one regards the
latter equation in the most general sense of solvation, in which "solvent"
could stand for a general polar and polarizable environment, such as an
enzyme-active site [5–7]. In this paper, we will largely focus on Eq. (2) in our
general discussion and on Eq. (1) in our discussion of water for our most
detailed and quantitative examination. Further, we will pay special attention
to the very first molecular step in either of Eqs. (1) and (2), i.e.,

$$HA \cdots B \rightarrow A^- \cdots HB^+, \tag{3}$$

in which we have now suppressed any notational reference to the identity
of the solvent, and by which we intend to indicate the elementary event of
the proton transfer from the acid to the base within a hydrogen-bonded
complex to produce a contact ion pair, similarly hydrogen bonded. It will be
argued that this first step can carry most of the free energy change associated
with the overall reaction to produce the completely separated ions, as in
Eqs. (1) and (2). The examples discussed will include HCl in water [8, 9], HF
in water [10, 11], and—in a less detailed fashion—the proton transfer
between the approximately modeled organic acid phenol and an amine base
in nonaqueous solution [12, 13]. The treatment of these reactions involves,
and has evolved from, a basic dynamical perspective presented some time
ago in [14, 15] and has points of inspiration arising from early work in the
Russian school [16]. One should note the existence of various other
theoretical approaches to the general problem of proton transfer [17–22],
but we refer to these only when they are particularly germane to the

reactions studied. For example, the electronic valence bond approach of [12] has some relation to the "empirical valence bond" approach of [21], and [19] is closely connected to some aspects of our discussion of phenol–amine.

In order to place acid ionization reactions in perspective vis à vis other reactions in solution, one can first observe [23], in connection with the aqueous version, Eq. (1), that the explicit involvement of a water molecule as a reaction partner introduces a new and essential feature compared with more generalized environmental solvent effects on reactions. One can also point to a number of special features of Eqs. (1)–(3): the variation of the solute electronic structure, which is itself coupled to the solvation, during the reaction; the special quantum character of the proton nuclear motion; and—for Eq. (1)—the special characteristics of proton transfer in water. These aspects all require particular attention and provide the basic perspective of the approach we describe. First, a problem relating to electronic structure in solution must be addressed (already a challenging problem in the absence of the solvent); this is a feature shared, for example, by unimolecular S_N1 ionizations [21, 24] $AC \rightarrow A^- + C^+$, although the acid ionization problem is bimolecular, since the proton must be attached to a base molecule. Second, the large polarity change in the acid ionization indicates a significant influence of the polar solvent, as is already well appreciated for the overall reaction thermodynamics [6]; indeed, for the isolated 1:1 hydrogen-bonded complex of HCl with H_2O [25], no acid ionization takes place, indicating at the least the essential importance of the polar water solvent in stabilizing the ionic products. More generally, solvent fluctuation and reorganization are accordingly expected to be critical factors for the reaction mechanism and the rate. This latter aspect [8–15, 21] is similar to that for electron-transfer reactions [26], excited electronic-state charge-transfer processes [27], and some unimolecular S_N1 ionizations [21, 24]. As will be seen throughout, the most closely related reaction class [28, 29] in some ways is electron transfer [12] between a donor and acceptor, $D + A \rightarrow D^+ + A^-$, but there are many significant differences as well, especially in the bond-making and -breaking aspect of proton transfer. In other ways, there are similarities to the S_N2 reaction class [28, 29], e.g., $A^- + BC \rightarrow AB + C^-$, especially in the bond-making and -breaking aspect, although typically, the charge redistribution pattern in S_N2 reactions is different, and attention to the quantum nuclear character of the transferring unit is not generally necessary.

Finally, we come to the special characteristics of the proton in aqueous media. In addition to the water molecule acting like a base in the aqueous version of Eq. (3), the solvent water molecules themselves can be involved further in the process of forming the separated ions in Eq. (1), via the "Grotthus mechanism" [6, 7]. A schematic version involving this mechanism, in which the solvent water molecules are themselves proton donors and acceptors, can be given as

$$
\text{ClH}' \cdots \text{H}_2\text{O} \cdots \text{H}_2\text{O} \rightleftharpoons \text{Cl}^- \cdots \text{H}'\text{OH}_2^+ \cdots \text{H}_2\text{O} \rightleftharpoons \text{Cl}^- \cdots \text{H}'\text{OH} \cdots \text{H}_3\text{O}^+
$$
$$
\mathbf{1} \qquad\qquad\qquad \mathbf{2} \qquad\qquad\qquad \mathbf{3}
$$

$$(4)$$

for the first two steps of the acid ionization of HCl in water, with step $\mathbf{2} \rightarrow \mathbf{3}$ being part of a Grotthus-like mechanism and the remaining steps producing the fully separated ions in Eq. (1) occurring by means of this mechanism. The "proton relay" among water molecules via a hydrogen-bonding network obviates the need for large displacements of an individual proton. A mechanism such as this is thought to be responsible for the large proton mobility in bulk water and ice [6, 30–36] and to play an important role in assorted acid–base catalyses in aqueous solution [5–7, 37]. We will pay some attention to the second step ($\mathbf{2} \rightarrow \mathbf{3}$) in Eq. (4), although our focus is on the first step. The presence of the anionic base molecule probably distinguishes this step in the acid ionization sequence given in Eq. (4) from the physical problem of proton transport in bulk water, a topic of much recent investigation [20] that is not of central concern here. Note that the PT character of the second step in Eq. (4), producing a "solvent-separated" ion pair from a contact ion pair, distinguishes this step fundamentally from other, more usual, ion-pair interconversion processes [10, 23], which are unimolecular processes involving rearrangement of the solvent as the ionic separation coordinate increases.

Finally, we will be concerned here only with acid ionization in the ground electronic state. There are a host of (largely experimental) studies [38] of excited electronic-state PT reactions in solution and cluster systems. These experiments are particularly attractive in that an initial "zero of time" is set for kinetic studies, especially in modern ultrafast experiments. While it is to be expected, or at least hoped, that many of the considerations of the present chapter will also be relevant in that context, we do not address the excited electronic-state problem here, since it requires the solution of an even more difficult electronic structure problem. Nonetheless, progress should be made in this theoretical area in the next few years.

The outline of the remainder of this chapter is as follows: In Section II, we address a number of general ideas and issues, largely in the context of model studies of phenol–amine PT reactions in solution [12, 13, 19, 39]. In Section III, we discuss the acid ionization of HCl in water [8, 9, 40, 41] and also deal more briefly with the related problem of HF ionization [10, 11, 42–44]. Concluding remarks are given in Section IV.

II.   GENERAL ISSUES AND MODEL STUDIES

A.   Some General Considerations

In a useful first division [45], the reaction type of proton transfer [45, 46] given in Eq. (3) is distinguished from transfer in an ion–molecule complex, i.e., $AH^+ \cdots B \rightarrow A \cdots HB^+$, which would not have a dramatic alteration of the polarity of the complex [30, 32, 47]. It is also distinguished from the intramolecular reaction class, as, for example, in malonaldehyde and other molecules [48], and from simultaneous proton transfer between molecules with mobile π-electrons, as, for instance, in acid dimer systems [49]; the latter two classes will generally differ in the magnitude of the proton coupling and thus the basic quantum character of the proton transfer.

The pronounced change in polarity in reaction (3) signals that the electronic structure of the neutral complex can be drastically affected by the solvent and, more generally, that attention must be paid to electronic structure issues as part of the description of the basic reaction; indeed, this last aspect is one of the central themes of the present chapter and is considered in general terms in the next few paragraphs.

Usually, $AH \cdots B$ complexes in the vacuum are predominantly neutral in character and have a single potential well in the proton coordinate [45, 46, 50]. By contrast, the ionic state in the vacuum is itself unstable or at least is of significantly higher energy, although the stable complex may have some partial ionic character. In the simplest two-valence-bond, electronically diabatic state perspective, the stable complex would be described by a neutral complex wave function ψ_n with only a small admixture, via electronic coupling, of an ionic wave function ψ_i to form the ground adiabatic electronic state.

But in solution, the description will change: The ionic state will be stabilized to some degree (compared with the neutral state), due to the electrostatic interactions with the surrounding polar solvent molecules [39, 46, 51, 52]. The influence could range from simply a slightly increased admixture of the ionic state in the ground adiabatic state—leaving the latter still largely neutral in character—all the way to the creation of a stable ionic complex, which is itself possibly in tautomeric equilibrium with a stable neutral complex. Effects such as these can lead to significant differences in gas-phase and solution acidity [54]. The same basic feature can occur in biochemical systems, in which the stabilizing role of the solvent is played instead by the ionic and polar environment in, for example, an enzyme-active site [5, 21, 54–56]. The solvent effect just described involves a competition between two factors. The first is the electronic coupling between, and the vacuum energetics of, the neutral and ionic VB states, which will typically favor a stable neutral complex with some admixture of the ionic state. The second factor is the solvation, which will favor or, indeed, completely establish an ionic character for the complex.

For simplicity, the previous paragraph has focused on the solvent's influence on the electronic structure in the determination of the electronic character of the stable complex. But in fact, precisely these same considerations indicate two essential features. First, in order to follow the mechanism of the proton-transfer reaction, one must deal with following the changing electronic structure of the system along the proton-transfer coordinate, in the field of the surrounding solvent molecules. This is because the electronic character of the PT system will change as the proton moves. Second, just how that electronic character changes will depend on the precise state of the surrounding solvent molecules; thus, certain configurations of the solvent molecules will be more effective in stabilizing the ionic VB state, while other configurations could instead stabilize the neutral VB state. Therefore, one must deal with the electronic structure of the PT system both as a function of the proton coordinate and as a function of the state of the solvent.

We will return later to details of how the electronic structure is to be described in practice, but the preceding discussion serves as an introduction to a preliminary, qualitative examination of the proton-transfer-potential characteristics as a function of the state of the solvent, which we here characterize by a solvent coordinate s, taken as a measure of the nuclear electrical polarization in the solvent. Figure 1 shows the general character of the proton potential, versus the proton coordinate, for three different values of the solvent coordinate s. It is assumed, as is typically the case, that the vacuum energy of the ionic VB state is noticeably above the vacuum energy of the neutral VB state. In (a), s corresponds to the situation in which the solvent is equilibrated to the neutral structure. The PT potential is

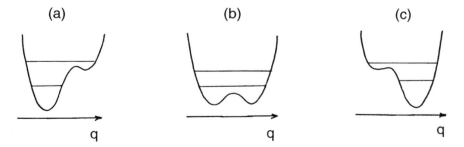

Figure 1. Schematic proton-transfer potential-energy curves versus the proton coordinate, for several different values of the solvent coordinate. The first two proton vibrational energy levels are indicated, in the proton adiabatic limit.

asymmetric favoring that structure, and PT to the base B is strongly endothermic, a consequence of the vacuum ionic–neutral energy gap and the solvent polarization being unfavorable for the proton-transferred ionic structure. In (c), we depict the corresponding situation in which the solvent polarization is equilibrated to the ionic structure. Again, the PT potential is asymmetric, but now in the sense favoring the ionic state; the solvation of the latter has overcome the unfavorable vacuum energy gap (in the example shown). Finally, in (b), we show the situation in which the solvent polarization is in an intermediate situation, in equilibrium with neither the neutral nor the ionic state, but such that the vacuum energy gap is counterbalanced by the differential solvation, so that a symmetric PT potential is obtained. In what follows, will be argued that it is in fact this solvent configuration that is critical, in that it governs the activation free energy for the PT reaction. Critical to that argument is the fact that the characteristic time scales for the high-frequency proton vibration are short compared with the relevant time scales for the nuclear motion of the solvent molecules that change the solvent polarization, and here we simply proceed on that premise.

If the solvent coordinate motion is slow on the proton time scale, then the attainment of situation (b) in Figure 1 from the reactant configuration (a) in the same figure requires that the solvent start to move before the proton does and that motion be towards configurations that are unfavorable for the initial neutral, non-proton-transferred complex. This indicates that, in fact, there will be a free-energy cost in the solvent coordinate in order to reach the situation in Figure 1(b). The situation is depicted in Figure 2, where the free energy in the solvent coordinate rises smoothly, from that appropriate to the situation in Figure 1(a) to a maximum value corresponding to the situation in Figure 1(b), before descending to the final stable value, corresponding to the situation in Figure 1(c). The solvent free energy is approximately parabolic in the neighborhood of the regions labeled reactant (R) and product (P), since there the solvent polarization is fluctuating in the presence of stable, non-proton-transferred neutral structures and proton-transferred ionic structures, respectively.

The foregoing discussion indicates that, in fact, the activation free energy for the reaction must involve the free-energy rise in Figure 2. We now go into the details of this mechanism and argue that the rise is essentially the activation free energy for the PT reaction and that the reaction coordinate for the PT is not the proton coordinate itself, but rather the solvent coordinate.

Figure 1 shows the first two quantized vibrational levels for the proton, the lowest, of course, being the zero-point energy, for what we will call the regime of adiabatic proton transfer. We return to this regime presently, but,

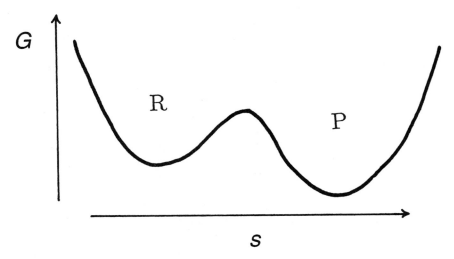

Figure 2. Schematic picture of the free energy versus the solvent coordinate for the adiabatic proton-transfer situation in Figure 1.

for simplicity (and contrast), begin the discussion of the nonadiabatic, or tunneling, regime, shown in Figure 3. In panel 3(a), corresponding to the reactant complex with its attendant equilibrium solvation, the first two proton levels correspond to the quantized vibration of the proton in the configuration $AH \cdots B$, while Figure 3(c) corresponds to the related situation in the product complex $A^- \cdots HB^+$. In the symmetric situation of the central panel 3(b), the vibrational motions of the proton in either configuration, $AH \cdots B$ or $A^- \cdots HB^+$, are located below the top of the proton barrier and are degenerate in energy. (We display the case where the second vibrational level in the R and P configurations remains well above the barrier in this symmetric situation.) This is the symmetric proton tunneling condition, and, as is well known in such a situation [8–22], the degenerate levels are split by the so-called tunnel splitting $2C$. From a dynamical point of view, the system prepared in the degenerate state corresponding to the proton being initially localized in the $AH \cdots B$ configuration will undergo tunneling oscillations to and from the $A^- \cdots HB^+$ configuration, with a tunneling frequency equal to $2C/\hbar$. (Tunneling is a nonadiabatic transition between the degenerate, nonadiabatic proton levels.) Now, it is well known that the ease of tunneling is a very sensitive function of the symmetry of the proton potential [8–22], dropping rapidly as the potential becomes asymmetric—i.e., as it departs from the case of the symmetric proton potential shown in Figure 3(c). The key consequence of this for our present purposes is that the tunneling, which

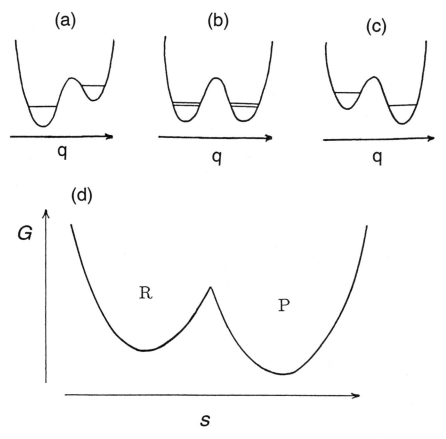

Figure 3.   Panels (a)–(c) correspond to those of Figure 1, but now in the proton nonadiabatic tunneling limit. Panel (d) corresponds to Figure 2, but again in the nonadiabatic limit.

allows the proton-transfer production of the P configuration $A^- \cdots HB^+$, can be ignored except when the proton-transfer potential is symmetric. (A rigorous treatment of the mechanism is complex, but the result is the same [14]). The picture for the tunneling proton-transfer reaction would then be that, starting from configurations appropriate to equilibrium with the neutral R complex, the solvent fluctuates, at a cost of free energy, into configurations that produce a symmetric proton-transfer potential and thence the proton tunnels, and the solvent fluctuates into configurations appropriate to the product. The free-energy profile of the solvent is shown in Figure 3(d), corresponding to the evolution of the proton potential (a) → (b) → (c) in Figure 3. The reaction coordinate is the solvent, and the activation free energy of the reaction is that along the solvent coordinate, attained in reaching the solvent transition state. Certainly, the height of the proton potential barrier does not determine the activation free energy: The proton tunnels through the barrier below that height. The characteristics of the proton potential, in the symmetric configuration of Figure 3(b), enter into a tunneling prefactor of the rate, proportional to the square of C, and not into the exponential activation free-energy factor in the rate [14, 16, 19].

We now return to the adiabatic proton transfer regime depicted in Figure 1, where, as the solvent reorganizes, the high-frequency proton motion adjusts adiabatically in the changing proton potential and remains at all times in the ground vibrational level, which, in the symmetric situation of Figure 1(b), is above the proton potential barrier. The vibrational splitting between the lowest level and the first excited level is $2C$, where we have retained the splitting notation from the tunneling regime, but it must be emphasized that C—which we call the proton coupling—has nothing to do with tunneling in the adiabatic regime, although it still depends on the characteristics of the proton potential. The special importance of the situation of the symmetric proton-transfer potential in Figure 1(b) in the adiabatic regime is evidently not of the fundamental quantum character, as it is in the nonadiabatic tunneling regime; however, as will become clearer subsequently, when we define the solvent coordinate in microscopic energy terms, it corresponds to an appropriate midway point for the overall proton-transfer reaction, even when that reaction is asymmetric, so that the transition state in the solvent coordinate should either be at, or very close to, this location.

To complete the qualitative picture just set forth, we need to discuss, in addition to the proton and solvent coordinates already considered, the vibrational coordinate Q between the proton donor and acceptor groupings A and B. Earlier, we noted that, in the nonadiabatic tunneling regime, the rate is proportional to C^2. The coupling C, which, in the tunneling regime, is exponentially sensitive to the width of the proton potential barrier [14, 16, 19], is an exponentially sensitive function of the Q coordinate, since Q variations modulate the proton potential, as shown schematically in Figure 4, for the relevant symmetric situation. For larger Q, a more complete breaking of the AH bond is required before the new HB^+ is formed, the proton barrier is higher, and a thicker barrier is presented to the transferring proton. Conversely, for smaller Q, the breaking of the original, and the formation of the new, bond for the proton is more concerted in character, the barrier is lower, and a more narrow feature of the potential is produced. Thus, the Q-coordinate is crucial for the tunneling regime. In the adiabatic PT regime, no under-the-barrier proton tunneling is occurring, and although the proton vibrational level splitting $2C$ depends on Q, since the proton potential is being modified, the dependence of the proton

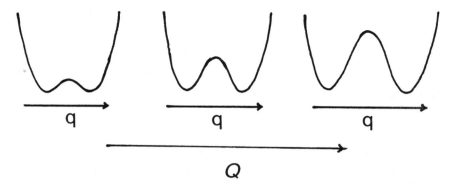

Figure 4. Illustration of the effect of the variation of the heavy-particle separation Q on the proton potential when the solvent coordinate has established a symmetric situation for the latter.

coupling C on Q will be much less than the exponential dependence characteristic of the tunneling regime; studies indicate that C depends approximately linearly on Q in the adiabatic regime [13, 19].

From the discussion in the last paragraph, it is clear that proton-transfer complexes $AH \cdots B$ that have a large equilibrium value Q_{eq} of the AB vibrational coordinate Q will have large proton-transfer barriers and will tend to be characterized exclusively by a tunneling mechanism of proton transfer. On the other hand, complexes that have smaller Q_{eq} values will tend to proceed via an adiabatic PT mechanism, since the barrier is lower and it is easier for the zero-point level of the proton to be above the proton barrier when the fluctuating solvent has established the symmetric scenario. These considerations give an extremely useful, albeit approximate, guide to what to expect in PT complexes of various sorts (although, of course, one has to confirm the expectations). In many intramolecular PT complexes, such as malonaldehyde and others [48], the structural architecture of the complex will result in a large Q_{eq} value, and a tunneling mechanism is indicated. In many intermolecular complexes, most especially those involving strong intermolecular hydrogen bonds, the resulting short equilibrium AB separations Q_{eq} will favor an adiabatic PT route.

A final point on the adiabatic PT regime should be emphasized. One gains the impression from many texts and much of the journal literature that it is widely thought that if a proton (or hydrogen atom or hydride ion) transfer does not proceed via quantum tunneling, then it must proceed via an over-the-proton-barrier classical mechanism. This thought is incorrect. In the adiabatic PT regime, the proton remains completely quantum in its motion. The reaction is not at all a classical passage of the proton over its barrier, in which the proton motion at the supposed transition state in the proton coordinate is an unstable motion. Instead, the proton motion at the reaction transition state, which is in the solvent coordinate, is a bound, high-frequency, quantized vibration.

B.   Valence Bond Study in a Dielectric Continuum Solvent

In the remainder of this section, we give a brief presentation of some highlights of several studies [12, 13] of assorted aspects of PT in hydrogen-bonded complex systems. These studies, involving models of varied sophistication of the ground electronic-state proton transfer between the organic acid phenol, C_6H_5OH, and a nitrogen molecular base, serve two purposes here. The first is to illustrate, in a comparatively simple context (especially in the quantum electronic structure aspects of the problem), some of the concepts we have just described quantitatively. The second is to address a few points—in particular, quantization of the electrons of the solvent quantization of the intermolecular AB vibrational Q coordinate, and dynamic reaction coordinate recrossing effects—that we will refer to, but not emphasize, in the subsequent discussion.

In the first study [12], the problem of the PT system electronic structure in solution was addressed via a two-VB state prescription (an approach with several antecedents in the literature [21, 39, 57–60], and a description of the solvent in terms of the dielectric continuum. Put simply, the functional forms of the vacuum energies of the two electronically diabatic states ψ_n and ψ_i for the neutral and ionic forms of the complex, respectively, together with the electronic coupling between them, were determined approximately, as a function of both the proton coordinate q and the AB (here, O–N) coordinate Q. This result was accomplished by adopting the empirical Lippincott–Schroeder potential [61, 62] for hydrogen-bonded complexes for the vacuum adiabatic potential-energy surface and extracting the VB components. The electronic structure of the PT system in solution was addressed via a theory [64] that simultaneously, takes into account the polarization effect of the solvent, which tends to localize the system into the ionic VB state, the electronic coupling H_{ni}, which tends to mix the states together, and the quantum character of the solvent electrons, whose participation in the solvation is a delicate issue [63, 64]. The solvent coordinate is related to the nuclear (i.e., nonelectronic) polarization of the solvent; it is an independent variable, and since it is not assumed that this polarization is in equilibrium with the instantaneous state of the PT complex system, nonequilibrium solvation is accounted for.

Initially, focus on some of the resulting electronic structure aspects of the theory. First, the magnitude of the electronic coupling H_{ni} between the VB states is in the range of 18–31 kcal/mol for Q in the range of 3.0–2.6Å [12]. The high value, which is associated with the fact that the PT reaction involves bond-making and -breaking, is to be strongly contrasted with the very much smaller electronic-coupling values dealt with in typical outer-sphere electron transfer reactions [26]. While we will see throughout this chapter that there are certainly points of contact and analogies between proton-transfer and electron-transfer reactions, there are also quite significant differences, and this is the first of those.

Second, the assumption that the PT system is described by an expansion over the two electronically diabatic neutral and ionic VB states—whose coefficients depend upon the two key variables q and Q of the PT solute system $AH\cdots B$ and the solvent coordinate—has a very definite physical, and actually somewhat nonconventional, meaning: Rather than being the more common picture of a simple transfer of the H^+ entity, such an expansion represents the Mulliken charge transfer picture [39, 58, 65, 66], depicted in Figure 5. The Mulliken description involves the transfer of electronic charge from the base-B lone-pair nonbonding orbital (n_B) to an antibonding molecular orbital (σ^*_{AH}) on A–H. This transfer produces a weakening of the AH bond, which lengthens in response, allowing the transfer of H to B. Note especially that the transferring hydrogen species is more like a hydrogen atom than a fully charged proton and in that sense is nonconventional. It also differs considerably from the current, widely used conception of hydrogen-bonded complexes in electrostatic terms [67, 68]. On the other hand, there is support for the Mulliken picture from various quarters [21, 39, 58–60, 66, 69], including ab initio considerations [70], and indeed, the picture is supported by the results regarding HCl ionization in water, to be discussed in Section III.

Third, the free energy G of the PT system in the nonequilibrium solvent, with quantized electronic polarization of the latter, has the form [64]

$$G = G_{AD} + \frac{\rho}{2c_N c_I + \rho}[G_{SC} - G_{AD}], \tag{5}$$

in which G_{AD} and G_{SC} are the free energies in, respectively, the adiabatic and self-consistent treatments to be explained subsequently, c_N and c_I are the

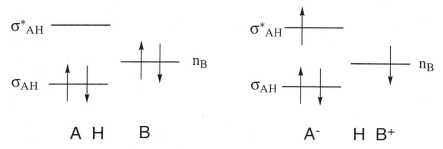

Figure 5.   Schematic illustration of the Mulliken charge transfer picture for a proton-transfer reaction.

expansion coefficients in the wave function over the diabatic basis sets ψ_N and ψ_I, respectively, and

$$\rho \equiv 2|H_{NI}|/\hbar\omega_e \tag{6}$$

is the ratio involving the invariant electronic coupling and the (high) characteristic frequency of the solvent electronic polarization, which is related approximately to the energy of the electronic absorption peak for the solvent. This ratio is an approximate measure of the frequency of the resonant quantum shuttling between the diabatic N and I states, compared with the frequency of the solvent electronic polarization. G_{SC} is the free energy of the system when $\rho \gg 1$; this self-consistent approximation is familiar, at least in the equilibrium solvation context, from many studies (see, e.g., [40]) of electronic structure in solution and assumes that the solvent electronic polarization is equilibrated to the charge distribution given by the full wave function ψ. G_{AD} is the free energy of the system when $\rho \ll 1$; this adiabatic description is most familiar in the context of electron-transfer-rate theory [26] and assumes that the solvent electronic polarization is equilibrated separately to the individual diabatic states ψ_N and ψ_I. For the proton-transfer problem, with ρ of order unity [12], neither limiting case generally applies, and one must deal with the full problem described by Eq. (5). In sum, this equation describes the free energy of the PT system in solution, in a two-VB state description, which accounts for the competition between the electronic coupling and the solvation. (Thus, the two states are not first coupled, as in the gas phase, and then solvated, but rather, the simultaneous influence of the electronic coupling and the solvent is accounted for.) Also, the solvent electronic polarization is described quantum mechanically, and the solvent nuclear polarization is not confined to its equilibrium value, so that nonequilibrium solvation is accounted for.

The general character of the results pertaining to the nonequilibrium solvation free-energy surfaces calculated with the model is illustrated schematically in Figure 6 for a fixed value of the AB separation Q. There, the neutral reactant well is evident at small q and at values of the solvent coordinate consistent with solvent polarization in equilibrium with, and in the near environs of equilibrium with, the neutral AH$\cdots$B structure. The second, ionic product complex well, located at larger q, is characterized by values of the solvent coordinate appropriate to, or not far from, equilibrium with that ionic complex. Also indicated in the figure is an equilibrium solvation path (ESP), which, in a widespread way of thinking, would represent the reaction path followed from the reactant complex to the product complex. Along the ESP, the solvent is equilibrated to the solute complex at each value of the proton coordinate q; that is, the free energy G is

ESP

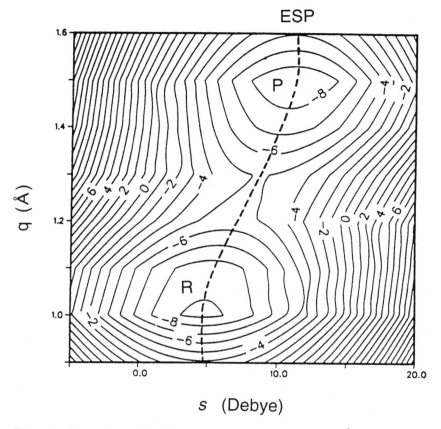

Figure 6. Free-energy surface for the OH$\cdots$N complex at $Q = 2.7\,\text{Å}$ in acetone. The contour spacing is 1 kcal/mol. The dashed line represents the equilibrium solvation path (ESP) in the two-dimenstional manifold.

a minimum with respect to the solvent coordinate at each q value. However, implicit in the assumption of the ESP is the condition that the solvent is fast compared with the proton, since the former is supposed to adjust to the latter in order to be equilibrated. Physically, this is just the opposite of the conditions that would be expected to apply for PT, where, typically, it is the high-frequency quantum proton that is fast compared with the motions of the solvent molecules. (Recall that the solvent coordinate represents the nuclear polarization of the solvent, e.g., the orientational motions of the solvent molecules, and not the fast electronic polarization of the solvent.) Accordingly, the ESP does not provide the correct picture of the reaction. Instead, one should consider the quantization of the high-frequency proton motion at each value of the comparatively slow solvent coordinate. Thus, at each value of s, the nuclear Schrödinger equation is to be solved to find the proton vibrational levels. When this is done, one obtains just the sort of pictures given in Figure 1 for the proton adiabatic regime, which, as previously discussed, is more likely for smaller Q values. At larger Q values, the proton nonadiabatic regime applies, and the lowest two levels would correspond to the tunnel-split levels in Figure 3.

The preceding discussion links the multidimensional free-energy surfaces that incorporate the solution to the problem of the PT solute system electronic structure in solution, as a function of the PT solute coordinates and in the presence of nonequilibrium solvation, to the schematic diagram of Figure 1, which, together with Figure 2, provides the basic paradigm for adiabatic proton transfer to produce a contact ion pair. Before turning to a study [13] that deals with some of these aspects in a more microscopic and mechanical context, we observe that nonequilibrium solvation is a general characteristic of reactions in solution that involve the motion of various nuclei in, for example, S_N2 [28, 29], S_N1 [24], and TICT [27] reaction dynamics. The quantum character of the proton motion in PT reactions brings this nonequilibrium aspect to the fore. In that sense, there is a similarity to electron transfer reactions [26], although one does not usually draw on the electron energy levels versus the electron coordinate. Still, the similarity is there: The quantized energies for the rapid electron motion, (i.e., the electronic energies) are to be determined at fixed values of the nuclear coordinates of the problem. Thus, there is a certain analogy between the electronic energies in the ET problem and the proton vibrational energies in the PT problem. We stress again the point that the electronic coupling in the PT problem is large, whereas typically it is not in the ET problem. As one might suspect from the analogy, the analogue of the electronic coupling in the ET problem is the proton coupling C in the PT problem. This point is pursued in more detail shortly.

C. Molecular Dynamics of Acid–Base Proton Transfer

The second study to be discussed also involves a model of phenol–amine PT, this time in a molecular dynamics simulation [13] of a modeled phenol–trimethyl amine system in methyl chloride solvent. In this study, the solvent was treated at the molecular level, the O–N vibrational coordinate Q was quantized in addition to the proton coordinate q, and the reaction rate constant was found, including the influence of dynamical recrossing of the reaction transition state. The electronic structure problem for the $AH\cdots B$ complex, on the other hand, was, in effect, treated at the level previously [12]; described as will be explained subsequently, only certain information from the electronic structure problem was, in fact, required, due to the basic formulation of the problem.

In the first step, the Hamiltonian $H(q, Q, \mathbf{S})$ for the system is written down as a function of all the system coordinates and momenta (the latter are suppressed in the notation used here); $\mathbf{S}$ stands for the set of all the solvent coordinates. The specification of the Hamiltonian involves all the short-range Lennard–Jones and electrostatic interactions for the system being modeled, and those details are not given here. In the next step—and this is key for the entire formulation—the proton was quantized in the proton adiabatic representation such that the quantum average over the fast proton motion is taken via the expectation value involving the proton wave function $\Psi_v(q; Q, \mathbf{S})$ for proton vibrational state v, where the notation indicates that the wave function is a function of the proton coordinate q and is parametrically dependent on the Q and $\mathbf{S}$ coordinates. This procedure, which is the proton analogue of the standard Born–Oppenheimer adiabatic approach to the electronic problem in isolated molecules, generates the Hamiltonians $H_v(Q, \mathbf{S})$, which are operators in Q and $\mathbf{S}$, although, since the nuclear degrees of freedom of the solvent are treated classically quantization in not a real issue. In essence, the $H_v(Q, \mathbf{S})$ represent the vibrational energy levels of the proton, plus the self-interaction energy of the solvent molecules, plus the kinetic energies of the solvent molecules and the Q vibration.

While the proton adiabatic formulation, in which the proton is de-localized, forms the basic core of the treatment, it was noted that, computationally, it is more convenient to use a diabatic, or localized, proton wave function perspective, in which the diabatic wave functions $\Phi_{R,P}(q)$ are related to the reactant (R) and product (P) nuclear configurations of the solute. The quantum averages of the total Hamiltonian $H(q, Q, \mathbf{S})$ give the reactant and product Hamiltonians,

$$H_{R,P}(Q, \mathbf{S}) = \langle \Phi_{R,P}(q) | H(q, Q, \mathbf{S}) | \Phi_{R,P}(q) \rangle, \qquad (7)$$

and also the proton coupling C between the proton vibrational diabatic states:

$$C(Q, \mathbf{S}) = \langle \Phi_R(q) | H(q, Q, \mathbf{S}) | \Phi_P(q) \rangle. \qquad (8)$$

In this diabatic representation, restricted with justification [13] to the two diabatic wave functions $\Phi_{R,P}$, the diagonalization of the 2×2 matrix of the diabatic Hamiltonians with the off-diagonal coupling C generates the representation of the proton adiabatic Hamiltonian for the lowest two proton vibrational levels as

$$H_{0,1}(Q, \mathbf{S}) = \frac{1}{2}[H_R(Q, \mathbf{S}) + H_P(Q, \mathbf{S})] \pm \frac{1}{2}\sqrt{\Delta E(Q, \mathbf{S})^2 + 4C(Q, \mathbf{S})^2}, \quad (9)$$

where

$$\Delta E(Q, \mathbf{S}) = H_R(Q, \mathbf{S}) - H_P(Q, \mathbf{S}), \qquad (10)$$

in which ΔE represents the difference in the solute–solvent electrostatic interaction energies, with the solute having both the reactant and product charge configurations. This collective solvent coordinate is similar to that introduced in a wide variety of charge transfer reactions in solution [21, 24, 26, 27], while the diabatic perspective has been employed in various H atom and PT studies [14, 18, 19] and is the molecular analogue of the solvent coordinate discussed earlier. Note that one is not dealing here with the electronically diabatic states, which would be coupled by a large electronic coupling, as has already been mentioned. Rather, the proton diabatic states are defined only in terms of the proton vibrational adiabatic states, and the coupling is the proton coupling C. This aspect is emphasized in the treatment by the determination of the coupling C via the inversion of Eq. (9), namely,

$$C(Q, \mathbf{S}) = \frac{1}{2} \sqrt{[H_1(Q, \mathbf{S}) - H_0(Q, \mathbf{S})]^2 - \Delta E^2}, \qquad (11)$$

in terms of the proton vibrationally adiabatic Hamiltonians H_0 and H_1. This was done by a certain finesse in which the vacuum electronically adiabatic potential for the PT system was taken from the Lippincott–Schroeder potential, as in Section IIB, and the interaction of this system, for each value of the proton coordinate q, was represented by a dipolar interaction [19] of the dipole moment of the PT complex, whose form was taken from the study described in Section IIB for the ground electronically adiabatic state [12]; with the electric field generated by the solvent molecules projected onto the solute internuclear axis. With this potential for the PT system to interact with the solvent molecules, with the electric field generated by MD, the one-dimensional nuclear Schrödinger equation for the proton was solved to generate the two lowest proton vibrationally adiabatic eigenvalues, giving H_0 and H_1 for use in Eq. (11). (See [13] for the details of the procedure.) The proton coupling was found to be a linear function of the O–N vibrational coordinate Q for separations less than about 2.8 Å, corresponding to the proton adiabatic regime, and was found to be exponential in Q for larger separations, corresponding to the proton diabatic, or tunneling, regime; these are explicit illustrations of the general features described in Section IIA. At the equilibrium value of Q (2.6 Å, see shortly), the calculated picture of the proton potentials and vibrational levels at the various ΔE values for the solvent coordinate is of precisely the character depicted in Figure 1 for an adiabatic PT reaction.

Together with umbrella sampling techniques [71, 72] to force the system into the range of ΔE values, the net output of the procedures just described was to generate a free-energy surface $G(Q, \Delta E)$ as a function of the Q vibrational coordinate and the collective solvent coordinate ΔEs, for each of the first two vibrational states of the proton. The surfaces are ones of free energy due to the reduction of the problem to only one coordinate, ΔE, for the solvent. An average was taken over the remaining degrees of freedom of the solvent.

The general appearance of the ground vibrational-state free-energy surface is that of two double wells at different ΔE's, but at about the same small Q value; at a fixed ΔE, as Q increases, the surface rises, corresponding to dissociation of either complex, for example. The free-energy wells for the neutral reactant complex and the proton-transferred contact-ion-pair complex are located at $Q = 2.6$ Å (such that the PT reaction is indeed expected to be proton adiabatic, with a large coupling C), with the latter at $\Delta E = 20$ kcal/mol, lying 1.2 kcal/mol lower in free energy than the former at $\Delta E = -18.5$ kcal/mol. The reaction is slightly exothermic and not far from symmetric, with a Q-dependent barrier separating the reactant and product wells.

The harmonic frequencies for the O–N Q vibration were found to be above $330\,\mathrm{cm}^{-1}$ at various values of the solvent coordinate ΔE (reaching $465\,\mathrm{cm}^{-1}$ in the barrier region of the $G(Q, \Delta E)$ surface). At the simulation temperature of $250\,\mathrm{K}$, this indicated that the Q vibrations should be quantized, which was effected, with justification, in the adiabatic approximation—that is, the solution of the nuclear Schrödinger equation for the Q vibration at fixed values of the solvent coordinate ΔE. One-dimensional free-energy surfaces were then generated for each of the Q vibrational levels $n = 0, 1, \ldots$. Although quite interesting effects were found in this connection—most notably, that Q vibrational excitation (such that, on average, the Q vibration is amplified) led to a deceleration of the reaction, in complete contrast to results in the proton diabatic tunneling regime [14, 16, 19]; we address this aspect no further here.

The calculated one-dimensional free-energy curve $G_0(\Delta E)$ versus the solvent coordinate, in the ground vibrational state of the Q vibration (and in the ground proton vibrational state), for the acid ionization to produce the contact ion pair, is an asymmetric double well, of barrier height $\Delta G^{\ddagger} = 3.37$ kcal/mol. As derived in [13] and [15], the reaction rate constant in the transition state theory (TST) approximation [1, 73], in which there is no dynamic recrossing of the barrier at the transition state $\Delta E^{\ddagger}$ in the reaction coordinate, is

$$k^{\mathrm{TST}} = \frac{\omega_R}{2\pi} \exp(-\Delta G^{\ddagger}/k_B T), \tag{12}$$

where ω_R is the harmonic frequency of oscillation of the solvent coordinate in the neutral pair $AH \cdots B$ reactant well. Actually, Eq. (12) is of a quite familiar general form for a reaction rate constant in a double-well context [1], but we again stress that this is the TST rate constant for a quantum adiabatic proton-transfer reaction, in the solvent reaction coordinate and not the proton coordinate. The activation free energy $\Delta G^{\ddagger}$ is determined here by the reorganization of the solvent and also contains the difference in the zero-point energies of the proton for the quantized proton bound vibrational motion at the transition state $\Delta E^{\ddagger}$ and at the reactant ΔE_R (cf. Figure 1), with the proton motion transverse to the reaction coordinate. In addition, $\Delta G^{\ddagger}$ contains the analogous difference of the zero-point energies for the Q vibration, but this is of less consequence here.

As noted earlier, the TST assumption for the rate constant ignores any dynamical recrossing of the transition state. In the present context, this assumption amounts to an equilibrium solvation requirement, in the sense that only the equilibrium probability of attaining the $\Delta E^{\ddagger}$ value, and no aspect of the dynamics of the solvent reaction coordinate ΔE, enters [13] into Eq. (12). However, such dynamical effects can arise for the ΔE coordinate and are related to the fact that the motion of each of the polar solvent molecules is hindered by its neighbors, causing a net dissipative damping or friction in the reaction coordinate and thus dynamical recrossings of the transition state.

These recrossings reduce the actual rate constant for the acid ionization from the TST approximation, such that the former is given by $k = k^{TST}\kappa$, where κ is the so-called transmission coefficient. The constant κ was determined via MD simulation to be 0.8, so that recrossing reduction of the rate constant is not very pronounced in this case. The κ value is well given from a wide variety of MD simulations of reactions [26, 29, 74, 75] by the Grote–Hynes theory [76], in which

$$\kappa_{GH} = \left[\kappa_{GH} + \omega_b^{-1} \int_0^{\infty} dt \exp(-\omega_b \kappa_{GH} t) \zeta^{\ddagger}(t) \right]^{-1}, \qquad (13)$$

where ω_b is the barrier frequency at the transition state and $\zeta^{\ddagger}(t)$ is the dynamical friction on the reaction coordinate. (Note that ω_b is the square root of the ratio of the magnitude of the free-energy curvature and the solvent coordinate mass at the transition state and $\zeta^{\ddagger}(t)$ is proportional to the time correlation function of the generalized force acting on the solvent coordinate at the transition state). This relation assumes that in the transition state neighborhood, the solvent coordinate satisfies the generalized Langevin equation

$$\Delta \ddot{E}(t) = \omega_b^2 \Delta E(t) - \int_0^t d\tau \zeta^{\ddagger}(t - \tau) t \Delta \dot{E}(\tau). \qquad (14)$$

The Kramers theory [77] prediction, which ignores the feature that the relevant friction for the barrier passage is time dependent and is determined on the very short time scale related to the barrier frequency, was found to be too small by about a factor of two. Before leaving this topic, it is again worth emphasizing that the recrossings, the transmission coefficient, and the time-dependent friction are those for the solvent reaction coordinate and not those for the motion of the proton.

Some alternative approaches to the TST rate were tested in [13], but at the fixed equilibrium values of the O–N stretch coordinate Q. The first, a Landau–Zener type of curve-crossing approach [15], in which one starts from the proton diabatic curves in the solvent coordinate and couples them by the proton coupling C, correctly predicts that the reaction is proton adiabatic. Further, the activation free energy in this formulation is $\Delta G^{\ddagger} = \Delta G^{\ddagger}_{na} - C$, where

$$\Delta G^{\ddagger}_{na} = \frac{(\Delta G_{rx} + E_r)^2}{4E_r}, \tag{15}$$

is the diabatic activation free energy involving the overall reaction free energy ΔG_{rx} and E_r is the solvent reorganization energy, all of which could be calculated via MD. Interestingly, Eq. (15), which is the proton non-adiabatic form of the Marcus expression familiar from electron transfer reactions [26], was found to hold rather well; this is an important observation. The overall activation free energy gives a reasonable, though not perfect, result compared to the MD value of $\Delta G^{\ddagger}$ and has an important subtractive contribution from the proton coupling, i.e., $-C$, whose electronic coupling analogue in the electron-transfer problem is usually neglected. These equations then provide a quick and useful way to make an estimate of the activation energy for an adiabatic PT reaction.

Finally, in Section IIB, it was emphasized that attention needed to be paid to the electronic polarization of the solvent in the acid ionization problem, via its influence on the electronic structure of the PT system and thus on the proton-transfer potential and the ultimate free-energy characteristics of the reaction. But in classical MD simulations, the solvent molecules are described by classical force fields, and the electronic polarization does not appear explicitly, and certainly not as a high-frequency quantum sort of motion; in a sense, the solvent electronic polarization has been effectively parameterized in interaction potentials and is no longer available as an explicit type of variable. This problem is present in many solution phase reaction MD simulations, but has received almost no attention. In [13], an analysis of the situation was presented in the context of the present topic of proton adiabatic acid–base reactions. The most important conclusion of that analysis was that a classical MD treatment of the solvent generally will produce proton potentials in the proton coordinate that are too high compared with what they should be; this will generally favor the proton adiabatic regime for the actual reaction. Such a situation would be of very great concern with regard to any proton transfer reaction that was calculated to be (and, in fact, was in reality) in the nonadiabatic, tunneling regime, where the rate is quite sensitive to the details of the proton potential. However, it also indicates that if one calculates a proton-transfer potential with the proton level slightly below the top of the barrier, there is a good chance that, in actuality, the level should be *above* that top and thus that the reaction should be in the proton adiabatic regime. For reactions in that regime, the consequence for the numerical value of the proton vibrational level, and thus for the resulting activation free energy of the reaction, will, fortunately, not be very pronounced; this is because that proton level is not so sensitive to the precise height of the barrier that is below it.

In Section II, we have covered many of the ideas and essential results needed to comprehend acid–base ionizations producing contact ion pairs in solution. On the other hand, the electronic structure problem was handled, while usefully, only very approximately. In the next section, we turn to acid ionization proton transfers in water, where the electronic structure problem is handled at a much higher level.

III. IONIZATION OF HYDROCHLORIC ACID IN WATER

A. Introduction

In this section, we describe the results of investigations into the acid–base reaction of hydrochloric acid in water (Eq. (4)) [8, 9]. These investigations, carried out at a detailed molecular level, involve many of the ideas discussed in the preceding sections. At the conclusion of the section, a very brief résumé of the analogous HF reaction is presented [10, 11].

In the work to be summarized, electronic structure calculations and Monte Carlo computer simulations were employed, not in a fully ab initio way, but rather within the framework of the ideas emphasized in Sections I and II. In fact, the calculations for HCl ionization PT to a base water molecule in aqueous solution produced numerical results for the step $1 \rightarrow 2$ in Eq. (4), whose general character is that of an adiabatic PT. The related solvent reaction coordinate free-energy profile was found to be generally similar in character to that of Figure 2, although details of the shape are different.

Before reviewing these and further results pertaining to HCl ionization, a brief overview of the calculation procedure will help to orient the discussion. Preliminary electronic structure calculations were first performed in HCl–water cluster systems to produce potentials suitable for use in Monte Carlo simulations involving many (classical) water molecules. Proton diabatic potential curves in solution were produced by special techniques, as a function of a microscopic solvent coordinate similar to that discussed in Section IIC; their appearance is somewhat similar to the two portions of Figure 3(d). Note that no proton nonadiabatic picture of the reaction is being assumed or implied by this step. Indeed, as detailed later and in [9], those curves are only an *intermediate* step toward accessing important regions of the solvent coordinate at which to perform the next step: the quantum chemical calculation of the proton potential-energy curves in the proton coordinate. The latter then serve as the base for quantizing the proton nuclear motion in order to find the proton vibrational levels as a function of the solvent coordinate. This then ultimately leads to the reaction free-energy profile in the solvent coordinate.

B.   Potentials and Computational Methods

In addressing the electronic structure issues, ab initio molecular orbital (MO) methods [78] were used (i) to optimize the nuclear geometries of small hydration clusters, (ii) to determine the model potential parameters used in subsequent Monte Carlo simulations, and (iii) to construct the potential-energy surfaces as a function of proton coordinates. The restricted Hartree–Fock (RHF) and the second-order Møller–Plesset (MP2) wave functions [79] were employed with the 3-21G* (Basis I) and 6-31G**(Cl+) (Basis II) basis sets [80]. The former set was used for nuclear geometry optimization, while the improved set was employed to determine various potentials— especially potentials for the transferring proton. In Basis II, a set of p-type diffuse functions [81] on the Cl atom with the exponent of 0.049 was added, to address the diffuse character of Cl^-. The adequacy of the MP2 method for the proton transfer reactions taking place in the HCl–water system was confirmed by checking the dominance of the Hartree–Fock configuration [82] in the single- and double-excitation configuration interaction (CI(SD)) expansion [83].

As a preliminary to the solution-phase simulations, the first step was the determination of the nuclear geometry of the reaction system **1**, hydrated by eight water molecules, shown schematically in Figure 7. These eight molecules are the nearest neighbors to the reacting molecules **1**, each of whose three members is coordinated by four molecules. The RHF analytic gradient method was then employed with Basis I. Because calculation of the

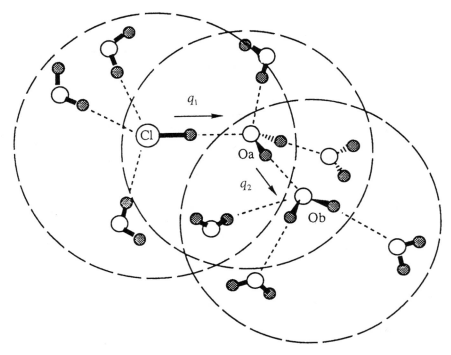

Figure 7. Schematic illustration of the primary reaction system considered (ClH$\cdots$OH$_2$ $\cdots$OH$_2$), with eight external water molecules displayed. The light (dashed) line enclosures are explained in the text.

nuclear geometry of entire system in Figure 7 was impractical, the structures of smaller four-hydrated clusters, designated by the dashed circles in the figure, were optimized. Details of the calculations are described in [9]. For example, for the state **1**, the resultant Cl$\cdots$O$_a$ and O$_a\cdots$O$_b$ distances, and the Cl–H (q_1) and O$_a$–H (q_2) bond lengths were 2.96, 2.83, 1.33, and 0.98 Å, respectively.

The many water molecules external to the reaction system **1–3** in solution were then accounted for via model potential functions, using information from the electronic structure calculations of the small-hydration clusters. The point charges on each atomic site of the reaction system **1–3**, used in the Monte Carlo simulations, were determined in order to reproduce the electrostatic potentials at approximately 500 points around the trimer **1–3** computed from the RHF wave function with Basis II. To include the solvent-induced polarization of the solute, the RHF wave functions were recalculated under the hypothetical influence of eight external water molecules reorganized to each charge distribution of the systems **1–3**.

The point charges on the external waters were then taken from the TIP3P model [84]. In the Monte Carlo simulations, the Lennard–Jones potential parameters were determined so as to approximate the interaction energies and the average heavy-atom distances for the clusters $HCl(H_2O)_4$, $Cl^-(H_2O)_4$, and $H_3O^+(H_2O)_3$. (A more detailed description and resultant parameters appear in [9].)

The next step was to compute the free-energy curves in the solvent coordinate (to be defined shortly) for the solution-phase reaction. For this purpose, Monte Carlo simulations were carried out using standard procedures [85] with Metropolis sampling, periodic boundary conditions, and the canonical (constant-NVT) ensemble at 298 K. Each cubic cell contains the reaction system **1**, **2**, or **3** and 248 TIP3P waters, with the box length 19.6 Å and the mass density therefore 0.997 g/cm^3. Each simulation started with an equilibration by generating more than 5×10^5 configurations, followed by several (4–8) sets of 1×10^6 configurations to compute the free-energy curves. Radial distribution functions were also computed for the chloride anion in water and were compared with X-ray and neutron diffraction experiments [9].

The calculations then proceeded in several steps, the first of which was to generate proton diabatic free-energy curves in the solvent coordinate, by a special sampling technique described next. (The second step, associated with generating proton-transfer potentials, is described subsequently.) The solvent coordinate was defined by the energy difference of the diabatic states,

$$\Delta E_{ij} \equiv V_i(\mathbf{S}; \mathbf{R}_i) - V_j(\mathbf{S}; \mathbf{R}_j) = \Delta V_{ij}(\mathbf{S}), \qquad (16)$$

where $V_i(\mathbf{S}; \mathbf{R}_i)$ denotes the total potential energy of the system at a fixed solute coordinate $\mathbf{R}_i$ as a function of the solvent configuration $\mathbf{S}$. (That is, ΔE gauges the relative energetics of two solute states at a given solvent configuration.) The subscripts i and j denote the diabatic states corresponding to **1–3**. $(i, j) = (1,2)$ and $(2,3)$ were considered for the first and the second proton-transfer steps, respectively, and $(1,3)$ was employed for the concerted pathway.

The proton diabatic free-energy curves were defined by

$$G_i(\Delta E_{ij}) = -k_B T \ln Q_i(\Delta E_{ij}), \qquad (17)$$

where $Q_i(\Delta E_{ij})$ is the partition function for ΔE_{ij}, computed via the probability distribution of ΔE_{ij}. To sample thermally inaccessible high free-energy regions, the free-energy perturbation method [72] was employed with an intermediate potential, i.e.,

$$V_\alpha(\mathbf{S}) = V_i(\mathbf{S}; \mathbf{R}_i) + \alpha(V_j(\mathbf{S}; \mathbf{R}_j) - V_i(\mathbf{S}; \mathbf{R}_i)), \qquad (18)$$

where the parameter α was varied from 0 to 1. The solvation effects from outside of the spherical truncation were estimated by the reaction field approximation and were on the order of 1 kcal/mol.

The proton diabatic free-energy curves for the first proton transfer $\mathbf{1} \to \mathbf{2}$ within a stepwise assumption in the solvent coordinate ΔE_{12} are displayed in Figure 8. The computed solvent reorganization energy for the first step, $\mathbf{1} \to \mathbf{2}$, was 12.6 kcal/mol, which indicates that the reactant ($\mathbf{1}$) and the contact ion-pair product ($\mathbf{2}$) equilibrium solvation states are well separated in the solvent coordinate (i.e., they are well-defined species).

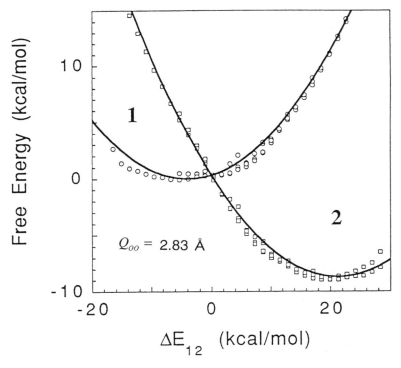

Figure 8. Calculated diabatic free-energy curves in the solvent coordinate ΔE_{12}. The assorted points are from Monte Carlo sampling, while the solid curves are parabolic fits, with the same curvatures.

So far, the free-energy curves are completely preliminary, since the problem of the proton potential has not been addressed. This was taken care of in the next step. Solvent configurations corresponding to the reactant ($\Delta E_{12} < 0$), the product ($\Delta E_{12} > 0$), and the intersection $\Delta E_{12} = 0$ in Figure 8 were then used as configurations from which to generate the proton-transfer potentials in the fixed field of the solvent in the following way. First, the solvent configurations were sampled in the course of the Monte Carlo simulations. Then, the MP2 wave function with Basis II was employed to compute the proton potential surfaces. Figure 9, displaying the resulting proton potential curves, confirmed the anticipated adiabatic scheme of Figure 1 for the HCl ionization in water. The variance of the proton barrier height among different solvent configuration samplings at $\Delta E_{12} = 0$ was checked and was found to be as small as 0.1 kcal/mol, indicating that the solvent polarization is well characterized by the ΔE_{12} coordinate, and in fact, the proton asymmetry is directly connected to ΔE_{12}.

The picture for the first transfer, $\mathbf{1} \rightarrow \mathbf{2}$, resulting from the calculations was thus determined to be as follows. The acid–base proton transfer is quantum adiabatic, rather than tunneling, with the proton adiabatically following the solvent rearrangement to configurations with $\Delta E_{12} = 0$. This rearrangement takes place at a slight cost of free energy. The precise cost of reaching the transition state in the solvent coordinate was determined by computing the vibrational energy levels of the proton at various points of ΔE_{12}. These final results, including the solvent self-free energy, are designated by small crosses in Figure 10. The process was found to be nearly barrierless: $\Delta G_{12}^{\ddagger} = 0.1$ kcal/mol. A reaction free energy of $\Delta G_{12} = -6.7$ kcal/mol was also estimated in a similar way. Thus, the adiabatic and almost activationless $\mathbf{1} \rightarrow \mathbf{2}$ transfer was computed to be markedly downhill

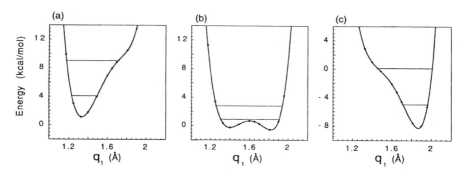

Figure 9. Proton-transfer potentials, with the Cl$\cdots$O distance at 2.91 Å, evaluated at negative, zero, and positive ΔE_{12} values. (Points are calculated values.) The ground and first excited proton vibrational levels are displayed.

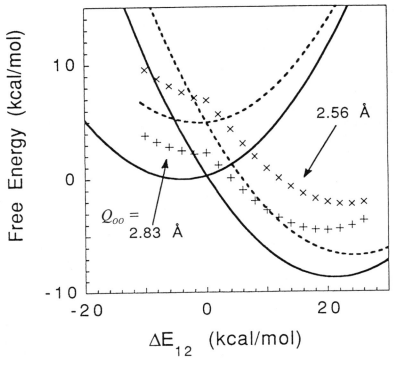

Figure 10. Diabatic free-energy curves in the solvent coordinate ΔE_{12}, with the ground proton vibrational levels (including the solvent self-free-energy) designated by $(+ + +)$ and $(\times \times \times)$ marks. The latter govern the adiabatic PT reaction. The O–O distance is (——, $+ + +$) 2.83 Å and (- - -, $\times \times \times$) 2.56 Å. The solid curves are taken from Figure 9.

in free energy. Note that the computational simulation uncertainty in $\Delta G_{12}^{\ddagger}$ was estimated to be much smaller than that of ΔG_{12}, which is about ± 0.3 kcal/mol. The experimental estimate of the overall reaction free energy for the reaction $HCl(aq) \rightleftharpoons Cl^-(aq) + H_3O^+(aq)$ is approximately -8 to 10 kcal/mol [86], the major portion of which was thus calculated to occur in the very first step.

The foregoing calculations were performed at an O–O distance of 2.83 Å in the reaction system, the equilibrium distance between two neutral water molecules in $(H_2O)_5$. When the calculations were repeated at an O–O distance of 2.56 Å, the H_3O^+–H_2O equilibrium distance in $H_3O^+(H_2O)_3$, almost the same basic picture emerged. However, the compression of the

O–O distance was computed to involve a significant cost in free energy, 4.9 kcal/mol in the neutral state **1** in the solution phase, and thus is strongly disfavored (Figure 10).

A contrasting equilibrium solvation view of the reaction was also considered. (See also Section IIB.) Figure 11 shows the calculated equilibrium solvation path (ESP), with the free energies minimized in the ΔE_{12} coordinate at each fixed proton position q_1, which assumes that the solvent motion is fast compared with the proton motion. The profile of the path along q_1 is a mean-force (mf) potential for the proton. But, as we have emphasized, the free-energy surface and the quantized character of the proton motion indicate that the *dynamical mechanism* of the proton transfer naturally involves a nonequilibrium solvation situation in which the proton is fast compared with the solvent [9, 12].

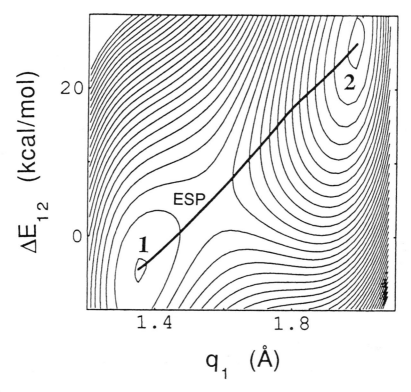

Figure 11. A two-dimensional contour plot of the free energy as a function of the proton coordinate and the solvent coordinate, with the equilibrium solvation path indicated. The contour spacing is 1 kcal/mol.

The question of what solvent rearrangements are involved in the aqueous solvation reorganization in the first proton-transfer step was also addressed. A significant guide here was the importance of solvent reorganization in the immediate vicinity of the hydrated excess proton in water, stressed early on by Newton and Ehrenson [32] for the hydrated clusters of H_3O^+ and proton transfer within them.

The observed characteristics of the structure change from **1** to **2** in an initial cluster calculation with eight external water molecules are summarized as follows: (i) The oxygen atoms in the external waters shrank slightly with respect to the solute, by 0.1–0.2 Å, but the hydrogen atoms in the external waters had large displacements, in such a way that either the molecular orientational dipole or the OH local dipole reorganized to stabilize the ion pair; (ii) in particular, the external water molecule that initially had been hydrogen bonded to the oxygen atom of the neutral H_2O_a turned away from the hydronium ion, $H_3O_a^+$ (Figure 12). The latter characteristic can be comprehended in terms of the local electrostatic repulsion between the net positive charge of H_3O^+ and the local partial charge on the H atom of that particular water molecule. In brief, the motion of the identified solvent water molecule is critical in the **1 → 2** proton transfer, playing a primary role in the change in coordination from 4 for H_2O in **1** to 3 for H_3O^+ in **2**. This is the key point.

Using an ab initio RHF method, Newton and Ehrenson [32] found that the first solvation shell of H_3O^+ can accommodate only three H_2O molecules, a result consistent with much chemical evidence [6, 23, 87] and subsequent calculations [88]; neither the "hydrogen-bonded" nor the "charge–dipole" configuration of the fourth water molecule resulted in

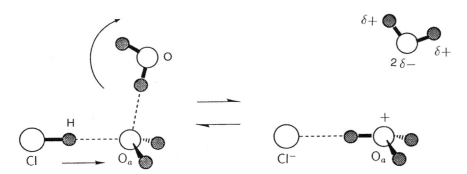

Figure 12. Schematic illustration of the reaction-promoting water molecule motion for the first proton-transfer step, **1 → 2**, associated with the change in the coordination number of the O_a-containing species from 4 to 3.

stabilization. Karlström, however, reported a stable (by about 4 kcal/mol) charge–dipole configuration [34]. The model potential of [9] gave a bound charge–dipole configuration. Common to all three works is the local electrostatic repulsion between H_3O^+ and the local OH dipole of the fourth H_2O molecule, resulting in a primary coordination number of 3 for H_3O^+.

The drastic changes in the local electrostatic interactions in the nearest neighbor solvation included in the preceding cluster calculation play the significant role, but long-range interactions and correlations in the aqueous solvent were explored by the Monte Carlo solution simulations. For a specific water molecule in the vicinity of the oxygen electron lone pair of H_2O_a or $H_3O_a^+$ in the neutral state 1, a linear hydrogen bonding was obtained, whereas a charge–dipole type of configuration of $H_3O_a^+$–H_2O with a binding energy distributed around -4 kcal/mol was found for the state 2. At the crossing point $\Delta E_{12} = 0$, that specific water molecule was very weakly bound in a configuration similar to the charge–dipole type, but noticeably *away* from the O_a atom. The solute–solvent interaction energy is smaller in the intermediate state, and consequently, that particular water molecule is pulled out from the reacting system into hydrogen-bond interaction with other external water molecules in the transition-state region; it clearly is a key player in the reaction coordinate and the solvent activation process, a fact that is consistent with the coordination change we have emphasized.

C. Second Proton-Transfer Step

Figures 13 and 14 display the corresponding solvent free-energy curves and proton-transfer potentials calculated for the second proton transfer, $2 \rightarrow 3$. The appropriate solvent coordinate is now ΔE_{23} [Cf. Eq. (16).] Here, the O–O distance is taken to be 2.56 Å, which is the equilibrium distance for H_3O^+–H_2O in the $H_3O^+(H_2O)_3$ cluster. As will be seen shortly, this separation in solution is slightly higher in free energy, by 0.5 kcal/mol, than the H_2O–H_2O equilibrium value (in an $(H_2O)_5$ cluster) of 2.83 Å. We believe that this difference reflects the influence of the bulk water solvent favoring a larger H_3O^+–H_2O separation than in an $H_3O^+(H_2O)_3$ cluster.

In Figure 14, the second proton transfer is seen to be adiabatic: The ground proton vibrational level is above the proton barrier in part (b). But now, in contrast to the first transfer, $1 \rightarrow 2$ (cf. Figure 10), activation in the solvent is required, and an estimate of the activation energy that takes account of the quantized proton vibrational level is $\Delta G_{23}^{\ddagger} = 0.9$ kcal/mol, with a solvent reorganization of 11.0 kcal/mol. This second transfer is nearly

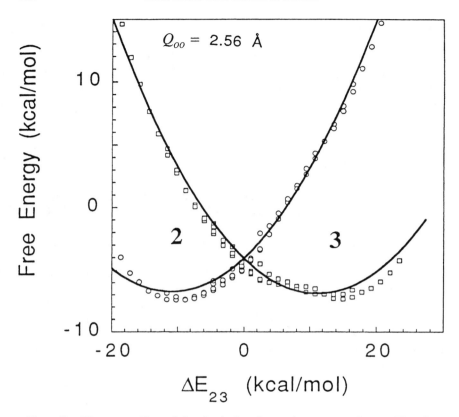

Figure 13. The same as Figure 8, but for the **2** → **3** second proton-transfer step. The solvent coordinate is ΔE_{23}, and the O–O distance is 2.56 Å.

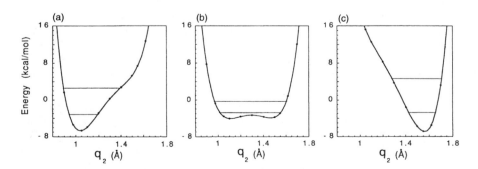

Figure 14. The analogue of Figure 9 for the situation of Figure 13.

thermodynamically neutral, with an estimated reaction free energy ΔG_{23} of only -0.2 kcal/mol.

We see in Figures 13 and 14 that the adiabatic scheme of Figure 1 is appropriate also for the second proton transfer between H_3O^+ and H_2O, and the full free-energy profile in the solvent coordinate is shown in Figure 15. In this case, the proton affinities of the two proton sites are almost the same. The reaction is thus not usefully regarded in "covalent versus ionic" terms, as is the first proton transfer, but is more similar to a (symmetric) charge-shift reaction. The polar aqueous solvation "prefers" a proton localized at either of the H_3O^+ sites, giving asymmetric proton potential wells at either site (Figure 14(a,c)) in the equilibrium reactant or product states. The transition between the two equilibrium polarization states requires activation in the solvent coordinate (Figure 13), and there is an

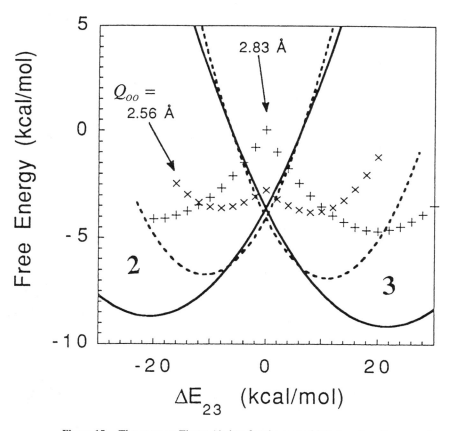

Figure 15. The same as Figure 10, but for the second PT step, $2 \rightarrow 3$.

intermediate crossing point that gives a symmetric proton potential (Figure 14(b)).

A rearrangement of the water molecule in the vicinity of H_2O_b or $H_3O_b^+$, which is similar to the one previously discussed for the first step, was found for the second proton-transfer step in the nearest neighbor solvation cluster of Figure 7: The water molecule hydrogen bonded to the neutral H_2O_b in state **2** turns away from $H_3O_b^+$ in state **3**, due to the critical change of the local electrostatic interaction. Further, the water molecule in the vicinity of $H_3O_a^+$ in a charge–dipole type of configuration in state **2** returns to hydrogen bond to H_2O_a in state **3** (Figure 16). These water rearrangement motions accommodating the primary change in hydration number from 4 for H_2O to 3 for H_3O^+ are expected to play an important role in step **2** → **3** in the solution-phase reaction and significantly contribute to the reaction coordinate.

An instructive carried out for comparison purposes, calculation, revealed the role of solute heavy-atom reorganization in the proton transfer. If the O–O separation is *not* allowed to "reorganize" from the $H_2O–H_2O$ equilibrium value of 2.83 Å, the proton transfer is found instead to be non-adiabatic tunneling. The small (0.27 Å) change in the separation of the heavy atoms affects the potential of the transferring proton significantly, and thus, nuclear reorganization plays a key role in this second transfer. The transition state for the small O–O distance is lower in energy than that for the larger O–O distance by 2.7 kcal/mol (Figure 15), providing a strong bias

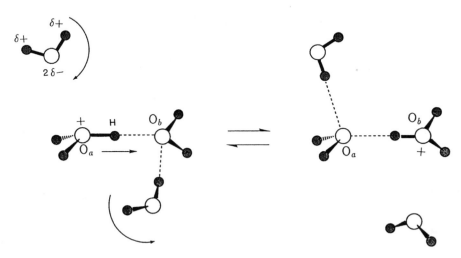

Figure 16. The analogue of Figure 12, but for the second proton transfer step, **2** → **3**.

against reaction at the larger O–O separation. In nearly thermoneutral processes such as these, the (diabatic) solvent activation energy is roughly proportional to the solvent reorganization energy; hence, these energy costs are smaller for the smaller proton shift. The adiabatic curves in [9] show that the activation energy for the smaller O–O distance is even smaller compared with the larger O–O case. Finally, the larger O–O separation was also found to require tunneling, an additional source of disfavor.

The issue of a stepwise ($1 \rightarrow 2 \rightarrow 3$) transfer versus a concerted ($1 \rightarrow 3$) proton transfer was also examined by calculating the free-energy curves for the concerted double transfer, now in the solvent coordinate ΔE_{13}. [See Eq. (16).]

When the O–O distance was taken to be the H_2O–H_2O equilibrium value of 2.83 Å, the activation energy for $1 \rightarrow 3$ in the ΔE_{13} coordinate was found to be 2.6 kcal/mol, which is noticeably higher than that for the stepwise $1 \rightarrow 2$ transfer discussed earlier. The $1 \rightarrow 3$ reorganization energy was computed to be 25.2 kcal/mol, which is larger than that for either $1 \rightarrow 2$ or $2 \rightarrow 3$. This (first) strong bias against a concerted transfer is general and robust against the numerical uncertainty in the simulation. Its qualitative origin is the larger reorganization free energy for the solvent required for the large charge separation in $1 \rightarrow 3$, versus the smaller reorganization costs for the less drastic charge separations in $1 \rightarrow 2$ and $2 \rightarrow 3$. Moreover, other biases against the concerted pathway were found, including an extra O–O compression energy cost absent in the stepwise process.

The strong conclusion is that the concerted pathway for the double proton transfer in Eq. (4) is unfavorable compared with the stepwise mechanism [89]. We believe that the qualitative arguments given in [9] should apply for general proton-transport processes in water, even though such a stepwise proton-transfer scenario contrasts with descriptions in which, for example, the existence of a solitonic motion [35] is suggested.

The calculations of [9] take no explicit account of the solvent electronic polarization, neglecting the correlation between the electrons in the solvent molecules and the change in the electronic structure of the solute during the proton transfer. In particular, the way the solvent electronic polarization solvates the intermediate electronic structure at the proton barrier region may alter the barrier height and thus the proton adiabaticity. This is the problem mentioned in Sections IIB and IIC.

As in Section IIC, the computational procedure employed in [9] roughly corresponds to the self-consistent (SC) limit in the sense that the fixed classical field of the solvent molecular point charges "sees" the delocalized electronic distribution of the (quantum chemically treated) solute molecules in the proton barrier region. In the opposite, Born–Oppenheimer (BO), limit, on the other hand, the solvent electronic polarization fully solvates the individual component solute electronic VB states at every point of the proton coordinate and would lead to greater stabilization in the proton barrier region than an SC treatment would. Aqueous proton transfers are expected to be somewhat closer to the BO limit [9]. Thus, the qualitative trend would be that a full account of the solvent electronic polarization lowers the proton barrier compared with the results we have described. A rough estimate suggests, for example for the first step, $1 \to 2$, a 1.4 kcal/mol change in the proton barrier height and a smaller effect on the proton vibrational levels [9, 13, 63], which is what is important for the reaction barrier. As in Section IIC, this indicates that the proton transfers would, in fact, be even more adiabatic (or nontunneling) than is displayed for the first and the second transfer steps shown in Figures 9 and 14, respectively.

In the foregoing discussion, the location of the zero-point proton vibrational level above the proton transfer barrier was taken as the indicator of adiabaticity. This issue was examined from a dynamical viewpoint in [9]. For the tunneling case, the proton-transfer mechanism is well characterized by the "proton coupling" C, defined as half the energy splitting of the lowest two proton vibrational levels held in each double well. In this case, the reaction rate constant is described by a perturbation expansion in C, giving a Golden Rule rate formula [14, 18b], which has a form similar to that of outer-sphere electron transfer reactions [26]. The idea is extended [13, 15] to the large coupling (adiabatic) limit by employing a semiclassical Landau–Zener curve-crossing model [90], also discussed in Section IIC. The key quantity governing the reaction adiabaticity and mechanism is the Landau–Zener transition probability per crossing, P_{LZ}. If one neglects quantum interference and friction effects [90, 91], the net probability for the passage to the product well—when all energetic reaction requirements are satisfied—is

$$P = 2P_{LZ}/(1 + P_{LZ}). \qquad (19)$$

Interestingly, P_{LZ} and P are slightly smaller than unity, even when the proton zero-point level is above the proton barrier, which means that this type of proton transfer is not purely "adiabatic" according to the curve-crossing concept. This is an interesting feature of the solvent water with a relatively fast fluctuation of the solvent librational reorganization [9], which tends to suppress P_{LZ} and P. Feature can be compared with, the polar CH_3Cl solvent, which gives $P_{LZ} \sim 1$ for an adiabatic acid ionization, with a comparable value of the proton coupling for the system in Section IIC [13]. However, this small dynamic nonadiabaticity for the HCl system would be expected to disappear for the reasons discussed in [9]. For instance, the preceding discussion and that in Section IIC indicate that an explicit account of the solvent electronic polarization will lower the proton barrier and, consequently, increase both the adiabaticity parameter P_{LZ} and P. Indeed, an assortment of possible impacts of going beyond the modeling of [9] indicates that the proton transfer between HCl and H_2O is most likely to be fully adiabatic, although small dynamical nonadiabatic effects could not be absolutely excluded.

The mechanistic picture that emerged from the work just described is that HCl ionization is a consecutive set of quantum adiabatic, nontunneling proton transfers. The first transfer involves an almost activationless solvent reorganization to a contact $Cl^- $–$H_3O^+$ ion pair, with an accompanying reaction free energy of -6.7 kcal/mol. The second transfer, producing a solvent-separated ion pair, is slightly activated in a solvent coordinate and is approximately thermoneutral. A key component in the required solvent reorganization in each step is a solvent water molecule rearrangement to accommodate the change in the coordination number of the proton-accepting water from 4 (for H_2O) to 3 (for H_3O^+). The two proton transfers occur in a stepwise fashion; there are assorted strong biases against a concerted pathway. The transfers follow a nonequilibrium solvation path in their respective solvent coordinate.

The overall experimental reaction free energy of HCl ionization in water given by Eq. (4) is approximately -8 to -10 kcal/mol [86]. While no simulations have been performed for the remaining passage from the solvent-separated ion-pair structure 3 in Eq. (4) to the fully separated ions, a dielectric continuum (screened Coulomb) estimate suggests that that process involves only a very small (less than 1 kcal/mol) overall change in free energy in aqueous solution with the high dielectric constant (~ 80). When a proton of $H_3O_b^+$ in 3 is further transferred to the next adjacent water molecular, the first solvation shells for each of Cl^- and H_3O^+ will be completed. The dielectric screening for these "dressed" hydration clusters, $Cl^-(H_2O)_n$ ($n = 4$–6) and $H_3O^+(H_2O)_3$, should be very efficient in aqueous solution. We expect, however, that there can be small barriers in a solvent coordinate—much like that seen earlier in the second step $2 \rightarrow 3$ —along the pathway, as well as nuclear reorganization of O–O distances to assist the adiabatic proton transfer. As we have argued, we anticipate these processes to be stepwise—rather than concerted and solitonic—in character.

Since the solvent barriers for the first two steps are either negligible or small, there could well be considerable barrier recrossing effects that would need to be taken into account in the computation of reaction rate constants [1, 9, 26]. In particular, the $1 \rightarrow 2$ proton transfer step could be diffusive in character, with multiple solvent barrier recrossings associated with the dynamics of the fourth H_2O molecule rearrangement noted in Figure 12. Also, since the $1 \rightarrow 2$ transfer evidently has a negligible barrier, time-dependent rate constants will likely be necessary to describe the non-exponential kinetics, with the solvent *dynamics* playing a significant role [26, 92]. This feature—like the dynamic curve-crossing aspects discussed in Section IIC—enriches, but would not alter, the energetic pathways. If one neglects these considerations, the time scales for the first and the second PT steps are estimated to be 0.1 and 0.4 ps, respectively, by using a semiclassical rate formula, $k = 2P_{LZ}/(1 + P_{LZ})(\Omega/2\pi)\exp(-\Delta G^{\ddagger}/k_B T)$, bearing in mind that these times should be lower limits. A rapid ionization was also found for nonquantized proton motion in [41].

Finally, the various site charges in the trimer solute for the transfer $\mathbf{1} \rightarrow \mathbf{2}$ in solution at the approximate transition state configuration $\Delta E_{12} = 0$ were examined, and the following pattern in the $\mathbf{1} \rightarrow \ddagger \rightarrow \mathbf{2}$ process emerged: Cl($-0.417 \rightarrow -0.654 \rightarrow -0.846$), H_1 ($0.298 \rightarrow 0.321 \rightarrow 0.327$), O_a ($-0.975 \rightarrow -0.672 \rightarrow -0.593$), $H_2 = H_3$ ($0.542 \rightarrow 0.485 \rightarrow 0.543$). This pattern reflects a non-conventional, Mulliken-type charge-transfer picture [12, 21, 39, 57–59, 65, 70], discussed in Section IIB. Its generality is presently under investigation.

D. Ionization of Hydrofluoric Acid in Water

That the acid ionization proton-transfer reaction of hydrofluoric acid (HF) in water must differ, at least in detail, from that of HCl follows from the fact that HF is a weak acid [3], with $pK_a = +3$, while HCl is strong. In terms of free energy, ΔG for the overall reaction required to produce completely separated and solvated ions is approximately $+4$ kcal/mol [3]; that is, the reaction is "endothermic". One major reason for the current interest in the HF system is that the first step could be slower than the corresponding step for HCl and thus more directly amenable to experimental scrutiny. Accordingly, the corresponding study of HF in water was undertaken [10, 11]. The focus here will be only on the first step,

$$FH' \cdots H_2O \cdots H_2O \rightleftharpoons F^- \cdots H'OH_2^+ \cdots H_2O \rightleftharpoons F^- \cdots H'OH \cdots H_3O^+.$$

$$(20)$$

Because the methodology is identical to that for HCl described in the preceding sections, with some minor differences, it is not detailed here. In fact, everything is so similar in its general features to the case of HCl, that we give only a very brief description.

The basic picture for the first ionization step of HF, yielding a contact ion pair, is given in Figure 17, in which the $F \cdots O$ distance is 2.57 Å. The essential story is the now-familiar one: The quantum proton transfer is adiabatic, rather than tunneling, with the proton adiabatically following the slower solvent rearrangement to configurations with $\Delta E_{12} = 0$. The precise

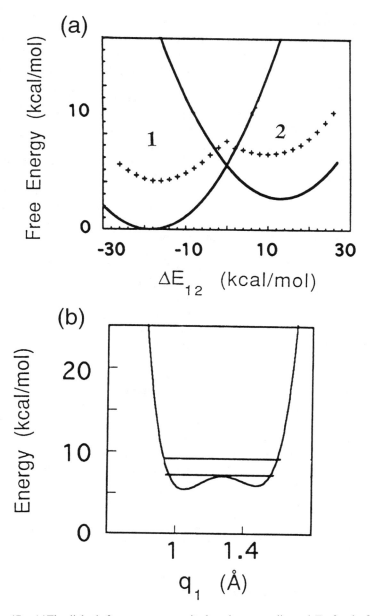

Figure 17.   (a)The diabatic free-energy curves in the solvent coordinate ΔE_{12} for the first PT step of HF in water. The crosses $(+ + +)$ show the ground vibrational proton levels, including the solvent self-free energy, and are the relevant curves for the adiabatic PT. (b) The proton transfer potential calculated at $\Delta E_{12} = 0$ and with $F \cdots O$ fixed at 2.57 Å. The first two proton vibrational levels are indicated.

cost of reaching that configuration is again calculated by computing the proton transfer potentials at various ΔE_{12} values and quantizing the nuclear motion of the proton in those potentials. The results, which also include the self-free energy of the solvent, are indicated by the crosses in Figure 17(a).

So, once again, the reaction coordinate is in the solvent, and the activation free energy was estimated to be 3.3 kcal/mol, whereas the change in the reaction free energy for the contact ion pair production $1 \rightarrow 2$ was estimated to be $\Delta G_{12} = +2.2$ kcal/mol. We do not discuss the second step, which produces the solvent-separated ion pair from the contact ion pair, but the calculation of ΔG for that step, plus a dielectric continuum estimate of the remaining change in free energy required to produce the free ions in water, gives [10, 11] an estimate of the overall change in free energy of approximately +3 kcal/mol. This value is to be compared with the +4 kcal/ mol experimental estimate, which can be fairly regarded as good agreement for such a complex system and calculation.

Several remarks need to be made about the numbers just presented. The first is that, as in the case of HCl, most of the overall reaction free energetics, which here are unfavorable for acid ionization, are associated with the first contact ion pair production step, which is uphill in free energy. This statement is in direct contradiction to the interesting suggestion [43] that the overall weakness of HF as an acid is due to the difficulty of separating the contact ion pair in water, whereas in the view presented here the first step in producing the contact ion pair is supposed to be favorable—i.e., negative— in free-energy terms. This issue is discussed elsewhere [93], where the results of [10] and [11] are supported. A second remark is that in a Car–Parrinello type of simulation [41] of HF in water (without any proton nuclear quan- tization), the acid ionization could not be studied on the relatively short time scale of the simulation, presumably again due to its "endothermic" character. Third, the results we have described indicate that, when the reaction is viewed in the reverse, ion combination sense, there will be a slight barrier to the final formation of the neutral acid molecular pair from the contact ion pair, with the contact ion pair itself being lower in free energy than the free ions at large separation. This relationship suggests that the short-range ion combination dynamics have interesting features that were inaccessible to prior experimental techniques [6], but that are accessible to modern ultrafast spectroscopy methods.

Finally, the HF acid ionization system presents an interesting candidate for infrared induced proton transfer in solution, in which the HF molecule would absorb an infrared photon and the contact ion pair would be produced. It would take us too far afield to discuss this reaction here, but we remark that the basic ideas about the process and the formula for calculating the appropriate rates pertaining to the process are greatly assisted by the picture of the PT reaction in proton adiabatic terms with a solvent reaction coordinate [44].

IV. CONCLUDING REMARKS

In the preceding sections, we have summarized several investigations into the fundamental reaction step of acid ionization in solution to produce a contact ion-pair product via proton transfer to a base molecule. While the description of the process is, to a degree, complex, involving significant aspects of electronic structure variation, nuclear quantum dynamics of the proton, and strong coupling to the surrounding solvent molecules, the overall picture that emerges is nonetheless fairly simple. Typically, the quantum proton will adiabatically follow the rearrangement of the surrounding solvent molecules, such that the reaction coordinate is in a collective solvent coordinate; the activation free energy is largely determined by the solvent coordinate, together with a difference in zero-point vibrational energies for the bound proton motion. This picture has been contrasted with other views in which the reaction either would be considered as the classical motion of the proton over a barrier in its own coordinate or would take place via quantum tunneling of the proton through the barrier in its coordinate. The former does not adequately address the key quantum features of the reaction, while the latter is, in general, more appropriate for intramolecular protons transfers or, even more generally, for those proton transfers in which the reactant acid and base molecules do not have a short internuclear separation induced by their mutual hydrogen bonding. For the special cases of HCl and HF in water, the free energetics involved in the contact ion pair formation were estimated to be a significant fraction of the overall thermodynamic change in free energy required to produce the completely separated and independently solvated anion and hydronium ion. It is clearly of interest to extend such studies to other acids, including organic acids, for which the considerations of Section IIB represent only a beginning.

The second step in the aqueous-phase reaction, in which the hydronium ion in the contact ion pair with the anion transfers its proton to another water molecule, has been discussed in some detail for the case of HCl. (That of HF is similar [11].) This process has features in common with, as well as important differences from, the first chemical step. The studies we have described indicate that the second step produces a solvent-separated ion pair for a hydronium ion and the anion. Clearly, at some point, as the proton-bearing water molecule becomes sufficiently removed from the perturbing influence of the anion, the process must change to that of the physical tran-sport of a proton in water, a topic of much current interest. This transition deserves further study.

The issue of H/D isotope effects for the contact ion-pair formation has not been addressed. This is an important area to clarify, since it offers a potential experimental probe of the reaction mechanism. The formal theory is under development, but it is clear that one will require accurate measurements of proton potential surfaces to do full justice to the area. There could be surprises [13]. For example, if the H^+ transfer reaction is adiabatic because the proton level at the solvent transition state lies above the proton barrier, the D^+ transfer could be nonadiabatic, due to the increased mass acting so as to reduce the corresponding D^+ level below the barrier. Thus, the deuteron would tunnel, but the proton would not, which is perhaps a startling statement if one has the incorrect view that a particle must either tunnel or execute a classical over-the-barrier motion. In any event, one clearly requires an accurate account of proton potentials for predictions of kinetic isotope effects.

Experimental study of the primary initial proton-transfer step is clearly desirable. This will be difficult, however, for the HCl reaction in water if the reaction is essentially barrierless, as is predicted. One possibility for slowing the reaction time scale is to examine the reaction in the significantly less polar supercritical water medium [94–96]. The slower aqueous-phase HF reaction offers possibilities, especially when it is viewed in the reverse recombination sense, so that the short-range features of the reaction dynamics, inaccessible to prior experimental techniques [6], might be probed. The significant population of the un-ionized HF acid in water could also offer possibilities for infrared-induced, mode-specific proton transfer to produce the contact ion pair; when viewed from an adiabatic proton, solvent coordinate perspective, this theory has some strong analogies to an electronic transition to an excited state followed by radiationless transition to the ground state; the theory is under development [44, 97].

Another reaction medium of interest for ground-electronic-state acid ionization is that of water clusters, and some theoretical work in this area has begun [9, 98–101]. Of special importance here will be the nature of the approach to the liquid-phase process, since small clusters presumably have a solvation power inferior to that of, for example, bulk liquid water, and in addition, geometric bonding patterns in clusters could be quite special. On the other hand, it is conceivable that the study of not-too-large clusters (e.g., 10 water molecules) may already reveal many aspects of the solution problem [9, 102]. Much remains to be discovered here.

Perhaps surprisingly, the issue of the acid ionization of HCl is one of direct relevance to the well-known phenomenon of ozone depletion in the Antarctic stratosphere. In the Antarctic winter, the low temperatures (about 190 K) and atmospheric flow patterns produce ice particles in polar stratospheric clouds. It is now believed that on these particles, the heterogeneous net reaction

$$HCl + ClONO_2 \rightarrow Cl_2 + HNO_3, \tag{21}$$

involving HCl and chlorine nitrate, formerly thought to be inactive reservoir species safely sequestering chlorine, to produce molecular chlorine and nitric acid, is the key source of Cl_2, which is then photolyzed to radicals to participate in gas-phase ozone destruction processes [103]. Whether HCl is ionized at the surface of ice under stratospheric conditions is important, since this would open up the possibility of rapid ionic reaction pathways for Eq. (21), rather than molecular routes (such as those followed in the gas phase, which are slow). It has been suggested in this connection that the surface of ice at 190 K might involve a liquidlike layer [104], in which case the HCl ionization would be thermodynamically feasible (and, according to Section III , kinetically rapid).

The issue of HCl acid ionization at the surface of ice was studied in [105] using methods somewhat similar to those described in Section III. But first an argument was presented that HCl could not ionize atop the surface of the ice, but instead needed to be dynamically incorporated slightly below the surface—"at" the surface, rather than "on" it. In the sites occupied by the HCl that is likely to be produced in such a dynamic incorporation, it was found [105, 106] that the HCl proton transferred to a nearby water molecule at the ice surface to produce the contact ion pair, with reaction free energetics and barrier heights not so very different from the aqueous-phase solution results described in Section III. Various tests of the degree of order of the ice surface region indicated that no liquidlike layer was required. Subsequent calculations [107, 108] supported the initial argument that HCl could not ionize atop the surface. A number of issues remain to be investigated here, such as the precise ionic mechanism of Eq. (21), which in turn can involve the issue of whether there is transport of the proton away from an initially produced contact ion, such that Eq. (21) would involve only the Cl^- ion, or whether the H_3O^+ ion is involved as well [107, 109].

While the acid ionization of the weaker HF is not likely to be important at the surface of the ice [110], the acid ionization of the stronger HBr at or on the ice surface is important for ozone depletion mechanisms in the Arctic troposphere [111]. This notion is under investigation, but it is interesting to note that in a cluster study, HBr was found [100] to form, as the most stable species over a wide temperature range, a "solvent-separated" ion pair—the Br^- ion and the H_3O^+ ion, bridged by three water molecules in a trigonal bipyramidal structure—in a four-water-molecule cluster. This suggests a certain ease of ionization at, or perhaps even atop, the ice surface, but that remains to be seen.

After the preceding few paragraphs, it is perhaps fitting to close this contribution with a paraphrase of a recent remark: It is interesting to note that questions of a very fundamental nature in chemical physics can have importance in socially relevant contexts [112].

Acknowledgments

The proton-transfer work of JTH has been supported over the years by grants from the NSF, most recently, NSF CHE97-00419 and NSF ATM96-13802, and by the NIH, including a Shannon award. In addition, acknowledgment is made to the donors of the Petroleum Research Fund, administered by the American Chemical Society, for the partial support of the work described herein. Acknowledgment is also made to the Pittsburgh Supercomputer Center and the National Center for Atmospheric Research, supported by the NSF, for computer time. Writing of the present article was supported by a faculty fellowship from the University of Colorado Council on Research and Creative Work. JTH acknowledges with pleasure and gratitude other former and present members of the research group who have worked, or are working, on various aspects of acid ionization proton-transfer reactions: Domenic Ali, Roberto Bianco, Daniel Borgis, Peggy Bruehl, Brad Gertner, Sunhee Jung, Phil Kiefer, Sangyoub Lee, Thomas Schroeder, Arnulf Staib, Jesus Timoneda, and Ward Thompson; we especially recognize that our proton-transfer work was started in collaboration with Drs. Ali, Borgis, and Lee. Related collaborations with Sally Chapman and with Dario Beksic, Juan Bertrán, and José María Lluch, the latter supported by the Iberdrola Foundation, are gratefully acknowledged. KA was partly supported by a JSPS Fellowship during his stay in the group with JTH. He is grateful to former colleagues in the group for helpful and fruitful discussions. Recent support by the Grants in Aid for Scientific Research from the Ministry of Education (Nos. 09740415 and 10120203) is also acknowledged.

References

[1] J. T. Hynes, in *The Theory of Chemical Reaction Dynamics*, M. Baer, Ed., CRC, Boca Raton, Vol. 4, 1985; D. G. Truhlar, W. L. Hase, and J. T. Hynes, *J. Phys. Chem.* **87**, 2664 (1983); R. M. Whitnell and K. R. Wilson, *Adv. Comp. Chem.* **4**, 67 (1993); P. Hänggi, P. Talkner, M. Borkovec, *Rev. Mod. Phys.* **62**, 250 (1990); A. Nitzan, *Adv. Chem. Phys.* **70**, 489 (Part 2, 1988).

[2] D. G. Truhlar, B. C. Garrett and S. J. Klippenstein, *ibid.* **100**, 12771 (1996); G. A. Voth and R. M. Hochstrasser, *ibid.* **100**, 13034 (1996); R. M. Stratt, M. Maroncelli, *ibid.* **100**, 12981 (1996); J. T. Hynes, in *Solvent Effects and Chemical Reactivity*, O. Tapia and J. Bertran, Eds., Kluwer, Amsterdam, (1996), p. 231; J. D. Simon, Ed., *Ultrafast Dynamics of Chemical Systems*, Kluwer Academic Publishers, Dordrecht, The Netherlands, 1994; Y. Gauduel and P. J. Rossky, Eds., *Ultrafast Reaction Dynamics and Solvent Effects*, AIP Press, New York, 1994.

[3] R. P. Bell, *The Proton in Chemistry*, Chapman and Hall, London, 1973; E. F. Caldin and V. Gold, Eds., *Proton Transfer Reactions*, Chapman and Hall, London, 1975; R. P. Bell, *The Tunnel Effect in Chemistry*, Chapman and Hall, London (1980); E. Buncel and C. C. Lee, Eds., *Isotope Effects in Organic Chemistry*, Elsevier, Amsterdam Vol. 2, 1976; F. Hibbert, *Adv. Phys. Org. Chem.* **22**, 113 (1986); F. Hibbert, *ibid.* **26**, 255 (1990); F. H. Westheimer, *Chem. Rev.* **61**, 265 (1961).

[4] (a) E. M. Kosower, D. Huppert, *Annu. Rev. Phys. Chem.* **37**, 127 (1986); M. M. Kreevoy and D. G. Truhlar, in *Rates and Mechanisms of Reactions*, 4th ed., C. F. Bernasconi, Ed., Wiley, New York, 1986, Chap. 1; H. Ratajczak, in *Electron and Proton Transfer Processes in Chemistry and Biology*, A. Müller, H. Ratajczak, W. Junge, and E. Diemann, Eds., Elsevier, Amsterdam, 1992. (b) T. Bountis, Ed., *Proton Transfer in Hydrogen-Bonded Systems*, Plenum, New York, 1993. (c) See also the special issues on proton transfer: *Chem. Phys.* **136**, No. 2 (1989); *J. Phys. Chem.* **95**, No. 25 (1991), and *Ber. Bunsenges. Phys. Chem.* **102**, No. 3 (1998).

[5] W. P. Jencks, *Catalysis in Chemistry and Enzymology*, McGraw-Hill, New York, 1969; M. L. Bender, *Mechanisms of Homogeneous Catalysis from Protons to Proteins*, Wiley, New York, 1971. A. Fersht, *Enzyme Structure and Mechanism*, W. H. Freeman, New York, 1985; R. D. Gandour and R. L. Schowen, *Transition States of Biochemical Processes*, Plenum, New York, 1978; J. P. Klinman, *CRC Crit. Rev. Biochem.* **10**, 39 (1981); F. Menger, *Acc. Chem. Res.* **26**, 206 (1993); M. Gutman and E. Nachliel, *Biochim. Biophys. Acta.* **391**, 1015 (1990).

[6] M. Eigen, W. Kruse, and L. De Maeyer, *Prog. React. Kin.* **2**, 285 (1964); M. Eigen, *Angew. Chem. (Int. Ed. Engl.)* **3**, 1 (1964); *Pure Appl. Chem.* **6**, 97 (1963).

[7] W. J. Albery, *Progr. React. Kinet.* **4**, 353 (1967); Y. Maréchal, in [4]b.

[8] K. Ando and J. T. Hynes, in *Structure and Reactivity in Aqueous Solution: Characterization of Chemical and Biological Systems*, C. J. Cramer and D. G. Truhlar, Eds., ACS Books, Washington, DC, 1994; *J. Mol. Liq.* **64**, 25 (1995).

[9] K. Ando and J. T. Hynes, *J. Phys. Chem. B* **101**, 10464 (1997).

[10] K. Ando and J. T. Hynes, *Faraday Discuss.* **102**, 435 (1995).

[11] K. Ando and J. T. Hynes, "Molecular Mechanism of HF Ionization in Water", to be submitted.

[12] J. Juanós i Timoneda and J. T. Hynes, *J. Phys. Chem.* **95**, 10431 (1991).

[13] A. Staib, D. Borgis, and J. T. Hynes, *J. Chem. Phys.* **102**, 2487 (1995).

[14] D. Borgis, S. Lee, and J. T. Hynes, *Chem. Phys. Lett.* **162**, 19 (1989); D. Borgis and J. T. Hynes, *J. Chem. Phys.* **94**, 3619 (1991); *Chem. Phys.* **170**, 315 (1993).

[15] D. Borgis and J. T. Hynes, *J. Phys. Chem.* **100**, 1118 (1996).

[16] E. D. German, A. M. Kuznezov, and R. R. Dogonadze, *J. Chem. Soc. Faraday Trans. 2* **76**, 1128 (1980); L. I. Krishtalik, *Charge Transfer Reactions in Electrochemical and Chemical Processes*, Plenum, New York, 1986.

[17] M. J. Gillan, *Phys. Rev. Lett.* **58**, 563 (1987); *J. Phys. C* **20**, 3621 (1987); G. A. Voth, D. Chandler, and W. H. Miller, *J. Phys. Chem.* **93**, 7009 (1989); *J. Chem. Phys.* **91**, 7749 (1989); D. H. Li and G. A. Voth, *J. Phys. Chem.* **95**, 10425 (1991); J. Lobaugh and G. A. Voth, *Chem. Phys. Lett.* **198**, 311 (1992); *J. Chem. Phys.* **100**, 3039 (1994); *ibid.* **104**, 2056 (1995).

[18] (a) D. Laria, G. Ciccotti, M. Ferrario, and R. Kapral, *J. Chem. Phys.* **97**, 378 (1992); M. Ferrario, D. Laria, G. Ciccotti, and R. Kapral, *J. Mol. Liq.* **61**, 37 (1994). (b) R. I. Cukier and M. Morillo, *J. Chem. Phys.* **91**, 857 (1989). (c) T. N. Truong, J. A. McCammon, D. J.

Kouri, and D. K. Hoffman, *J. Chem. Phys.* **96**, 8136 (1992); N. Makri and W. H. Miller, *J. Chem. Phys.* **91**, 4026 (1989); G. K. Scenter, M. Messina, and B. C. Garrett, *J. Chem. Phys.* **99**, 1674 (1993).

[19] H. Azzouz and D. Borgis, *J. Chem. Phys.* **98**, 7361 (1993); *J. Mol. Liq.* **61**, 17 (1994). D. Borgis, G. Tarjus, and H. Azzouz, *J. Phys. Chem.* **96**, 3188 (1992); *J. Chem. Phys.* **97**, 1390 (1992).

[20] M. E. Tuckerman, K. Laasonen, M. Sprik, and M. Parrinello, *J. Chem. Phys.* **103**, 150 (1995); *J. Phys. Chem.* **99**, 5749 (1995); M. E. Tuckerman, P. J. Ungar, T. von Rosenvinge, and M. L. Klein, *ibid.* **100**, 12878 (1996). See also N. Agmon, *Chem. Phys. Lett.* **244**, 456 (1995); *J. Chem. Phys.* (Paris) **93**, 1714 (1996).

[21] A. Warshel and R. M. Weiss, *J. Phys. Chem.* **102**, 6218 (1980); A. Warshel and S. Russel, *J. Am. Chem. Soc.* **108**, 6569 (1986); A. Warshel, *Computer Modeling of Chemical Reactions in Enzymes and Solutions*, Wiley, New York, 1991.

[22] S. Hammes-Schiffer and J. C. Tully, *J. Chem. Phys.* **103**, 8525 (1995); H. Decornez, K. Drukker, M. M. Hurley, and S. Hammes-Schiffer, *Ber. Bunsenges. Phys. Chem.* **102**, 533 (1998).

[23] G. W. Robinson, *J. Phys. Chem.* **95**, 10386 (1991); G. W. Robinson, P. J. Thistlethwaite, and J. Lee, *ibid.* **90**, 4224 (1986).

[24] J. R. Mathis and J. T. Hynes, *J. Phys. Chem.* **98**, 5445, 5460 (1994); H. J. Kim and J. T. Hynes, *J. Am. Chem. Soc.* **114**, 10508, 10528 (1992); J. R. Mathis, H. J. Kim, and J. T. Hynes, *ibid.* **115**, 8248 (1993); W. P. Keirstead, K. R. Wilson, and J. T. Hynes, *J. Chem. Phys.* **95**, 5256 (1991).

[25] A. C. Legon and L. G. Willoughby, *Chem. Phys. Lett.* **95**, 449 (1983); G. B. Bacskay, *Mol. Phys.* **77**, 61 (1992); G. B. Bacskay, D. I. Kerdraon, and N. S. Hush, *Chem. Phys.* **144**, 53 (1990).

[26] (a) R. A. Marcus, *Annu. Rev. Phys. Chem.* **15**, 155 (1964); M. D. Newton and N. Sutin, *ibid.* **35**, 437 (1984); B. Bagchi, *Ann. Rev. Phys. Chem.* **40**, 115 (1989); P. F. Barbara, T. J. Meyer, and M. A. Ratner, *J. Phys. Chem.* **100**, 13148 (1996). (b) J. Ulstrup, *Charge Transfer Processes in Condensed Media*, Springer-Verlag, Berlin, 1979. For representative simulation studies, see, e.g., (c) A. Warshel and J.-K. Hwang, *J. Chem. Phys.* **84**, 4938 (1986); J.-K. Hwang and A. Warshel, *J. Am. Chem. Soc.* **109**, 715 (1987); R. A. Kuharski, J. S. Bader, D. Chandler, M. Sprik, M. L. Klein, and R. W. Impey, *J. Chem. Phys.* **89**, 3248 (1988); C. L. Kneifel, M. D. Newton, and H. L. Friedman, *J. Mol. Liq.* **60**, 107 (1994); T. Fonseca and B. M. Ladanyi, in *Ultrafast Reaction Dynamics and Solvent Effects: Experimental and Theoretical Aspects*, Y. Gauduel and P. J. Rossky, Eds., AIP Press, New York, 1994; D. A. Zicchi, G. Ciccotti, J. T. Hynes, and R. Kapral, *J. Phys. Chem.* **93**, 6261 (1989); K. Ando, *J. Chem. Phys.* **101**, 2850 (1994); *ibid.* **106**, 116 (1997).

[27] T. Fonseca, H. J. Kim, and J. T. Hynes, *J. Mol. Liq.* **60**, 161 (1994); S. Kato and Y. Amatatsu, *J. Chem. Phys.* **92**, 7241 (1990); H. J. Kim and J. T. Hynes, *J. Photochem. Photobiol. A: Chemistry* **105**, 337 (1997); S. Hayashi, K. Ando, and S. Kato, *J. Phys. Chem.* **99**, 955 (1995).

[28] W. M. Olmstead and J. I. Brauman, *J. Am. Chem. Soc.* **99**, 4219 (1977); M. J. Pellerite and J. I. Brauman, *ibid.* **105**, 2672 (1983); J. Chandrasekhar, S. F. Smith, and W. L. Jorgensen, *ibid.* **107**, 154 (1985); R. A. Chiles and P. J. Rossky, *ibid.* **106**, 6867 (1984).

[29] J. P. Bergsma, B. J. Gertner, K. R. Wilson, and J. T. Hynes, *J. Chem. Phys.* **86**, 1356 (1987); B. J. Gertner, J. P. Bergsma, K. R. Wilson, and J. T. Hynes, *ibid.* **86**, 1377 (1987).

[30] S. Scheiner, *Acc. Chem. Res.* **18**, 174 (1985); *J. Chem. Phys.* **77**, 4039 (1982); *J. Am. Chem. Soc.* **103**, 315 (1981).

[31] T. Komatsuzaki and I. Ohmine, *Chem. Phys.* **180**, 239 (1994).

[32] M. D. Newton, and S. Ehrenson, *J. Am. Chem. Soc.* **93**, 4971 (1971); M. D. Newton, *J. Chem. Phys.* **67**, 5535 (1977).

[33] B. Halle and G. Karlstrom, *J. Chem. Soc. Faraday Trans. 2* **79**, 1047 (1983); F. H. Stillinger, in *Theoretical Chemistry: Advances and Perspectives,* H. Eyring and D. Henderson, Eds., Academic Press, New York, Vol. 3, 1978.

[34] G. Karlstrom, *J. Phys. Chem.* **92**, 1318 (1988).

[35] V. Ya. Antonchenko, A. S. Davydov, and A. V. Zolotariuk, *Phys. Stat. Sol. (b)* **115**, 631 (1983); A. S. Davydov, *Solitons in Molecular Systems*, Naukovaja Dumka, Kiev, 1984; A. Godzik, *Chem. Phys. Lett.* **171**, 217 (1990); E. S. Kryachko and V. P. Sokhan, in [4]b; J. F. Nagle, in [4]b.

[36] E. Hückel, *Z. Elektrochem.* **34**, 546 (1928); J. D. Bernal and R. H. Fowler, *J. Chem. Phys.* **1**, 515 (1933); B. E. Conway, J. O'M. Bockris, and H. Linton, *J. Chem. Phys.* **24**, 834 (1956).

[37] R. D. Gandour, G. M. Maggiora, and R. L. Schowen, *J. Am. Chem. Soc.* **96**, 6967 (1974).

[38] The literature here is enormous. See, for example, 9–11 in references [13]. Some other recent references can be found in J. Syage, *J. Phys. Chem.* **99**, 5772 (1995); A. Douhal, F. Lahmani, and A. H. Zewail, *Chem. Phys.* **207**, 477 (1996); and [4]c.

[39] H. Ratajczak, *J. Phys. Chem.* **76**, 3000, 3991 (1972); H. Ratajczak and W. J. Orville-Thomas, *J. Phys. Chem.* **58**, 911 (1973); M. Ilczyszyn, H. Ratajczak, and K. Skowronek, *Magn. Reson. Chem.* **26**, 445 (1988).

[40] C. Chipot, L. G. Gorb, and J.-L. Rivail, *J. Phys. Chem.* **98**, 1601 (1994); J.-L. Rivail, S. Antonczak, C. Chipot, M. F. Ruiz-López, and L. G. Gorb, in *Structure, Energetics, and Reactivity in Aqueous Solution*, C. J. Cramer and D. G. Truhlar, Eds., ACS Books, Washington, DC, 1994.

[41] K. Laasonen and M. L. Klein, *J. Am. Chem. Soc.* **116**, 11620 (1994).

[42] P. A. Giguere, *J. Chem. Ed.* **56**, 571 (1979); *Chem. Phys.* **60**, 421 (1981).

[43] K. Laasonen and M. L. Klein, *Mol. Phys.* **88**, 135 (1996).

[44] T. Schroeder, A. Staib, and J. T. Hynes, work in progress; K. Ando, A. Staib, and J. T. Hynes, in *Femtochemistry: Ultrafast Chemical and Physical Processes in Molecular Systems*, M. Chergui, Ed., World Scientific, Singapore, 1996, p. 534; in *Fast Elementary Processes in Chemical and Biological Systems*, A. Tramer, Ed., AIP, New York, 1996, p. 326.

[45] P. Schuster, W. Jakubetz, W. Meyer, and B. M. Rode, in *Chemical and Biochemical Reactivity*, E. D. Bergman and B. Pullman, Eds., Jerusalem, 1974.

[46] Th. Zeegers-Huyskens and P. Huskens, in *Molecular Interactions*, H. Ratajczak and W. J. Orville-Thomas, Eds., Wiley, New York, Vol. 2, 1980; H. Ratajczak and W. J. Orville-Thomas, *ibid.* Vol. 1, 1980.

[47] L. Jaroszewski, B. Lesyng, J. J. Tanner, and J. A. McCammon, *Chem. Phys. Lett.* **175**, 282 (1990); J. L. Andres, M. Duran, A. Lledos, and J. Bertran, *Chem. Phys. Lett.* **24**, 177 (1986); M. A. Muniz, J. Bertran, J. L. Andres, M. Duran, and A. Lledos, *J. Chem. Soc. Faraday Trans. 1* **81**, 1547 (1985).

[48] N. Shida, P. F. Barbara, and J. Almlof, *J. Chem. Phys.* **92**, 4061 (1989); T. Carrington, and W. H. Miller, *ibid.* **84**, 4364 (1986); P. Schuster, *Monatsh. Chem.* **100**, 2084 (1969); *Chem. Phys. Lett.* **3**, 433 (1969).

[49] E. Clementi, J. Mehl, and W. J. von Niessen, *J. Chem. Phys.* **54**, 508 (1971); P. Schuster, *Int. J. Quantum Chem.* **3**, 851 (1969); N. Shida, P. F. Barbara, and J. Almlof, *J. Chem. Phys.* **94**, 3633 (1991); P. Bosi, G. Zerbi, and E. Clementi, *J. Chem. Phys.* **66**, 3376 (1977).

[50] M. D. Joesten and L. J. Schaad, *Hydrogen Bonding*, Marcel Dekker, New York, 1974; P. Schuster, in *The Hydrogen Bond*, P. Schuster, G. Zundel, and C. Sandorfy, Eds., North Holland, Amsterdam, Vol. I, 1976; M. S. Gordon, D. E. Tallman, C. Monroe, M. Steenbach, and J. Armbrust, *J. Am. Chem. Soc.* **97**, 1326 (1975); R. R. Lucchese, H. F. Schaefer, III, *ibid.* **97**, 7205 (1975); E. Clementi, *J. Chem. Phys.* **47**, 2323 (1967).

[51] S. Nagakura, *J. Chim. Phys.* **60**, 217 (1964).

[52] J. Jadzyn and J. Malecki, *Acta Phys. Pol.* **A41**, 599 (1972); P. Kollman and I. Kuntz, *J. Am. Chem. Soc.* **98**, 6820 (1976).

[53] E. M. Arnett, *Acc. Chem. Res.* **6**, 404 (1973).

[54] P. Th. van Duijnen, *Enzyme* **36**, 93 (1986); B. T. Thole and P. Th. van Duijnen, *Biophys. Chem.* **18**, 53 (1983); O. Tapia and G. Johannen, *J. Chem. Phys.* **75**, 3624 (1981); P. Th. van Duijnen and B. T. Thole, in *Quantum Theory of Chemical Reactions*, R. Daudel, A. Pullman, L. Salem, and A. Veillard, Eds., Reidel, Dordrecht, The Netherlands, Vol. 3, 1982, p. 85; O. Tapia, C.-I. Branden, and A.-M. Armbruster, *ibid.* p. 97.

[55] D. Borgis and J. T. Hynes, in *The Enzyme Catalysis Process*, A. Cooper, J. L. Houben, and L. C. Chien, Eds., Plenum, New York, 1989.

[56] D. Hadži, in *Spectroscopy of Biological Molecules, NATO-ASI Series C 139*, C. Sandorfy and T. Theophanides, Eds., Reidel, Holland, 1984, p. 61; T. Solmayer and D. Hadži, *Int. J. Quantum. Chem.* **23**, 945 (1983); J. H. Wang, *Science* **161**, 137 (1968).

[57] C. A. Coulson, in *Hydrogen Bonding*, D. Hadži and H. W. Thompson, Eds., Pergamon London, 1959.

[58] S. Bratož, *Adv. Quant. Chem.* **3**, 209 (1967).

[59] M. Hasegawa, K. Daiyasu, and S. Yomosa, *J. Phys. Soc. Jpn.* **27**, 999 (1969); *ibid.* **28**, 275, 1304 (1970).

[60] P. C. McKinney and G. M. Barrow, *J. Chem. Phys.* **31**, 294 (1959).

[61] E. R. Lippincott, J. N. Finch, and R. Schroeder, in *Hydrogen Bonding*, D. Hadži and H. W. Thompson, Eds., Pergamon, London, 1959; E. R. Lippincott and R. Schroeder, *J. Chem. Phys.* **23**, 1099, 1131 (1955); R. Schroeder and E. R. Lippincott, *J. Phys. Chem.* **61**, 921 (1957); C. Reid, *J. Chem. Phys.* **30**, 1982 (1959).

[62] W. J. Hehre, R. F. Steward, and J. A. Pople, *J. Chem. Phys.* **51**, 2657 (1969).

[63] H. J. Kim and J. T. Hynes, *J. Chem. Phys.* **96**, 5088 (1992).

[64] J. N. Gehlen, D. Chandler, H. J. Kim, and J. T. Hynes, *J. Phys. Chem.* **96**, 1748 (1992); J. N. Gehlen and D. Chandler, *J. Chem. Phys.* **97**, 4958 (1992); H. J. Kim, R. Bianco, B. J. Gertner, and J. T. Hynes, *J. Phys. Chem.* **97**, 1723 (1993).

[65] R. S. Mulliken, *J. Phys. Chem.* **56**, 801 (1952); *J. Chem. Phys.* **20**, 20 (1964).

[66] K. Szczepaniak and A. Tramer, *J. Phys. Chem.* **71**, 3035 (1967); P. G. Puranik and V. Kumar, *Proc. Indian Acad. Sci.* **29**, 58, 327 (1963).

[67] A. C. Legon and D. J. Millen, *Acc. Chem. Res.* **20**, 39 (1987); C. E. Dykstra, *ibid.* **21**, 355 (1988).

[68] H. Umeyama and K. Morokuma, *J. Am. Chem. Soc.* **99**, 1316 (1977); K. Kitaura and K. Morokuma, *Int J. Quantum Chem.* **10**, 325 (1976); K. Morokuma, *Acc. Chem. Res.* **10**, 294 (1977); K. Morokuma and K. Kitaura, in *Molecular Interactions*, H. Ratajczak and W. J. Orville-Thomas, Eds., Wiley, New York, Vol. 1, 1980.

[69] B. A. Zwilles and W. B. Parson, *J. Chem. Phys.* **79**, 65 (1983); G. B. Bacskay and N. S. Hush, *Chem. Phys.* **82**, 303 (1983).

[70] A. E. Reed, L. A. Curtis, and F. Weinhold, *Chem. Rev.* **88**, 899 (1988).

[71] S. H. Northrup, M. R. Pear, C.-Y. Lee, J. A. McCammon, and M. Karplus, *Proc. Natl. Acad. Sci. U.S.A.* **79**, 4035 (1982).

[72] K. Ando and S. Kato, *J. Chem. Phys.* **95**, 5966 (1991).

[73] H. Eyring, *J. Chem. Phys.* **3**, 107 (1935); E. Wigner, *Trans. Faraday Soc.* **34**, 29 (1938).

[74] B. B. Smith, A. Staib, and J. T. Hynes, *Chem. Phys.* **176**, 521 (1993).

[75] J. P. Bergsma, J. R. Reimers, K. R. Wilson, and J. T. Hynes, *J. Chem. Phys.* **85**, 5625 (1986); B. J. Gertner, K. R. Wilson, and J. T. Hynes, *ibid.* **90**, 3537 (1989); G. Ciccotti, M. Ferrario, J. T. Hynes, and R. Kapral, *ibid.* **93**, 7137 (1990); R. Rey and E. Guardia, *J. Phys. Chem.* **96**, 4712 (1992); S. B. Zhu, J. Lee, and G. W. Robinson, *ibid.* **92**, 2401 (1988); B. J. Berne, M. Borkovec, and J. E. Straub, *ibid.* **92**, 3711 (1988); B. Roux and M. Karplus, *ibid.* **95**, 4856 (1991); S. Tucker and D. Truhlar, *J. Am. Chem. Soc.* **112**, 3347 (1990).

[76] R. F. Grote and J. T. Hynes, *J. Chem. Phys.* **76**, 2715 (1980).

[77] H. A. Kramers, *Physica* **7**, 284 (1940).

[78] M. Dupuis, J. D. Watts, H. O. Villar, and G. J. B. Hurst, *HONDO Ver. 7.0, QCPE* **544** (1987).

[79] C. Møller and M. S. Plesset, *Phys. Rev.* **45**, 618 (1934); R. Krishnan, M. J. Frisch, and J. A. Pople, *J. Chem. Phys.* **72**, 4244 (1980).

[80] J. S. Binkley, J. A. Pople, and W. J. Hehre, *J. Am. Chem. Soc.* **102**, 939 (1980); W. J. Hehre, R. Ditchfield, and J. A. Pople, *J. Chem. Phys.* **56**, 2252 (1972).

[81] T. H. Dunning and P. J. Hay, in *Methods of Electronic Structure Theory*, H. F. Schaefer, III, Ed., Plenum, New York, 1977.

[82] M. Urban, I. Cernusak, V. Kellö, and J. Noga, in *Methods in Computational Chemistry*, S. Wilson, Ed., Plenum, New York, 1987.

[83] I. Shavitt, in *Methods of Electronic Structure Theory*, H. F. Schaefer, III, Ed., Plenum, New York, 1977.

[84] W. L. Jorgensen, J. Chandrasekhar, J. Madura, R. W. Impey, and M. L. Klein, *J. Chem. Phys.* **79**, 926 (1983).

[85] M. P. Allen and D. J. Tildesley, *Computer Simulation of Liquids*, Clarendon, Oxford, 1987.

[86] L. Ebert, *Naturwiss* **13**, 393 (1925); R. A. Robinson, *Trans. Faraday Soc.* **32**, 743 (1936).

[87] B. E. Conway, in *Mod. Aspects Electrochem.*, J. O'M. Bockris and B. E. Conway, Eds., Butterworths, London, 1964; Y. K. Lau, S. Ikuta, and P. Kebarle, *J. Am. Chem. Soc.* **104**, 1462 (1982).

[88] J. E. Del Bene, M. J. Frisch, B. T. Luke, and J. A. Pople, *J. Phys. Chem.* **87**, 3279 (1983); M. J. Frisch, J. A. Pople, and J. E. Del Bene, *ibid.* **89**, 3664 (1985); J. E. Del Bene, M. J. Frisch, and J. A. Pople, *ibid.* **89**, 3669 (1985); E. Clementi, *ibid.* **89**, 4426 (1985).

[89] A concerted double PT was favored for a special structural arrangement of the $HCl(H_2O)_4$ cluster by ab initio density functional theory calculations; see M. Planes, C. Lee, and J. J. Novoa, *J. Phys. Chem.* **100**, 16495 (1996).

[90] E. E. Nikitin, *Theory of Elementary Atomic and Molecular Processes in Gases*, Clarendon, Oxford, 1974.

[91] For example, J. N. Onuchic and P. G. Wolynes, *J. Phys. Chem.* **92**, 6495 (1988); G. E. Zahr, R. K. Preston, and W. H. Miller, *J. Chem. Phys.* **62**, 1127 (1975); Y. Hurwitz, Y.

Rudich, R. Naaman, and R. B. Gerber, *J. Chem. Phys.* **98**, 2941 (1993); P. G. Wolynes and H. Frauenfelder, *Science* **229**, 337 (1985); J. E. Straub and B. J. Berne, *J. Chem. Phys.* **87**, 6111 (1987), and references therein.

[92] T. Fonseca, *J. Chem. Phys.* **91**, 2869 (1989); S. H. Northrup and J. T. Hynes, *J. Chem. Phys.* **69**, 5246 (1978).

[93] T. Schroeder and J. T. Hynes, in preparation.

[94] P. B. Balbuena, K. P. Johnston, and P. J. Rossky, *J. Phys. Chem.* **100**, 2716 (1996).

[95] K. Heger, M. Uematsu, and E. U. Franck, *Ber. Bunsenges. Phys. Chem.* **84**, 758 (1980); W. L. Marshall and E. U. Franck, *J. Phys. Chem. Ref. Data* **10**, 295 (1980).

[96] J. Gao, *J. Phys. Chem.* **98**, 6049 (1994); S. T. Cui and J. G. Harris, *Chem. Eng. Sci.*, in press.

[97] H. J. Kim, A. Staib, and J. T. Hynes, in *Ultrafast Reaction Dynamics at Atomic-Scale Resolution Femtochemistry and Femtobiology*, V. Sundstrom, Ed., Nobel Symposium 101, Imperial College Press, London, 1998.

[98] M. J. Packer and D. C. Clary, *J. Phys. Chem.* **99**, 14323 (1995); C. Lee, C. Sosa, M. Planas, and J. J. Novoa, *J. Chem. Phys.* **104**, 7081 (1996).

[99] R. Kapral and S. Consta, *J. Chem. Phys.* **101**, 10908 (1994); *ibid.* **104**, 4581 (1996); D. Laria, R. Kapral, D. Estrin, and G. Ciccotti, *J. Chem. Phys.* **104**, 6560 (1996).

[100] B. J. Gertner and J. T. Hynes, "Acid Ionized HBr in a Four-Water Cluster," submitted.

[101] B. J. Gertner and J. T. Hynes, "Hydrogen Halide Acid Ionization in Water Clusters: (a) 1. HBr"; (b) 2. HCl,"; (c) 3. HF," to be submitted.

[102] T. Schindler, C. Berg, G. Niedner-Schatteburg, and V. E. Bondeybey, *J. Chem. Phys.* **104**, 3998 (1996); Controlled temperature studies of HCl in protonated water clusters will soon be available (A. W. Castleman, Jr., Penn. State Univ., personal communication).

[103] S. Solomon, R. R. Garcia, F. S. Rowland, and D. J. Wuebbles, *Nature* **321**, 755 (1986); C. E. Kolb, D. R. Worsnop, M. S. Zahniser, P. Davidovits, L. F. Keyser, M.-T. Leu, M. J. Molina, D. R. Hanson, A. R. Ravishankara, L. R. Williams, and M. Tolbert, in *Progress and Problems in Atmospheric Chemistry*, J. R. Barker, Ed., World Scientific, New York, 1995.

[104] J. P. Abbatt, K. D. Beyer, A. F. Fucaloro, J. R. McMahon, P. J. Woodbridge, R. Zhang, and M. J. Molina, *J. Geophys. Res.* **97**, 15819 (1992).

[105] B. J. Gertner and J. T. Hynes, *Science* **271**, 1563 (1996).

[106] B. J. Gertner and J. T. Hynes, "Model Molecular Dynamics Simulation of Hydrochloric Acid Ionization at the Surface of Stratospheric Ice," *Faraday Discuss.* **110**, xxxx (1998), in press.

[107] R. Bianco, B. J. Gertner, and J. T. Hynes, *Ber. Bunsenges. Phys. Chem.* **102**, 518 (1998).

[108] D. A. Estrin, J. Kohanoff, D. H. Laria, and R. O. Weht, *Chem. Phys. Lett.* **280**, 280 (1997).

[109] R. Bianco and J. T. Hynes, work in progress.

[110] S. H. Robertson and D. C. Clary, *Faraday Discuss.* **100**, 309 (1995).

[111] H. Niki and K. H. Becker, Eds., *The Tropospheric Chemistry of Ozone in the Polar Regions*, Springer-Verlag, New York, 1993; D. R. Hanson and A. R. Ravishankara, *J. Phys. Chem.* **96**, 9441 (1992); S.-M. Fan and D. J. Jacobs, *Nature* **359**, 522 (1992); J. P. D. Abbatt, *Geophys. Res. Lett.* **21**, 665 (1994); D. J. Lary, M. P. Chipperfield, R. Toumi, and T. Lenton, *J. Geophys. Res.* **101**, 1489 (1996).

[112] D. C. Clary, *Science* **271**, 1509 (1996).

CHAPTER 7

STRUCTURES, SPECTROSCOPIES, AND REACTIONS OF ATOMIC IONS WITH WATER CLUSTERS

KIYOKAZU FUKE

Department of Chemistry, Kobe University, Kobe, 657-8501 Japan

KENRO HASHIMOTO

Computer Center, Tokyo Metropolitan University, Minami-Ohsawa, Hachioji, 192-0397 Japan

SUEHIRO IWATA

Institute for Molecular Science, Okazaki, 444-8585 Japan

CONTENTS

Advances in Chemical Physics, Volume 110, Edited by I. Prigogine and Stuart A. Rice.
ISBN 0-471-33180-5. © 1999 John Wiley & Sons, Inc.

I. INTRODUCTION

Hydrated metal ions play an important role in many aspects of chemical and biological phenomena and have been a subject of numerous investigations for the last several decades. Hydrated halogen anions also are ubiquitous in various chemical species, from aerosols to ionic crystals. These ions are bound to ligating water molecules and are further stabilized by long-range interaction with solvent molecules. Hydrated electrons could be regarded as a kind of hydrated ion. Although many experimental and theoretical efforts have been made to understand the nature and dynamics of solvation for these species, their microscopic aspects are not yet fully understood (Dogonadge et al., 1988). Recent advances in molecular beam techniques, combined with mass spectroscopy (Bower et al., 1996), allow us to prepare various kinds of clusters in the gas phase and open new approaches to a microscopic investigation of metal cations and halogen anions, as well as excess electrons in solutions (Kebarle, 1977; Castleman et al., 1986). The study of successively larger clusters containing ions is a model study of the solvation process in solution and provides us with a microscopic view of solvation structure and dynamics, in addition to reactions in solution. Cluster research also offers an opportunity to bridge the gap between the gas and condensed phases (Castleman and Bowen, 1996).

Great advances in quantum chemical methods and computer technology in the last few decades have made accurate quantum chemical calculations possible for clusters. With a systematic examination of the approximation levels for small to medium-sized clusters, the calculations can now be extended to larger clusters. Water clusters and their complexes with atoms and their ions are particularly suitable systems for quantum chemical calculations, because electrostatic interaction and local covalent (or electron-donor–acceptor) interactions are dominant in binding clusters. Van der Waals interaction which is more difficult to estimate, is less important. It is also true in theoretical studies that cluster research offers an opportunity to bridge the gap between the isolated molecules and the molecules in solution, on surfaces and in solids. Thus, the experimental and theoretical studies of such heteroclusters allow us to investigate specific cluster size dependencies, such as the buildup of the solvation shell in small and medium-sized clusters, and to explore the gradual transition from a finite system to an infinite bulk medium in large clusters.

As in metal–ion solvation, gas-phase cluster ions have been used to study solvent shell structure. An accurate characterization of properties of cluster ions will help to develop the fundamental interaction potentials needed for structural models of solvated ions. Especially as regards the structural aspects of solvated metal and halogen ions, high-pressure mass spectroscopy (HPMS) was perhaps the first experimental method that proved successful. Investigations employing this technique have yielded a wealth of thermo-dynamic information, for example, on enthalpy, entropy, and free-energy changes of association for ion-neutral complexes (Kebarle, 1977; Castleman et al., 1986; Hiraoka et al., 1988). Sharp changes in the enthalpy of associa-tion as a function of cluster size are used to infer the number of molecules in the first solvent shell. Since the temperature of solvated ions is well characterized in the HPMS technique, the results can be readily compared with solution studies or theoretical simulations. However, with successively larger cluster ions, the enthalpy of association rapidly approaches the value of the enthalpy of vaporization of the bulk solvent. This has limited the application of HPMS to solvation studies of ions with small ionic radii and low solvation numbers (typically, $n \leq 6$). A particular concern in these studies is where the atom (or ion) is located in the cluster — on the surface of the cluster (Surface(S) type) or in the middle of the cluster surrounded by water molecules (Interior(I) type). The size dependence of enthalpy changes should be reflected in the structural difference.

Although spectroscopic studies are expected to provide much more detailed knowledge of the properties of solvated cluster ions, the studies are yet limited in number (Duncan, 1997). Regarding cluster anions, photo-detachment studies have had a direct impact on questions concerning the nature of the solvated anions. With respect to hydrated metal anions, it is only recently that measurements of the photoelectron spectra (PES) of metal ions (Li^-, Na^-, and Cu^-) embedded in water clusters have been reported (Misaizu et al., 1995; Takasu et al., 1996, 1997). Similarly, only recently have

PES been reported for halogen anions with water clusters and their size dependence examined (Markovich et al., 1991, 1994). IR spectroscopy has also been used to study the gas-phase solvation of alkali–metal cations (Weinheimer and Lisy, 1996). In particular, combined with theoretical calculations, IR spectroscopy of halogen–water cluster anions has proven to be a powerful technique in determining the structure of the clusters. The electronic spectra of water clusters containing monovalent metal ions have been examined by photodissociation excitation spectroscopy (Shen and Farrar, 1991; Misaizu et al., 1992b). The shift of electronic transitions can be used as a probe for metal ion–solvent interaction. Besides, the photodissociation products and their yield spectra are informative on the potential-energy surfaces not only of the excited state, but also of the ground state, of the cluster.

The water clusters containing group-2 metal ions afford us with other interesting chemical processes, such as redox reactions of metal ions in a finite-sized clusters (Misaizu et al., 1992b). Since these metal ions are stable as divalent species in aqueous solution, monovalent metal ions can be oxidized to divalent ions at a certain cluster size. Thus, we can trace this transition by probing the electron-transfer reaction between metal ions and water molecules as a function of cluster size. These studies will give us information on the energetics and dynamics of the reduction and oxidation reactions of metal ions in aqueous solution, which is extremely important in chemistry and biology.

One of the highlights in the studies of metal–water clusters in recent years is a finding of ionization energies in $M(H_2O)_n$ converging on a constant value for $n \geq 4$, independently on the alkali metal atom (M = Li, Na, and Cs) (Hertel et al., 1991; Misaizu et al., 1992a; Takasu et al., 1998). The value converged on nearly coincides with an estimated photoelectric threshold of bulk water. The result strongly suggests the formation of an ion-pair state $M^+-(H_2O)_n^-$. The structure of the negatively charged unit, $(H_2O)_n^-$, is not strongly dependent on the counterion M^+, and the $(H_2O)_n^-$ could be a precursor of hydrated electrons in liquid. More structural information on the ion-pair state is required to understand the convergence of the ionization energy even at as small a value as $n = 4$.

In relation to hydrated electrons, water cluster anions, $(H_2O)_n^-$ (for $n = 2$, 6, 7, ≥ 11), have been prepared by localization during the cluster nucleation process or via the capture of very low energy electrons by cold water clusters (Haberland et al., 1984, 1885; Mark, 1991; Kondow, 1987; Kondow et al., 1989). The photoelectron spectra of $(H_2O)_n^-$ were reported, and vertical detachment energies (VDEs) were determined from the spectra (Coe et al., 1990). The VDEs of water cluster anions increase smoothly with cluster sizes and plot approximately linearly with $n^{-1/3}$, extrapolating to a value that is very close to the photoelectric threshold energy for the corresponding condensed-phase hydrated electron system. Besides, the absorption spectra of $(H_2O)_n^-$ (for $n \leq 60$), which might correspond to the well-known electronic transition of hydrated electrons in the condensed phase, have been investigated by photodepletion spectroscopy (Ayotte and Johnson, 1997). The IR spectrum of $(H_2O)_n^-$ is also reported (Bailey et al., 1996). The comparison of spectroscopic studies of the water cluster anions with those of $M(H_2O)_n$ could reveal a new insight into hydrated electrons in a bulk medium.

In this review, we will discuss the experimental and theoretical studies of atom–water clusters and their ions containing only a single metal atom or halogen anion. Even with this restriction, there is an enormous literature on the subject. We have decided not to expand the scope, and concentrate mostly on the experimental works by Fuke and his coworkers. All of us have been working closely for last few years. The review is a product of our collaborative efforts. The related experimental and theoretical work will be covered, but not exhaustively.

II. GROUP 1 (ALKALI) METALS

A. Size Dependence of the Ionization Energy

1. Experimental Study

Alkali (group 1) metals have been known to ionize spontaneously when dissolved in polar solvents, forming solvated metal ions and solvated electrons (Dogonadze et al., 1988). Clusters of solvent molecules containing an alkali atom might serve as a good model for linking the macroscopic properties of alkali metal-solvent systems with microscopic properties and for elucidating the early stage of solvated electron formation. As in the case of bulk solution, a valence electron of an alkali metal atom might be transferred to the clusters of water molecules at a sufficiently large n. In other words, the ground state of alkali metal–water clusters might have an ion-pair character if the number of water molecules in the cluster exceeds a certain number n. The photoionization threshold as a function of n provides information on this transition of the electronic state in the clusters. Hertel and coworkers first reported experimental studies on sodium–water clusters, (Schulz et al., 1988; Hertel et al., 1991) and subsequently, Fuke and his coworkers studied both cesium–water (Misaizu et al., 1992a) and lithium–water (Takasu et al., 1998) clusters. In this section, we discuss the results of experiments on $Cs(H_2O)_n$ and $Li(H_2O)_n$ by Fuke's group, together with those of experiments on $Na(H_2O)_n$ by Hertel et al.

Lithium—water clusters are produced by a laser vaporization method. The second harmonic of an Nd:YAG laser is focused onto a Li rod (≈ 5 mm) that is rotating and translating in an aluminum block. Argon gas of 2 atm mixed with water vapor is expanded through the block from a pulsed valve. Cesium—water clusters are produced with a pickup type of cluster source, which is composed of two pulsed valves arranged at right angles. Solvent molecular clusters are produced by supersonic expansion of 2 atm Ar gas mixed with the sample gas from the first pulsed valve at room temperature. A pulsed Cs atom beam is formed by the expansion of neat vapor from the second valve, which is heated to about 350°C. The Cs atom beam is injected 10–15 mm downstream from the first nozzle. Both clusters are skimmed by a skimmer, introduced into the ionizing region of a time-of-flight mass spectrometer, and then photoionized by crossing with an excimer-pumped dye laser. Photoionization mass spectra are recorded at various photon energies with an interval of 0.03 eV in the region 5.45–2.14 eV (228–580 nm). The ionization thresholds are determined to an accuracy of ± 0.06 eV by analyzing the mass spectra.

Figure 1 shows the ionization threshold energies (IE) of Li(H$_2$O)$_n$ and Cs(H$_2$O)$_n$ as a function of $(n+1)^{-1/3}$, which is approximately proportional

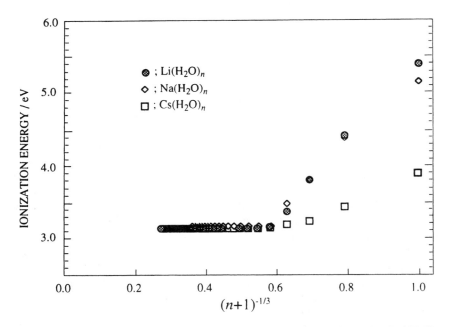

Figure 1. The size dependence of the ionization threshold energies (IE) of Li(H$_2$O)$_n$, Na(H$_2$O)$_n$ and Cs(H$_2$O)$_n$. The abscissa is $(n + 1)^{-1/3}$ (From Takasu et al., 1997).

to the inverse of the cluster radius (Takasu et al., 1998; Misaizu et al., 1992a). The results of experiments on $Na(H_2O)_n$ reported by Hertel and coworkers (1991) are also plotted in the figure. The IEs of these clusters exhibit anomalous size dependence. For $Li(H_2O)_n$, the IE decreases with cluster size up to $n = 4$ and then becomes constant at 3.12 eV for larger clusters. Similar results have also been observed for $Na(H_2O)_n$ and $Cs(H_2O)_n$, with constant IEs of 3.10 and 3.17 eV ($n \geq 4$), respectively. Interestingly, these limiting values of IEs [$IE(\infty)$] are nearly equal to the estimated photoelectric threshold of bulk ice (3.3 eV) and the vertical detachment energy (VDE) of $(H_2O)_n^-$ reported by Haberland and coworkers (Lee et al., 1991). These results are in marked contrast to the size dependence of IEs for $Li(NH_3)_n$ and $Cs(NH_3)_n$. The IEs of the latter clusters decrease almost linearly with $(n + 1)^{-1/3}$, and the intercept at $n = \infty$ gives an $IE(\infty)$ of 1.47 eV and 1.45 eV, respectively. The difference in the size dependence of IE of water– and ammonia–alkali metal clusters clearly indicates that the solvent molecules, and not the metal atoms, determine $IE(n)$ at $n \geq 4$ in water–alkali metal clusters.

2. Theoretical Study

The interesting findings on the characteristic size dependence of the ionization energies for $M(H_2O)_n$ clusters have inspired several theoretical studies. The main questions are (i) What are the most stable hydration structures of alkali–metal atoms, and are they similar to those of their corresponding cluster cations? (ii) What kind of interactions is important in stabilizing these hydration clusters? (iii) Do the most stable neutral structures of the clusters reproduce the characteristic size dependence in the ionization energies? and (iv) What is the character in the electronic state of the clusters, behind this peculiar behavior in IEs? The last question is, in particular, related to the localization mode of the excess electron and the formation of a two-center ion-pair state in the cluster.

Hydration Structures. The structures of $Li(H_2O)_n$ ($n = 1–6$ and 8) have recently been studied by Hashimoto and Kamimoto (1998) with ab initio molecular orbital (MO) methods. The selected sets of optimized structures and the total binding energies of $Li(H_2O)_n$ ($n = 1–6$ and 8) are shown in Figure 2.

For a 1:1 complex, the optimized structure with the HF level is planar, becoming nonplanar with the MP2 level. However, the potential energy surface is very flat for the out-of-plane coordinate. The binding energy of $Li–(H_2O)$ is 51.1 kJ/mol at the $MP2/6–31 ++ G(d, p)//HF/6–31 ++ G(d, p)$ level (Hashimoto and Kamimoto, 1998a). The corrections of BSSE and zero-point vibrations reduce the figure to 38.9 kJ/mol. The calculated

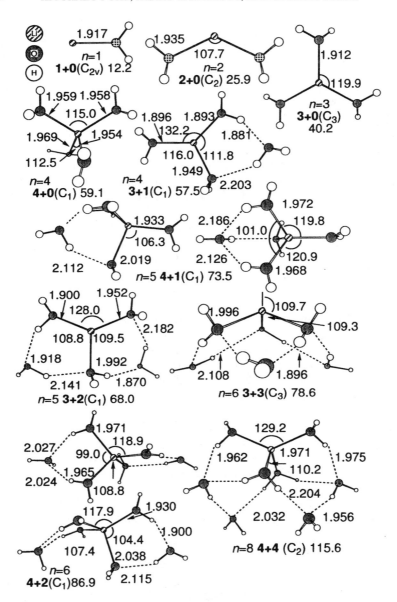

Figure 2.   The structures of Li(H₂O)ₙ ($n = 1$–6 and 8) at the HF/6–31 + + G(d, p) level. Geometrical parameters are given in angstroms and degrees. The total binding energies (kcal/mol, 1kcal/mol = 4.1868 kJ/mol) at the MP2/6–31 + + G(d, p)//HF/6–31 + + G(d, p) level are also given (From Hashimoto and Kamimoto, 1998).

binding energy of a water dimer with a similar level of calculations ranged from 27.2 (MP2/6–311 + + G(d, p)) to 20.9 (HF/6–31 + + G(d, p))kJ/mol; these two values bracket the experimental one of 22.8 ± 2.9 kJ/mol (Curtiss et al., 1979). The Li–O bond is stronger than the hydrogen bonds among water molecules, and therefore, it is expected that the *interior* structures, in which the metal atom is surrounded by water molecules, are "preferred", in terms of energy, to the *surface* structures, wherein the metal atom sits on the hydrogen-bond network of water molecules. The threefold $Li(H_2O)_n$ is the global minimum for $n = 3$, but is the high-energy isomer for larger n. For clusters with $n \geq 4$, two types of structures are seen in Figure 2. One is the *interior* structure, in which a Li atom is surrounded tetrahedrally by four water molecules, forming the first shell, and further water molecules are hydrogen bonded to the water molecules in the first shell. The other structure is a threefold one in which three water molecules are around the metal atom and further hydration forms the second shell; we may call this the *surface* structure. As is expected, the interior structure is more stable than the surface one for $n \geq 3$. Thus, the maximum hydration number of the first shell of the Li atom is four, which is the same as that of the Li^+ cation. In addition, the O–Li–O angles and Li–O bond lengths become close to those in the corresponding hydrated cations for $n = 3$ and 4. The formation of the second shell starts at $n = 5$, for which there are many interior structures with different hydrogen-bond networks of water molecules. For instance, the relative energy among several isomers of $n = 8$ lies within approximately 6 kcal/mol. The *surface* structures of $Na(H_2O)_n$, as will be discussed, are similar to those of the corresponding Li clusters, but the O–M–O angle in the Li clusters is much larger than in the Na clusters, because the strong and short Li–O bonds cause a repulsion among water molecules in the first shell.

The hydration structures of the Li^+ cation were studied by ab initio MO methods (Glendening et al., 1995; Glendening, 1996; Feller et al., 1994, 1995; Hashimoto and Kamimoto, 1998). Similarly to the neutral clusters, four water molecules can be bound directly to the central Li cation from oxygen sides, and the other water molecules form the second shell. Consequently, the most stable structure of $Li^+(H_2O)_n$ is similar to the interior structure of the corresponding neutral clusters. It is, however, interesting to note that in the cations, all water molecules in the second shell of the lowest energy clusters for $n = 5$ and 6 are bound to water molecules in the first shell by oxygen atoms as $O^I H \cdots O^{II} \cdots HO^I$, where O^I and O^{II} are oxygen atoms of the first and second shells, respectively. This is in contrast to the neutral clusters, whose structures with the networks of $O^I H \cdots O^{II} H \cdots O^I$ are the most stable. Furthermore, the intershell hydrogen bonds in $Li^+ (H_2O)_8$ are in most cases elongated or broken. Thus, only the first shells of the interior structures in the neutral and cationic clusters are similar to each other; the structures of the intershell hydrogen bonds are different, which results from the interaction between the excess electron and the hydrogen atoms in the neutral clusters.

The structures of $Na(H_2O)_n$ have been reported by Barnett and Landman (1993) and by Hashimoto et al. (1993, 1994a,b). The former employed local spin-density functional theory and latter conventional ab initio molecular orbital methods. Some of the optimized structures by Hashimoto et al. are shown in Figure 3. The interaction energy between a sodium atom and a water molecule is weaker (30.1(24.7) kJ/mol at the MP2/6–31 + G(d) (HF/6–31 + G(d)) level) than that between a lithium atom and a water molecule and is comparable to the hydrogen-bonding energy among water molecules (28.9(22.6) kJ/mol). The stable structure for small n has the form in which a sodium atom interacts with the hydrogen-bonded water clusters. One typical example is found in the most stable structure of $Na(H_2O)_3$: A sodium atom sits on a hydrogen-bonded water trimer of C_3 symmetry. The Na–O bond distances are longer than those in its cation by approximately 0.2 Å, and the O–Na–O angles are 72.1°. This form of $Na(H_2O)_3$ is the smallest member of the *surface* structures. For $n \geq 4$, the *surface* clusters are close in energy to the *interior* forms, in contrast to the situation for $Li(H_2O)_n$. The relative stability of this form of the $Na(H_2O)_3$ depends on the method used, and the approximation levels employed for $n \geq 5$ clusters are not so high as in the case of Li–water systems.

The most stable structure of $Na^+(H_2O)_n$ is essentially same as that of the corresponding lithium clusters, $Li^+(H_2O)_n$ (Bauschlicher et al., 1991a; Hashimoto and Morokuma 1994b; Kim et al., 1995; Feller et al., 1995). The Na–O distances are longer than Li–O bonds. Because there is no valence electron around Na^+ and the electrostatic interaction between Na^+ and O is large, all water molecules are directly bound to Na^+ by oxygen atoms, and no hydrogen bonds among ligand water molecules are found for $n \leq 4$. The repulsive interaction among water molecules is overcome by strong ion–water interaction. The maximum hydration number 4 is also consistent with the saturation of the sp^3 hybrid orbitals of Na^+, which accept the lone-pair electrons from the oxygen atoms.

The hydration structures of other neutral alkali atoms have not been reported yet. As we have seen, they are determined by the balance between the strengths of metal–O bonds and of the hydrogen bonds of water molecules. Since the interaction energy between the alkali metal and oxygen atoms is expected to become weaker as the size of the metal atom increases, the competition between the *interior* and *surface* structures for the larger alkali atom–water clusters is important. In addition, electron correlation could play an essential role in describing the structures, and there might

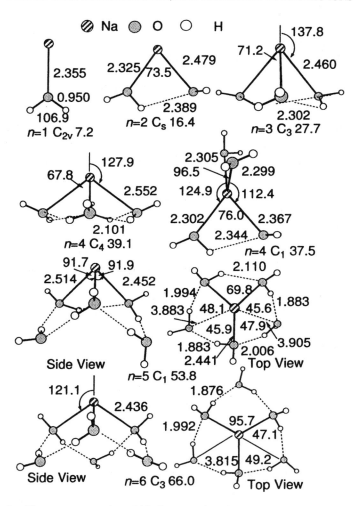

Figure 3. The structures and total binding energies of $Na(H_2O)_n$ by a $6-31 + G(d)$ basis. Geometrical parameters are in angstroms and degrees and are calculated at the HF level. Total binding energies in kcal/mol ($1 \, kcal/mol = 4.1868 \, kJ/mol$) are at the $MP2/6-31 + G(d)//HF/ 6-31 + G(d)$ level (From Hashimoto and Morokuma, 1994b).

exist new types of stable isomers other than the *interior* and *surface* clusters. Because the ionization energies for $n \geq 4$ show no dependence on the kind of alkali metal atoms, it is important to accumulate more structural information and examine whether or not the electronic states of the hydrated alkali atom clusters can be related to their structures.

Interaction energy. Total binding energies ($\Delta E(n)$) for the most stable structures of $Li(H_2O)_n$ and for the corresponding cations are listed in Table 1. The total binding energy $\Delta E(n)$ is defined by

$$-\Delta E(n) = E[M(H_2O)_n] - E(M) - nE[H_2O] \quad (M = Li, Li^+). \quad (1)$$

It is interesting to note that $\Delta E(n)$ for neutral clusters is almost additive for n, even after the first shell is completed. We have further analyzed $\Delta E(n)$ by decomposing it into contributions from the solute–water cluster interaction and that of the water–water interaction (Hashimoto and Kamimoto, 1998), defined, respectively, as

$$-\Delta E_M(n) = E[M(H_2O)_n] - E(M) - E[(H_2O)_n^{\#}] \quad (2)$$

$$-\Delta E_S(n) = E[(H_2O)_n^{\#}] - nE[H_2O]. \quad (3)$$

Here, $E[(H_2O)_n^{\#}]$ is the energy of an $(H_2O)_n$ cluster whose structure is fixed at that of the $Li(H_2O)_n$ or $Li^+(H_2O)_n$ complex; thus, $\Delta E_M(n)$ is the interaction energy between the prepared $(H_2O)_n^{\#}$ cluster and the Li atom or ion. On the other hand, $\Delta E_S(n)$ is the interaction energy among the H_2O molecules in the hydrated Li or Li^+ complex. The sum of the two components gives the total binding energy

$$\Delta E(n) = \Delta E_S(n) + \Delta E_M(n). \quad (4)$$

The energies $\Delta E_M(n)$ and $\Delta E_S(n)$ are also listed in Table 1. In the most stable neutral structures, the Li–water cluster interaction ($\Delta E_M(n)$) is a main

TABLE 1
Total binding energy, $\Delta E(n)$, and contributions of solute–water interaction, $\Delta E_M(n)$, and water–water interaction, $\Delta E_S(n)$, for $Li(H_2O)_n$ clusters and their cations at MP2/ 6–31 + + G(d, p)//HF/6–31 + + G(d, p) level.
Values are in kcal/mol (1 kcal/mol = 4.1868 kJ/mol) and are without BSSE and zero-point corrections

n	Interior $Li(H_2O)_n$			Interior $[Li(H_2O)_n]^+$		
	$\Delta E(n)$	$\Delta E_M(n)$	$\Delta E_S(n)$	$\Delta E(n)$	$\Delta E_M(n)$	$\Delta E_S(n)$
1	12.2	12.1	0.1	36.1	35.8	0.3
2	25.9	27.4	−1.5	69.3	70.6	−1.3
3	42.8	47.5	−4.7	94.7	99.6	−4.9
4	59.1	67.7	−8.6	115.1	125.0	−9.9
5	73.5	71.0	2.5	131.7	135.6	−3.9
6	86.9	80.7	6.2	147.5	147.2	0.3
8	115.6	99.2	16.4	172.8	158.1	14.7

component of the total binding energy for all n. The water–water interaction ($\Delta E_S(n)$) is very small or negative for $n \leq 4$ and is positive for larger n. Thus, the stabilization energies in the $n \leq 4$ clusters are gained by strong Li–O bonds that overcome the repulsive interaction among the water molecules, and the hydrogen-bond interaction for the formation of the second shell becomes important for $n \geq 5$.

Interestingly, $\Delta E_M(n)$ still increases for $n \geq 5$, although it has a smaller slope than for $n \leq 4$, which implies that the formation of the second hydration shell enhances the interaction between the metal atom and the water molecules in the first shell. This augmentative aspect becomes clearer when $\Delta E_M(n)$ and $\Delta E_S(n)$, for $n \geq 4$, are further decomposed by the following formulas:

$$\Delta E_S(4 + q) = \Delta E(W^{\#}_{4\,1\text{st}}) + \Delta E(W^{\#}_{q\,2\text{nd}}) + \Delta E(W^{\#}_{4\,1\text{st}} - W^{\#}_{q\,2\text{nd}}), \qquad (5)$$

$$\Delta E_M(4 + q) = \Delta E(\text{Li} - W^{\#}_{4\,1\text{st}}) + \Delta E(\text{Li} - W^{\#}_{q\,2\text{nd}})$$
$$+ \Delta E(\text{Li} - W^{\#}_{4\,1\text{st}} - W^{\#}_{q\,2\text{nd}}). \qquad (6)$$

Here, we regard $\text{Li}(\text{H}_2\text{O})_{4+q}$ as a cluster consisting of three parts: a lithium atom, a first-shell $(\text{H}_2\text{O})_4$ part, and a second-shell $(\text{H}_2\text{O})_q$ part. The energies $-\Delta E(W^{\#}_{4\,1\text{st}})$ and $-\Delta E(W^{\#}_{q\,2\text{nd}})$ are the energies of $(\text{H}_2\text{O})^{\#}_4$ and $(\text{H}_2\text{O})^{\#}_q$, respectively, relative to the energies of the isolated water monomers. $\Delta E(W^{\#}_{4\,1\text{st}})$ is the sum of all intrashell interactions among the first-shell water monomers up to four body terms, and $\Delta E(W^{\#}_{q\,2\text{nd}})$ is that among the second-shell waters up to q body terms. The two terms also include the deformation energy of each water monomer from a free H_2O molecule. $\Delta E(W^{\#}_{4\,1\text{st}} - W^{\#}_{q\,2\text{nd}})$ is the interaction energy between the first-shell $(\text{H}_2\text{O})_4$ and second-shell $(\text{H}_2\text{O})_q$ water molecules, including all intershell water interactions. On the other hand, $\Delta E(\text{Li} - W^{\#}_{4\,1\text{st}})$ is the interaction energy between Li and the first-shell $(\text{H}_2\text{O})_4$ part, and $\Delta E(\text{Li} - W^{\#}_{q\,2\text{nd}})$ is that between Li and the second-shell $(\text{H}_2\text{O})_q$ part. In other words, the former is the sum of all interactions among the Li atom and the first-shell water molecules up to five body terms, and the latter is that among the Li atom and the second-shell molecules up to $q + 1$ body terms. Therefore, the overall intershell interaction energies among Li and the first-shell and second-shell water molecules including up to $5 + q$ body terms are included in $\Delta E(\text{Li} - W^{\#}_{4\,1\text{st}} - W^{\#}_{q\,2\text{nd}})$, which we call the intershell Li–water interaction energy in the discussion that follows. The terms in Eqs. (5) and (6) are summarized in Table 2.

In the neutral clusters, $\Delta E(\text{Li} - W^{\#}_{4\,1\text{st}})$ is a main contributor in $\Delta E_M(n)$, while $\Delta E(\text{Li} - W^{\#}_{q\,2\text{nd}})$ is almost negligible. Both $\Delta E(\text{Li} - W^{\#}_{4\,1\text{st}})$ and

TABLE 2

Detailed components of solute–water and water–water interactions in $4 + q$ type of Li $(H_2O)_n$ and Li$^+(H_2O)_n$ clusters ($n \geq 4$) at MP2 6–31 + +G(d, p)//HF/6–31 + +G(d, p) level. Values are given in kcal/mol (1kcal/mol = 4.1868 kJ/mol) and without counterpoise and zero-point corrections

		$\Delta E_M(n)$			$\Delta E_S(n)$		
n	q	ΔE (Li–$W^{\#}_{4\ 1st}$)	ΔE (Li–$W^{\#}_{q\ 2nd}$)	ΔE (Li–$W^{\#}_{4\ 1st}$–$W^{\#}_{q\ 2nd}$)	ΔE ($W^{\#}_{4\ 1st}$)	ΔE ($W^{\#}_{q\ 2nd}$)	ΔE ($W^{\#}_{4\ 1st}$–$W^{\#}_{q\ 2nd}$)
				Li$(H_2O)_n$			
4	0	67.7	0.0	0.0	−8.6	0.0	0.0
5	1	66.5	0.6	3.9	−10.4	0.2	12.7
6	2	68.5	1.3	10.9	−10.6	0.2	16.6
8	4	69.2	2.3	27.7	−12.7	9.1	20.0
				Li$^+ (H_2O)_n$			
4	0	125.0	0.0	0.0	−9.9	0.0	0.0
5	1	124.3	10.8	0.6	−11.5	0.2	7.4
6	2	124.8	20.9	1.5	−14.1	0.0	14.3
8	4	124.5	31.4	2.2	−14.5	11.8	17.5

$\Delta E(W^{\#}_{4\ 1st})$ change only slightly, from $n = 4$ to $n = 8$, which implies that the addition of second-shell water molecules does not affect the core Li$(H_2O)_4$ very much. On the other hand, ΔE(Li $- W^{\#}_{4\ 1st} - W^{\#}_{q\ 2nd}$) becomes larger as n grows. This intershell Li–water interaction, which becomes more than 25 percent of the total ΔE_M for $n = 8$, is responsible mainly for the increase in $\Delta E_M(n)$. The interaction energy among the second-shell water molecules is almost zero for $n = 5$ and 6, but for $n = 8$ it becomes about twice the hydrogen-bond energy of a water dimer, which reflects the network structure of the second shell in $n = 8$. The intershell water–water interaction $\Delta E(W^{\#}_{4\ 1st} - W^{\#}_{q\ 2nd})$ also increases as n does, reaching as large as 83.7 kJ/mol for $n = 8$. Thus, although the interaction between the Li and the first-shell ligands is important, the intershell Li–water, as well as intershell water–water, interactions become essential in forming the most stable structures for $n \geq 5$, especially $n = 8$.

The total binding energies of the interior cations also increase monotonically with n, having a much larger slope than the neutral complexes, which reflects the strong electrostatic interaction between the Li$^+$ and solvent water molecules, as is seen in large ΔE(Li $- W^{\#}_{4\ 1st}$) and ΔE(Li– $W^{\#}_{q\ 2nd}$). The Li$^+$–water cluster interaction is important even for the second-shell water molecules. The interaction among the first-shell water molecules is repulsive, and the n dependence of $\Delta E(W^{\#}_{q\ 2nd})$ is similar to that of the neutral clusters. On the other hand, the $\Delta E(W^{\#}_{4\ 1st} - W^{\#}_{q\ 2nd})$ value for the cation is a little smaller than that for the corresponding neutral, showing weaker intershell hydrogen-bond interactions. The intershell Li–water interaction makes almost no contribution to the total binding energy and is not so essential in stabilizing the cation complexes as it is in the corresponding neutrals.

Ionization Energies and the Electronic State.   The observed size dependence of IEs shows at least two interesting features: the convergence of the IEs for $n \geq 4$ and the approximate convergence of the asymptotic IE value to the electron affinity of the bulk water, independently of the metal atom. The former property has been examined mainly by MO calculations, the latter by the bulk models.

Hertel et al. (1991) have tried to explain the n dependence of IEs in $Na(H_2O)_n$ and $Na(NH_3)_n$ by solving the one-electron Schrödinger equation in a dielectrically screened Coulomb potential. They succeeded in reproducing IEs of $Na(NH_3)_n$ for n up to around 10, but it was impossible to obtain any explanation for the measured IEs or even a resemblance of the experimentally found trend for $Na(H_2O)_n$ for $n > 4$. From this fact, they suggested the existence of a two-center state in these clusters, although they could not show any calculated IEs for the two-center state.

Barnett and Landman (1993) calculated the IEs of $Na(H_2O)_n$ ($1 \leq n \leq 8$) on the basis of spherical structure models and found that the addition of water molecules to the Na atoms results in a successive decrease in the IEs, with a reduced variation for $n > 4$. The authors concluded that the reduction in the successive IE change reflects the formation of a molecular shell around Na and that the electronic state for $Na(H_2O)_n$ for $n \geq 4$ is the surface Rydberg state, in which the electron is delocalized and spread rather equally around water molecules; this might be called the one-center state.

Makov and Nitzan (1994) applied a continuum dielectric theory for ions and neutral atomic solutes near planar and spherical surfaces, and found that the size dependence of the IEs was insensitive to the location of the solute in the cluster. Although their computed asymptotic behavior of the vertical detachment energies of hydrated electron and the hydrated I^- ion agreed well with the experiment, the authors failed to explain the observed behavior in the IEs of the Na–water clusters.

Stampfli and Bennemann (1994), and Stampfli (1995), however, indicated the importance of the polarization effect by a polarizable electropole model and pointed out that the IEs of the spherically symmetric structure of the hydrated sodium atom in the surface state strongly decrease with increasing cluster size.

Hashimoto et al. (1994b) showed that the IEs of the surface-type $Na(H_2O)_n$ reproduce the observed IEs well for $n \geq 4$. They also evaluated the IEs of the interior-type structures for $n \geq 4$, but their calculated IEs differed from the experimental values. Hashimoto and Kamimoto (1998a) extended the study to $Li(H_2O)_n$, and their calculated IEs are plotted against n in Figure 4. The theoretical values of the IEs of both the most stable interior and the high-energy threefold forms of $Li(H_2O)_n$ are in good agreement with the experiment. They decrease monotonically until $n = 4$ and become nearly constant for larger n, converging to the bulk limit of the vertical detachment energies of $(H_2O)_n^-$.

In the surface type of $Na(H_2O)_n$, only three or four water molecules interact with a sodium atom, and the excess electron distribution is rather localized around the sodium in the surface region of the clusters, extending widely on one side of the space, avoiding hydrating H_2O molecules. The excess electron distribution is almost unchanged for $n \geq 4$. The indirect hydration in the second and further outer shells does not very much affect the local electronic state around the sodium atom. Therefore, the IEs of these complexes remain nearly constant, at around $n \approx 4$. The excess electron distribution of the threefold $Li(H_2O)_n$ resembles that of the corresponding surface-type $Na(H_2O)_n$, and thus, the same argument about the n dependence of the IEs of the $Li(H_2O)_n$ may be made. These surface and threefold structures are candidates for the asymmetric structures suggested by Stampfli and Benneman. The excess electron is distributed asymmetrically around the cluster, but the electronic state is not the two-center one in these clusters.

The electron distribution of typical examples for $Li(H_2O)_2$ and $Li(H_2O)_4$ are shown in Figures 5(a) and (b) (Hashimoto and Kamimoto, 1998), where, inside the cloud, 50% of the odd electron is contained. The most striking result has been obtained for the interior clusters of $Li(H_2O)_n$, which are the most stable for $n \geq 4$. Figures 5(c) and (d) show the excess electron distribution for $n = 6$ and $n = 8$ (Hashimoto and Kamimoto, 1998). The figures clearly show that the odd electron is far separated from the metal ion both in $Li(H_2O)_6$ and $Li(H_2O)_8$. The metal ion is hydrated with four water molecules, and the excess electron is trapped by the H–O bonds of waters. In particular, in $Li(H_2O)_8$, the excess electron is supported by two water molecules that have no direct interaction with the metal ion. This is the first theoretical finding relating to the formation of the two-center ion-pair state in hydrated alkali atom clusters. The electron distribution of $Li(H_2O)_4$, shown in Figure 5(b), is an intermediate case. The excess electron is well separated from the metal ion, but the water molecules that support the excess electrons are all hydrating the metal ion. The electronic nature of the hydrated Li atom clusters changes from the one-center atomic state for $n \leq 3$ to the two-center ionic state at $n = 4$. It is worth noticing that two second-shell H_2O molecules for $n = 8$ point their free OH bonds toward the excess

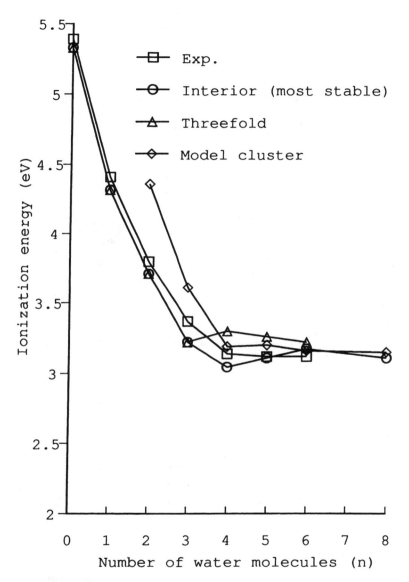

Figure 4. Size dependence of experimental and calculated ionization energies (IE) for $Li(H_2O)_n$ (Hashimoto and Kamimoto, 1998).

electron in the surface region of the cluster. It is the dangling hydrogens, which do not take part in the hydrogen-bond network, that support the nearly spherical excess electron. While studying pure water cluster anions

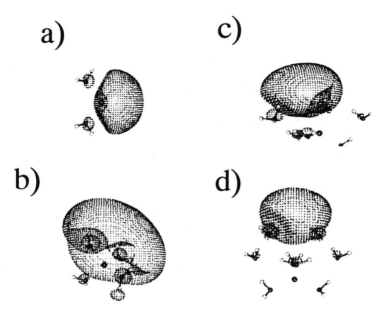

Figure 5. The excess electron distribution of $Li(H_2O)_n$. (The structures are determined by Hashimoto and Kamimoto, 1998: (a) $n=2$, (b) $n=4$ $(4+0$ structure), (c) $n=6$ $(4+2$ structure), (d) $n=8$ $(4+4$ structure). Inside the sphere, 50% of the electron is contained in the singly occupied molecular orbital (SOMO). The Li–O bond distances in the first shell are (a) 1.94 Å for $n=2$, (b) 1.95–1.96 Å for $n=4$, (c) 1.93–2.04 Å for $n=6$, and (d) 1.94–1.99 Å for $n=8$.

$(H_2O)_n^-$ with ab initio MO calculations, Tsurusawa and Iwata (1998) found a similar excess-electron distributions for $n=2$ and 3, and the structures of some of $(H_2O)_n^-$ are shown in Figure 6. The similarity of the structures and the electron distributions between pure water cluster anions and $Li(H_2O)_n$ $(n \geq 4)$ can easily be noticed in the figure. In both cases, HO bonds of water molecules support the excess electron, and no hydrogen bonds exist between the water molecules that surround the excess electron. Tsurusawa and Iwata called the interaction between the excess electron and HOs of water molecules the {e}–HO bonds (1999).

Hashimoto and Kamimoto (1998) also studied the electronic state of $Li(H_2O)_n$ by examining the model clusters in which the Li atom is replaced by a point charge. The model clusters can be regarded as the negatively charged $(H_2O)_n^-$ clusters, interacting with a cation center in which the structure of the water molecules is fixed at the structure in the corresponding

a)   b)

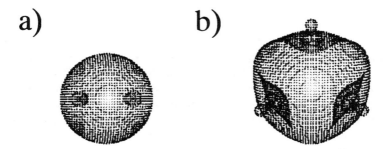

Figure 6. The interior type of water cluster anions $(H_2O)_n^-$ ($n = 2$ and $n = 3$). Inside of the sphere, 50% of the electron in the singly occupied molecular orbital (SOMO) is contained. The O–O distance is 9.91 Å for $n = 2$ and 5.39 Å for $n = 3$.

$Li(H_2O)_n$. The vertical IEs of the model clusters almost coincide with those of the true Li–water complexes for $n \geq 4$, and their excess electron distribution is similar to that of the corresponding true Li–water cluster. These facts indicate that a lithium atom in the hydration system releases its valence electron and acts as a cation center with more than four water molecules. In photoionization, the electron is ejected from the electron cloud surrounded by water molecules for $n \geq 4$. This picture is consistent with the similarity in structure of the first shell between the neutral $Li(H_2O)_n$ and cationic $Li^+(H_2O)_n$ clusters ($n \geq 4$). Because of the interaction between HO bonds and excess electrons (the $\{e\}$–HO bonds) in the neutral clusters, the structure of the outer shell in the neutral clusters differs from that of the corresponding cation. Once the neutral clusters form an ion pair of $[Li(H_2O)_4]^+$ and a hydrated electron, their IEs behave insensitively to the cluster size (n) and, probably, to the metal atom.

However, as mentioned earlier and shown in Figure 3, the structures determined for $Na(H_2O)_n$ by Hashimoto et al. (1994b) are mostly the surface type, and their excess electron is distributed on one side of, but near, the metal ion. The strong hydrogen-bond networks grow on the other side of the metal ion. These researchers found no such water molecules supporting the excess electron. Tsurusawa and Iwata (1999) found $Na(H_2O)_n$ clusters whose structures are similar to the corresponding $Li(H_2O)_n$ clusters. Further careful studies might be required on the other alkali metal–water clusters.

The convergence of the IE values to the vertical detachment energy of $(H_2O)_n^-$ might not be surprising if one assumes that the anion unit of the ion pair is formed by $\{e\}$–HO bonds, as in pure water cluster anions (Tsurusawa and Iwata, 1999).

B. Photoelectron Spectroscopy of Solvated Metal Anions

1. Experimental Study

Photoionization studies offer a clue to understanding the microscopic solvation process of alkali atoms in finite clusters. However, such studies provide information only on the energy difference between the neutral and ionic states, and thus, it is difficult to obtain a deeper understanding of the solvation state of metal atoms. The photoelectron spectroscopy of negatively charged solvent clusters containing an alkali atom may afford further information on the electronic structure of these clusters, both in the neutral ground state ^2S and the excited states derived from the ^2P states of the alkali atom. The cluster anions are successfully prepared with a laser vaporization technique coupled with a supersonic expansion method (Misaizu et al., 1995). The photoelectron spectra (PES) of mass-selected $Li^-(H_2O)_n$ and $Na^-(H_2O)_n$ are studied with a magnetic-bottle type of photoelectron spectrometer.

Figure 7 shows the photoelectron spectra of $Na^-(H_2O)_n$ ($n \leq 7$) obtained at the detachment energy of 3.50 eV for $n = 1$ and $n = 2$, and at 4.66 eV for $n \geq 3$ (Takasu et al., 1997). In contrast to the Li–water system shown in Figure 8, both the $Na(3^2S)\leftarrow Na^-(^1S)$ and $Na(3^2P)\leftarrow Na^-(^1S)$ types of transitions shift to the higher electron binding energy by almost the same amount as n. The spectral shifts of the former transition from that of the bare anion ($^2S-^1S$, 0.55 eV) are 0.21, 0.44, and 0.58 eV for $n = 1$, 2, and 3, respectively. A similar blue shift in vertical detachment energy has also been observed for $Cu^-(H_2O)_n$ (Misaizu et al., 1995) and $I^-(H_2O)_n$, (Markovich et al., 1991, 1994), but the amounts of the shift for these cluster ions are much larger than those for the Na system. For Na^-–water clusters, Hashimoto et al. (1997) have also calculated the structures, total binding energies, and vertical detachment energies for $n \leq 3$, which are shown in Figure 9 and Table 3. Two isomers of $Na^-(H_2O)$ were found, one with symmetrical Na^-–H (0.377 nm) bonds and the other an Na^-–O (0.233 nm) bond. The former is more stable by 4 kJ/mol. For larger clusters, the isomers with the Na^-–H bonds and hydrogen bonds among water molecules are more stable by 12–21 kJ/mol than the isomers having only Na–O bonds. These optimized structures of $Na^-(H_2O)_n$ are quite different from those for the neutral $Na(H_2O)_n$ clusters, in which the Na–O bonds are predominant. The calculations at the CCSD//MP2 level with the extended basis sets reproduce the observed vertical detachment energies with sufficient accuracy, as is seen in Table 3. These theoretical results suggest that the observed photoelectron bands are due mainly to the isomers with Na–H interactions; the blue shift in vertical detachment energy is caused by much larger electrostatic interaction in the anion state than in the neutral state. The

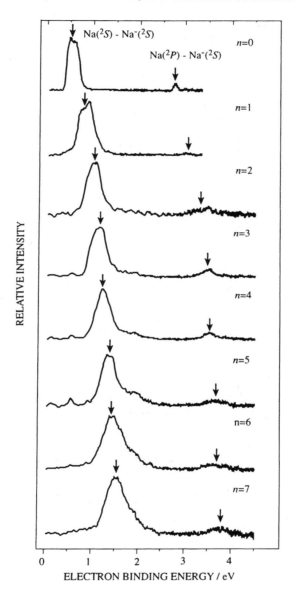

Figure 7. The photoelectron spectra of $Na^-(H_2O)_n$ ($n \leq 7$). The detachment energy is 3.50 eV for $n = 1$ and $n = 2$, and 4.66 eV for $n \geq 3$ (From Takasu et al., 1997).

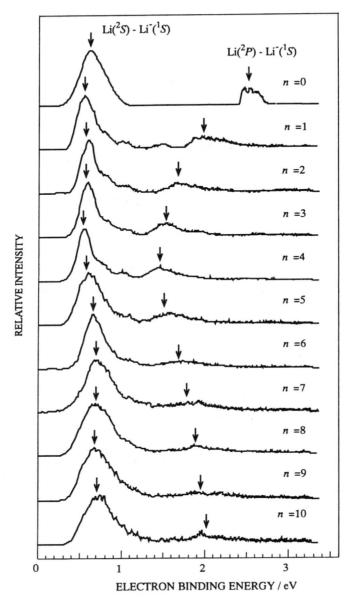

Figure 8. The photoelectron spectra of $Li^-(H_2O)_n$ $(n \leq 10)$. The detachment energy is 3.50 eV.

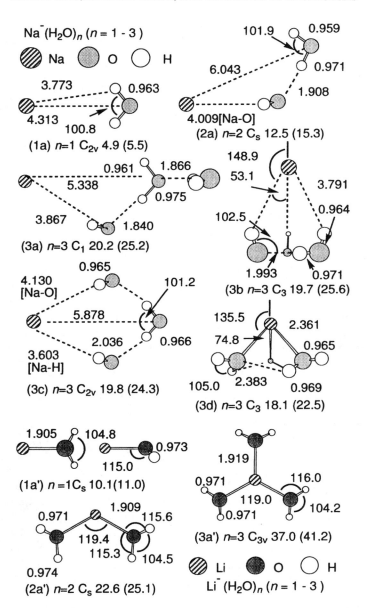

Figure 9. Selected sets of optimized structures of $[Na(H_2O)_n]^-$ and $[Li(H_2O)_n]^-$ and their total binding energies in kcal/mol (1 kcal/mol = 4,1868 kJ/mol). The structures and energies are at the MP2/6–311 + + G(d, p) levels of approximation. The values in parentheses are without any zero-point correction (From Hashimoto et al., 1997).

TABLE 3
Vertical electron detachment energy (eV) of $[Na(H_2O)_n]^-$ and $[Li(H_2O)_n]^-$

	$[Na(H_2O)_n]^-$			$[Li(H_2O)_n]^-$		
n	Isomer in Figure 9	CCSD	Exp.	Isomer in Figure 9	CCSD (CCSD(T))	Exp.
1	1a	0.74	0.76	1a′	0.45 (0.51)	0.56
2	2a	0.94	0.99	2a′	0.42 (0.50)	0.57
3	3a	1.09	1.13	3a′	0.41 (0.51)	0.56
	3b	1.00				
	3c	1.19				
	3d	0.31				

experimental and theoretical results together imply that the potential energy minimum of $Na(H_2O)_n$ cannot be probed by photoelectron spectroscopy of its anions, because of the large difference in equilibrium structures of the anion and corresponding neutral clusters. As seen in Figure 7, the energy separation between the 3^2S and 3^2P states is almost constant for $n \leq 7$.

Photoelectron spectroscopy of $Na^-(H_2O)_n$ is thus less informative about the origin of the anomalous IE behavior. On the other hand, the size dependence of the photoelectron spectra of $Li^-(H_2O)_n$ may offer deep insight into the electronic structure of the clusters. This is because, as shown in Figure 9, theoretical calculations by Hashimoto et al. (1997) predict that the most stable $Li^-(H_2O)_n$ isomers have similar Li–O bonds, as have the corresponding neutral $Li(H_2O)_n$ clusters. Figure 8 shows the PES of $Li^-(H_2O)_n$ ($n \leq 10$), recorded at a photon energy of 3.50 eV (Takasu et al., 1998). The spectrum of Li^- exhibits a strong band at 0.62 eV and a weak band at 2.47 eV, corresponding, respectively, to the Li $(2^2S) \leftarrow Li^-(^1S)$ and Li $(2^2P) \leftarrow Li^-(^1S)$ transitions. For the 1:1 complex $Li^-(H_2O)$ shown in Figure 8, the spectrum consists of two peaks at 0.56 and 1.97 eV, corresponding to two transitions of Li^-. In contrast to $Na^-(H_2O)$, these bands are shifted to the lower binding energy with respect to those of Li^-, by 0.06 eV for the 2^2S transition and by 0.50 eV for the 2^2P transition. For $Li^-(H_2O)$, we can also expect two isomers, the calculated total binding energies of which are 42.3 for the Li–O isomer and 22.2 kJ/mol for the Li–H bonds at the MP2/6–311 + + G(d, p) level with zero-point vibrational correction (Hashimoto et al., 1997). The calculated vertical detachment energies of the 2^2S-type state are 0.45 and 0.85 eV for the stable and less stable isomers, respectively, at the CCSD/6–311 + + G(d, p)//MP2/6–311 + + G(d, p) level. Thus, the anionic, neutral, and cationic 1:1 complexes have structures that are similar to each other. By assuming the same equilibrium structures, we estimate the binding energies of the neutral and anion complexes, using the observed IEs, vertical detachment energies, and experimental binding energy of the cation complex [1.42 eV ($= 136.9$ kJ/mol)]. The binding energies are determined to be 0.44 and 0.38 eV for the neutral and anionic forms, respectively, which values are consistent with the calculated energies of 0.43 and 0.44 eV, respectively. From these results, as well as the observed and calculated vertical detachment energies, the bands at 0.56 and 1.97 eV in Figure 8 are safely ascribed to the 2^2S and 2^2P transitions of the stable isomer having an Li–O bond. The large shift of the second transition (about 0.5 eV) from the bare atom indicates a large increase in the binding energy between the Li atom and solvent molecule in the first excited state; indeed, the shift is as large as 0.90 eV, probably due to a much larger ionic interaction.

For $Li^-(H_2O)_2$, the first band does not shift appreciably from $n = 1$, but the second band shifts further, to 1.72 eV, as shown in Figure 8. The first bands of $n = 3$ and $n = 4$ are found to be shifted to the red by 0.06 and 0.09 eV, respectively, from that of Li^-, while the second band is shifted extensively to the red, as shown in the figure. The calculations reveal that the most stable isomers of these clusters also have Li–O bonds, and the other isomers with Li–H bonds become increasingly unstable with n (Hashimoto et al., 1997). The calculated vertical detachment energies of the first band of the most stable isomers are close to 0.50 eV, but those of the other local-minimum isomers become larger with n. Thus, these experimental and theoretical results indicate that the observed PES could be explained by the most stable clusters for $n \leq 10$. A large decrease in the ^2S–^2P energy separation is consistent with a rapid decrease in IEs of Li $(H_2O)_n$ $(n \leq 4)$, as shown in Figure 1, and the electronic structure around the Li atom is changed extensively, even for $n \leq 4$. The observed red shifts in the vertical detachment energies of the transition to the ^2S state for $Li^-(H_2O)_n (n \leq 4)$ with respect to that of Li^- clearly indicate that the solvation energies of the neutral clusters are larger than those of the anionic forms. The large solvation energy in the neutral Li–water clusters is due to a large deformation of the SOMO orbital, as was shown in Section II.A. The s orbital of Li is strongly polarized in the direction opposite that of the water molecules and is destabilized. The ^2P–^2S energy separation becomes small.

The spectral trend observed for $Li^-(H_2O)_n$ $(n \leq 4)$ is in marked contrast to that of $Na^-(H_2O)_n$, shown in Figure 7. The structural difference in the anionic clusters of two metals explains the change in the direction of the shift.

As indicated in Figure 8, both transitions in $Li^-(H_2O)_5$ shift to the higher binding energy with respect to those of $n = 4$. For larger clusters, these transitions monotonically shift further, to high binding energy. This change in the shift at $n = 5$ indicates that the first solvation shell of Li is closed with four water molecules, both in the neutral and anionic clusters. Besides, the $^2P-^2S$ gap becomes slightly broader for $n \geq 5$. The neutral clusters for $n \geq 5$ become ion-pair states, as discussed in Section II.A. Although the theoretical results for the larger anionic clusters are not available at present, the electronic and geometric structures of the anions are expected to differ from those of the corresponding neutral clusters, and thus, the $^2P-^2S$ energy separation starts to change at $n = 5$, from 0.95 eV to about 1.2 eV ($n = 10$). If the ground state becomes the two-center ion-pair state, the anionic center might be similar to that of water cluster anions. Ayotte and Johnson (1997) examined the electronic absorption spectra of water cluster anions $(H_2O)_n^-$ ($n = 6-50$) and observed the band maxima shift to higher photon energy as a function of n, from 0.81 eV ($n = 15$) to 1.19 eV ($n = 50$). However, a direct comparison of the $^2P-^2S$ energy separation with the absorption spectrum of water clusters is not possible, because in the photoabsorption spectrum of water clusters, the $^2P\leftarrow^2S$ transitions cannot be separated from the other transitions, such as bound-free transitions.

2. Theoretical Study

The photoelectron spectra (PES) of the negatively charged $Na^-(H_2O)_n$ and $Li^-(H_2O)_n$ clusters, as has been described, have shown the different size dependence in the vertical detachment energies. The dependence on n for $Na^-(H_2O)_n$ is similar to that for halogen–water cluster anions (Markovich et al., 1991, 1994), while the PES for $Li^-(H_2O)_n$ are similar to those for ammoniated Li^- and Na^-, which will be discussed in Section II.C for comparison. The difference in the vertical detachment energies is related to that of the geometric structure of the anionic and neutral clusters. Hashimoto et al. (1997) examined this question by ab initio MO calculations.

Structures.   The selected sets of optimized structures of $Na^-(H_2O)_n$ and $Li^-(H_2O)_n$ ($n = 1-3$), together with their total binding energies, are shown in Figure 9. For $Na^-(H_2O)$, the C_{2v} structure 1a, in which the H_2O molecule is bound to Na from the hydrogen side, is more stable than the C_s structure with an Na–O bond by 4.6 kJ/mol. The structure 1a is similar to that of $Cu^-(H_2O)$ studied by Zhan and Iwata (1995) and that of $X^-(H_2O)$ for $X=F$, Cl, Br, and I, as is discussed in Section V. The most stable structure of $Na^-(H_2O)_2$ is a C_s structure 2a, wherein one H_2O molecule is bound to 1a via a hydrogen bond. This structure is more stable than that of a C_{2v} isomer having two equivalent Na–O bonds, by 9.6 kJ/mol. The structures 3a–c are the lowest energy clusters among the 18 optimized structures for $n = 3$. They are almost isoenergetic, and their calculated binding energies are

84.5–82.5 kJ/mol. In the C_1 structure 3a, the third water molecule is bound to 2a through a hydrogen bond, while three H_2O molecules are bound equivalently to Na by hydrogen atoms and are bound to one another through hydrogen bonds in structure 3b. On the other hand, two H_2O molecules are bound to Na by H atoms and are bridged by the third H_2O molecule through hydrogen bonds in structure 3c. Another C_3 structure, 3d, with three equivalent Na–O bonds and three hydrogen bonds, is also at a local minimum on the potential surface, but is less stable than 3a by 8.8 kJ/mol. Thus, the structures with the Na–H and hydrogen bonds among water molecules are more stable than those having Na–O bonds. The structures of the most stable isomer of Na–water anions differ from those of the corresponding neutral clusters. Therefore, in the photoelectron spectroscopy of $Na^-(H_2O)_n$ ($n = 1-3$), the global minimum of the neutral clusters cannot be accessed.

The structures of $Li^-(H_2O)_n$ shown in Figure 9 are different from those of the corresponding $Na^-(H_2O)_n$ and are apparently similar to those of the neutral $Li(H_2O)_n$. The Li–O bonds dominate the geometric structure. The isomers having Li–H bonds, as in $Na^-(H_2O)_n$, are local-minimum structures. The binding energy gained by the Li–O bonds are more than 42 kJ/mol per water molecule. The structures of $Li^-(H_2O)_n$ ($1 \leq n \leq 3$) are unique among $X^-(H_2O)_n$ ($X =$ Li, Na, Cu, F, Cl, Br, and I).

Vertical Detachment Energies.   The calculated vertical detachment energies of $Li^-(H_2O)_n$ and $Na^-(H_2O)_n$ ($n = 1-3$) are summarized in Table 3 (Hashimoto et al., 1997) and are compared with the experimental values (Takasu et al., 1997, 1998). The vertical detachment energy of the atomic Na anion is computed to be 0.52 eV at the CCSD level, which differs from the experimental value by only 0.03 eV. For $Na^-(H_2O)_n$ ($n = 1-3$), the calculated vertical detachment energies of the most stable isomers, 1a, 2a, and 3a in Figure 9, are 0.74, 0.94, and 1.09 eV, respectively. These values are in excellent agreement with the experimental values of 0.76 ($n = 1$), 0.99 ($n = 2$), and 1.13 ($n = 3$) eV. The theoretical calculations for the most stable isomer reproduce the experimentally seen gradual shift of the vertical detachment energies to higher levels with n. The vertical detachment energies for the structures with the maximum numbers of the Na–O bonds, which are less stable, are 0.41 ($n = 1$), 0.40 ($n = 2$), and 0.31 ($n = 3$) eV. They are much smaller than the experimental values, and their shifts with n are opposite to what has been experimentally observed. Thus, the photoelectron spectra are safely attributed to the most stable isomers having Na–H bonds with hydrogen-bonded water molecules.

The calculated vertical detachment energy for the atomic Li^- anion is 0.62 eV at the CCSD//6–311 + + G(d, p) level, and that for the most stable

$Li^-(H_2O)$ is 0.51 (0.45) at the CCSD (CCSD(T)) level, which is in good agreement with the experimental energies (0.62 and 0.56 eV, respectively). The corresponding calculated vertical detachment energies for the most stable structures of $n = 2$ and 3 become 0.50 (0.42) and 0.51 (0.41) eV, respectively. Therefore, the observed size dependence of the first band is well reproduced in the calculations if the most stable isomers having the maximum number of Li–O bonds are assumed. On the other hand, the calculated vertical detachment energies for the isomers having Li–H bonds are 0.85 eV for $n = 1$ and more than 1.0 eV for $n > 1$. Therefore, the observed size dependence of these energies in $Li^-(H_2O)_n$ is consistent with the calculated results only if the isomers with Li–O bonds are assumed, which is in sharp contrast to $Na^-(H_2O)_n$.

Excited States and Electronic Nature.   Hashimoto (1999a) recently extended the theoretical calculations for the excited states to analyze the second band in PES. Table 4 lists the excitation energies of the $Na(H_2O)_n$ and $Li(H_2O)_n$ clusters at the most probable anionic geometries. The excitation energies for the neutral clusters correspond to the energy separation between the first and second bands in the observed PES. For all clusters, the calculated values are in good agreement with experiment. As with the $Na(H_2O)_n$ clusters, the energy separation Na (3^2P)–Na (3^2S) is nearly constant around the atomic value (2.1 eV) with n. On the other hand, the energy separation in the PES of $Li^-(H_2O)_n$ decreases rapidly with n, both in the experiments and in the theoretical calculations. Thus, the observed second bands are safely assigned to the transitions to the excited states of the most stable hydrated clusters of each hydrated anion of Na and Li.

   The estimated expectation values $\sqrt{< r^2 >_{SOMO}}$ of the radial distribution of the unpaired electron are also presented in Table 4. They are evaluated by placing the origin at the alkali atom. In an isolated Na atom, it is 4.5 bohr (1 bohr = 52.9 pm) for the ground ^2S state and 6.5 bohr for the first excited ^2P state. In the $Na(H_2O)_n$ clusters, these values are almost unchanged from the atomic value for both the ground and the excited states, even when n grows. The excess electron is located around the Na atom. On the other hand, as is seen in the table, the radial distribution of the unpaired electron in $Li(H_2O)_n$ increases stepwise by hydration and becomes nearly double the atomic value in $Li(H_2O)_3$ for both the ground and the excited states. This indicates that the valence electron of the Li atom becomes extended widely over the solvent molecules in the Li–water clusters as the number of the water molecule increases. Therefore, the electronic state of $Li(H_2O)_n$ is quite different from that of $Na(H_2O)_n$ at their anion geometries.

TABLE 4

Vertical excitation energy (EE) of $^2P \leftarrow\ ^2S$ transition for [Na(H$_2$O)$_n$] and [Li(H$_2$O)$_n$] and expectation value $\sqrt{<r^2>}_{\text{SOMO}}$ (RD) of unpaired electron. The origin is at the metal atom (Hashimoto, 1999a)

	[Na(H$_2$O)$_n$]				[Li(H$_2$O)$_n$]		
Label in Figure 9	EE (eV)		RD (a.u.)	Label in Figure 9	EE (eV)		RD (a.u.)
	Experiment	Calculated	Calculated		Experimental	Calculated	Calculated
Na Atom	0.0	0.0 (^{2}S)	4.5	Li Atom	0.0	0.0 (^{2}S)	4.2
	2.1	2.1 (^{2}P)	6.5		1.9	1.8 (^{2}P)	5.2
1a (C$_{2v}$)	0.0	0.0 (^{2}A$_1$)	4.5	1a' (C$_s$)	0.0	0.0 (^{2}A')	4.9
	2.3	2.0 (^{2}A$_1$)	6.1		1.4	1.5 (^{2}A')	7.1
		2.0 (^{2}B$_1$)	6.3			1.7 (^{2}A')	7.0
		2.1 (^{2}B$_2$)	6.3			2.6 (^{2}A')	6.3
2a (C$_s$)	0.0	0.0 (^{2}A')	4.5	2a' (C$_s$)	0.0	0.0 (^{2}A')	6.1
	2.3	2.1 (^{2}A')	6.0		1.2	1.0 (^{2}A')	7.8
		2.1 (^{2}A')	6.2			1.4 (^{2}A')	8.2
		2.1 (^{2}A')	6.1			1.2 (^{2}A')	8.6
3a (C$_1$)	0.0	0.0 (^{2}A)	4.5	3a' (C$_{3h}$)	0.0	0.0 (^{2}A$_1$)	7.4
	2.4	2.1 (^{2}A)	6.0		1.0	0.9 (^{2}E)	9.0
		2.1 (^{2}A)	6.0				
		2.2 (^{2}A)	6.1			1.3 (^{2}A$_1$)	9.9

C.   Ammoniated Metal Anions for Comparison with Hydrated Clusters

1.   Experimental Study

As mentioned previously, IEs for the clusters consisting of a group 1 atom with ammonia molecules exhibit a characteristic size dependence; the metal atom does not change the size dependence much, and its limiting value, IE(∞), coincides with that of negatively charged ammonia clusters and the photoelectric threshold of liquid ammonia. In contrast to IEs in metal–water systems, the IEs in clusters with a group 1 atom and ammonia molecules decrease almost linearly with $(n+1)^{-1/3}$. By analyzing the size dependence of IEs for water clusters and ammonia clusters, both containing a group 1 metal atom, we may be able to explore the electronic structures and the solvation states of the atoms. Therefore the photoelectron spectra of ammoniated metal anions are also investigated using the magnetic-bottle type of photoelectron spectrometer (Takasu et al., 1996, 1997).

Figure 10 shows the photoelectron spectra of the $Li^-(NH_3)_n$ clusters recorded at the photodetachment energy of 3.50 eV. The transitions to the Li (^2S) and Li (^2P) states are observed at 0.62 and 2.47 eV, respectively. For $Li^-(NH_3)$, one strong band and one weak band are also observed at 0.56 and ca. 2.05 eV, which are shifted by 0.06 and 0.42 eV, respectively, from the corresponding bands in the bare anion. As described in Section II.C.2, calculations predict that $Li^-(NH_3)$ has two isomers. The most stable isomer has a structure similar to that of the neutral complex, in which an ammonia molecule is bound to Li^- by the N atom, while, in the less stable isomer, NH_3 is bound to Li^- by the H atoms, as in the case of $Li^-(H_2O)$. The calculated transition energy to the ^2S state for the stable isomer is 0.4 eV, while that for the less stable isomer is much higher (Hashimoto et al., 1997). Based on these theoretical results, the 0.56- and 2.05-eV bands of $Li^-(NH_3)$ are assigned to the transitions to 2^2S and 2^2P states of the stable isomer. Although its spectra are not shown, $Na^-(NH_3)$ exhibits similar spectral trends (Takasu et al., 1997). For $Li^-(NH_3)$, the spectrum reveals a third band at 2.78 eV (see Figure 10); similar weak bands are observed at 1 eV above the 2^2P-type transition for $Li^-(NH_3)_n$ ($n \leq 11$). These bands are assigned to the transition to the 3^2S state, shifted from 3.99 eV for the isolated atomic anion Li^-.

For large Li^-– and Na^-–ammonia clusters, the experimental and theoretical results suggest that the isomer having the maximum number of metal–N bonds is increasingly favored with n, as in the case of the neutral clusters. Thus, the photodetachment process of the clusters with $n \geq 2$ corresponds to the transition from the negative state to near the potential minimum of the neutral states, as in the case of $Li^-(H_2O)_n$ mentioned in Section II.B. For $Li^-(NH_3)_n$ ($2 \leq n \leq 9$), the 2^2P transition shifts to as low

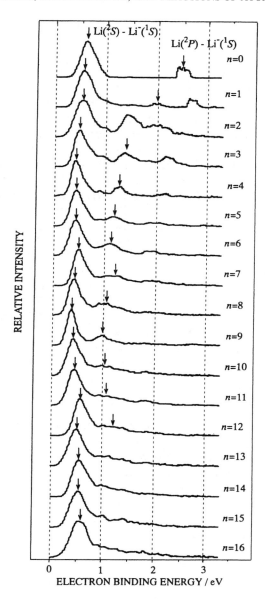

Figure 10.   Photoelectron spectra of the $Li^-(NH_3)_n$ clusters ($n \leq 16$). The detachment energy is 3.50 eV (From Takasu et al., 1997).

as 1.4 eV and almost becomes degenerate with the transition to the neutral ground state 2^2S for $n \geq 10$. (See Figure 10). The $3^2S \leftarrow {}^1S$ transition at 2.78 eV for $n = 1$ also gradually shifts further to the lower energy up to $n = 11$ and then smears out for $n \geq 12$. For $Na^-(NH_3)_n$, the 3^2P transition rapidly shifts to the lower electron binding energy for $n = 2$–4 and then shifts at a much slower rate for $n \geq 5$. (The amount of shift for $n = 5$ is more than 1.4 eV with respect to that of Na^-). A large change between $n = 4$ and $n = 5$ may be ascribed to the formation of a solvation shell around Na^-. Moreover, the bandwidth of the 3^2P transition monotonically increases with n up to 12, although the shift in the band position is almost the same for $n \geq 5$. The similar large spectral change in the ^2P–^2S type transition has also been observed for $Na(NH_3)_n$ by Schulz and coworkers (Schulz et al., 1997; Brockhaus et al., 1999). The other interesting feature in the PES of $Li^-(NH_3)_n$ is that the 2^2S transition is shifted to lower binding energy with respect to that of Li^- even with $n \geq 5$. The binding energy of the second-shell NH_3 in the neutral clusters is larger than that in their anions. Because the theoretical calculations indicate that the first solvation shell around the Li and Na atoms is closed with four or five NH_3 molecules (Hashimoto et al., 1995, 1999b), further changes in PES for $n \geq 5$ imply that the electronic structure of the metal atom is affected by the second-shell ammonia molecules. Therefore, the observed PES suggest that the alkali metal atom is spontaneously ionized, and its valence electron is delocalized over the clusters and interacts directly with the second-shell ammonia molecules, as in the case of the Li^-–H_2O system described in Section II.B.

2. Theoretical Study

The observed size dependence (or size independence) of PES bands of both $Na^-(NH_3)_n$ and $Li^-(NH_3)_n$ is somehow similar to that of $Li^-(H_2O)_n$, but not of $Na^-(H_2O)_n$. The spectral features of these clusters have been analyzed in relation to their structures (Hashimoto et al., 1997). The calculated structures and total binding energies of $Na^-(NH_3)_n$ and $Li^-(NH_3)_n$ for $n = 1$, 2, and 3 are shown in Figure 11. In contrast to the $Na^-(H_2O)_n$ clusters, the most stable isomers have ammonia molecules directly bound to the Na atom by N atoms. The binding energy gained by the Na-N bond is about 17 kJ/mol for $n = 1$ and increases in a nonadditive manner as n increases. On the other hand, the Na–H interaction is about 13 kJ/mol per NH_3 molecule, which is almost constant for $n = 1$–3. Other isomers with fewer Na–N bonds and some hydrogen-bonding ammonia molecules are found to be local-minimum structures. The most stable structures of the neutral $Na(NH_3)_n$ ($n = 1$–3) and their cation complexes are similar to those of the corresponding anion clusters (Hashimoto and Morokuma, 1995). This structural similarity between the neutral and

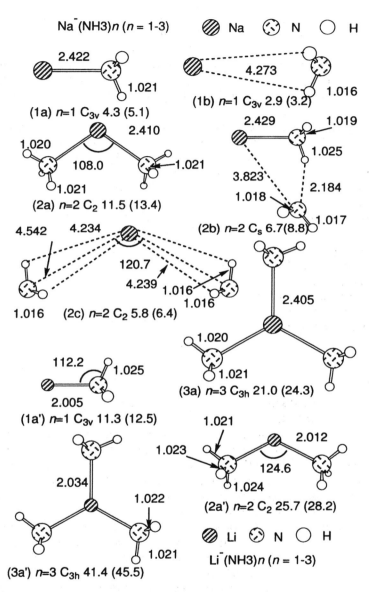

Figure 11. Structures and total binding energies in kcal/mol (1 kcal/mol = 4.1868 kJ/mol) of $[Na(NH_3)_n]^-$ and $[Li(NH_3)_n]^-$ ($n = 1-3$).

anionic states of the small clusters of Na–NH$_3$ is a remarkable difference from what is exhibited by hydrated Na atom clusters.

The most stable structures of Li$^-$(NH$_3$)$_n$ ($n = 1$–3) resemble the corresponding Na$^-$(NH$_3$)$_n$. The isomers having fewer Li–N bonds are less stable, and the energy difference between the most stable isomer and the local-minimum structures of Li$^-$(NH$_3$)$_n$ for each n increases gradually with n.

The vertical detachment energies for the most stable structures of Li$^-$(NH$_3$)$_n$ ($1 \leq n \leq 3$) are calculated to be around 0.4 eV, which is in reasonable agreement with experimental results. The corresponding vertical detachment energies for Na$^-$(NH$_3$)$_n$ are 0.35 for $n = 1$ and 0.32 for $n = 2$. The calculated vertical detachment energies for complexes in which all NH$_3$ molecules are bound to Li by H atoms are 0.76 (0.74) eV for Li$^-$(NH$_3$) and 0.91 (0.88) for Li$^-$(NH$_3$)$_2$ at the CCSD(T) (CCSD) level. The corresponding energies for the similar isomers of Na$^-$(NH$_3$) and Na$^-$(NH$_3$)$_2$ are 0.65 eV and 0.78 eV, respectively, at CCSD level. As seen in Figure 10, the band around 0.5 eV in the photoelectron spectrum of Li$^-$(NH$_3$)$_n$ ($n = 1$ and 2) is symmetric, but the corresponding band in the photoelectron spectrum of Na$^-$(NH$_3$)$_n$ is asymmetric and has a shoulder at the high-energy side, which suggests that less stable isomers coexist with the more stable ones in Na$^-$(NH$_3$)$_n$ ($n = 1, 2$) (Takasu et al., 1997).

III.   GROUP 2 ELEMENTS

In Section II, we discussed the solvation process of alkali (group 1) atoms whereby solvated electrons are generated in finite clusters. The chemical reactivity of alkali metals with a water molecule and with water clusters is also an interesting target in water cluster chemistry, which might be related to microscopic aspects of the well-known explosive reaction of solid sodium with liquid water. However, recent studies indicate that no reaction products are observed in the scattering of an Na atom (and its clusters) with water clusters and that the observed species in the mass spectra are the hydrated metal Na(H$_2$O)$_n$ (Bewig et al., 1998). The chemical reactivity of group 2 and group 3 metal ions in aqueous solution has been studied for many years. In the gas phase, the clusters consisting of singly charged ions of group 2 and group 3 metals with water molecules are readily produced, and, in contrast to group 1 atoms, it turned out that the chemical reactions proceed in the metal–water clusters. In bulk aqueous solutions, these metal atoms are oxidized to produce ions of the form $[M(OH)_n]^{+s}$, in which M is formally doubly or triply charged. Therefore, with water clusters, metal ions may undergo a charge transfer, oxidation, or charge disproportion reactions to produce ions with more charge as the cluster size increases. Thus, the experimental and theoretical studies of these cluster ions may provide us with information on the microscopic aspect of redox reactions of metal ions with water molecules in bulk solution.

A. Collision Reactions of Metal Ions with Water Clusters

1. Product Switches

Let us examine collision-induced reactions of metal ions of Mg^+ and Ca^+ generated by laser vaporization with water clusters, $M^+ + (H_2O)_n$. From an experimental perspective, one cannot completely specify the size of the neutral target water clusters. However, studying these reactions is still informative and lends insight into the dynamics of metal–ion solvation in gas-phase clusters.

Figure 12 shows a typical mass spectrum of the nascent cluster ions produced by the collision of the Mg^+ ions with water clusters (Misaizu et al., 1992a; Fuke et al., 1993). Both $Mg^+(H_2O)_n$ and $MgOH^+(H_2O)_{n-1}$ ions are produced by the two reactions

$$Mg^+ + (H_2O)_n \rightarrow Mg^+(H_2O)_m + (n - m)H_2O \tag{7}$$

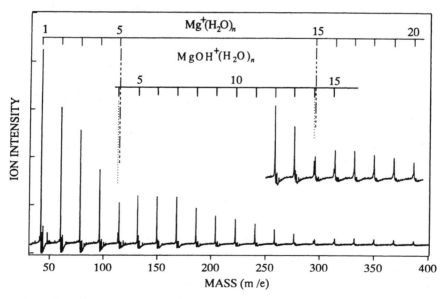

Figure 12. TOF mass spectrum of the product ions for the reaction of singly charged ions Mg^+ with water clusters $(H_2O)_n$.

and

$$Mg^+ + (H_2O)_n \rightarrow Mg\ OH^+(H_2O)_m + H\ +\ (n - m - 1)H_2O \qquad (8)$$

The relative abundance of $Mg^+(H_2O)_n$ and $MgOH^+(H_2O)_{n-1}$, obtained from the mass spectrum, is shown in Figure 13(a). These cluster ions are produced with a characteristic size distribution: $Mg^+(H_2O)_n$ clusters are produced predominantly for $1 \leq n \leq 5$ and $n \geq 15$, while $MgOH^+\ (H_2O)_{n-1}$ clusters are almost exclusively observed for $6 \leq n \leq 14$. The results indicate that the product ions switch rapidly at two critical cluster sizes: $n = 5$ and $n = 15$. These features are not influenced by experimental conditions, such as the stagnation pressure of expansion or the intensity of the vaporization laser. For the Mg^+–D_2O system, we also find similar product switchings, with a slight shift of the critical sizes, as shown in Figure 13(b): $Mg^+(D_2O)_n$ are dominant for $1 \leq n \leq 6$ and $n \geq 14$, while $MgO^+(D_2O)_{n-1}$ are exclusively observed for $7 \leq n \leq 13$. Castleman and coworkers have studied the reaction of Mg^+–H_2O and –D_2O systems at room temperature using a flow tube method and found a similar H atom elimination reaction (Harms et al., 1994). As in the case of the Mg^+ ions, we also found both $Ca^+(H_2O)_n$ and $CaOH^+(H_2O)_{n-1}$ ions in Ca^+– $(H_2O)_n$ reactions with similar product switchings; the critical sizes are $n = 4$ and $n = 13$. However, the product distribution is considerably influenced by the deuterium substitution, as shown in Figures 14(a) and (b). Although the hydrated $CaOD^+$ ions are formed for $5 \leq n \leq 12$, $Ca^+(D_2O)_n$ ions persist in this size range. To understand these "anomalous" product distributions, Fuke and his co-workers also examined the reaction for the Mg^+–H_2O system at several collision energies using a simple crossed-beam apparatus. The results indicated that the critical sizes, where the product switchs, remain unchanged within the KE range of 0–60 eV (Sanekata et al., 1995).

In Sections III.A.2 and III.A.3, the mechanism of the observed first product switching are interpreted with the aid of the observed photodissociation reactions of $M^+(H_2O)_n$ and with the aid of ab initio MO calculations.

2. Ab initio MO studies

There are several ab initio MO studies of complexes of group 2 atoms with water clusters. Hashimoto, Iwata, and their coworkers investigated a beryllium atom and its ion with water clusters (Hashimoto et al., 1987a,b; Hashimoto and Iwata, 1989; Hashimoto et al., 1990). Although almost no experimental studies for Be–water clusters and their ions have been reported, theoretical results provided some relevant results to the aqueous solution chemistry of beryllium ions. Bauschlicher and coworkers (1992)

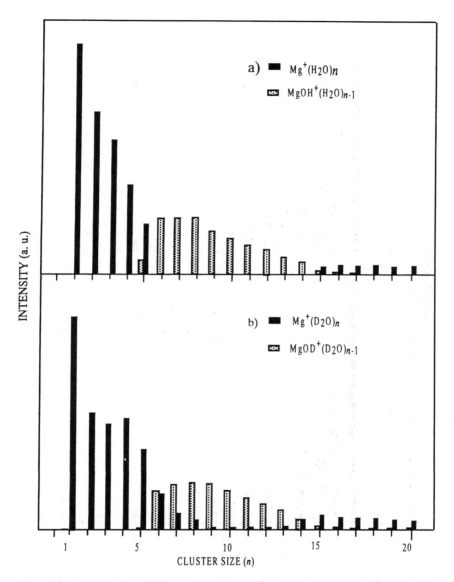

Figure 13. (a) Relative abundance of $[Mg(H_2O)_n]^+$ and $[MgOH(H_2O)_{n-1}]^+$. (b) Relative abundance of $[Mg(D_2O)_n]^+$ and $[MgOD(D_2O)_{n-1}]^+$ (From Sanekata et al., 1995). The filled (open) bars in the figures correspond to the $[Mg(H_2O)_n]^+$ ($[MgOH(H_2O)_{n-1}]^+$) ions.

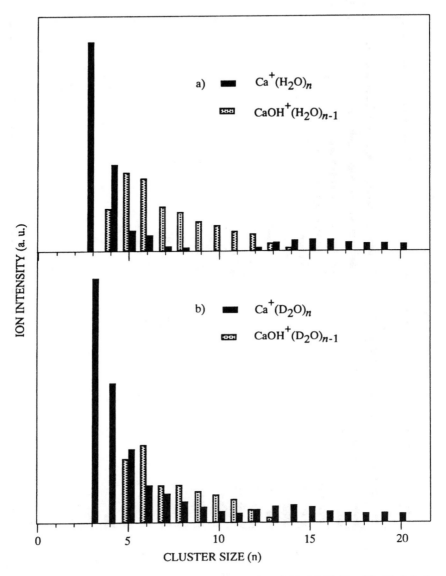

Figure 14. (a) Relative abundance of $[Ca(H_2O)_n]^+$ and $[CaOH(H_2O)_{n-1}]^+$. (b) Relative abundance of $[Ca(D_2O)_n]^+$ and $[CaOD(D_2O)_{n-1}]^+$. The filled (open) bars in the figures correspond to the $[Ca(H_2O)_n]^+$ ($[CaOH(H_2O)_{n-1}]^+$) ions.

examined the geometric and electronic structures of Mg^+- and Ca^+-water clusters ($n \leq 4$). To analyze the first product switch mentioned in the preceding subsection, Watanabe et al. determined the geometric structures of the reactants $Mg^+(H_2O)_n$ and $Ca^+(H_2O)_n$, and products $[MgOH(H_2O)_{n-1}]^+$ and $[CaOH(H_2O)_{n-1}]^+$ of the hydrogen elimination reaction (Watanabe et al., 1995; Watanabe and Iwata, 1997b). The optimized structures by Bauschlicher et al. and by Watanabe et al. are

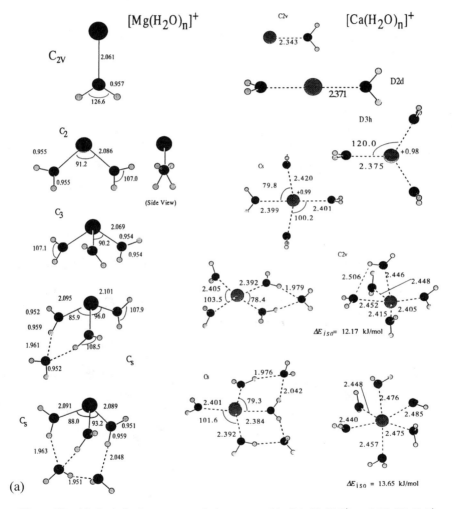

Figure 15. (a) Optimized structures of the most stable $[Mg(H_2O)_n]^+$ and $[Ca(H_2O)_n]^+$. (b) Optimized structures of the most stable $[MgOH(H_2O)_{n-1}]^+$ and $[CaOH(H_2O)_{n-1}]^+$.

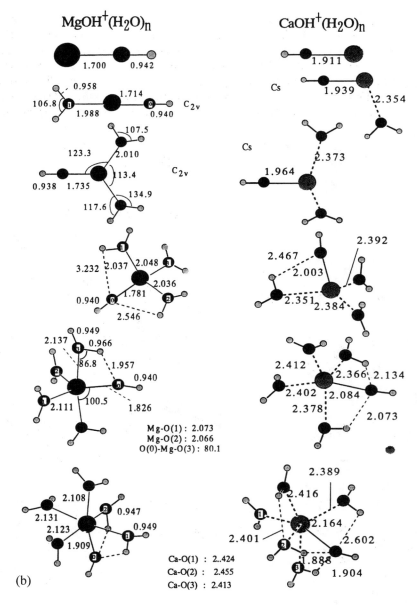

Figure 15. (continued)

similar to each other, but, in some cases, not exactly equal. Figures 15(a) and (b) are the optimized structures of some of the most stable isomers of the reactants and products. The structures are optimized with the unrestricted self-consistent field (UHF) method for the reactants and with the closed-shell self-consistent field (HF) method for the products. For smaller clusters, the effect of electron correlation on geometry was also examined, but turned out to be small for structures around the metal atom. The basis sets used for Mg ion clusters are $6–31G^*$ in most cases, except for large clusters; for Ca^+, the $[12s9p5d/9s7p3d]$ set was used. In energy calculations, MP2 and MP4 levels of electron correlation are taken into account for Mg ion complexes, but only HF calculations are performed for Ca^+ ion complexes. It has been confirmed for Mg ion complexes that semi-quantitative results are attained at the HF level of approximation, because of the strong ion–water interaction in these complexes.

The structures are classified, in terms of the number of water molecules in the first shell, as $[M^+(H_2O)_s](H_2O)_{n-s}$ and $[MOH(H_2O)_s]^+(H_2O)_{n-s-1}$. For a given n, there are a few isomers with different s. The largest s is 5 for $[Mg^+(H_2O)_s](H_2O)_{n-s}$ and 6 for $[Ca^+(H_2O)_s](H_2O)_{n-s}$. As in hydrated group 1 elements, the balance between strengths of the metal–O bond and the hydrogen bonds among water molecules determines the relative energies among the isomers. But another factor also influences the structure of the clusters. The difference in the structures of Mg^+ and Ca^+ ion complexes with water clusters is easily noticed in Figure 15(a). In $M^+(H_2O)_2$, the angle of O–Mg–O is nearly a right angle. On the other hand, O–Ca–O is linear. In $Mg^+(H_2O)_3$, the Mg^+ ion complex has a pyramidal form with a nearly O–Mg–O right angle, and this form persists as the structure of the first shell in the most stable isomer of larger clusters. On the other hand, in $Ca^+(H_2O)_3$ and $Ca^+(H_2O)_4$, all of the oxygen atoms and the calcium ion lie in a plane. For larger clusters of Ca^+ ion complexes, the structure of $Ca^+(H_2O)_4$ is the first shell in the most stable isomers, but isomers having more water molecules in the first shell are close in energy to the most stable one, as is also shown in Figure 15(a).

The maximum number of water molecules in the first shell in the most stable isomers of the Ca^+ complexes is larger than that in the Mg^+ ion complexes; a fact that cannot be explained by the competition between the metal ion–water interaction and the hydrogen-bonding interaction among water molecules. The calculated binding energy ΔE_{hyd} for $Ca^+(H_2O)$ is $-124.5\,kJ/mol$ and for $Mg^+(H_2O)$ is $-166.6\,kJ/mol$, at the $MP4//UHF/6–31G^*$ level of approximation. Since the metal ion–water interaction in the Ca^+ ion is weaker than in the Mg^+ ion, the hydrogen-bonding network of water molecules in the first and second shells is expected to determine the structure of the metal ion complexes with water clusters. The other factor we have to take into account is the ionic radius of the metal ions: R_{ion} is 1.18 Å for Ca^+ and 0.82 Å for Mg^+. Thus, more water molecules can directly bond to the Ca^+ ion with less steric hindrance among themselves than to the Mg^+ ion. But the difference in the ionic radii of Mg^+ and Ca^+ does not explain the structural difference around the metal ion. The orbital character of the singly occupied molecular orbital (SOMO) is different for the Mg^+ and Ca^+ ion complexes. For instance, the SOMO of $Mg^+(H_2O)_3$ is localized at the end opposite the water molecules, in a manner somewhat similar to that of the lone-pair orbital of the ammonia molecule; in other words, the SOMO is an sp^3 hybridized orbital, as was first pointed out by Bauschlicher et al. (1991a, 1991b). The other three hybrid orbitals are coordinated by the non-bonding orbitals of water molecules. The SOMO of $Ca^+(H_2O)_n$, on the other hand, is almost a pure $4s$, which is diffuse and spherical. The water molecules hydrate to the Ca^+ ion as they do to the bare alkali metal ion, such as Na^+ or Li^+, forming the interior structure. But this simple analogy does not account for the planar structures of the first shell. Interestingly, in the natural population analysis, the $3d$ electron population in $Ca^+(H_2O)_n$ increases from 0.03 at $n = 1$ to 0.15 at $n = 4$, while the $4s$ population decreases from 0.97 to 0.85, and almost no population in the $4p$ orbitals is found. Obviously, the electron donation to the d orbitals from the non-bonding orbitals of water molecules contributes the structure around the metal ion. To confirm the role of the d orbitals, Watanabe and Iwata (1997b) optimized the structures of $Ca^+(H_2O)_2$ and $Ca^+(H_2O)_3$ by excluding the d functions from the basis set and obtained the similar structures of $Mg^+(H_2O)_2$ and $Mg^+(H_2O)_3$; a slight difference is that in these model $Ca^+(H_2O)_2$ and $Ca^+(H_2O)_3$ complexes, a stronger hydrogen-bond network within the first shell of the water molecule is present than in the corresponding Mg^+ ion complexes.

3. Analysis of the Product Switches

As mentioned previously, $Mg^+(H_2O)_n$ ions are produced exclusively for n up to 5, while $MgOH^+(H_2O)_{n-1}$ ions are the unique products for n larger than 5 (Misaizu et al., 1992a; Sanekata et al., 1995). These results indicate that the formation of $MgOH^+(H_2O)_n$ requires at least five water molecules. Moreover, they confirmed that the critical size is not affected by the reaction conditions, which implies that the product switching might be caused by the thermodynamic stability of the product ions and not by the reaction dynamics. On the basis of the results of the photodissociation studies discussed in Section III.B, the incremental binding energies ΔH for $MgOH^+-H_2O$ and $MgOH^+(H_2O)-H_2O$ were determined to be 188 kJ/ mol (1.95 eV) and 184 (1.91), respectively, which are in reasonably good agreement with the calculated results of 238 kJ/mol (2.47 eV) and 184 (1.91), respectively (Watanabe et al., 1995). These binding energies are much larger than the theoretical results for Mg^+-H_2O (159 kJ/mol = 1.65 eV) and $Mg^+(H_2O)-H_2O$ (130 kJ/mol = 1.35 eV), due to a large polarization of the $MgOH^+$ ions; the effective charge on the Mg atom is more than +1.5. The hydrogen elimination reaction is nothing but a redox reaction. The large polarization in the Mg–O bond enhances the bonding with water, and the $MgOH^+(H_2O)_n$ ions are stabilized much faster than those for $Mg^+(H_2O)_n$ with n. To explore the relative energy of the hydrated ions and reaction products more clearly, energy diagrams for the Mg^+ ion and Ca^+ ion complexes are shown in Figure 16. As is clearly seen in the figure, the reaction products become lower in energy at $n = 5$, for both the Mg^+ ion and Ca^+ ion complexes. The product switching observed for the $Ca^+(H_2O)_n$ systems at $n = 4$ is also almost reproduced by the theoretical calculations — in particular, when the entropy factor is taken into account by evaluating the thermal equilibrium constants. Hence, as a result of the difference in hydration energies for the M^+ and MOH^+ ions, the metal-oxidation reaction

$$M^+(H_2O)_n \rightarrow MOH^+(H_2O)_{n-1} + H \qquad (9)$$

becomes thermodynamically feasible for clusters larger than the critical size. As we shall see in Section III.B, this same reaction proceeds by photoexcitation for smaller clusters.

Upon deuterium substitution, the critical size for the product switching for Mg^+ shifts to $n = 6$. However, weak signals for reactants are observed for $n = 6$–15. A similar deuterium effect is also observed in the Ca^+–water reaction. Watanabe et al. (1995) evaluated the equilibrium constant and its temperature dependence by assuming thermal equilibrium. Although under real experimental conditions thermal equilibrium is never attained, the first product switch takes place at $n = 5$ over a broad temperature range. Besides, if high temperature — in other words, high internal energy — is assumed, the observed deuterium effect and metal dependence are partly reproduced.

The observed deuterium effect on the product distribution is much larger for Ca^+ than for Mg^+. This is because the enthalpy changes from $Ca^+(H_2O)_n$ to $CaOH^+(H_2O)_{n-1}$ ($5 \leq n \leq 12$) are smaller than those from $Mg^+(H_2O)_n$ to $MgOH^+(H_2O)_{n-1}$, whereas the enthalpy changes due to deuterium substitution are almost the same for both systems.

For $n \geq 15$, the $Mg^+(H_2O)_n$ ion signals again become the main ones, as shown in Figure 13. Similar product switchings are also observed for $Mg^+(D_2O)_n$ at $n = 14$, for $Ca^+(H_2O)_n$ at $n = 13$, and for $Ca^+(D_2O)_n$ at $n = 12$ (Sanekata et al., 1995). The experimental conditions do not affect the critical sizes of product switching. The cause of this second product switch has been discussed (Sanekata et al., 1995; Watanabe et al., 1995; Berg et al.,

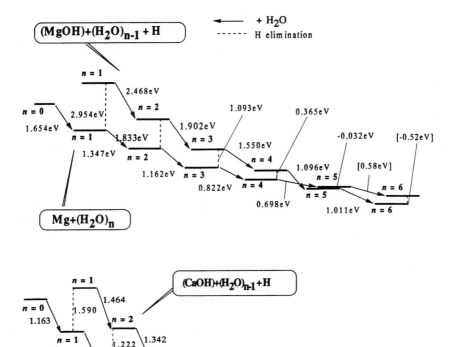

Figure 16. (a) Relative energies of $[Mg(H_2O)_n]^+$ and $[MgOH(H_2O)_{n-1}]^+$ + H. The numbers are evaluated for the most stable isomer with the MP4SDTQ method at the HF optimized structure. (b) Relative energies of $[Ca(H_2O)_n]^+$ and $[CaOH(H_2O)_{n-1}]^+$ + H. The numbers are evaluated with the HF level of approximation. In both figures, the zero-point vibration correction, estimated with HF calculations, is included. (1 eV = 96.48 kJ/mol.)

1997, 1998), but seems not yet resolved. Recently, Berg et al. (1997, 1998) reported Fourier-transform–ion cyclotron resonance studies of reactions similar to Eq. (9). In their resonance cell, they isolated complexes of a particular size n and followed the photofragmentation by blackbody radiation. As with the collision reactions, they found the size dependence of the reaction to be

$$Mg^+(H_2O)_n + h\nu \rightarrow Mg^+(H_2O)_{n-1} + H_2O \qquad \text{(for } n > 21 \text{ and } n \le 4\text{)},$$

$$\text{(10)}$$

$$Mg^+(H_2O)_n + h\nu \rightarrow MgOH^+(H_2O)_{n-2} + H_2O + H \quad \text{(for } n = 16 - 21\text{)},$$

$$\text{(11)}$$

and showed that $MgOH^+(H_2O)_{n-2}$ with smaller n is the product of the successive evaporation of a water molecule [Eq. (11)]. Thus, again, the apparently stable species $Mg^+(H_2O)_{n-1}$ is found for large n. Berg et al. (1997) advocated the model of the ion-pair state $M^{2+}(H_2O)_n^-$, as Farrar and coworkers used in explaining the photodissociation spectra of $Sr^+(NH_3)_n$ (Shen and Farrar, 1988, 1991; Donnelly and Farrar, 1993). To ionize Mg^+ and Ca^+ one step further requires 15.0 eV (1,450 kJ/mol) and 11.9 eV (1,150 kJ/mol), respectively. Farrar and colleagues argued that, because of its high positive charge, the hydration energy of the doubly charged ion exceeds a certain stabilization energy, forming the ion pair; thus, the intracluster reaction is prohibited. This model might explain the metal dependence of the onset size of the reaction—$n = 15$ for Mg^+ complexes and $n = 13$ for Ca^+ complexes.

Other hypotheses may explain the apparent nonreactivity in the clusters. The intracluster reactions might be taking place even in larger clusters, as in the first product switch. In the product $MgOH^+$ of that switch, the metal is not fully oxidized, but full oxidization does not mean the production of a fully charged ion Mg^{2+}. The chemical bonds, in place of the solvation, can stabilize the cluster. One of the candidates for those chemical species is, for instance, $MgOH-H_3O^+(H_2O)_{n-2}$. One of the important findings in the experiments by Berg et al. (1997) is that the reactive species are limited only for $n = 16-21$; in other words, the precursor of the hydrogen elimination reaction is present only for these sizes of clusters. Under the assumption of $MgOH-H_3O^+(H_2O)_{n-2}$, the oxonium ion H_3O^+ and the MgOH radical are well separated in larger clusters, and both are solvated independently, with no chance of the electron or proton transfer. But in a limited size of clusters, such as $n = 16-21$, the ion and the radical are not far away from each other, and successive electron and proton transfers may induce the hydrogen elimination reaction. Obviously, the discussion is based on no experimental or theoretical data; however, we cannot automatically rule out this mechanism as the origin of the second product switching. In Section IV.D, a model study of such electron- and proton-transfer reactions in B^+- water clusters is discussed. At present, it is difficult to conclude which is responsible for the second product switch, the ion-pair state formation or the unexpected reaction products. We will revisit this issue when we discuss a similar switch in the reaction of the Al^+ ion with water clusters in Section IV.C.

B. Photodissociation Spectra and the Dissociation Process

As mentioned in Section III.A, solvated metal ions such as $Mg^+(H_2O)_n$ and $Ca^+(H_2O)_n$ are easily synthesized by the laser vaporization technique coupled with supersonic expansion. A reflection-type time-of-flight (TOF) mass spectrometer can be utilized to examine the photodissociation products, as well as the dissociation spectra, of mass-selected cluster ions. The spectral data obtained here may include information on the size-dependent geometric and electronic structures of hydrated metal ions. Moreover, since the photodissociation process of these ions is a half-collision analogue of the reaction described in Section II.A, it provides us with a wealth of insight into the intrinsic nature of the cluster reactions.

1. Photodissociation Spectra of $Mg^+(H_2O)_n$

Figure 17(a) shows the photodissociation spectra of $Mg^+(H_2O)$ ($n = 1$–5) (Misaizu et al., 1994). The photoexcitation of these ions induces the evaporation of water molecules, as well as the hydrogen elimination process [Eq. (9)]. The photochemical reaction will be discussed in more detail in Section III.B.2. The spectra are recorded by monitoring the total product ions as a function of excitation wavelength. In the energy region of the Mg^+ ($^2P \leftarrow ^2S$) transition (35,700 cm^{-1}), a shoulder at 28,300 and peaks at 30,500 and 38,500 cm^{-1} are observed. [See Figure 17(a)]. In Figure 17(b), the calculated stick spectra of $Mg^+(H_2O)_n$ are also shown. The $Mg^+(H_2O)$ ion, shown in Figure 15(a), has a planar structure with C_{2v} symmetry in the ground state. With this conformation, the degeneracy of the p orbital of the Mg^+ ion is removed by interaction with the nonbonding orbital of the water molecule. The schematic orbital interaction diagram is drawn in Figure 18. The excitation of an electron to these orbitals generates three low-lying excited electronic states, $1^2B_2(x)$, $1^2B_1(y)$, and $2^2A_1(z)$. The calculated vertical excitation energies are 28,000 cm^{-1} ($1^2B_2 \leftarrow 1^2A_1$), 29,800 cm^{-1} ($1^2B_1 \leftarrow 1^2A_1$), and 37,700 cm^{-1} ($2^2A_1 \leftarrow 1^2A_1$), respectively, with relative transition probabilities of approximately 1:1:1 (Misaizu et al., 1992a, Watanabe and Iwata, 1998). Bauschlicher et al. (1992) have also reported similar theoretical results (29,200, 31,200 and $< 38,900$ cm^{-1}, respectively). Based on these theoretical results, the shoulder and two peaks are safely assigned to the transitions from the 1^2A_1 ground state to the 1^2B_2, 1^2B_1, and 2^2A_1 excited states, respectively. Duncan and coworkers have recorded a high-resolution spectrum of this complex and confirmed the symmetry of the two low-lying excited states by rotational analysis (Yeh et al., 1992; Willey et al., 1992). The calculations also predict that these excited states are all bound

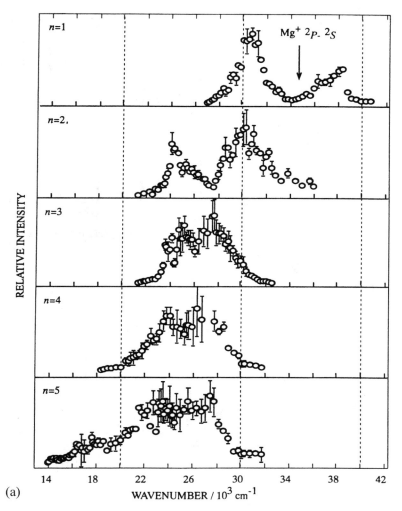

RELATIVE INTENSITY

n=1

Mg⁺ $^2P_-$ ^2S

n=2,

n=3

n=4

n=5

14   18   22   26   30   34   38   42

(a)

WAVENUMBER / 10^3 cm^{-1}

Figure 17. (a) The experimental photodissociation spectra of $Mg^+(H_2O)_n$ ($n = 1$–5). (b) The theoretical photoabsorption spectra for isomers of $Mg^+(H_2O)_n$ ($n = 1$–5).

with respect to the Mg–O stretching coordinate (Misaizu et al., 1992a). These results are important for an understanding of the mechanism of the photodissociation process of $Mg^+(H_2O)_n$, as mentioned later.

For $Mg^+(H_2O)_2$, the preceding transitions are shifted to the red and peak at 25,000 and 30,000 cm^{-1}, respectively. The relative intensities of the two bands are reversed. As shown in Figure 15(a), the most stable structure of

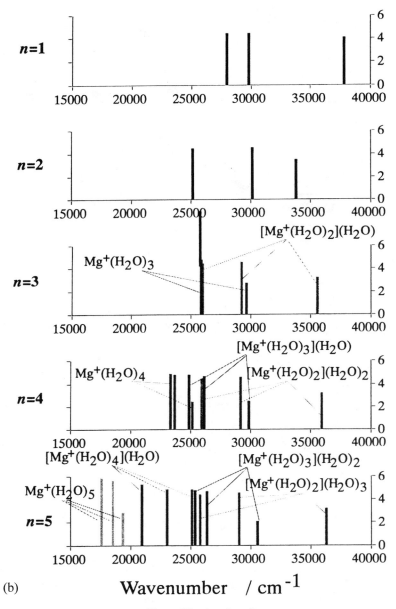

(b)   Wavenumber / cm^{-1}

Figure 17. (continued)

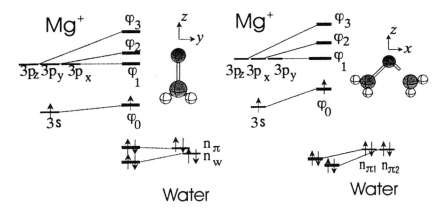

Figure 18.   The schematic orbital diagrams of $Mg^+(H_2O)_n$ ($n = 1, 2$).

$n = 2$ in the ground state has a bent form (C_2 symmetry), with two waters staggered (Misaizu et al., 1992a; Watanabe et al., 1995). The observed excitation energies are found to be 25,330 ($1^2B(y) \leftarrow 1^2A$), 30,060 ($2^2B(x) \leftarrow 1^2A$), and 33,740 cm^{-1} ($2^2A(z) \leftarrow 1^2A$), respectively, whereas the calculated vertical energies by Watanabe and Iwata (1998) are 25,100 (26,500), 30,100 (31,400), and 33,800 (<35,000) cm^{-1}, respectively. (The numbers in parentheses are those of Bauschlicher et al., 1992). Thus, the observed spectrum of $n = 2$ agrees very well with these theoretical predictions, including the transition probabilities. A broad, intense band at about 30,000 cm^{-1} shown in Figure 17 is due to overlapped transitions to the higher two electronic states (2^2B and 2^2A). The schematic orbital diagrams in Figure 18 can qualitatively explain the spectral shifts from $n = 1$ to $n = 2$.

The spectrum of $Mg^+(H_2O)_3$ exhibits two closely spaced bands at 25,000 and 28,000 cm^{-1}. The total width of the spectrum is substantially narrower than that of either $n = 1$ or $n = 2$. The spectrum of $Mg^+(H_2O)_4$ is somewhat similar to that of $Mg^+(H_2O)_3$, but the tail extending to the lower energy is pronounced. This tail becomes a shoulder in the spectrum of $Mg^+(H_2O)_5$ and extends to as low as 15,000 cm^{-1}. The analysis of these spectral changes requires systematic ab initio MO calculations for the excited states of the complexes.

2.   Ab initio MO Studies of the Vertical Excitation Energies

The structures of the most stable isomers of $Mg^+(H_2O)_n$ were shown in Figure 15(a). But the observed absorption spectra for $n = 4$ and $n = 5$ suggest that the spectra could be attributed to a mixture of the isomers. Figure 19 shows the structures of isomers for $n = 3$, 4, and 5, optimized at the UHF/6–31 + G* level of approximation. The isomers are classified by the number of

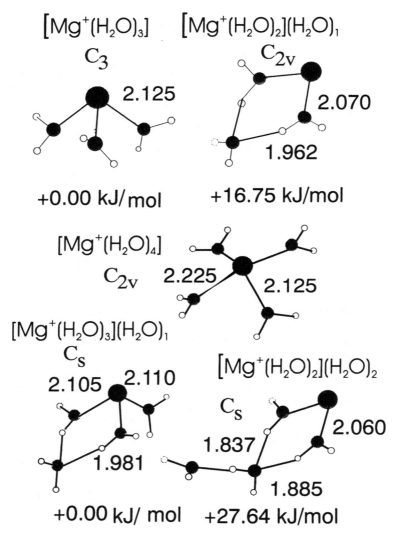

Figure 19. Optimized structures of several isomers and relative energy of $Mg^+(H_2O)_n$ for $n = 3, 4,$ and 5.

first-shell water molecules, s, as $[Mg^+(H_2O)_s](H_2O)_{n-s}$. The number can be as large as 5 in the Mg^+ ion–water complexes, but the structures are quite different from the corresponding Ca^+ ion–water complexes, as shown in Figure 15(a). In the structure of $[Mg^+(H_2O)_5]$, all of the water molecules are located at one side, indicating that the SOMO is localized at the other side

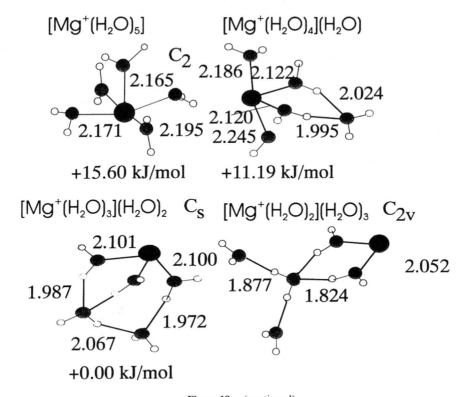

$[Mg^+(H_2O)_5]$ $[Mg^+(H_2O)_4](H_2O)$

C_2 2.186 2.122

2.165   2.024

2.171   2.195   2.120   1.995
                   2.245

+15.60 kJ/mol   +11.19 kJ/mol

$[Mg^+(H_2O)_3](H_2O)_2$ C_s $[Mg^+(H_2O)_2](H_2O)_3$ C_{2v}

2.101   2.100   2.052

1.987   1.877   1.824

2.067   1.972

+0.00 kJ/mol

Figure 19.   (continued)

and has a nonspherical *sp* mixing. The electronic excitations result from the excitation from this orbital to $3p$ orbitals, one of which is destabilized by the *sp* mixing, as depicted in Figure 18. The shape of the SOMO is determined by the number and structure of the water molecules in the first shell, which also determine the pattern of the calculated vertical excitation energies, seen in Figure 17(b).

The vertical excitation energies are calculated with the multireference configuration interaction method (Watanabe and Iwata, 1998). The molecular orbitals are determined with the state averaged complete active space self-consistent field method. The active orbitals in the complete active space are the $3s$, $3p_x$, $3p_y$, and $3p_z$ orbitals of Mg. Equal treatment of these orbitals is essential to an accurate evaluation of the excitation energies. As described in Section III.B.1, the calculated vertical excitation energies for $n=1$ and $n=2$ are in very good agreement with the observed peaks. The same level of calculation is performed for most of the isomers, up to $n=5$.

We first discuss the spectrum of $n = 5$, because it most clearly demonstrates the contributions from more than one isomer. The spectrum in Figure 17(a) exhibits the characteristic shoulder in the low-energy region. The structure of $[Mg^+(H_2O)_5]$, with five water molecules on one side of the first shell, indicates that the SOMO is unstable, and accordingly, the excitation energy shifts to lower frequencies. The calculated vertical excitation energies of this isomer, which is less stable by 15.6 kJ/mol than the most stable isomer of $[Mg^+(H_2O)_3](H_2O)_2$, are 17,600, 18,600, and 19,400 cm^{-1}, and thus, the observed shoulder around 16,000 to 20,000 cm^{-1} is unequivocally assigned to the isomer $[Mg^+(H_2O)_5]$. The slightly more stable isomer $[Mg^+(H_2O)_4](H_2O)_1$ has vertical transitions at 20,900, 23,000 and 25,300 cm^{-1}. Comparison with the experimental spectrum clearly points out the presence of the contribution from this isomer. But the mixture of the two isomers alone cannot account for the intense peak around 27,000, which can be ascribed to the electronic transitions of the most stable isomers $[Mg^+(H_2O)_3](H_2O)_2$, calculated at 25,300, 26,400, and 30,600 cm^{-1}. Thus, to explain the observed photodissociation spectrum of $n = 5$, we have to assume the coexistence of at least three isomers in the molecular beam. From the spectrum, the exact population ratio among the isomers cannot be estimated. Roughly speaking, the ordering of the population is as in the calculated relative energy. The numbers given in the figure are the total energy difference with the zero-point energy correction. If the entropy factor is taken into account, the difference becomes slightly smaller.

As in the case of $n = 5$, the observed spectrum of $n = 4$ is reproduced theoretically by assuming the coexistence of at least two isomers, $[Mg^+(H_2O)_4]$ and $[Mg^+(H_2O)_3](H_2O)_1$, and possibly also $[Mg^+(H_2O)_2]$ $(H_2O)_2$. For $n = 3$, the pyramidal isomer $[Mg^+(H_2O)_3]$ is the most stable, and the first excited state at 25,800 cm^{-1} is degenerate because of C_3 symmetry of the isomer. The observed main peak at 25,000 cm^{-1} is assigned to this transition. The calculated second transition is at 29,600 cm^{-1}, and the ratio of the calculated transition intensities is 3.4:1. So again, the most stable isomer alone cannot explain the observed peak at 28,000 cm^{-1}, which might be assigned to the isomer $[Mg^+(H_2O)_2](H_2O)_1$.

The coexistence of the isomers strongly suggests a high internal energy of the clusters. In the analysis of the photodissociation processes, the coexistence of isomers and a high internal energy of the clusters should be considered.

3.   Photodissociation Spectra of $Ca^+(H_2O)_n$

The ^2D level of Mg^+ lies 35,800 cm^{-1} above the first excited state, ^2P, and a perturbative mixing of these two states is negligibly small. On the other hand, the ^2D level of Ca^+ lies $11,600\,cm^{-1}$ below the ^2P level (about $25,300\,cm^{-1}$), and as a result, the electronic structure of the states originating from the ^2P states in the complex is expected to be significantly altered by the $3d$–$4p$ mixing. Figure 20(a) shows the photodissociation

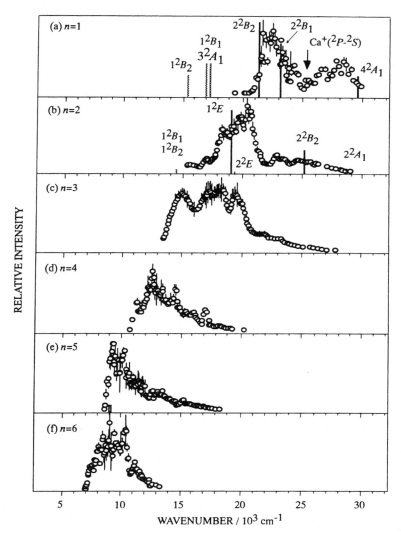

Figure 20.   The experimental photodissociation spectra of $Ca^+(H_2O)_n$ ($n = 1$–6). The bars in (a) and (b) are the calculated stick spectrum by Bauschlicher et al. (1992).

spectra of $Ca^+(H_2O)$. The spectrum consists of two distinct bands: an intense band at 22,400 cm^{-1} and a weaker band at 28,000 cm^{-1}, with the approximate intensity ratio of 1.7:1. The structure of the ground state of $Ca^+(H_2O)$ is shown in Figure 15(a). Bauschlicher et al. (1992) calculated the transition energies of the $4^2P \leftarrow 4^2S$ type of transitions, resulting in figures of 21,300 (2^2B_2, in-plane $p\pi$), 23,100 (2^2B_1, out-of-plane $p\pi$), and < 29,600 cm^{-1} (4^2A_1, $p\sigma$), in good agreement with the observed band positions as shown by the bars in Figure 20(a). The characters of these transitions are similar to those of the excited states of $Mg^+(H_2O)$. Bauschlicher et al. also predicted that three 3^2D-4^2S types of transitions lay just below the $4^2P \leftarrow 4^2S$ transitions between 14,000 and 17,000 cm^{-1} and that they had substantially large transition moments, which suggests large $d-p$ mixings, because the $3^2D \leftarrow 4^2S$ transition is forbidden in an isolated atom. On the contrary, in the observed photodissociation spectrum, no distinct bands are found below 20,000 cm^{-1}. Thus, the observed bands can be assigned to the transitions to the ^2P-type excited states having a much smaller $3d$ population than that predicted by theory. One of the most probable reasons that the $3^2D \leftarrow 4^2S$ transitions are not found is that the calculations overestimate the $p-d$ mixing and that the transition probability is not that large.

The spectrum of $Ca^+(H_2O)_2$ exhibits an intense band (20,000 cm^{-1}) and a weak, broad band (24,700 cm^{-1}), as shown in Figure 20(b). The bands are redshifted with respect to the Ca^+-ion transition by 5,300 and 600 cm^{-1}, respectively. The structure of $Ca^+(H_2O)_2$ has a D_{2d} form, as shown in Figure 15(a). Bauschlicher et al. (1992) estimated the optically allowed transitions at 19,100 cm^{-1} for 1^2E ($4p$) and at < 25,200 cm^{-1} for 2^2B_2 ($4p$) and reproduced the observed band positions fairly well. In contrast to the $Ca^+(H_2O)$ ion, $Ca^+(H_2O)_2$ has an appreciable photodissociation cross section in the energy region below 16,000 cm^{-1}, as seen in Figure 20(b). These results may indicate that metal hybridization is substantially enhanced for $Ca^+(H_2O)_2$.

The photodissociation spectra for $Ca^+(H_2O)_n$ ($n \leq 6$) in Figure 20 (Sanekata et al., 1996) show significant redshifting as large as 16,000 cm^{-1} relative to the atomic resonance lines of Ca^+. In contrast to the main bands of the spectrum of $Mg^+(H_2O)_n$, shown in Figure 17, the main bands for $Ca^+(H_2O)_n$ shift monotonically to the red with n up to 6. The spectral width does not change much with n, or it becomes slightly narrower, which suggests that the spectrum of each n corresponds to one isomer. As shown in Figure 15(a), the most stable isomer has four water molecules in the first shell for $n \geq 4$ (Watanabe and Iwata, 1997), but the isomers having five or six water molecules in the first shell also are close in energy. The character of the SOMO in these isomers is different from that of the SOMO of the corresponding Mg^+ complexes. The spherical s character remains in the complexes and is substantially destabilized by interaction with the ligating waters. The $3d$ orbitals are also destabilized by the bonding interaction with the lone-pair orbitals of water molecules; the blue shifts of the $3^2D \leftarrow 4^2S$ transitions are found in the calculated results of Bauschlicher et al. (1992). On the other hand, the diffuse $4p$ orbitals are relatively unchanged in the complex, and thus, the $4^2P \leftarrow 4^2S$ transitions shift to the red. The monotonic shift reflects the destabilization of the $4s$ orbital.

The observed monotonic shift up to $n=6$ suggests that the water molecule sequentially hydrates directly to the Ca^+ ion, although in the ab initio MO calculations it is not the most stable isomer for $n=5$ and $n=6$. The spectral shifts suggest the shell closing with six water molecules. Preliminary results on the photodissociation spectra for $Ca^+(D_2O)_n$ ($n = 7-9$) show that no appreciable intensity is observed in the energy region below $7,000\,cm^{-1}$, and, importantly, the band positions for $n = 7$ and $n = 8$ barely shift from the bands for $n = 6$ (Sanekata and Fuke, unpublished). These results are consistent with the foregoing arguments.

On the other hand, Farrar and his coworkers proposed another interpretation for the similar large red shifts observed for the $Sr^+(NH_3)_n$ transitions (Shen and Farrar, 1988; Shen and Farrar, 1991; Donnelly and Farrar, 1993). These researchers carried out a moment analysis of the absorption spectra and found a large increase in the electronic radial distribution in the ground state with increasing cluster size. Based on the results, they proposed the involvement of an increasing Rydberg-type ion-pair character, $Sr^{2+}(H_2O)_n^-$, in both the ground (5^2S) and first excited (5^2P) electronic states. The argument is somewhat similar to that for the alkali atom–NH_3 systems discussed in Section II.C.2. Because a neutral alkali atom is isoelectronic with a singly charged alkali earth metal ion, we might argue that the large decrease in vertical detachment energy for the Li (^2P)-type state seen in Figure 10 corresponds to the large red shifts in the absorption bands for $Ca^+(H_2O)_n$ and $Sr^+(NH_3)_n$. However, because the ionization energies of Ca^+ (11.87 eV) and Sr^+ (11.0 eV) are much larger than those for alkali atoms, the application of a simple analogy might require a more careful analysis.

4.   Photodissociation Process of $Mg^+(H_2O)_n$ and $Ca^+(H_2O)_n$

In the photodissociation process of $M^+(H_2O)_n$ ($M = Mg, Ca$), photo-induced intracluster reactions to produce the $MOH^+(H_2O)_{m-1}$ ions [similar to Eq. (11)] are observed, as is the evaporation of water molecules [in a reaction similar to Eq. (10)]. The two reactions are competitive with each other, and the branching ratio depends strongly on both the photolysis energy $h\nu$ and the cluster size n.

Evaporation Process. Figures 21 and 22 show the branching ratios of the fragment ions in the photodissociation of $Mg^+(H_2O)_n$ and $Ca^+(H_2O)_n$, respectively (Misaizu et al., 1994; Sanekata et al., 1996). As with $Mg^+(H_2O)$, the evaporation process that produces Mg^+ is the minor

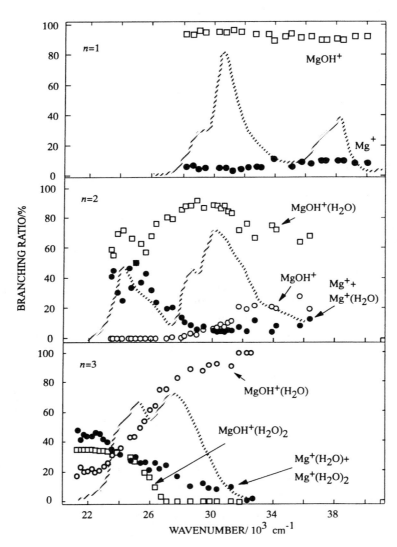

Figure 21. The branching ratios of the fragment ions in the photodissociation of $Mg^+(H_2O)_n$ ($n = 1-3$). The blurred line is the photodissociation spectrum.

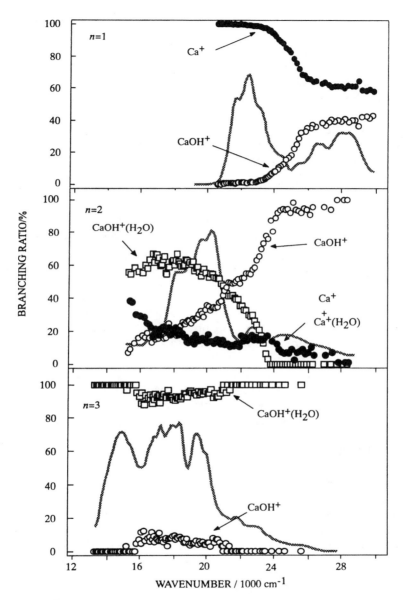

Figure 22. The branching ratios of the fragment ions in the photodissociation of $Ca^+(H_2O)_n$ ($n = 1$–3). The blurred line is the photodissociation spectrum.

process, and its branching ratio is almost constant with photolysis energy. These results are in marked contrast to those for $Ca^+(H_2O)$: The evaporation process to produce Ca^+ is predominant at least in the first and second excited states ($2\,^2B_2$ and $2\,^2B_1$, respectively) of the $Ca^+(H_2O)$ ion. Moreover, the dissociation cross section of the $Ca^+(H_2O)$ complex is found to be at least one order of magnitude larger than that for $Mg^+(H_2O)$. The difference in the excited-state electronic structures of these ions leads to different evaporation behaviors. Since the optically accessible states of these complexes are all bound along the metal ion–oxygen coordinate, the dissociation dynamics are predissociative in character: Internal conversion to lower electronic states induces the dissociation process. In $Mg^+(H_2O)$, the ground state lies far below the states derived from $Mg^+(^2P)$, and as a result, the de-excitation rate of $Mg^+(^2P)(H_2O)$ is rather slow (on the order of 10^{-8} s^{-1}), and the radiative process is expected to take part in the decay process. Through internal conversion, the higher vibrational levels of the ground state, which are far above the threshold energy of the $MgOH^+$ formation, are populated efficiently. Hence, the evaporation reaction becomes a minor process compared with the hydrogen elimination reaction for $Mg^+(H_2O)$. On the other hand, in $Ca^+(H_2O)$, the ^2D levels of Ca^+ lie just below the ^2P levels, and as a result, internal conversion to ^2D-type states is strongly enhanced. Because these states are expected to lie close to the threshold (1.59 eV in Figure 16) of the hydrogen elimination reaction, the reaction might be suppressed. Direct internal conversion to the ground state, as in the case of $Mg^+(H_2O)$, may also play some role. However, the de-excitation process via the ^2D-type states may be the predominant channel for evaporation of a water molecule. Thus, fast internal conversion in $Ca^+(H_2O)$ may be responsible for the large dissociation cross section of this ion.

In $Mg^+(H_2O)_n$ for $n \geq 2$, the evaporation process is also observed, and this channel becomes slightly larger than for $n = 1$ (Misaizu et al., 1994). On the other hand, the evaporative fraction of $Ca^+(H_2O)_2$ becomes much smaller than for $n = 1$ and is almost zero for $n \geq 3$. The persisting evaporation processes of $Mg^+(H_2O)_n$, even for large n, may be due to the presence of a relatively high energy barrier for the ground-state reaction that leads to the product $MgOH^+(H_2O)_{n-1}$. The difference in the evaporation process between the Mg^+ and Ca^+ systems is due to a large enhancement of intracluster reactions for the former ions.

Intracluster Metal–Oxidation Reaction.   The most interesting feature of the $M^+(H_2O)_n$ ($M = Mg$, Ca) photodissociation process is the presence of intracluster reactions that produce the hydrated MOH^+ ions (Misaizu et al., 1994; Sanekata et al., 1996). The hydrogen elimination reaction in the

$Mg^+(H_2O)$ ion is the predominant process and does not depend on photolysis energy, as shown in Figure 21. As discussed in Section III.B.2, the excitation energy for the ^2P-like states for $Mg^+(H_2O)$ is well above the threshold energy (2.95 eV, 23,800 cm^{-1} in Figure 16) for the $MgOH^+ + H$ reaction. Moreover, internal conversion to the ground state is the sole channel for the deactivation of the excited state of $Mg^+(H_2O)$. Through this mechanism, excitation energy is redistributed directly in the ground state to produce the highly vibrationally excited complexes, leading to a reaction product. On the other hand, the hydrogen elimination reaction in the photodissociation of $Ca^+(H_2O)$ is the minor process, as seen in Figure 22, and its branching ratio exhibits a threshold behavior with an onset energy of about 24,000 cm^{-1} (288 kJ/mol, 2.98 eV) (Sanekata et al., 1996). Duncan and coworkers have also studied the photodissociation process of cooled $Ca^+(H_2O)$ ions (Scurlock et al., 1996). They found that the evaporation process producing the Ca^+ ion is predominant in the origin region of the $^2B_1$ state near 23,300 cm^{-1}, but both channels are comparable in the lower $^2B_2$ state near 22,000 cm^{-1}. The power-dependence study exhibits a linear behavior for the $^2B_1$ state, but a nonlinear dependence is found for the $^2B_2$ state, indicating more complex kinetics at the lower excitation energy. These results may be consistent with those of $n = 1$, shown in Figure 22. The apparent hydrogen elimination reaction increases around 24,000 cm^{-1}, and the branching ratio becomes nearly constant above 26,000 cm^{-1}. The onset of the increase coincides with a new electronic transition, which might be assigned to the 4^2A_1 ($p\sigma$) state. As is shown in Figure 16, the required energy for the reaction $Ca^+(H_2O) \rightarrow CaOH^+ + H$ is 1.6 eV (12,900 cm^{-1}), a figure much smaller than that for the observed onset of the reaction. Thus, the threshold behavior implies an opening of a state-specific reaction channel to produce $CaOH^+$. The character of the excited state 4^2A_1 ($4p\sigma$) is similar to that of the third excited state $3p_z \leftarrow 3s$ of $Mg^+(H_2O)$ shown in Figure 18, although mixing with $3d$ orbitals may change the details of the potential energy surface. To distinguish the reaction mechanism and to understand the characteristic change in the branching ratio around 24,000 cm^{-1}, extensive theoretical studies are required.

In the photolysis of $Mg^+(H_2O)_2$, the branching fraction of the $MgOH^+(H_2O)$ formation in Figure 21 increases with increasing photon energy, up to 29,500 cm^{-1}, and becomes almost unity at energies close to the peak of the broad band, indicating a rapid reaction (Misaizu et al., 1994). Since the higher energy side of this broad band at 30,000 cm^{-1} corresponds to the excitation of a $3s$ electron to the $3p_z$ orbital (see Figure 18), the features in the photodissociation of $Mg^+(H_2O)_2$ might be similar to those in the 4^2A_1 state of $Ca^+(H_2O)$. The linear dependence on laser fluence in this energy range, which suggests a significant shortening of the lifetime of the relevant state, also supports the state-specific reaction for $n = 2$, at least in the energy region above $28,600 \, \text{cm}^{-1}$. On the other hand, the ground state channel, as discussed for $n = 1$, may be responsible for $\text{MgOH}^+(\text{H}_2\text{O})$ production in the energy region below $27,000 \, \text{cm}^{-1}$, where the 1^2B (out-of-plane, $3p_y$) state is located and the orbital alignment is quite unfavorable for the insertion reaction. At this moment, there is no clear answer as to why an excited-state channel is opened for $\text{Mg}^+(\text{H}_2\text{O})_2$ and not for $n = 1$.

In the photodissociation of $\text{Ca}^+(\text{H}_2\text{O})_2$, the hydrogen elimination reaction is enhanced dramatically, surpassing the evaporation process, as shown in Figure 22, although both internal conversion and evaporation processes are expected to become faster than for $n = 1$, due to the closer separations among the low-lying states for $n = 2$. The drastic change in reactivity may be ascribed to the large hydration energy of the reaction products. As is shown in Figure 16, the energy difference between the evaporation product $\text{Ca}^+(\text{H}_2\text{O})_{n-1} + \text{H}_2\text{O}$ and the reaction product $\text{CaOH}^+(\text{H}_2\text{O})_{n-2} + \text{H}$ becomes only $0.13 \, \text{eV}$ for $n = 2$ and is even negative for $n > 2$ in theoretical calculations. These energetic situations cause the large enhancement in reactivity for $n = 2$.

For larger clusters, two metal ions also exhibit different photodissociation features, as is seen in Figures 21 and 22. Although the results of the branching ratios are not shown, the fractions of intracluster reaction for $\text{Mg}^+(\text{H}_2\text{O})_n$ ($n = 4$ and $n = 5$) exhibit no photolysis-energy dependence, and the evaporative fraction is twice as large as that for $\text{Mg}^+(\text{H}_2\text{O})$. On the other hand, for $\text{Ca}^+(\text{H}_2\text{O})_n$ ($n \geq 3$), the intracluster reaction is further enhanced and no evaporation product is detected, which is compatible with the energy ordering of the evaporation product and reaction product. But a similar switch in the energy ordering is also found for Mg^+ ion complexes, and so, to explore the persistence of the evaporation process for $\text{Mg}^+(\text{H}_2\text{O})_n$ ($n \geq 4$), the existence of an energy barrier along the reaction coordinate has to be assumed, as was mentioned previously for $\text{Ca}^+(\text{H}_2\text{O})$.

IV.   GROUP 13 ELEMENTS

In this section, the solvation processes of group 13 metal atoms in the gas phase are examined. The single-photon ionization of Al–water clusters, as well as photodissociation spectroscopies of $\text{Al}^+(\text{H}_2\text{O})_n$ ions, is discussed (Misaizu et al., 1993). Aluminum atom vapor is produced by irradiating a rotating Al rod with the second harmonic of an Nd:YAG laser. The vapor is co-expanded by 10 atm of He gas mixed with water vapor at room temperature from a pulsed valve. To produce hydrated aluminum atom clusters containing a single aluminum atom, the conically diverging channel is connected in front of the pulsed valve. In photoionization experiments, the resulting cluster beams are ionized at the source region of the TOF mass spectrometer by irradiation with an output of an ArF laser at 193 nm or with the second harmonic of an output of an XeCl excimer-pumped dye laser. In the photodissociation studies, techniques similar to those described in Section III.B are used.

A.   Photoionization of $Al(H_2O)_n$

Figures 23(a) and (b) show typical photoionization mass spectra of $Al(H_2O)_n$ clusters, recorded at the ionization wavelengths of 193 and 213 nm, respectively (Misaizu et al., 1993). The observed peaks are those for Al^+, Al_3O^+, and a series of $Al^+(H_2O)_n$ cluster ions. At wavelengths longer than 213 nm, a characteristic size distribution is clearly observed. The intensities of the hydrated cluster ions $Al^+(H_2O)_n$ with $n \leq 4$ are one order of magnitude larger than those with $n \geq 5$. On the other hand, the intense ion signals around $n = 5$–10 are observed in the mass spectrum of the ions, which emerge directly from the laser vaporization source (not shown). Moreover, at 193-nm ionization, the ion signals of $n \geq 5$ decrease more clearly than those of $n \leq 4$, as the laser fluence decreases. These findings suggest that most of the ion signal intensities for $n \geq 5$ produced at 193-nm ionization originate through the dissociation of larger clusters. In other words, the observed intensity gap at $n = 5$ may reflect a change in the abundance of neutral clusters.

Equilibrium structures and hydration energies of neutral $Al(H_2O)_n$ clusters have been determined with ab initio MO calculations for $n \leq 4$ (Watanabe et al., 1993). The optimized structures are shown in Figure 24. Calculations predict that up to $n = 2$ water molecules are bound directly to the Al atom. For $n = 2$, however, the most stable isomer has only one ligating water molecule, and the second water molecule hydrogen-bonds to the ligating water as $[Al(H_2O)](H_2O)$. This is because of weaker neutral metal–water bonds than water–water hydrogen bonds. For $n \geq 3$, there are at least two types of isomers, such as $[Al(H_2O)](H_2O)_{n-1}$ and $[Al(H_2O)_2]$ $(H_2O)_{n-2}$, but the former is more stable than the latter. Three water molecules cannot be directly bound to a neutral aluminum atom, because of the repulsion interaction with three valence electrons in the s^2p^1 electron configuration of Al. The structures in Figure 24(d), (f), and (g) suggest attractive interaction between a hydrogen atom of the end water molecule of the hydrogen bond network and an odd electron on the Al atom. The calculated incremental hydration energies peak at $n = 3$ for $[Al(H_2O)]$ $(H_2O)_{n-1}$ and at $n = 3$ or $n = 4$ for $[Al(H_2O)_2](H_2O)_{n-2}$. Thus, the characteristic cluster distribution, as shown in Figure 23(b), may be partially due to the difference in hydration energy in the neutral clusters.

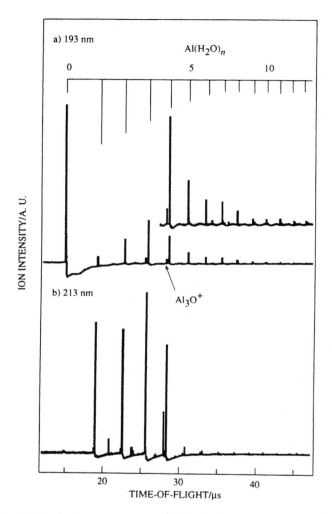

Figure 23. Photoionization mass spectra of $Al(H_2O)_n$ clusters. The ionization wavelengths are (a) 193 and (b) 213 nm.

Ionization energies (IEs) of $Al(H_2O)_n$, determined by scanning the ionization laser wavelength, are 5.07, 4.68, 4.47, and 4.41 eV, for $n = 1$–4, respectively. The numbers decrease monotonically with cluster size and become almost constant at $n = 4$ (Misaizu et al., 1993). The IEs calculated with the configuration interaction method for $[Al(H_2O)](H_2O)_{n-1}$ for $n = 1$, 2, and 3 are 5.10, 4.39, and 4.25 eV, respectively, and agree well with the

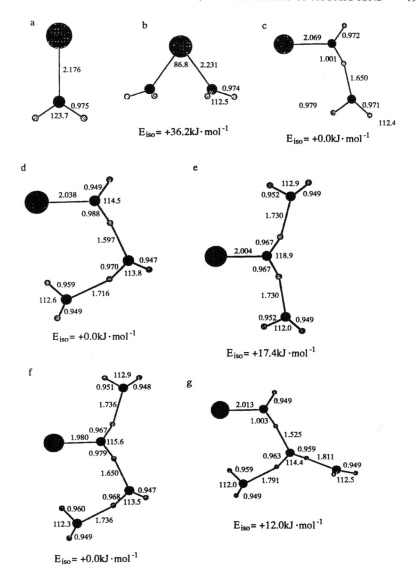

Figure 24. Optimized structures of $Al(H_2O)_n$ at UHF/6–31G level. The relative energy without correction for zero-point vibration is given.

experimental values (Watanabe et al., 1993). On the other hand, if the structure $[Al(H_2O)_2](H_2O)_{n-2}$ is assumed for $n=2$ and 3, the calculated IEs with the configuration interaction method slightly underestimate the experimental IEs. But the comparison of the theoretical and experimental IEs alone cannot determine the structures of the clusters.

B.   Photodissociation of $Al^+(H_2O)_n$

With the use of an experimental apparatus similar to that employed in the case of the group 2 metal–water systems, the photodissociation process of the mass-selected $Al^+(H_2O)_n$ ions was examined for $n = 1$–10 with photons at wavelengths of 193–308 nm (Misaizu et al., 1993). Figure 25 shows photofragment ion mass spectra taken at 248-nm excitation. At this wavelength, efficient dissociation is observed for the parent ions with $n = 3$–5. No photoinduced intracluster reaction processes are observed, as are found in the photoexcitation of $Mg^+(H_2O)_n$. (See Section III.B.4.) Only the evaporation process

$$Al^+(H_2O)_n + h\nu \rightarrow Al^+(H_2O)_{n-k} + kH_2O \tag{12}$$

is observed in the region of wavelengths examined. The most efficient process is $k = 1$ (the loss of one H_2O molecule) for $n \leq 4$, $k = 2$ for $n = 5$, and $k \geq 3$ for $n \geq 6$ at 248 nm. In particular, the process producing the

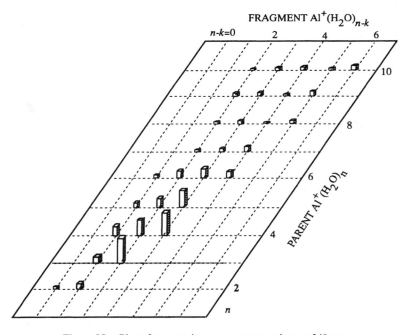

Figure 25.   Photofragment ion mass spectra taken at 248 nm.

$Al^+(H_2O)_3$ daughter ion is dominant in the dissociation of the ions with $n = 4–6$. One possible explanation for these results is that the $Al^+(H_2O)_3$ ion is stable in comparison with those of $n = 4$ and 5. As will be shown in Figure 28(c), the most stable ion for $n = 3$ has an $[Al^+(H_2O)_2](H_2O)_1$ structure (Watanabe and Iwata, 1996). The isomer of $[Al^+(H_2O)_3]$ is less stable, but is at a local minimum on the potential energy surface. For $n \geq 4$, the $[Al^+(H_2O)_3](H_2O)_{n-3}$ isomers become the most stable. These theoretical predictions appear to be consistent with the observed photofragment yields.

Photodissociation spectra of $Al^+(H_2O)_2$ and $Al^+(H_2O)_3$ are obtained as shown in Figure 26 by measuring the total yield of the fragment ions (Misaizu et al., 1993). For $n = 2$, only an absorption edge is observed, while the spectrum of $n = 3$ exhibits an absorption peak at about $41,500\ cm^{-1}$. On the analogy of the photodissociation excitation spectra in $Mg^+(H_2O)_n$ (Misaizu et al., 1993), these bands are assigned to the transitions localized on the Al^+ ion. The lowest allowed transition of Al^+, $3^1P \leftarrow 3^1S$, lies at $59,850\ cm^{-1}$ (Moore, 1971). This transition is shifted to red and is split into three transitions as a result of interaction with water molecules. The observed absorption edge of $n = 2$ is assigned to the lowest component among three, and the spectrum for $n = 3$ is further redshifted, as in the case of $Mg^+(H_2O)_n$.

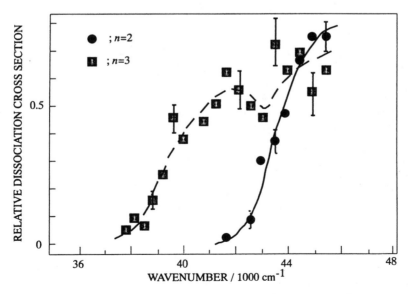

Figure 26. Photodissociation excitation spectra of $Al^+(H_2O)_2$ and $Al^+(H_2O)_3$.

C. Unimolecular Fragmentation Reactions

Using the ion cyclotron resonance (ICR) technique, Bondybey and his coworkers recently examined the unimolecular fragmentation process of $Al^+(H_2O)_n$ ($n = 3$–50) (Beyer et al., 1996; Berg et al., 1997). The cluster ions produced by a laser vaporization source are stored in a collision-free ion trap, and the fragmentation reactions induced by blackbody infrared radiation at room temperature are examined with a holding time up to 120 sec. Figure 27 shows the mass spectra of the Al^+–water clusters recorded with various delay times. Two series of cluster ions, $Al^+(H_2O)_n$ for $n = 6$– 50 and $AlO^+(H_2O)_{n-1}$ for $9 \leq n \leq 25$, are observed in the different mass regions. The latter ions are assigned to the reaction products, hydrated aluminum hydroxides ($[Al(OH)_2]^+(H_2O)_m$). Apparently, a hydrogen mole- cule is produced. With increasing holding times, the $Al^+(H_2O)_n$ ions disappear by absorbing the IR radiation from the cell wall, as shown in the figure. The distribution of the $Al(OH)_2^+(H_2O)_m$ ions also shifts toward smaller clusters. About 85% of the hydrated hydroxide clusters eventually reach a reaction product, $Al(OH)_2^+(H_2O)_3$, and the other reaction products are $Al(OH)_2^+(H_2O)_2$ and $Al(OH)_2^+(H_2O)_4$. The distribution of the mass spectrum in the earlier holding time is similar to those of the group 2 metal ion–water systems described in Section III.A.3, in which the products of elimination of a hydrogen atom, but not of a hydrogen molecule, are detected. In the group 2 metal systems, the partial oxidation of M^+ to produce the stable, closed-shell MOH^+ ions takes place in the limiting size of $6 \leq n \leq 15$. On the other hand, in the Al^+–water clusters, the apparent oxidation of Al^+, leading to the loss of an H_2 molecule, and the formation of the closed-shell hydroxide $Al(OH)_2^+$, are induced by the IR background radiation. This intracluster reaction is found to be strongly size dependent and proceeds only in the region of $12 \leq n \leq 24$, which is a little broader than that for Mg^+ complexes, wherein the hydrogen elimination reactions start only for $16 \leq n \leq 21$. For $n < 12$, only the sequential evaporation of a water molecule proceeds, and eventually, a long-lived $Al^+(H_2O)_4$ is formed (Berg et al., 1997). No intracluster reactions for smaller n in the ICR experiments are consistent with the aforementioned results on the photodissociation of $Al^+(H_2O)_n$, which also exhibits no intracluster reactions. In these experiments, no intracluster reactions are detected in larger clusters of size $n > 24$. The product switches in Al^+ complexes with water molecules are very similar to those found in the Mg^+ complexes described in Section III.A.3.

Berg et al. (1997) provided three explanations for the observed size dependence of the reactions in Al^+ ion complexes with water clusters. The

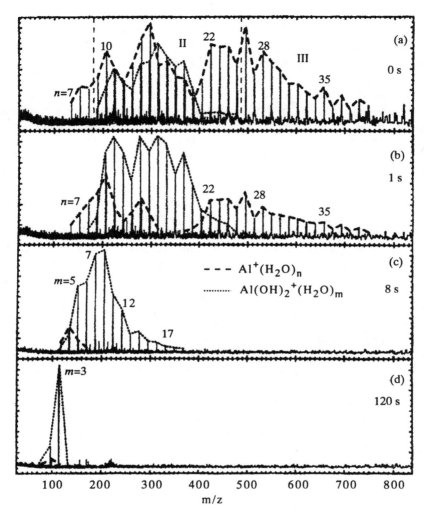

Figure 27.   Mass spectra of a cluster ion distribution of $Al^+(H_2O)_n$, taken with different time delays before detection (From Beyer et al., 1996). (a) The initial distribution is observed after accumulating the cluster ions for 2 s in the cell. A pure $Al^+(H_2O)_n$ progression is observed in region III ($n > 25$) and $n = 6–8$. In region II ($n = 9–25$), each peak of $Al^+(H_2O)_n$ is accompanied by a peak of $[Al(OH)_2]^+(H_2O)_{n-2}$. (b) After 1 s. Large clusters ($n > 25$) evaporate water molecules. The reaction products $[Al(OH)_2]^+(H_2O)_{n-2}$ become dominant for $n = 8–22$. (c) After 8 s. The reaction products further evaporate water molecules. (d) After 120 s. The fairly stable clusters $[Al(OH)_2]^+(H_2O)_3$ and $[Al(OH)_2]^+(H_2O)_2$ are reached.

first is just the same as they gave for the reactions in the Mg^+ ion complexes (see Section III.A.3): The ion-pair state $Al^{3+}(H_2O)_n^{2-}$ is formed in larger complexes. But they had to admit the difficulty in this model for the case of Al^+, because as large as 47.26 eV(4,560 kJ/mol) is required to doubly ionize an Al^+ ion. Therefore, that much solvation energy has to be gained by water clusters of size $n \approx 24$. Besides, in their model, two excess electrons have to be held in the water clusters of size $n \approx 24$. The second explanation is solvation shell effects. Berg et al. assumed that the first and second shells around the Al^+ ion are completed with $n=4$ and 12, respectively, and speculated that the third shell is closed at $n \approx 24$. Once this shell is closed, the ion complexes become very stable, so no reactions proceed. In this model, the water molecules in the third coordination sphere are assumed to play a key role in the intracluster reactions. The third explanation Berg et al. gave in their paper is similar to that given in Section III.A.3 as an alternative model for the reactions in the Mg^+ ion complexes. In this model, even in large clusters, the intracluster reactions are actually taking place, and the reaction products simply have the same mass number as that of $Al^+(H_2O)_n$. With $n > 24$, the recombination reactions of the products are prohibited, but in the intermediate size $16 \leq n \leq 21$, a series of electron and proton transfer reactions becomes possible, and the hydrogen formation reaction and the oxidation reaction of Al^+ to $[Al(OH)_2]^+$ simultaneously take place. One of the candidates of the intermediate reaction products is $(HAlOH)^+(H_2O)_{n-1}$. In fact, Watanabe and Iwata (1996) found the stable reaction products of this form in their ab initio MO study, and some of these products for $n=4$ are shown in Figure 28 (h)–(k). One of them has a form of $HAl(OH)_2$ $(H_3O)^+(H_2O)$, which is the most stable among the chemical species of {Al, $(H_2O)_4$}$^+$ they examined, and is 116 kJ/mol more stable than the hydrated Al^+ ion, $[Al^+(H_2O)_3](H_2O)$. The hydrogen atom bound to the Al atom is negatively charged (-0.38 e in the natural population analysis). This negative hydrogen atom reminds us of the reaction of a hydrated beryllium ion,

$$[(H_2O)_3BeH]^+ + H_3O^+ \rightarrow [(H_2O)_4Be]^{2+} + H_2, \qquad (13)$$

modeled and examined some time ago by Hashimoto et al. (1990). The H atom bound to a Be atom is also negatively charged (-0.23 e in the Mulliken population analysis). The hydrogen and oxionium ion $(H_3O)^+$ are spatially separated in $HAl(OH)_2(H_3O)^+(H_2O)$, but as more water molecules hydrate to the ion, an oxionium ion may be able to approach AlH, and the hole migration and hydrogen molecule formation may simultaneously take place as in Eq. (13). To verify this model, more realistic calculations are required.

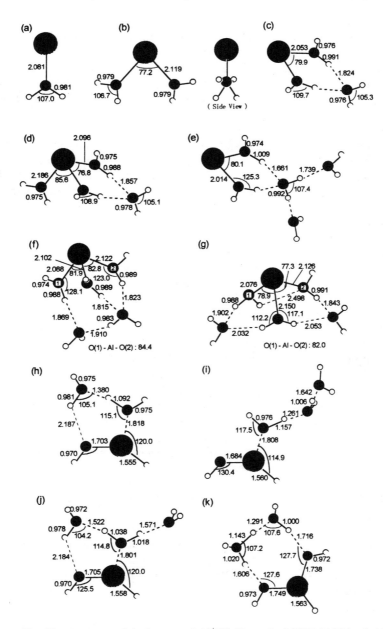

Figure 28. The structures of the isomers of $Al^+(H_2O)_n$ at the MP2/6–31G* level. (a)–(g): $[Al^+(H_2O)_{n-s}](H_2O)_s$. (h)–(k): The isomers of $HAlOH^+(H_2O)_{n-1}$, in which the Al atom has sp^2 hybridization.

D. Intracluster Reactions of $[B(H_2O)_n]^+$ Studied by a Hybrid Procedure of Ab Initio MO Calculations and Monte Carlo Samplings

In Sections III.A and IV.C, we discussed the size-dependent intracluster reactions in the Mg^+, Ca^+, and Al^+ ions. A common feature of these reactions is two types of product switches as the size of the solvent water clusters increases. In large clusters, the intracluster redox reactions are apparently suppressed. Only in the intermediate size of water clusters, reaction products are observed in the mass spectrum in the collision experiments as well as in the visible–UV photoexcitation. In both cases, the metal–water clusters have a certain amount of internal energy. As FT–ICR experiments show, similar reactions occur through excitation by IR blackbody radiation (Berg et al., 1997). As discussed in Section IV.C, there are a few plausible explanations, but none of them are yet convincing. To examine the models proposed, extensive theoretical calculations for the reactive clusters are needed.

The simulation of the intracluster reactions demands a large computational resources because both the dissociation and rearrangement of bonds are taken into account. Empirical potential energy functions are almost impossible to apply to the reactions that we are interested in. Asada and Iwata (1996) developed a hybrid procedure of ab initio MO calculations and Monte Carlo (MC) samplings to study the cluster $Mg^+(H_2O)_n$. To increase the sampling efficiency, they combined the rigid and nonrigid moves; the latter allow any of the atoms to move at a step and, thus, the intracluster reaction to proceed. Asada and Iwata (1996 and unpublished results) examined the cluster ions up to $n = 6$ at $T = 300$ K. Although charge fluctuations on the Mg atom become large as n increases, reactions do not occur in the simulation. In retrospect, the results are consistent with the FT–ICR experiments, which showed that the reactions start only for $16 \leq n \leq 21$. So for Mg^+ ion complexes, the simulation of larger clusters is required.

Watanabe et al. (1997) extended Asada and Iwata's study to the $B^+(H_2O)_n$ complexes as a model system and found that intracluster reactions can occur in the simulation. An example is shown in Figure 29. The simulation starts at the structure $[B^+(H_2O)_3]$ and ends up at the product $(HBOH)^+(H_2O)$ of an apparent insertion reaction and an evaporating water molecule. The basis set in the ab initio MO calculations is 6–31 + G*, and at every MC step a closed-shell SCF calculation is performed. The strategy for moving the atoms is the same as that adopted by Asada and Iwata (1996). The assumed temperature is $T = 300$ K. Figure 29(a) shows the energy profile along the MC samplings. A series of snapshots is shown in Figure 29(b). Starting from this structure, one of the B^+–O bonds immediately becomes longer [see the 501st step (i)], and then another B^+–O bond starts

a: Energy

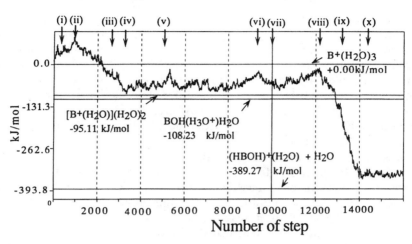

b: Snapshots

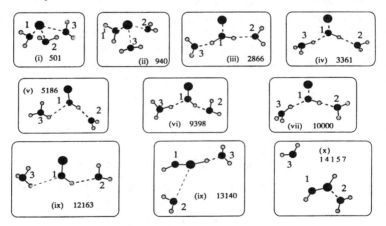

Figure 29. Monte Carlo samplings with ab initio MO calculations along the reaction path of $B^+(H_2O)_3$ to $HBOH^+(H_2O) + (H_2O)$. (a) Energy profile along the path. In the Metropolis sampling, $T = 300\,K$ is assumed. (b) Snapshots at some intermediate steps.

to get longer [see the 940th step (ii)]. The isomerization reaction proceeds to structure (iv), in which an oxionium ion is already formed. The highest point of the profile in this run is at the 940th step (ii), which might be close to the transition-state structure from the starting structure to the structure with the proton transferred (iv). Between steps 2,900 and 12,000, the system stays at configurations similar to either (iv) or (iii). (In most steps, at structure (iv) the system has a BOH molecule as a core.) The structures are entropy favored, because of the floppy motion of one or more water molecules and an oxionium ion. Eventually, after several changes of the hydrogen-bond pairs and proton transfers (note that in the figure the numbering of atoms is kept by assuming classical motion), an ion HOBH$^+$ is formed at (ix). Then, a water molecule starts to leave the ion HOBH$^+$(H$_2$O). So the final product is a hydrated ion HOBH$^+$(H$_2$O), plus a water molecule. In terms of energy, this is quite conceivable, because the exothermic energy of the reaction to the product HOBH$^+$(H$_2$O) + H$_2$O from the initial isomer [B$^+$(H$_2$O)$_3$] is as large as -389.27 kJ/mol.

Interestingly, the product ion HOBH$^+$(H$_2$O) has the chemical formula [B(H$_2$O)$_2$]$^+$. In other words, if mass spectrometry is employed, no intracluster reactions can be noticed, and there is no way to distinguish the reaction from mere evaporation of a water molecule. The simulation predicts that the intracluster reaction proceeds for smaller n in B$^+$ ion complexes with water molecules. It is this kind of reaction that takes place, according to the alternative reaction model we discussed in Section III.A.3 for Mg$^+$ and Ca$^+$ ion complexes and in Section IV.C for Al$^+$ complexes. For these metal ions, larger water clusters are required to make the reaction possible. To confirm this model, we have to perform ab initio MO–MC simulations for $n \approx 21$, which at this moment demands a prohibitively large computational resource. Also, further development of the computational methods is requisite to the point where they allow intracluster reactions during the simulation.

V. HALOGEN–WATER CLUSTER ANIONS

A. Incremental Enthalpy Changes and Structures of the Clusters

Another interesting target for exploring the microscopic details of ion–water interaction is a halogen anion complex with water clusters. The structure and energetics of halide–water clusters have been extensively studied by mass spectrometric methods. The studies provide a wealth of information on the thermodynamics of clusters, as well as on intrasolvent and solvent–solvent interactions. The incremental enthalpy change $\Delta\Delta H^0$ and the entropy change $\Delta\Delta S^0$ have been determined via high-pressure mass spectroscopy (HPMS) (Arshadi et al., 1970; Keesee and Castleman, 1980; Hiraoka et al., 1988). Recently, a new technique, called zero-pressure thermal-radiation-induced dissociation (ZTRID), was introduced to determine the enthalpy changes (Dunbar et al., 1995). Cheshnovsky's group (Markovich et al., 1991,1994) reported the photoelectron spectra (PES) for these clusters. With a few assumptions, the difference between the vertical detachment energy and the electron affinity of the halogen atom is related to the enthalpy change as

$$E_{\text{stab}}(n) \equiv \text{VDE}(n) - \text{EA}(X) = -\Delta H_s^i(n) + \Delta H_s^n(n) + \Delta H_n(n). \quad (14)$$

The quantities in this equation are defined in Figure 30. By taking the difference as

$$\begin{aligned}
\Delta E_{\text{stab}}(n) &\equiv E_{\text{stab}}(n) - E_{\text{stab}}(n-1) = \text{VDE}(n) - \text{VDE}(n-1) \\
&= -\{\Delta H_s^i(n) - \Delta H_s^i(n-1)\} + \{\Delta H_s^n(n) - \Delta H_s^n(n-1)\} \\
&\quad + \{\Delta H_n(n) - \Delta H_n(n-1)\} \\
&\approx -\{\Delta H_s^i(n) - \Delta H_s^i(n-1)\} \equiv \Delta\Delta H_s^i,
\end{aligned} \quad (15)$$

we can estimate the incremental enthalpy change from the shift in the vertical detachment energy determined by photoelectron spectra. Here, we assume that the size dependence of the solvation energy $\Delta H_s^n(n)$ of neutral

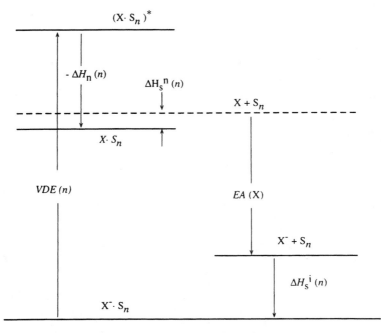

Figure 30. An energy-level diagram of neutral clusters $X \cdot S_n$ and anion clusters $X^- S_n$. $\Delta H_s^i(n)$ and $\Delta H_s^n(n)$ are the changes in enthalpy of the anionic and neutral clusters. $\Delta H_n(n)$ is the geometric relaxation energy from the neutral cluster to the anionic cluster.

species and the geometric relaxation energy $\Delta H_n(n)$ are small. In Tables 5–8, incremental enthalpy changes or corresponding approximate energies for F^-, Cl^-, Br^-, and I^- are summarized, along with some theoretically estimated values. Care should be taken in these tables, because the entity in each row is not exactly equivalent in experiment and theory. Yet, from the tables, we can see the size (n) and anion (X^-) dependence on the incremental hydration energy of halogen–water clusters. From the n dependence of the hydration energy, we may infer the structures of the clusters. In particular, there has been controversy on where the halogen anion is located — at the surface of the water clusters or in the interior of the clusters.

More recently, vibrational spectroscopy has also become feasible for studying these clusters (Johnson et al., 1996; Bailey et al., 1997; Choi et al., 1998; Cabarcos et al., 1998). By comparing the observed infrared bands with the harmonic frequencies and their intensities calculated with various levels of ab initio methods, the structures of the clusters might be identified. A comparison between these species will be discussed in Section V.C.

As summarized in Tables 5 through 8, the general trends in halogen and size dependencies of incremental hydration energy are consistent with various experiments and theoretical calculations. As expected, with an

TABLE 5

The incremental enthalpy change $\Delta\Delta H^{298}$ (n) and hydration energy $\Delta E_{solv}(n)$ of F^- in kcal/mol (1 kcal/mol = 4.1868 kJ/mol)

n	1	2	3	4	5	6	7	8	9	10
AYK70	23.3	16.6	13.7	13.5	13.2					
HMY88		19.2	15.3	13.9	12.3	10.9	10.4	11.2	11.1	~11
XD94	26.7	19.5	16.9							
298 K	25.5	19.0	15.9							
XDa96	26.5	21.6	18.3	13.8	10.9	10.5	10.4	8.7	13.6	13.1
CK 94	24.5	20.0	17.4	14.0 ss	13.3 ss	13.4 ss				
		19.6 i		14.9 is	13.3 is	11.1 is				
		s			13.4 i's	11.1 's				
					12.5 i'i					

AYK70: $\Delta\Delta H^{298}$ (n) determined by HPMS (Arshadi et al., 1970).

HMY88: $\Delta\Delta H^{298}$ (n) determined by HPMS (Hiraoka et al., 1988).

XD94: $\Delta\Delta H^{298}$ (n) and $\Delta\Delta H^0$ (n) evaluated with MP4 level of ab initio MO calculations (Xantheas and Dunning, 1994).

XDa96: $\Delta\Delta H^{298}$ (n) evaluated by MD simulations with a new set of polarizable potential functions (Xantheas and Dang, 1996).

CK94: $\Delta\Delta E(n)$ evaluated with MP2/HF level of approximation (Combariza and Kestner, 1994).

The subscript "is" stands for the energy difference between surface S-type anion isomer and interior-I type neutral isomer. Similarly, the subscripts "ss" and "is" stand for the energy differences between S-type and S-type isomers and between I-type and S-type isomers, respectively.

TABLE 6

The incremental enthalpy change $\Delta\Delta H^{298}$ (n) and hydration energy ΔE_{solv} (n) of Cl$^-$ in kcal/mol

n	1	2	3	4	5	6	7
AYK70	13.1	12.7	11.7	11.1			
KC	14.9	12.6	11.5	0.9			
HMY88	14.7	13.0	11.8	10.6	9.5	8.8	~8.1
DMTTSW 95		11.6	9.5				
MPGC94	17.5	13.8	12.2	9.7	6.7	8.5	6.9
X96 ΔH	13.4	12.0	12.4 ss 10.5 s's	10.2 ss 9.0 s's			
ΔE	13.8	13.7	14.5 ss 12.2 s's	12.1 ss 10.7 s's			
CKJ94	12.9	13.1	15.4	12.5 ss	9.0 ss	14.6 ss	
ΔH		11.6 is		9.8 is	9.2 is	11.7 s's 10.7 is (10.4 ii)	

AYK70: $\Delta\Delta H^{298}$ (n) determined by HPMS (Arshadi et al., 1970).

KC: $\Delta\Delta H^{298}$ (n) determined by HPMS (Keesee and Castleman, 1980).

HMY88: $\Delta\Delta H^{298}$ (n) determined by HPMS (Hiraoka et al., 1988).

DMTTSW95: $\Delta\Delta H^{298}$ (n) determined by ZPTRID (Dunbar et al., 1995).

MPGC94: E_{stab} (n) estimated by PES (Markovich et al., 1994).

X96: $\Delta\Delta H^{298}$ (n) and $\Delta\Delta E$ (n) estimated with MP4 level of ab initio MO calculations (Xantheas, 1996). The subscript "is" stands for the energy difference between surface S-type anion isomer and interior-I type neutral isomer (s: pyramidal, s': 3 + 1 structure).

CKJ95: $\Delta\Delta H^0$ (n) estimated with MP2 level of ab initio MO calculations (Combariza, 1994).

TABLE 7

The incremental enthalpy change $\Delta\Delta H^{298}$ (n) and hydration energy ΔE_{solv} (n) of Br$^-$ in kcal/mol

n	1	2	3	4	5	6	7	8	9	10	11
AYK70	12.6	12.3	11.5	10.9							
HMY88	11.7	11.6	11.4	11.0	10.8	10.3	~10				
MPGC94	12.7	12.2	10.8	6.9	6.2	6.2	6.0	2.5	1.6	0.7	1.6
CKJ94	11.1	11.8	14.6	12.2	8.8	13.4 v					
ΔH		10.2 is		9.0 is	8.6 is	S 11.1 9.3 is					

AYK70: $\Delta\Delta H^{298}$ (n) determined by HPMS (Arshadi et al., 1970).

HMY88: $\Delta\Delta H^{298}$ (n) determined by HPMS (Hiraoka et al., 1988).

MPGC94: ΔE_{stab} (n) estimated by PES (Markovich et al., 1994).

CKJ94: $\Delta\Delta H^0$ (n) estimated with MP2 level of ab initio MO calculations (Combariza et al., 1994).

The subscript "is" stands for the energy difference between surface S-type anion isomer and interior-I type neutral isomer.

TABLE 8

The incremental enthalpy change $\Delta\Delta H^{298}$ (n) and hydration energy $\Delta E_{solv}(n)$ of I^- in kcal/mol

n	1	2	3	4	5	6	7	8	9	10
AYK70	10.2	9.8	9.4	11.1						
KC80	11.1	9.9	9.3							
HMY88	10.3	9.5	9.2	9.2	~9					
MPGC94	10.4	9.4	8.5	6.9	4.1	7.8	2.1	0.7	3.9	1.4
CKJ94a	9.3	10.5	14.1	11.9	8.7	12.1 v				
ΔH		8.5		7.6	6.9	S				
		is		is	is	10.4				
						7.2 is				
CKJ94b						4.9 vs	2.3 ss	3.9 ss		
ΔVDE						11.1 s		14.1 is		
						7.8 ii	12 is			
						5.1 ii		4.1 ii		

AYK70: $\Delta\Delta H^{298}$ (n) determined by HPMS (Arshadi et al., 1970).

KC80: $\Delta\Delta H^{298}$ (n) determined by HPMS (Keesee and Castleman, 1980).

HMY88: $\Delta\Delta H^{298}$ (n) determined by HPMS (Hiraoka et al., 1988).

MPGC94: E_{stab} (n) estimated by PES (Markovich et al., 1994).

CKJ94a: $\Delta\Delta H^0$ (n) evaluated with HF level of ab initio MO calculations (Combariza et al., 1994a). The subscript "is" stands for the energy difference between surface S-type anion isomer and interior-I type neutral isomer. (s: pyramidal, s′: 3 + 1 structure).

CKJ94b: ΔVDE evaluated with MP2 level of ab initio MO calculations (Combariza et al., 1994b).

increase in the ionic radius, the hydration energy decreases. The empirical ionic radii $R(X^-)$ are 1.33 Å for F^-, 1.81 Å for Cl^-, 1.96 Å for Br^-, and 2.20 Å for I^- (Elsevier, 1987). As the size increases, the incremental hydration energy monotonically decreases, with a few exceptions, which might be related to structural changes in the clusters. Among the reported data, however, several systematic differences are found.

For fluoride anions, the incremental changes determined by the earlier HPMS experiments of Kebarle's group (Arshadi et al., 1970) are substantially smaller than those determined by Hiraoka et al. (1988). The former data were used to parametrize the nonadditive interaction potential for use in the MD simulations (Dang and Smith, 1993; Perera and Berkowitz, 1994). Xantheas and Dunning (1994) determined the structures of clusters for $n = 1,2$, and 3, with larger basis sets. The estimated incremental enthalpy change for $n = 2$ and $n = 3$ with an MP4 level of approximation agrees well with Hiraoka's larger energy values. In these clusters, one of the hydrogen atoms of each water molecule is bound to the anion atom. So the clusters are of surface type (S). Interestingly, with the MP2 optimization, two water molecules in $F^-(H_2O)_2$ become nonequivalent because they form a hydrogen bond and one of them is a proton donor. The existence of water–water hydrogen bonds is characteristic of surface (S) structures. Encouraged with this success, Xantheas and Dang (1996) reparametrized the polarizable interaction potential by fitting it to the accurate ab initio energies and performed the MD simulations for $n = 1$–10. The estimated enthalpy change up to $n = 7$ agrees with Hiraoka's values. The authors attributed the discrepancy at $n \geq 8$ to the low-temperature ($T = 200$ K) simulations, which they performed for larger clusters in order to avoid evaporation. It is, nevertheless, suggestive that both in the MD simulations and in Hiraoka's experimental data, larger incremental enthalpy changes are observed for $n = 9$ and $n = 10$ than for $n = 7$. In the MD simulations for larger clusters, a competition between surface (S) and interior (I) structures is found, the latter being dominant for clusters with six or more water molecules. As we have already discussed in the case of metal–water cluster cations, the S-versus-I competition is determined by the interplay between the strengths of the ion–water and water–water interactions. Combariza and Kestner (1994) also carried out ab initio calculations for $F^-(H_2O)_n$ ($n = 1$–6) with a slightly smaller basis set. In their optimized geometry, at the HF level of calculations the I structure for $n = 4$ is more stable than the S structure in terms of the MP2 energy. For $n = 5$ and $n = 6$, the S structure is again more stable than two I structures. The authors argued that the entropy factor favors the I structure for $n = 5$ and $n = 6$. To unambiguously determine the structures of the clusters, vibrational spectroscopic studies were recently reported (Cabarcos et al., 1999). For $F^-(H_2O)_n$ ($n = 3$–5), the spectra and supporting ab initio MO calculations show that the fluoride anion is bound as an interior species.

Among the experimental data of the incremental hydration energies for chloride cluster anions, the estimation from the shift in vertical detachment energy (Markovich et al., 1994) is slightly different from the others: The hydration energy for $n = 1$ is much larger than the HPMS data. Dunbar et al. (1995) introduced a new technique, zero-pressure thermal-radiation-induced dissociation (ZPTRID), to determine enthalpy changes, and the incremental enthalpy changes they found for $n = 2$ and $n = 3$ are substantially smaller than Hiraoka's values. Dunbar et al. repeated the HPMS experiment and obtained values consistent with ZPTRID. As the ion–water interaction becomes weaker from the fluoride to the chloride anion, the O–H–X angle in the optimized structure of $X^-(H_2O)$ deviates from a linear configuration to somewhere from 177° to 169° and eventually becomes 148° in $I^-(H_2O)$ (Johnson et al., 1996). For $n = 2$, Combariza et al. (1994a) found a linear type I structure, which has no water–water hydrogen bonds, but they also found a typical S structure similar to that determined by Xantheas (1996) with a better basis set. For $n = 3$, Xantheas found two S structures; one has three water molecules directly bonded to Cl, while the other has two such water molecules, and the other water molecule is in the second shell. The latter is less stable, and these types of structures are more likely in larger water clusters. The theoretical enthalpy changes for $n = 2$ and $n = 3$ can support both experimental data if the reliability of the theoretical model used is taken into account. But the 39.8 kJ/mol for $n = 3$ found by Dunbar et al. (1995) might be a little small for an S structure. For $n = 4$, Xantheas found two S structures, and in addition, Combariza et al. located an I structure $(3 + 1)$, in which a water ring trimer and a single water molecule sandwich a Cl^- anion. A similar I structure $(4 + 1)$ is found for $n = 5$. For $n = 6$, the I structure has a $(3 + 3)$ form, in which two water trimers face each other, holding the Cl^- anion at the center. In these calculations, the S structures are more stable than the I structure, but when the entropy factor is taken into account, the energy difference between the two types of structure becomes small. Experimentally, a contradictory change is found from $n = 5$ to $n = 6$. In the HPMS data, the change is monotonic. On the other hand, in estimates from photoelectron spectra, the incremental hydration energy increases at $n = 6$. As is seen in Table 6, if the S structure persists at $n = 6$, an increase in the incremental hydration energy is expected; this is because, in the S structures, stable hydrogen bond networks start to form in the second shell. Thus, the S structure is consistent with the estimation of vertical detachment energy. On the other hand, if the large entropy factor in the I structure is taken into account at 298 K, the incremental enthalpy change may continue to decrease at $n = 6$. (Note that in Table 6 $\Delta\Delta H$ is evaluated at $T = 0$ by Combariza et al., 1994a.) The importance of the entropy factor in $Cl^-(H_2O)_n$ clusters was first noticed by Asada et al. (1994) in their MD simulation. These researchers observed the transition from an S to an I structure and analyzed it by using a simple double-well model potential.

The size dependence of incremental hydration energies of bromide–water anions is similar to that of chloride–water anions. The exception is $\Delta\Delta H$ in the estimation of the vertical detachment energy between $n = 5$ and $n = 7$. For bromide anions, this energy is nearly constant, while for chloride anions, a sudden jump at $n = 6$ is found, as mentioned before. It is, however, true that the vertical detachment energy for both chloride and bromide anions does not monotonically change in this size range. Because of the large electron affinity of the bromine atom, Markovich et al. (1994) succeeded in extending the measurement of the photoelectron spectra up to 16. The incremental change in vertical detachment energy monotonically decreases up to 10 and then increases slightly. So convergence is not reached at $n = 16$.

The ionic radius of the iodide anion is very large, and the interaction between the ion and water molecules becomes weak. As already mentioned, the angle O–H–I in $I^-(H_2O)$ becomes as small as 148°, and the barrier height for the H–X bond switch through a C_{2v} structure of $I^-(H_2O)$ is only 0.8 kJ/mol (Johnson et al., 1996). The corresponding height is 5.4 for $Cl^-(H_2O)$ (Xantheas, 1996) and 31.4 kJ/mol for $F^-(H_2O)$ (Xantheas and Dunning, 1994). The floppy nature of the H–I^- bond may be reflected in the structures of larger clusters. The calculated size dependence of $\Delta\Delta H$ by Combariza et al. (1994a), even for $n < 4$, shows a marked difference from the observed dependence, which is contrary to the other halides. Markovich et al. (1994) noticed a sudden jump in the incremental hydration energy at $n = 6$ in their analysis of the photoelectron spectra and suggested a shell closing at $n = 6$. A similar change is found for chloride anion clusters, but the change at $I^-(H_2O)_6$ is more distinct. As was just mentioned, the ab initio MO studies for even small clusters cannot properly predict the size dependence of the stabilization energy. Combariza et al. (1994b) found three isomers for $n = 6$; two of them (a V-shaped one and a distorted hexagonal pyramid) are of S type, and the other is a typical I structure of a distorted octahedral. Their conclusion was that the I structure was selected for $n = 6$ and was prevalent for $n = 7$ and $n = 8$. But, according to the data in Table 8, this statement is not yet conclusive. The accuracy of the calculations is not high, as it is for the other halide–water anions, and most importantly, the entropy factor is expected to play a more important role in $I^-(H_2O)_n$ than in the other halide–water anions.

B. Photoelectron Spectra of $X^-(H_2O)_n$

Cheshnovsky and his collaborators succeeded in observing the photoelectron spectra of $Cl^-(H_2O)_n$ ($n = 0$–7), $Br^-(H_2O)_n$ ($n = 0$–16), and $I^-(H_2O)_n$ ($n = 0$–60) (Markovich et al., 1994). From the shift in the vertical detachment energy, the incremental hydration energies are estimated and have already been discussed. The vertical detachment energies increase substantially, from 3.61 eV ($n = 0$) for the bare ion to 6.88 eV ($n = 7$) for $Cl^-(H_2O)_n$, from 3.36 eV ($n = 0$) to 6.70 eV ($n = 16$) for $Br^-(H_2O)_n$, and from 3.06 eV ($n = 0$) to 6.58 eV ($n = 60$) for $I^-(H_2O)_n$. Markovich et al. made an attempt to analyze the halide solvation in clusters by using a classical continuous dielectric model and found that the inverse dependence of the solvation energy on the cluster radius in the model apparently agrees with the observed dependence of $E_{stab}(n)$. However, the limiting value, $E_{stab}(\infty)$ converges to a higher value than that measured for the bulk solution. In order to locate an anion X^- in the cluster, they also analyzed the results by an analogous electrostatic model recently developed by Makov and Nitzan (1994) for the vertical detachment of an anion solvated at the surface of the solvent cluster. Makovich et al. found that the experimental E_{stab} and $E_{stab}(\infty)$ fit somewhat better to the model of the surface solvated ion.

However, since the fitting quality of these models is very sensitive to the experimental parameters used, such as the ionic volume, water density, and dielectric constants, a final conclusion about the location of the anion in the clusters could not be deduced from the PES data alone.

The metal atoms used in many coins — Cu, Ag, and Au — have electronic configurations of ^2S $((n-1)d^{10}ns^1)$ in their ground state and thus have an electronic character analogous to that of the alkali metal atoms. They also have positive electron affinities because of the stabilization of the negative ions as a result of the shell closing, as in halide anions. Therefore, Fuke and his coworkers (Misaizu et al., 1995) investigated the photoelectron spectra of Cu^-–water systems to obtain information on the metal–water interaction analogous to the information obtained from hydrated halide clusters. The Cu^-–water clusters are produced by the laser vaporization technique. In the mass spectrum of the nascent negative-ion beam, two series of ion signals, which can be assigned to those of $^{63}Cu^-(H_2O)_n$ and $^{65}Cu^-(H_2O)_n$, are observed. In addition, another series of cluster ions, $CuOH^-(H_2O)_{n-1}$ and the anion of copper dimer solvated with water molecules, $Cu_2^-(H_2O)_n$, also appear in the mass spectrum.

Figure 31 shows the photoelectron spectra of $Cu^-(H_2O)_n$ for $n = 0$–4, detached at 266 nm. The spectrum of Cu^- consists of an intense band at 1.23 eV and two weak bands at 2.6 and 2.8 eV. The first band corresponds to the transition $Cu^2S(3d^{10}4s^1) \leftarrow Cu^{-1}S(d^{10}s^2)$ and the remaining bands to $Cu^2D_{5/2}$ $(d^9s^2) \leftarrow Cu^{-1}S$ and $Cu\ ^2D_{3/2}$ $(d^9s^2) \leftarrow Cu^{-1}S$, respectively.

The spectra for $Cu^-(H_2O)_n$ with $n = 1$ and $n = 2$ exhibit more complicated features than that of Cu^-, due to spectral contamination by $CuOH^-(H_2O)_{n-1}$. For $n = 1$, the relative intensity of $CuOH^-$ is estimated to be about 30% of $Cu^-(H_2O)$ from the mass spectrum. Since the latter ions are only a single atomic mass unit smaller than $Cu^-(H_2O)_n$, they could not be discriminated by the pulsed-mass gate. The rest of the bands observed in the photoelectron spectra of $Cu^-(H_2O)$ and $Cu^-(H_2O)_2$ are all assigned to the bands originating from the $Cu^2S \leftarrow Cu^{-1}S$ and $Cu^2D \leftarrow Cu^{-1}S$ transitions: 1.58 and ~ 2.9 eV for $n = 1$, and 1.95 and ~ 3.2 eV for $n = 2$. Although the 3d electron detachment should produce two spin-orbit states, as in the case of a Cu^- anion, these states are not resolved in the spectra for $n = 1$ and $n = 2$ because of the band broadening. The broadening is probably due to large structural differences between the excited states and the negative ions, as well as to the congestion of the vibrational modes. The PES results of $Cu^-(H_2O)_3$ are found to be more complicated than those for $n = 1$ and $n = 2$, as shown in the figure. The observed spectrum exhibits bands with peaks at 0.88, 1.71, 2.28, and 2.79 eV. The band at 1.71 eV is attributed to electron detachment from $CuOH^-(H_2O)_2$. The complexity of the spectrum is also due to the incapability of mass selection of $Cu^-(H_2O)_3$ from other ions, such

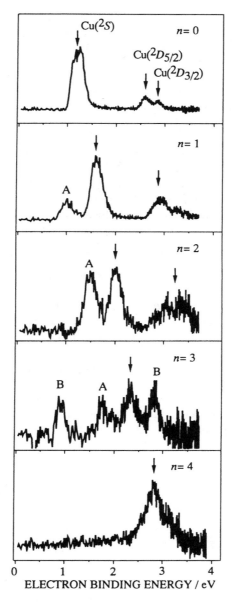

Figure 31.   Photoelectron spectra of $Cu^-(H_2O)_n$ for $n = 0$–4 detached at 266 nm. The bands A and B are assigned to the contaminated species (see text).

as Cu_2^- ions in the pulsed-mass gate. The photoelectron spectrum of Cu_2^- is known to exhibit two bands at 0.92 ($Cu_2 X^1\Sigma_g^+ \leftarrow Cu_2^- X^2\Sigma_u^+$) and 2.73 eV ($Cu_2 a^3\Sigma_u^+ \leftarrow Cu_2^- X^2\Sigma_u^+$) from the separate experiments (Misaizu et al., 1995). The bands at 0.88 and 2.79 eV are thus ascribed to the detachment from Cu_2^-. Therefore, the band at 2.28 eV can be assigned to the detachment from $Cu^-(H_2O)_3$, which corresponds to the transition to the neutral cluster state derived from the ^2S state of the Cu atom. This transition in $Cu^-(H_2O)_4$ is observed at 2.77 eV, as shown in the figure.

The foregoing results indicate that the vertical detachment energies of the ^2S-type state in $Cu^-(H_2O)_n$ increase monotonically with increasing number of water molecules. The average stabilization energy with an increase of one water molecule is determined to be about 0.38 eV. The bandwidths of the series are almost constant (about 300 meV) up to $n = 3$. These results seem to imply that the excess electron in the cluster is localized on the Cu atom, at least up to $n = 4$ in the ground state, and that successive solvation by water simply results in the stabilization of the states. The vertical detachment energies of the ^2D-type state also show a linear dependence with stepwise hydration. The increase in energy for this series is determined to be about 0.2 eV per water molecule.

From the vertical detachment energies of $Cu^-(H_2O)_n$, we estimated the successive hydration energies for the anion states. Unfortunately, there are no experimental and theoretical data on the hydration energies of the neutral $Cu(H_2O)_n$, except for that of $Cu(H_2O)$ calculated by Blomberg et al. (1986) and by Curtiss and Bierwage (1991). Hence, we adopted the theoretical result (0.2 eV). We also assume that the hydration energies of $Cu(H_2O)_n$ are equivalent to that of Cu. The estimated incremental hydration energies of the $Cu^-(H_2O)_n$ cluster, $n = 0-4$, are thus found to be almost constant, about 0.6 eV. These results indicate that the first solvation shell around the Cu^- ion is still not filled at $n = 3$. It is worth noticing that the results are consistent with those of the hydrated halide negative ions, $X^-(H_2O)_n$ ($X = Cl$, Br, and I), described in the previous paragraphs. The simplest view of the interaction between X^- and H_2O is that it is electrostatic and occurs between the excess electron on the halogen atom and the dipole of the ligands. The energy of this interaction is inversely proportional to the distance between the halide ion and the dipole. In fact, the crystal ionic radii of the ions, 1.81 for Cl^-, 1.96 for Br^-, and 2.20 Å for I^-, are qualitatively consistent with the order of the hydration energies determined for the three halide negative ions. As for the Cu^- ion, its ionic radius has not been reported. However, assuming a correlation between the ionic radii and the van der Waals radii, the ionic radius of Cu^- is estimated to be 1.86 Å (Zhan and Iwata, 1995). The hydration energy of about 0.6 eV for Cu^- is qualitatively explained by this simple electrostatic consideration: Since the

Cu$^-$ ion has a closed-shell configuration, as in the cases of the X^- and H$^-$ ions, the hydrated Cu$^-$ ions are expected to have geometric structures similar to those of the latter ions. The entropy factor is similarly important, as in hydrated I$^-$.

C. Vibrational Spectroscopy

To examine the structures of solvated halides, Okumura and coworkers (Choi et al., 1998) carried out the vibrational photodissociation spectroscopy of Cl$^-$(H$_2$O)$_n$. Figure 32 shows the IR spectra obtained for $n = 1$–5 in the region from 3,100 to 3,800 cm^{-1}, recorded by vibrational predissociation with the loss of one water molecule. For $n = 1$, intense and weak bands are observed at 3,285 and 3,156 cm^{-1}, respectively. The 3,285-cm^{-1} band is assigned to the OH stretching mode of the ionic hydrogen bond, which is shifted downward by 365 cm^{-1} from the symmetric OH band (at 3,650 cm^{-1}) of the free water molecule. The researchers ascribed the peak at 3,156 cm^{-1} to the bending overtone (at 3,151 cm^{-1}) of H$_2$O, because of a near coincidence with the overtone of a free water molecule, although the intensity is much higher in the complex. From a large frequency shift of the intense band and through comparison with an ab initio MO study, Okumura and his colleagues concluded that, in Cl$^-$(H$_2$O), one of the OH bonds in the water molecule strongly bonds with the Cl$^-$ anion. The band corresponding to the free OH vibration is detected at 3,698 cm^{-1} in the photodissociation spectra of Cl$^-$(H$_2$O)(CCl$_4$); the shift from the antisymmetric OH band (at 3,755 cm^{-1}) of free water is only 57 cm^{-1}. The IR spectra for I$^-$(H$_2$O) were also observed by Johnson et al. (1996) and Bailey et al. (1997). The band assigned to the OH stretching mode of the ionic hydrogen bond is centered at 3,400 cm^{-1} and consists of a quartet of almost equal intensity. Bailey et al. found that the complex structure of the band persists in I$^-$(H$_2$O)Ar and I$^-$(H$_2$O)N$_2$. The band for a free OH is observed at 3,713 cm^{-1}. The shifts for both bands are smaller for I$^-$(H$_2$O) than for Cl$^-$(H$_2$O), which is consistent with the weaker ion–water interaction in the former. Johnson and his coworkers (1998) extended the IR photodissociation spectra to X^-(H$_2$O)Ar$_3$ for $X = $ I, Br, and Cl and found that the quartet bands become a single sharp band. They concluded that the previously observed quartets are assigned to the hot bands due to high internal energy of the clusters.

As seen in Figure 32, the spectrum of $n = 2$ exhibits a strong doublet, with maxima at 3,245 and 3,317 cm^{-1}, but no significant peak in the region of 3,500–3,650 cm^{-1}, where water-bound OH stretching modes (O $\cdots$ H–O) are expected to appear. These spectral features indicate that $n = 2$ has a structure with two nearly equivalent ionic bonds, $X \cdots$ (H–OH)$_2$, and that no water–water hydrogen bonds exist in the cluster. But the structures

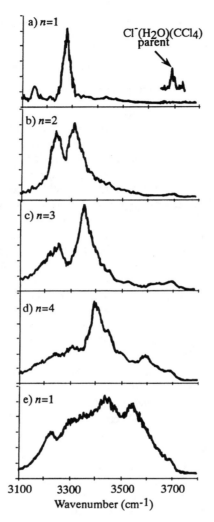

Figure 32. The IR spectra of $Cl^-(H_2O)_n$ ($n = 1$–5) in the region from 3,100 to 3,800 cm^{-1} (From Choi et al., 1998).

determined with the ab initio MO calculations, both by Choi et al. (1998) and by Xantheas (1996), have hydrogen-bonding waters and a nearly linear $O-H \cdots Cl \cdots HO$ structure that corresponds to a third-order transition state. This apparent inconsistency between the ab initio MO studies and the observed spectrum is also not resolved. One suggestion, given by Choi et al.

(1998) and supported by Johnson's (1998) new experiments, relates to the high temperature of the clusters. In their classical simulation, Asada et al. (1994) demonstrated that the cluster undergoes a kind of phase transition if its internal energy is high.

The spectra of larger clusters exhibit distinct features that indicate the existence of hydrogen bonds among the water molecules: The water-bound OH stretching bands appear in the $3,500–3,650 \, cm^{-1}$ region, and they are increasingly prominent for $n = 5$. These bands are almost as intense as the Cl^--bound OH stretching bands. The spectra also contain an underlying broad absorption indicating the possible existence of multiple isomers due to rather high internal energies. In addition, the spectra show a clear peak at $3,696 \, cm^{-1}$, corresponding to the free OH. From a comparison of these results with the results of ab initio calculations by Xantheas, Okumura and coworkers (Choi et al., 1998) suggested that the dominant structures of $n = 3$ and $n = 4$ are pyramidal-like and $3 + 1$ geometries, in which Cl^- is surface solvated.

More recently, Lisy and his colleagues studied the IR photodissociation spectra for $F^-(H_2O)_n$ ($n = 3$, 4, 5) (Cabarcos et al., 1999). They could not detect spectra for $n = 1$ and $n = 2$, probably because no dissociation takes place under the IR excitation due to strong ion–water bonds. The ion-bound OH stretching modes shift downward as low as $2,800 \, cm^{-1}$. In Lisy et al.'s analysis, the spectra up to $n = 5$ do not contain contributions from the OH bands with water–water hydrogen bonds, which suggests that all water molecules directly bond to the F^- ion and that internal (I) structures dominate, even at $n = 5$. This conclusion seems to contradict one previously derived by Xantheas and Dang (1996), who predicted that the interior states start to appear with six or more water molecules. Their new calculations support the experimental findings (Cabarcos et al., 1999).

D.   MD and MC Simulations of the Clusters

A halogen ion in water clusters and in aqueous solution has been the target of molecular dynamics (MD) and Monte Carlo (MC) simulations. Some results were already mentioned in the previous sections. Here, we briefly discuss two aspects of the simulations. First, we emphasize the importance of simulations in exploring the structures, dynamics, and even spectra of the halogen–water cluster anions. In their MC study of $Cl^-(H_2O)_n$, Asada et al. (1993,1994) showed that the interaction potential between the ion and water molecules is highly anharmonic and that entropy-favored structures (interior structures) become dominant at high temperatures. By evaluating the entropy contribution with the harmonic approximation, Combariza et al. (1994a) also noted systematic differences in the entropy changes for interior and surface structures. Some inconsistencies among experimental data and between such data and theoretical calculations might be due to thermal effects. To take these effects into account, simulations with realistic potential energy functions are required.

Crucial to the MD and MC simulations are the potential energy functions used to evaluate the energies and forces. Probably, the current consensus is that both ions and water must be polarizable (Carignano et al., 1997; Xantheas and Dang, 1996) for halide–water clusters. But in all simulations reported to date, the structure of the water molecule is frozen. Sato and Iwata (1999) examined the importance of the structural change in water molecules by carrying out ab initio MO calculations for $X^-(H_2O)$ at every MC step. The radial distributions for the intra-molecular parameters, together with their temperature and halide dependencies, are shown in Figure 33. Because the ion–water interaction is not symmetric for two OH bonds, the bond lengths are not equal, which is expected by the frequency shift of the OH bands. The O–H bond bound to the ion is substantially lengthened, and the width of the distribution is much broader than for the free OH bond. Quantum effects are expected to be more pronounced. In particular, for $X = F$, the effect cannot be neglected. When Xantheas and Dang (1996) fitted the ab initio energies to the polarizable ion–water interaction potential functions, they properly included the effect of non-equivalent OH bonds. Similar treatments are required for the other halides.

VI. SUMMARY AND CONCLUDING REMARKS

This review has presented experimental and theoretical studies on the geometric structures, electronic structures, and reactivity of water clusters containing a single atom or an atomic ion. The results may form a basis for understanding the fundamental aspects of hydration phenomena of a single atom (ion), as well as oxidation–reduction reactions in metal ion–water systems. The review has demonstrated that studies of atom–water clusters and their ions provide crucial information on the structures of hydrated electrons. We believe that we have proven the importance of collaborative work between experimental and theoretical groups in cluster chemistry and physics.

In Section II, we described photoionization and photoelectron spectroscopic studies, as well as ab initio studies of alkali atom–water clusters. Intriguingly, photoionization experiments revealed that the ionization energies (IEs) of the alkali atom–water clusters converge to a constant value for $n \geq 4$, which is independent of the kind of metal. To clarify the origin of this anomalous n dependence of IEs, the photoelectron spectra (PES) of $Li^-(H_2O)_n$ and $Na^-(H_2O)_n$ have also been investigated. Surprisingly, the n dependencies of the experimental PES for $Li^-(H_2O)_n$ and

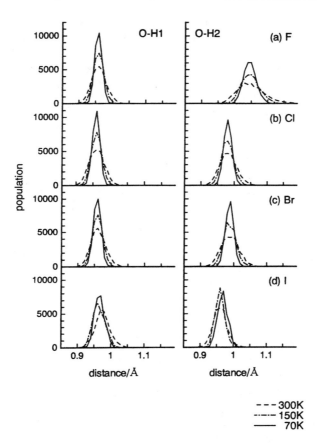

Figure 33. The temperature dependence of the radial distribution of the intramolecular OH bond distance in the water molecule.

$Na^-(H_2O)_n$ were very different from each other. Theoretical studies established that the structures of $Li^-(H_2O)_n$ had forms similar to those of the corresponding neutral clusters, while in $Na^-(H_2O)_n$ the water molecules in the first shell were bound to the metal atom by hydrogen atoms, completely opposite the situation in the corresponding neutral clusters. Accurate calculations of vertical detachment energies using optimized structures of Li and Na anion clusters agree with experimental values. So the PES of $Na^-(H_2O)_n$ could not resolve the mystery of the convergence of IE. On the other hand, the PES of $Li^-(H_2O)_n$ seemed consistent with the convergence of IE. Theoretical studies of neutral Li–water clusters predicted that in $Li(H_2O)_4$, the center of the odd electron would starts to be separated from the atomic center, Li. For one of the isomers of $Li(H_2O)_8$, the odd electron was trapped between two water molecules in the outer shell, located far from a hydrated Li^+ cation. For comparison, we also presented the PES results for $Li^-(NH_3)_n$ and $Na^-(NH_3)_n$. Both experimental and theoretical results indicated that in the metal atom–ammonia clusters with $n \leq 10$, the metal atom was gradually ionized to form the one-center ionic state because of a weaker interaction between the ejected electron and ammonia molecules.

In Section III, we presented the characteristic size distributions in TOF spectra of the group 2 (alkaline-earth) metal ions Mg^+ and Ca^+ with the $(H_2O)_m$ and $(D_2O)_m$ clusters. Apparent metal-oxidation reactions to form $MOH^+(H_2O)_{n-1}$ displayed a characteristic size dependence, with two critical sizes ($n \sim 5$ and $n \sim 14$). With the help of ab initio MO calculations of the structure of the $M^+(H_2O)_n$ and $MOH^+(H_2O)_{n-1}$ ions, the origin of the first product switching for $n \sim 5$ was ascribed to the difference in the successive hydration energies of M^+ and MOH^+. As for the second product switching at $n \sim 15$, the cause was not resolved yet. We discussed two possible mechanisms: the participation of the ion-pair state $M^{2+}(H_2O)_n^-$ and the formation of new reaction products such as $MgOH \cdot H_3O^+(H_2O)_{n-2}$. To confirm the nature of the mechanism, experimental and theoretical studies of larger clusters are indispensable. We also presented the results of photodissociation spectra and dissociation processes of $M^+(H_2O)_n$, which are a half-collision analogue of the foregoing reaction. Comparison of the observed spectra of $Mg^+(H_2O)_n$ with theoretical calculations indicated the presence of geometrical isomers having different numbers of water molecules in the first shell for $n > 2$. The absorption bands for $Ca^+(H_2O)_n$ redshift monotonically with increasing n and exhibited a trend to converge at $n \sim 6$. The photodissociation processes of $M^+(H_2O)_n$ were found to include dehydrogenation reactions that produced the hydrated MOH^+ ions, as well as simple evaporation. The features of the dissociation processes for these systems were quite different for $M = Mg$ and $M = Ca$. The difference was ascribed to the absence and the presence of the ^2D-type states in between the ground (n^2S) and optically allowed excited (n^2P) states for Mg^+ and Ca^+, respectively. The other interesting difference in the photo dissociation processes of these cluster ions with a large n was the extensive enhancement of the hydrogen elimination reaction for the latter system. These studies were one of the first experimental studies that unveiled the microscopic solvation dynamics of metal ions, as well as redox reactions in the gas-phase clusters.

In Section IV, we presented experimental and theoretical results pertaining to equilibrium structures, hydration energies, and ionization energies of neutral $Al(H_2O)_n$ clusters. We also presented results of photodissociation experiments on $Al^+(H_2O)_n$. As in the case of alkaline-earth metal ion–water systems, the absorption band derived from the lowest allowed transition of Al^+, $3^1P \leftarrow 3^1S$, was found to be shifted extensively to lower energy in clusters. However, no photoinduced intracluster reaction was found in the photoexcitation of $Al^+(H_2O)_n$; only the evaporation process was observed in the wavelength region examined. On the other hand, recent ICR experiments showed that large $Al^+(H_2O)_n$ clusters ($n \geq 16$) underwent a photoinduced intracluster reaction; the apparent oxidation of Al^+, leading to the loss of an H_2 molecule and the formation of the closed-shell hydroxide $Al(OH)_2^+$, was induced by IR background radiation from the cell wall.

Intimately related to the preceding reaction, the intracluster reactions of $B^+(H_2O)_n$ were theoretically examined with a hybrid procedure of ab initio MO calculations and Monte Carlo samplings. When a simulation started at the structure $[B^+(H_2O)_3]$, it ended up at the product $(HBOH)^+(H_2O)$ of an apparent insertion reaction with an evaporating water molecule. These results may provide us with a clue to understanding the mechanism of the oxidation reaction in large Mg^+– and Al^+–water clusters.

In Section V, we reviewed recent experimental and theoretical studies on the structure and energetics of halide–water clusters. Although the thermodynamic properties of these clusters have been studied extensively for the last three decades, the structures of the clusters are still a subject of investigation. The solvation structure of halide anions is the interior type in liquid water. But in small water clusters, the surface structure was expected to be dominant, except in the case of F^-. Advances in ab initio and MC calculation methods, as well as in spectroscopic techniques such as photoelectron and vibrational dissociation spectroscopy, have allowed us to shed new light on this problem and to explore the size-dependence of the transition from surface to interior structures and its dependence on halogen.

These spectroscopic studies, coupled tightly with ab initio MO calculations, now allow us to understand details about the structure and energetics of the gas-phase solvation of atoms in the anionic, neutral, and cationic forms. The elements we have dealt with are in groups 1, 2, 11, 13, and 17. The studies on group 1 atoms have succeeded in answering the fundamental questions of what kind of interaction is most important and how many solvent molecules are required to dissolve alkali metal atoms in polar solvent molecules. The gas-phase studies reviewed here were the first experimental and theoretical efforts to characterize the one-center and two-center ion-pair states, which are the counterpart of the presolvated state of electrons and metal ions in bulk solution. The studies on group 2 and group 13 atoms also revealed details about the fundamental interaction of metal ions with water molecules. In particular, these studies have led to a new approach to investigating the microscopic aspect of the redox reaction of metal–water systems. The studies may further serve as a guide to exploring the gas-phase redox reaction of transition-metal atom–water systems, which have more complex electronic structures than those discussed here. The studies on the anions of the group 11 and group 17 atoms have elucidated the existence of the interior- and surface-type structures in atomic anion–water clusters. Entropy effect was found to play an important role. The other important issue to be solved is the dynamical aspect of the afore-mentioned gas-phase solvation and reactions. In solution chemistry, under-standing the dynamical role of solvent molecules is a paramount issue. The cluster approach may provide us with a wealth of information on the microscopic aspects of this subject. Experimental findings reviewed in this paper have already suggested some clues to extending research to the dynamical case. Information on the rate and dynamics of the inter conversion between various isomers will help us understand the role of statistical fluctuation in solution reactions. Also, examining the relaxation processes in metal ion–water clusters will give us a microscopic view of a stratified structure of the energy flow from solute to solvent molecules. Research along these lines, which may require the development of new experimental and theoretical techniques, is important in bridging the gap between gas-phase and bulk solvation.

Acknowledgments

Most of the sections of this review are based on collaborative works with our colleagues for last few years. They are Prof. Keiji Morokuma (Emory University), Prof. Fuminori Misaizu (University of Tohoku), Dr. Masaoni Sanekata (Keio University), Dr. Ryozo Takasu (Kobe University), Dr. Hidekazu Watanabe (RIKEN), Dr. Katsuhiko Sato (IMS), Mr. Takeshi Tsurusawa (IMS), and Mr. Tetsuya Kamimoto (Tokyo Metropolitan University). Many aspects of the research have been supported by Grants-in-Aid for Scientific Research awarded by the Ministry of Education, Science, Sports, and Culture in Japan. We have made extensive use of resources of the computer center of the Institute for Molecular Science. We thank Prof. J. Lisy for critically reading the manuscript.

References

Arshadi, M., Yamdagni, R., and Kebarle, P., *J. Phys. Chem.* **74**, 1475 (1970).

Asada, T. and Iwata, S., *Chem. Phys. Lett.* **260**, 1 (1996).

Asada, T., Nishimoto, K., and Kitaura, K., *J. Molec. Struct. (THEOCHEM)* **310**, 149 (1994).

Asada, T., Nishimoto, K., and Kitaura, K., *J. Phys. Chem.* **97**, 7724 (1993).

Asada, T. and Iwata, S., unpublished results.

Ayotte, P. and Johnson, M. A., *J. Chem. Phys.* **106**, 811 (1997).

Bailey, C. G., Kim., J., Dessent, C. E. H., and Johnson, M. A., *Chem. Phys. Lett.* **269**, 122 (1997).

Bailey, C. G., Kim, J., Johson, M. A., *J. Phys. Chem.* **100**, 16782 (1996).

Barnett, R. N. and Landman, U., *Phys. Rev. Lett.* **70**, 1775 (1993).

Bauschlicher, C. W., Jr., and Partridge, H., *J. Phys. Chem.* **95**, 3946 (1991b).

Bauschlicher, C. W., Jr., and Partridge, H., *J. Phys. Chem.* **95**, 9694 (1991c).

Bauschlicher, C. W., Jr., and Sodupe, M., and Partridge, H., *J. Chem. Phys.* **96**, 4453 (1992).

Bauschlicher, C. W., Jr., Langhoff, S. R., Partridge, H., Rice, J. E., and Komoronicki, A., *J. Chem. Phys.* **95**, 5142 (1991a).

Berg, C., Achatz, U., Beyer, M., Joos, S., Albert, G., Schindler, T., Niedner-Schatteburg, G., and Bondybey, V. E., *Int. J. Mass. Spctrom. Ion Proc.* **167**, 723 (1997).

Berg, C., Beyer, M., Achatz, U., Niedner-Schatteburg, G., and Bondybey, V. E., *J. Phys. Chem.* (1999), forthcoming.

Berg, C., Beyer, M., Achatz, U., Joos, S., Niedner-Schatteburg, G., Bondybey, V. E., *Chem. Phys.* **239**, 379 (1998).

Bewig, L., Buck, U., Rakowsky, S., Reymann, M., and Steinbach, C., *J. Phys. Chem.* **A102**, 124 (1998).

Beyer, M., Berg, C., Gorlitzer, H. W., Schindler, T., Achatz, U., Albert, G., Niedner-Schatteburg, G., and Bondybey, V. E., *J. Am. Chem. Soc.* **118**, 7386 (1996).

Blomberg, M. R. A., Brandemark, U. B., and Siegbahn, P. E. M., *Chem. Phys. Lett.* **126**, 317 (1986).

Bower, M. T., Marshall, A. G., Lafferty, F. W., *J. Phys. Chem.* **100**, 12897 (1996).

Brockhaus, P., Hertel, I. V., Schulz, C. P., *Chem. Phys.* **110**, 393 (1999).

Cabarcos, O. M., Weinheimer, C. J., Lisy, J. M., and Xantheas, S. S., *J. Chem. Phys.* **110**, 5 (1999).

Carignano, M. A., Karlstrom, G., and Linse, P. *J. Phys. Chem.* **B101**, 1142 (1997).

Castleman, Jr., A. J., Keesee, R. G., *Chem. Rev.* **86**, 589 (1986).

Castleman, Jr., A. J., Bowen, Jr., K. H., *J. Phys. Chem.* **100**, 12911 (1996).

Choi, J.-H., Kuwata, K. T., Cao, Y.-B., and Okumura, M., *J. Phys. Chem.* **A102**, 503 (1998).

Coe, J. V., Lee, G. H., Eaton, J. G., Arnold, S. T., Sarkas, H. W., Bowen, K. H., Ludewigt, C., Haberland, H., and Worsnop, D. R., *J. Chem. Phys.* **92**, 3980 (1990).

Combariza, J. E. and Kestner, N. R., *J. Phys. Chem.* **98**, 3513 (1994).

Combariza, J. E., Kestner, N. R., and Jortner, J., *Chem. Phys. Lett.* **203**, 423 (1993).

Combariza, J. E., Kestner, N. R., and Jortner, J., *J. Chem. Phys.* **100**, 2851 (1994a).

Combariza, J. E., Kestner, N. R., and Jortner, J., *Chem. Phys. Lett.* **221**, 156 (1994b).

Curtiss, L. and Bierwage, E., *Chem. Phys. Lett.* **176**, 417 (1991).

Curtiss, L. A., Frurip, D. L., and Blander, M., *J. Chem. Phys.* **71**, 2703 (1979).

Dang, L. X., and Smith, D. E., *J. Chem. Phys.* **99**, 6950 (1993).

Dogonadze, R. R., Kalman, E., Kornyshev, A. A., and Ulstrup, J. Eds., *The chemical Physics of Solvation,* Part C, Elsevier, Amsterdam, 1988.

Donnelly, S. G. and Farrar, J. M., *J. Chem. Phys.* **98**, 5450 (1993).

Dunbar, R. C., McMahon, T. B., Tholmann, D., Tonner, D. S., Salahub, D. R., and Wei, D., *J. Am. Chem. Soc.* **117**, 12819 (1995).

Duncan, M. A., *Ann. Rev. Phys. Chem.* **48**, 69 (1997).

Elsevier's Periodic Table of Elements, Elsevier, Amsterdam, Elsevier, 1987.

Feller, D., Glendening, E. D., Kendall, R. A., and Peterson, K. A., *J. Chem. Phys.* **100**, 4981 (1994).

Feller, D., Glendening, E. D., Woon, D. E., and Feyereisen, M. W., *J. Chem. Phys.* **103**, 3526 (1995).

Fuke, K., Misaizu, F., Sanekata, M., Tsukamoto, K., and Iwata, S., *Z. Phys.* **D26**, S180 (1993).

Glendening, E. D., *J. Am. Chem. Soc.* **118**, 2473 (1996).

Glendening, E. D., Feller, D., *J. Phys. Chem.* **99**, 3060 (1995).

Haberland, H., Ludewigt, C., Schindler, H. -G., and Worksnop, D. R., *Surf. Sci.* **156**, 157 (1985).

Haberland, H., Schindler, H.-G., and Worksnop, D. R., *Ber. Bunsenges. Phys. Chem.* **88**, 270 (1984).

Harms, A. C., Khanna, S. N., Chen, B., and Castleman, A. W., Jr., *J. Chem. Phys.* **100**, 3540 (1994).

Hashimoto, K., He, S., and Morokuma, K., *Chem. Phys. Lett.* **206**, 297 (1993).

Hashimoto, K. and Iwata, S., *J. Phys. Chem.* **93**, 2165 (1989).

Hashimoto, K., Kamimoto, T., and Fuke, K., *Chem. Phys. Lett.* **266**, 7 (1997).

Hashimoto, K. and Morokuma, K., *Chem. Phys. Lett.* **223**, 423 (1994a).

Hashimoto, K. and Morokuma, K., *J. Am. Chem. Soc.* **116**, 11436 (1994b).

Hashimoto, K. and Morokuma, K., *J. Am. Chem. Soc.* **117**, 4151 (1995).

Hashimoto, K., Osamura, Y., and Iwata, S., *J. Molec. Struct. (THEOCHEM)* **152**, 101 (1987a).

Hashimoto, K., Yoda, N., and Iwata, S., *Chem. Phys.* **116**, 193 (1987b).

Hashimoto, K., Yoda, N., Osamura, Y., and Iwata, S., *J. Am. Chem. Soc.* **112**, 7189 (1990).

Hashimoto, K. and Kamimoto, T., *J. Am. Chem. Soc.* **120**, 3560 (1998).

Hashimoto, K. and Kamimoto, T. (1999a), in preparation.

Hashimoto, K. and Kamimoto, T. (1999b), in preparation.

Hertel, I. V., Huglin, C., Nitsch, C., and Schulz, C. P., *Phys. Rev. Lett.* **67**, 1767 (1991).

Hiraoka, K., Mizuse, S., and Yamabe, S., *J. Phys. Chem.* **92**, 3943 (1988).

Johnson, M. S., Kuwata, K. T., Wong, C-K., Okumura, M., *Chem. Phys. Lett.* **260**, 551 (1996).

Johnson, M. A., International conference on "Water in Gases", Paris, June 1998.

Keesee, R. G. and Castleman, A. W., Jr., *Chem. Phys. Lett.* **74**, 139 (1980).

Kim, J., Lee, S., Cho, S. J., Mhin, B. J., and Kim, K. S., *J. Chem. Phys.* **102**, 839 (1995).

Kebarle, P., *Ann. Rev. Phys. Chem.* **28**, 445 (1977).

Kondow, T., *J. Phys. Chem.* **91**, 1307 (1987).

Kondow, T., Nagata, T., and Kuchitsu, K., *Z. Phys.* **D12**, 291 (1989).

Lee, G. H., Arnold, S. T., Eaton, J. G., Sarkas, H. W., Bowen, K. H., Ludewigt, C., and Haberland, H., *Z. Phys.* **D20**, 9 (1991).

Lisy, J. M., *Int. Rev. Phys. Chem.* **16**, 267 (1997).

Makov, G., Nitzan, A., *J. Phys. Chem.* **98**, 3459 (1994).

Mark, T. D., *Int. J. Mass. Spctrom. Ion Proc.* **107**, 143 (1991).

Markovich, G., Giniger, R., Levin, M., and Cheshnovsky, O., *J. Chem. Phys.* **95**, 9416 (1991).

Markovich, G., Pollack, S., Giniger, R., and Cheshnovsky, O., *J. Chem. Phys.* **101**, 9344 (1994).

Misaizu, F., Sanekata, M., Fuke, K., and Iwata, S., *J. Chem. Phys.* **100**, 1161 (1994).

Misaizu, F., Sanekata, M., Tsukamoto, K., Fuke, K., and Iwata, S., *J. Phys. Chem.* **96**, 8259 (1992a).

Misaizu, F., Tsukamoto, K., Sanekata, M., and Fuke, K., *Chem. Phys. Lett.* **188**, 241 (1992b).

Misaizu, F., Tsukamoto, K., Sanekata, M., and Fuke, K., *Z. Phys.* **D26**, S177 (1993).

Misaizu, F., Tsukamoto, M., Sanekata, M., and Fuke, K., *Laser Chem.* **15**, 195 (1995).

Moore, C. E. ed. "Atomic Energy Levels", Natl. Stand. Ref. Data Ser., NBS 35, Vol. 1. (National Bureau of Standards, Washington, D. C. 1971).

Perera, L. and Berkowitz, M. L., *J. Chem. Phys.* **100**, 3085 (1994).

Sanekata, M., Misaizu, F., and Fuke, K., *J. Chem. Phys.* **104**, 9767 (1996).

Sanekata, M., Misaizu, F., Fuke, K., Iwata, S., and Hashimoto, K., *J. Am .Chem. Soc.* **117**, 747 (1995).

Sanekata, M., Fuke, K., unpublished results.

Satoh, K. and Iwata, S., in preparation (1999).

Schulz, C. P., Haugstatter, R., Tittes, H.-U., Hertel, I. V., *Z. Phys.* **D10**, 279 (1988).

Schulz, C. P., Nitsch, C., *J. Chem. Phys.* **107**, 4794 (1997).

Scurlock, C. T., Pullins, S. H., Reddic, J. E., and Duncan, M. A., *J. Chem. Phys.* **104**, 4591 (1996).

Shen, M. H. and Farrar, J. M., *J. Chem. Phys.* **94**, 3322 (1991).

Shen, M. H. and Farrar, J. M., *J. Phys. Chem.* **93**, 4386 (1988).

Sodupe, M. and Bauschlicher, C. W., Jr., *Chem. Phys. Lett.* **195**, 494 (1991).

Stampfli, P., *Physics Rep.* **255**, 1 (1995).

Stampfli, P. and Bennemann, K. H., *Comp. Mater. Sci.* **2**, 578 (1994).

Takasu, R., Hashimoto, K., and Fuke, K., *Chem. Phys. Lett.* **258**, 94 (1996).

Takasu, R., Misaizu, F., Hashimoto, K., and Fuke, K., *J. Phys. Chem.* **A101**, 3078 (1997).

Takasu, R., Taguchi, T., Hashimoto, K., and Fuke, K., *Chem. Phys. Lett.* **290**, 481 (1998).

Tsurusawa, T. and Iwata, S., *Chem. Phys. Lett.* **287**, 553 (1998a).

Tsurusawa, T. and Iwata, S. (1999), *J. Phys. Chem.* in press.

Watanabe, H., Aoki, M., and Iwata, S., *Bull. Chem. Soc. Jpn.* **66**, 3245 (1993).

Watanabe, H., Iwata, S., Hashimoto, K., Misaizu, F., and Fuke, K., *J. Am. Chem. Soc.* **117**, 755 (1995).

Watanabe, H. and Iwata, S., *J. Phys. Chem.* **100**, 3377 (1996).

Watanabe, H., Asada, T., and Iwata, S., *Bull. Chem. Soc. Jpn.* **70**, 2619 (1997a).

Watanabe, H. and Iwata, S., *J. Phys. Chem.* **A101**, 487 (1997b).

Watanabe, H. and Iwata, S., *J. Chem. Phys.* **108**, 10078 (1998).

Weinheimer, C. J. and Lisy, J. M., *J. Chem. Phys.* **105**, 2938 (1996).

Willey, K. F., Yeh, C. S., Robbins, D. L., Pilgrim, J. S., and Duncan, M. A., *J. Chem. Phys.* **97**, 8886 (1992).

Xantheas, S. S., *J. Phys. Chem.* **100**, 9703 (1996).

Xantheas, S. S. and Dang, L. X., *J. Phys. Chem.* **100**, 3989 (1996).

Xantheas, S. S., Dunning, T. H., Jr., *J. Phys. Chem.* **98**, 13489 (1994).

Xantheas, S. S., Dunning, T. H., Jr., *J. Phys. Chem.* **96**, 7505 (1992).

Yeh, C. S., Willey, K. F., Robbins, D. L., Pilgrim, J. P., and Duncan, M. A., *Chem. Phys. Lett.* **196**, 233 (1992).

Zhan, C.-C. and Iwata, S., *Chem. Phys. Lett.* **232**, 72 (1995).

AUTHOR INDEX

Numbers in parentheses are reference numbers and indicate that the author's work is referred to although his name is not mentioned in the text. Numbers in *italic* show the pages on which the complete references are listed.

SUBJECT INDEX